Stem Cells and C
From the Essentials to Applic..

Stem Cells and Cancer in Hepatology
From the Essentials to Application

Edited by
Yun-Wen Zheng

Academic Press is an imprint of Elsevier
125 London Wall, London EC2Y 5AS, United Kingdom
525 B Street, Suite 1800, San Diego, CA 92101-4495, United States
50 Hampshire Street, 5th Floor, Cambridge, MA 02139, United States
The Boulevard, Langford Lane, Kidlington, Oxford OX5 1GB, United Kingdom

Copyright © 2018 Elsevier Inc. All rights reserved.

No part of this publication may be reproduced or transmitted in any form or by any means, electronic or mechanical, including photocopying, recording, or any information storage and retrieval system, without permission in writing from the publisher. Details on how to seek permission, further information about the Publisher's permissions policies and our arrangements with organizations such as the Copyright Clearance Center and the Copyright Licensing Agency, can be found at our website: www.elsevier.com/permissions.

This book and the individual contributions contained in it are protected under copyright by the Publisher (other than as may be noted herein).

Notices
Knowledge and best practice in this field are constantly changing. As new research and experience broaden our understanding, changes in research methods, professional practices, or medical treatment may become necessary.

Practitioners and researchers must always rely on their own experience and knowledge in evaluating and using any information, methods, compounds, or experiments described herein. In using such information or methods they should be mindful of their own safety and the safety of others, including parties for whom they have a professional responsibility.

To the fullest extent of the law, neither the Publisher nor the authors, contributors, or editors, assume any liability for any injury and/or damage to persons or property as a matter of products liability, negligence or otherwise, or from any use or operation of any methods, products, instructions, or ideas contained in the material herein.

Library of Congress Cataloging-in-Publication Data
A catalog record for this book is available from the Library of Congress

British Library Cataloguing-in-Publication Data
A catalogue record for this book is available from the British Library

ISBN: 978-0-12-812301-0

For information on all Academic Press publications visit our website at
https://www.elsevier.com/books-and-journals

Publisher: Mica Haley
Acquisition Editor: Mica Haley
Editorial Project Manager: Tracy I. Tufaga
Production Project Manager: Mohanapriyan Rajendran
Designer: Victoria Pearson

Typeset by Thomson Digital

I faithfully dedicate this book to:
My mother, Cuihua Gong, for her love to me, and her kindness and forgiveness, and for her great support, alive in cherished memory. I pledge my best to promote stem-cell and related cancer research and to help preserve the lives of mothers in the future from this kind of illness, consoling her great human spirit.

Yun-Wen Zheng
Tsukuba, Japan

Contents

List of Contributors	xv
Editor Biography	xvii
Foreword	xix
Preface	xxi
Acknowledgments	xxiii
Introduction	xxv

1. **Molecular Mechanisms of Liver Development: Lessons From Animal Models**   **1**

Norio Miyamura and Hiroshi Nishina

1	Introduction		1
2	Liver Development		2
	2.1	Chronology of Liver Bud Formation	3
	2.2	Genes Involved in Liver Bud Formation	3
	2.3	Hepatocyte and Cholangiocyte Differentiation	6
	2.4	Mesothelial Cell Differentiation	7
	2.5	Hepatic Stellate Cell Differentiation	8
	2.6	LSEC Differentiation	8
	2.7	Differentiation of Kupffer Cells and Pit Cells	8
3	Studies of Liver Formation Using Zebrafish		9
4	Studies of Liver Formation Using Medaka		10
	4.1	Group 1: Mutations Affecting Hepatoblast Development	11
	4.2	Group 2: Mutations Affecting Hepatoblast Proliferation	11
	4.3	Group 3: Mutations Affecting Liver Laterality	11
	4.4	Group 4: Mutations Affecting Hemoglobin-Bilirubin Metabolism	11
	4.5	Group 5: Mutations Affecting Lipid Metabolism	11
	4.6	Analysis of Medaka Mutants	12
5	Studies of Liver Size Regulation in Mice		12
6	Human Liver Diseases		13
	6.1	Hepatocellular Carcinoma	13
	6.2	Metabolic Disorders	13
	6.3	Alagille Syndrome	14
7	Conclusion		14
References			15

2. Molecular Mechanisms Regulating the Proliferation and Maturation of Hepatic Progenitor Cells During Liver Development   21

Hiromi Chikada and Akihide Kamiya

1	Introduction	21
2	Liver Regeneration and Normal Liver-Derived Stem/Progenitor Cells	23
3	Regulation of Transcription Factors Regulating the Proliferation and Maturation of Hepatic Progenitor Cells	23
4	Signaling Pathways Regulating Hepatoblast Proliferation and Differentiation	26
5	Hepatic Progenitor Cells Derived from Human Pluripotent Stem Cells	28
6	Conclusion	30
	References	31

3. Plasticity of Liver Epithelial Cells in Healthy and Injured Livers   35

Naoki Tanimizu and Toshihiro Mitaka

1	Introduction		35
2	Lineage Specification During Development		36
3	Lineage Plasticity and Heterogeneity of Cholangiocytes		37
	3.1	Lineage Plasticity of Cholangiocytes	37
	3.2	Heterogeneity of Cholangiocytes	38
	3.3	Molecular Mechanisms Regulating the Plasticity of Cholangiocytes	40
4	Lineage Plasticity and Heterogeneity of Hepatocytes		41
	4.1	Lineage Plasticity of Hepatocytes	41
	4.2	Heterogeneity of Hepatocytes	43
	4.3	Molecular Mechanisms Regulating Lineage Plasticity of Hepatocytes	45
5	Concluding Remarks		48
	References		50

4. Stem Cells in the Liver Cancer and the Organ Size Control   55

Amita Tiyaboonchai, Yulong Su and Bin Li

1	Introduction		55
2	Stem Cells in the Liver		55
3	Liver Cancer and Cancer Stem Cells		58
4	Liver Organ Size		59
5	Epigenetic Deregulation in HCC		60
	5.1	DNA Methylation	61
	5.2	Histone Modification and Chromatin Structure	62
	5.3	Noncoding RNAs	64

	6	Manipulate Liver Cancer with Over Expressing Oncogene	65
	7	Conclusions and Future Perspective	67
	References		67

5. **Exploration for Cell Sources for Liver Regenerative Medicine: "CLiP" as a Dawn of Cell Transplantation Therapy** — 77

   Kazunori Hosaka, Takeshi Katsuda, Shuji Terai and Takahiro Ochiya

1	Introduction	77
2	Hepatocytes as the Source of Liver-Repopulating Cells	78
3	Hepatocyte-Like Cells Induced from ESCs as the Source of Liver-Repopulating Cells	80
4	Hepatocyte-Like Cells Induced from iPSCs as the Source of Liver-Repopulating Cells	81
5	Induced Hepatocyte-Like Cells as the Source of Liver-Repopulating Cells	83
6	Induction of Hepatic Cells from Mesenchymal Stem Cells as the Source of Liver-Repopulating Cells	84
7	Liver Progenitor/Stem Cells as the Source of Liver-Repopulating Cells	85
8	Emergence of LPCs from MHs Using Small Molecules	87
9	LPC-Like Characteristics of YAC-Induced Proliferative Cells	89
10	YAC-Cell Bipotentiality	90
11	In Vitro Reprogramming of MHs to LPCs	90
12	Stable Long-Term Expansion of CLiPs	91
13	In Vivo Repopulation Capacity of CLiPs in Chronically Injured Liver	91
14	Diploid Hepatocytes as the Origin of CLiPs	92
15	Insights from CLiPs Into Regenerative Therapy and Basic Liver Biology	92
References		94

6. **Generation of Hepatocytes by Transdifferentiation** — 103

   Pengyu Huang and Qiwen Chen

1	Introduction		103
2	Generation of iHep Cells		104
3	Delivery Systems		105
	3.1	Integrative Delivery Approaches	105
	3.2	Nonintegrative Delivery Approaches	106
4	Donor Cell Types		106
	4.1	Fibroblasts	106
	4.2	Mesenchymal Stem Cells	107
	4.3	Urinary Cells	107
5	Evaluation of iHep Cell Quality		107
6	Applications and Challenges		109
References			111

7. Generation of Liver Organoids and Their Potential Applications 115
Li-Ping Liu, Yu-Mei Li, Ning-Ning Guo, Lu-Yuan Wang, Hiroko Isoda, Nobuhiro Ohkohchi, Hideki Taniguchi and Yun-Wen Zheng

1	Introduction	115
2	Liver Development and Microenvironment	125
3	Hepatic Cell Sources, Function, and Toxicity Assessment	127
4	3D vs. 2D Culture	129
5	Humanized Liver Animal Models Established by Hepatocyte Transplantation	132
6	Summary and Perspectives	135
	References	136

8. Methods for Engineering of Multicellular Spheroids to Reconstitute the Liver Tissue 145
Nobuhiko Kojima, Fumiya Tao, Hirotaka Mihara and Shigehisa Aoki

1	Introduction	145
2	Rapid Formation of MCS Using Methylcellulose Medium	146
3	Microchannel Formation in MCS	148
4	Regulation of Hepatic Functions by ECM Loading	151
5	Cell Polarity Formation by ECM Loading	155
6	Conclusion	156
	References	157

9. Role of Platelet, Blood Stem Cell, and Thrombopoietin in Liver Regeneration, Liver Cirrhosis, and Liver Diseases 159
Tomohiro Kurokawa and Nobuhiro Ohkohchi

1	Introduction	159
2	Blood Stem Cell	160
3	Thrombopoietin	163
4	Chronic Liver Disease	163
5	Platelets in Liver Regeneration	164
6	Platelets in Liver Cirrhosis	166
7	Platelet Transfusions on CLD and Cirrhosis	168
8	Effect of TPO Receptor Agonist Eltrombopag in CLD	168
9	Conclusion	170
	References	171

10. Dynamic Tissue Remodeling in Chronic Liver Diseases: Abnormal Proliferation and Differentiation of Hepatocytes and Bile Ducts/Ductules 179
Yuji Nishikawa

1	Introduction	179
2	Chronic Liver Diseases and Cirrhosis	180

	3	Ductular Reaction and its Possible Cellular Origins	182
	4	Phenotypic Plasticity of Mature Hepatocytes and its Involvement in the Ductular Reaction	188
		4.1 Ductular Differentiation of Hepatocytes In Vitro	188
		4.2 Ductular Differentiation of Hepatocytes In Vivo	194
		4.3 Transdifferentiation Versus Dedifferentiation of Mature Hepatocytes	196
	5	Remodeling of the Intrahepatic Bile Duct System in Chronic Liver Diseases	196
	6	Enhanced Hepatocarcinogenesis in Liver Cirrhosis	198
		6.1 Clonality of Regenerative Nodules in Liver Cirrhosis	198
		6.2 Experimental Evidence for the Selective Proliferation of Subsets of Hepatocytes in Chronic Liver Injury	199
		6.3 Dedifferentiation of Hepatocytes and Hepatocarcinogenesis	202
	7	Conclusions	204
	References		205
11.	The Role of Stem Cells in the Hepatobiliary System and in Cancer Development: A Surgeon's Perspective		211
	Naoto Koike		
	1	Introduction	211
	2	Anatomy and Development of the Hepatobiliary System	212
	3	Stem Cells in the Development and Regeneration of Hepatobiliary System	214
		3.1 Stem Cells in Liver Development	214
		3.2 Stem Cells in the Adult Liver	215
		3.3 Stem Cells in the Extrahepatic Biliary Tract	217
		3.4 The Role of Stem Cells in Hepatobiliary Surgery	218
	4	CSCs in Hepatobiliary Cancers	222
		4.1 Tumorigenesis and Aggressiveness of Hepatobiliary Cancers	222
	5	CSC Markers as Therapeutic Targets	224
		5.1 Side Population	224
		5.2 CD133	227
		5.3 OV6	228
		5.4 CD90	228
		5.5 CD44	229
		5.6 EpCAM	229
		5.7 CD13	230
		5.8 CD24	230
		5.9 CD47	231
		5.10 K19	231
		5.11 SOX9	232
		5.12 TLR4	232
		5.13 CD274	233
		5.14 $\alpha 2\delta 1$	233

	6	CSCs Markers as Roles of Prognostic Factors	233
	7	Other Novel Therapies for Hepatobiliary CSCs	236
		7.1 Immunotherapy for CSCs of Hepatobiliary Cancers	237
		7.2 Drug Repositioning	237
	8	Conclusions	238
	References		241

12. Stem/Progenitor Cells in Chronically Injured Liver and the Surrounding Microenvironment 255

Michitaka Matsuda and Minoru Tanaka

1	Introduction		255
2	Historical View of Liver Stem/Progenitor Cell Studies		257
3	Characterization of LPC		257
	3.1	The Deduced Origin of LPC	257
	3.2	Experimental Rodent Models and Human Liver Diseases Related to LPC	259
	3.3	Characterization of LPCs by Molecular Marker	261
4	Microenvironment Surrounding LPC in Injured Liver		263
	4.1	Signaling Pathways Related to LPC Regulation	263
	4.2	Contribution of LPC to Liver Regeneration	264
5	Conclusions and Future Directions		267
References			267

13. The Stem Cells in Liver Cancers and the Controversies 273

Hiroyuki Tomita, Tomohiro Kanayama, Ayumi Niwa, Kei Noguchi, Takuji Tanaka and Akira Hara

1	Introduction		273
2	Stem Cell Concept of Liver (Hepatocyte and Cholangiocyte)		273
	2.1	Liver Function and Architecture	273
	2.2	The Concept of Liver Stem/Progenitor Cell	274
	2.3	Fetal Liver Stem/Progenitor Cell	274
	2.4	Adult Liver Stem/Progenitor Cell in Homeostasis	275
	2.5	Liver Stem/Progenitor Cell and Regeneration after Injury	275
	2.6	Transdifferentiation between Hepatocyte and Cholangiocyte	276
3	Hepatocellular Carcinoma		276
	3.1	The Characteristics of HCC	276
	3.2	CSC Markers in HCC	277
	3.3	Heterogeneity in CSC of HCC	277
	3.4	Therapy for HCC	278
4	Cholangiocarcinoma (CCA)		279
	4.1	The Characteristics of CCA	279
	4.2	CSC Markers in CCA	279
	4.3	Therapy for CCA	280

		5	The Controversies in Stem Cells of HCC and CCA	280

       5   The Controversies in Stem Cells of HCC and CCA   280
    Conclusion   281
    References   282

14. Liver Cancer Stem Cells   289
Jin Ding and Wei-Fen Xie

    1   Characteristics of Liver Cancer Stem Cells   289
        1.1   Self-renewal   289
        1.2   Multiple Differentiation Capacity   289
        1.3   Constant Latency and Chemoresistance   290
        1.4   Heterogeneity of Liver Cancer Stem Cells   290
        1.5   Evasion of Immune Clearance   290
        1.6   Metastatic Potential   291
    2   Identification and Isolation of Liver Cancer Stem Cells   291
        2.1   Surface Biomarkers   291
        2.2   CSC Function-based Isolation   293
    3   Molecular Mechanism Underlying Regulation of LCSCs   294
        3.1   Transcription Factors   294
        3.2   Non-coding RNAs   295
        3.3   Epigenetic Factors   297
        3.4   Stemness-related Cascades   297
        3.5   Microenvironment   299
        3.6   Metabolism   300
    4   Putative Origins of Liver Cancer Stem Cells   300
        4.1   Liver Progenitor Cells   300
        4.2   Hepatocytes   301
        4.3   Liver Cancer Cells   301
    5   Therapeutic Strategy Targeting Liver Cancer Stem Cells   302
        5.1   Potential in Diagnosis   302
        5.2   Eliminating LCSCs   303
        5.3   Differentiation Strategy of LCSCs   303
        5.4   LCSC-targeted Immunotherapy   305
        5.5   Targeting the LCSC Microenvironment   305
    6   Current Controversy and Future Direction   306
    References   308

15. Clinical Application of Stem Cells in Liver Diseases: From Bench to Bedside   317
Yan Xu and Qi Zhang

    1   Introduction   317
    2   **PSCs in Liver Disease Study**   317
        2.1   PSC-Derived HPCs/Hepatocytes and Cholangiocytes in Cell Therapy of Liver Disease   318
        2.2   PSC in Liver Disease Modeling and Drug Screening   325
        2.3   Perspectives   330

3	Clinical Application of MSCs in Liver Diseases	331
	3.1 General Properties of MSCs	331
	3.2 Clinical Studies of MSC Therapy in Liver Diseases	332
	3.3 Potential Mechanisms of MSCs Therapy in Liver Disease	334
	3.4 Perspectives	338
References		338

Index	347

List of Contributors

Shigehisa Aoki Saga University, Saga, Japan

Qiwen Chen Shanghai Cancer Center, Shanghai Medical School, Fudan University, Shanghai, China

Hiromi Chikada Tokai University School of Medicine, Kanagawa, Japan

Jin Ding Changzheng Hospital, Second Military Medical University; The Eastern Hepatobiliary Surgery Hospital, Second Military Medical University, Shanghai, China

Jian-Yun Ge University of Tsukuba, Tsukuba, Japan

Ning-Ning Guo The Affiliated Hospital of Jiangsu University, Zhenjiang, Jiangsu, China

Akira Hara Gifu University Graduate School of Medicine, Gifu, Japan

Kazunori Hosaka National Cancer Center Research Institute, Tokyo; Niigata University Medical and Dental Hospital, Niigata, Japan

Pengyu Huang ShanghaiTech University, Shanghai, China

Hiroko Isoda University of Tsukuba, Ibaraki, Japan

Akihide Kamiya Tokai University School of Medicine, Kanagawa, Japan

Tomohiro Kanayama Gifu University Graduate School of Medicine, Gifu, Japan

Takeshi Katsuda National Cancer Center Research Institute, Tokyo, Japan

Naoto Koike Seirei Sakura Citizen Hospital, Sakura, Yokohama City University School of Medicine, Yokohama, Japan

Nobuhiko Kojima Yokohama City University, Yokohama, Japan

Tomohiro Kurokawa University of Tsukuba, Tsukuba, Ibaraki, Japan

Bin Li Oregon Stem Cell Center, Oregon Health and Science University, Portland, OR, United States

Yu-Mei Li The Affiliated Hospital of Jiangsu University, Zhenjiang, Jiangsu, China

Li-Ping Liu University of Tsukuba, Ibaraki, Japan; The Affiliated Hospital of Jiangsu University, Zhenjiang, Jiangsu, China

Michitaka Matsuda National Center for Global Health and Medicine, Tokyo, Japan

Hirotaka Mihara Yokohama City University, Yokohama, Japan

Toshihiro Mitaka Research Institute for Frontier Medicine, Sapporo Medical University School of Medicine, Sapporo, Japan

Norio Miyamura Medical Research Institute, Tokyo Medical and Dental University (TMDU), Tokyo, Japan

Yuji Nishikawa Asahikawa Medical University, Asahikawa, Hokkaido, Japan

Hiroshi Nishina Medical Research Institute, Tokyo Medical and Dental University (TMDU), Tokyo, Japan

Ayumi Niwa Gifu University Graduate School of Medicine, Gifu, Japan

Kei Noguchi Gifu University Graduate School of Medicine, Gifu, Japan

Takahiro Ochiya National Cancer Center Research Institute, Tokyo, Japan

Nobuhiro Ohkohchi University of Tsukuba, Tsukuba, Ibaraki, Japan

Yulong Su Oregon Health and Science University, Portland, OR, United States

Minoru Tanaka National Center for Global Health and Medicine, Tokyo, Japan

Takuji Tanaka Research Center of Diagnostic Pathology (RC-DiP), Gifu Municipal Hospital, Gifu, Japan

Hideki Taniguchi Yokohama City University, Yokohama, Japan

Naoki Tanimizu Research Institute for Frontier Medicine, Sapporo Medical University School of Medicine, Sapporo, Japan

Fumiya Tao Yokohama City University, Yokohama, Japan

Shuji Terai Niigata University Medical and Dental Hospital, Niigata, Japan

Amita Tiyaboonchai Oregon Stem Cell Center, Oregon Health and Science University, Portland, OR, United States

Hiroyuki Tomita Gifu University Graduate School of Medicine, Gifu, Japan

Lu-Yuan Wang The Affiliated Hospital of Jiangsu University, Zhenjiang, Jiangsu, China

Wei-Fen Xie Changzheng Hospital, Second Military Medical University, Shanghai, China

Yan Xu Biotherapy Center, The Third Affiliated Hospital, Sun Yat-sen University, Guangzhou, PR China

Qi Zhang Cell-gene Therapy Translational Medicine Research Center, The Third Affiliated Hospital, Sun Yat-sen University, Guangzhou, PR China

Yun-Wen Zheng University of Tsukuba, Tsukuba, Ibaraki, Japan; The Affiliated Hospital of Jiangsu University, Zhenjiang, Jiangsu, China; Yokohama City University, Yokohama, Japan

Editor Biography

Dr. Yun-Wen Zheng recently moved to the University of Tsukuba from Yokohama City University, where he was an Assistant Professor in the Department of Regenerative Medicine for ten years. He has joined the University of Tsukuba Faculty of Medicine as an Associate Professor in the Department of Advanced Gastroenterological Surgical Science and Technology for the researches on regenerative medicine. As a distinguished Medical Expert and Visiting Professor, he also serves for Research Center of Stem Cells and Regenerative Medicine, Jiangsu University Hospital, China as well as a Visiting Associated Professor for Yokohama City University School of Medicine.

Dr. Zheng received his B.S, M.S. in 1989 from Nanjing University and Ph.D. of Medical Science majoring in Physiology/Stem Cell Biology in 2003 from the University of Tsukuba Graduate School of Medicine. Following a postdoctoral fellowship of Japan Society for the Promotion of Science, he joined the medical faculty at Yokohama City University School of Medicine in the Department of Regenerative Medicine.

As a stem cell biologist, since 2003, he has been a team leader for liver and cancer stem cell researches as well as regenerative medicine. Dr. Zheng is interested in iPS and somatic stem cell developing into endodermal and hepatic direction as well the stem cells in normal and cancer tissue, especially for hepatic primordial development and the cellular microenvironment; Organoids and cell-cell interaction in 3D system; Immunodeficient animals, humanized liver as well humanized blood and drug development. Furthermore, he is also interested in patient-derived cellular reprogramming and transplant therapy. He holds eleven patents in Japan, United States, Europe, and China in the realm of embryonic stem cells, induced pluripotent stem cell, hepatic stem cells, and humanized animals.

Foreword

LIVER STEM CELLS: ARE THEY REAL OR ILLUSIONS?
The liver is a central organ for homeostasis and carries out a wide variety of functions, including digestion, metabolism, glycogen storage, plasma protein synthesis, hormone production, detoxification, and so on. Most of those liver functions are carried out by hepatocytes, the liver parenchymal cells. Another epithelial cell type in the liver is biliary epithelial cells (BECs), cholangiocytes. These two types of epithelial cells are derived from the common progenitor cell, which is also considered liver stem cell. Tissue stem cells are generally defined as the cells having the potentials to self-renew and differentiate to mature cells. While there are many reports on liver stem cell, the definitions of liver stem cell are rather ambiguous in many cases. During development, as hepatoblasts emerged from the foregut endoderm give rise to both hepatocytes and BECs, they may be considered liver stem cells. Although hepatoblasts proliferate, they appear transiently during development and there is no clear evidence showing that they persist after birth. Thus it would be more appropriate to describe hepatoblasts as a liver progenitor rather than liver stem cell. It has been known that adult livers harbor a specialized type of cells, which proliferate clonally in vitro and give rise to hepatocytes and BECs depending on culture conditions. However, it remains unclear whether there is a bona fide liver stem cell in the adult liver. It has been hypothesized that new hepatocytes are produced from the canal of Hering, the junction between the code of hepatocytes and the bile duct. However, due to the lack of specific marker of such cells, the nature of the cells remains unclear. In contrast, the hepatocytes next to the central vein were reported to be homeostatic liver stem cell that produces hepatocytes under normal conditions. Thus, there has been still a debate whether there is a liver stem cell that constitutively produces hepatocytes in a manner similar to hematopoietic stem cells in the bone marrow or the intestine stem cell in the intestinal crypt, which self-renew and differentiate to mature cells.

The liver is well known for its remarkable capacity to recover from injuries caused by various insults, such as surgical resection, viral infection, metabolic disorders, chemical and toxic stresses; the mechanisms underlying the recovery process are different depending on the type of injury. In the case of partial hepatectomy or acute liver injuries induced by chemicals, the remaining cells proliferate to repair the damage and liver stem/progenitor cells are unlikely to be required for the repair process. In contrast, severe/chronic injuries often induce proliferation of cells with bile duct markers. This phenomenon is known as ductular reaction and those proliferating cells are considered liver progenitor cells (LPCs), which are also known as oval cells in rodents. However, whether those LPCs contribute to newly developed hepatocytes has been intensively debated. Using lineage tracing, several reports showed that new hepatocytes are derived mostly from hepatocytes, but not from BECs or LPCs, whereas there are also several reports showing that BECs or LPCs become hepatocytes. Furthermore, the origin of LPCs is another subject of debate. While the controversy could be due to the injury models employed by different investigators, this is an issue that needs to be clarified experimentally.

Liver cancer is an aggressive disease with a poor outcome. Among primary liver cancers, hepatocellular carcinoma (HCC) accounts for 70%–85% of the cases and intrahepatic cholangiocarcinoma (ICC) is the second most frequent type of liver cancer. Both HCC and ICC are heterogeneous in their cellular morphology and clinical outcome. Mixed HCC-cholangiocellular carcinoma (HCC-CCC) is a rare form of liver cancer exhibiting both hepatocellular and cholangiocellular features, reminiscent of liver stem/progenitor cells. Because stem/progenitor cell markers are often expressed in HCC, mature hepatocytes may be dedifferentiated to an immature stage during tumorigenesis. Alternatively, there may be some immature hepatocytes such as LPCs that can become tumors. Intriguingly, recent studies on cell fate tracing of hepatocytes demonstrated that ICC can originate even from hepatocytes. Again there are rooms for investigating the origins of liver cancer. In this book, the authors discuss recent advance and understanding on liver stem/progenitor cells and cancer.

Atsushi Miyajima
Institute of Molecular and Cellular Biosciences,
University of Tokyo, Tokyo, Japan

Preface

Hepatocellular carcinoma (HCC) is the most popular disease of the liver in the world. It is well known that chronic deterioration and fibrosis of the liver have a strong relation for pathogenesis of HCC, especially in patients with hepatitis B and/or hepatitis C, but the precise pathophysiology is still not clarified yet. Resection, radio frequency ablation, trans-catheter-arterial chemo-embolization, and molecular targeted drug are carried out for the patients with HCC. In the patients with HCC after treatment we have sometimes experienced the recurrence of HCC, however, these treatments mentioned above, cannot prevent the recurrence of HCC. The most effective treatment is liver transplantation but not every patient can receive the liver transplantation because of shortage of the donor. For development of the effective treatment for liver disease, especially liver cirrhosis and HCC, the elucidation of the molecular pathogenesis is necessary.

In 21st century, researches about the stem cell have progressed remarkably, and embryology of the body and organs also have been elucidated precisely. In this book, entitled "Stem Cells and Cancer in Hepatology," the contents are as follows: development of the liver as the organ, generation and differentiation of various kind of cells of the liver, that is, parenchymal hepatocytes, endothelial cells and bile duct cells, remodeling of the liver in animal experiment and clinical setting. Chapters described on cancer stemness of HCC is also contained; therefore, this book is very instructive and informative for students, researchers, and clinicians of medical staff. This book is also helpful in developing powerful treatment for HCC.

Nobuhiro Ohkohchi
University of Tsukuba, Tsukuba, Japan

Acknowledgments

Editing this book was a great challenge for me.

I feel indebted to the international distinct experts, especially the members of Liver Biology Study Group in Japan, and my enthusiastic colleagues and partners in China and United States for sharing the same interests in liver and stem cells for this book as well in this field. I am sincerely grateful to these scientists for their strong commitment, cooperation, and excellent contributions.

I thank the colleagues Mrs. Dan Song of University of Tsukuba Faculty of Medicine, Dr. Yun-Zhong Nie of Yokohama City University School of Medicine to edit and proofread the manuscripts, and Dr. Jaejeong Kim of the University of Tsukuba Faculty of Medicine to design the elegant image of cover.

I thank the staff of Elsevier, especially Tracy Tufaga, Mohana Rajendran, Lisa Eppich, Mica Haley, Glyn Jones, and Tessa de Roo for their excellent help, cooperation, support, and editorial assistance for the timely publication of this book.

Lastly, I should remember that there would be no this book publication if none of the supports and assistants from the affiliated institutions, Department of Surgery, University of Tsukuba Faculty of Medicine, the Research Center of Stem Cells and Regenerative Medicine, the Affiliated Hospital of Jiangsu University for the supports and assistants.

I thank the staff of Elsevier, especially Tracy Tufaga, Mohana Rajendran, Lisa Eppich, Mica Haley, Glyn Jones, and Tessa de Roo for their excellent help, cooperation, support, and editorial assistance for the timely publication of this book.

Introduction

Yun-Wen Zheng, Jian-Yun Ge
University of Tsukuba, Tsukuba, Japan

The liver is the largest organ in the human body and plays important roles in homeostasis, including metabolism, glycogen storage, drug detoxification, production of various serum proteins, and bile secretion. In cases of acute liver injury, hepatic cells, especially hepatocytes, can rapidly proliferate to repair damage. However, when the proliferative capacity is impaired in the case of severe and/or chronic liver damage, a subset of the cell population—called liver stem/progenitor cells (LPCs)—is expected to appear and undergo a massive expansion. The shortages in donor supply and transplant rejections badly limit the clinical application of liver transplantation for disease treatment at present. The LPC has been considered as an alternative source to the traditional strategy. Therefore, the medical focus should be set to find, isolate, and use LPCs for making functional hepatocytes. Stem/progenitor cells have already been found in developing livers. Using fluorescence activated cell sorting (FACS), Zheng and his colleagues identified hepatic stem cells in developing murine [1,2] and human [3] livers, although it is still unknown whether human hepatic stem cells can be precisely isolated from a normal adult liver. Furthermore, despite numerous experiments, little is known about the stem cell inside the liver itself, and questions remain as to how hepatic stem cells differentiate into functional hepatocytes and cholangiocytes.

LPCs have been experimentally shown to be the cells of origin in liver cancer [4]. To understand the biology of stem cells and to exploit their use for therapy, it is critical to identify and characterize the factors that control the decision between their self-renewal and differentiation under normal physiological conditions and in disease. Study on stem cells can help us understand critical events during development, stem cell differentiation toward functional somatic cell fate as well as tumor initiation. Reprogramming, not limited to iPS cells, but also includes trans-differentiation between germ layer derived somatic cells. Moreover, reprogramming also enable mature cell de-differentiated into tissue stem cells. As a result, working on reprogramming would be helpful to understand cell plasticity and then regenerative medicine and cell therapy. Under persistent inflammatory processes, cell damage and high cellular turnover occur; the LPCs, including mature hepatocytes or cholangiocytes, are considered able to acquire stemness or de-differentiation and convert into liver CSCs. In fact, the heterogenicity of liver CSCs is well accepted and may therefore contribute to the observed morphological and biological heterogeneity characteristics of liver cancers, mainly hepatocellular carcinoma (HCC) and cholangiocarcinoma (CCA). However, the initiation of primary liver CSCs is still controversial.

It is becoming increasingly clear that discovering the mechanism of stem cell or CSC initiation and development in the liver are essential for solving the problems related to liver disease therapy, while developing a novel strategy to generate sufficient functional liver cells in vitro or in vivo is the major project for medical research and preclinical research. This unique book brings together the ideas and work of several talented individuals who are pioneering this exciting research field, and dealing with newly discovered molecular mechanisms concerning the role of the liver stem cells and CSCs involved in liver homeostasis and disease progression. It is written at a level designed to take the readers to the frontier of this specific research field. This book covers a variety of newly developed technologies for generating hepatocytes, which will lead to achievements of great clinical significance. It reflects the remarkable development in the field of liver regeneration in recent years, provides a great of exceptional chances and magnificent prospects, as well as offers insights into the new challenges of the future. Therefore, we sincerely hope this book will provide authoritative information and a new vision, for researchers in this field, of the exciting times that lie ahead for the future of clinically oriented human liver regeneration projects.

REFERENCES

[1] Suzuki A, Zheng YW, Kaneko S, Onodera M, Fukao K, Nakauchi H, et al. Clonal identification and characterization of self-renewing pluripotent stem cells in the developing liver. J Cell Biol 2002;156(1):173–84.

[2] Suzuki A, Zheng YW, Kondo R, Kusakabe M, Takada Y, Fukao K, et al. Flow-cytometric separation and enrichment of hepatic progenitor cells in the developing mouse liver. Hepatology 2000;32(6):1230–9.

[3] Zhang RR, Zheng YW, Li B, Nie YZ, Ueno Y, Tsuchida T, et al. Hepatic stem cells with self-renewal and liver repopulation potential are harbored in CDCP1-positive subpopulations of human fetal liver cells. Stem Cell Res Ther 2018;9(1):29. doi: 10.1186/s13287-017-0747-3.

[4] Chiba T, Zheng YW, Kita K, Yokosuka O, Saisho H, Onodera M, et al. Enhanced self-renewal capability in hepatic stem/progenitor cells drives cancer initiation. Gastroenterology 2007;133(3):937–50.

Chapter 1

Molecular Mechanisms of Liver Development: Lessons From Animal Models

Norio Miyamura, Hiroshi Nishina
Medical Research Institute, Tokyo Medical and Dental University (TMDU), Tokyo, Japan

1 INTRODUCTION

The liver plays a central role in organismal homeostasis due to its roles in metabolism and in the synthesis, storage, and redistribution of nutrients [1,2]. These functions are mostly undertaken by parenchymal hepatocytes, which in adult liver constitute about 70% of the total number of liver cells. Other liver cell types include cholangiocytes (1%), liver sinusoidal endothelial cells (LSECs; 12%), stellate cells (3%), Kupffer cells (15%), Pit cells and mesothelial cells [3–5]. A liver is functional only when the organ maintains the correct three-dimensional (3D) structure containing the appropriate numbers of the above cell types.

The liver has a unique dual blood supply mediated by the hepatic artery and the portal vein. The hepatic artery carries oxygen-rich blood from the aorta, whereas the portal vein carries blood rich in digested nutrients obtained from the gastrointestinal tract, spleen, and pancreas. Both the hepatic artery and portal vein subdivide within liver into small capillaries called liver sinusoids. These sinusoids are made up of LSECs, which cover the hepatic cords [4]. The sinusoids connect to the central vein within the liver, which is a branch of the hepatic vein that carries blood out of this organ.

One particularly important function of the liver is the maintenance of normal blood glucose levels. When blood glucose levels begin to decline, the liver activates the depolymerization of stored glycogen and the export of the resulting glucose monomers into the blood. Conversely, if excess glucose accumulates in the blood, it is rapidly taken up by the liver and sequestered as glycogen. The liver is also responsible for detoxification, removing waste and xenobiotics through metabolic conversion and biliary excretion. Heavy metals, toxic byproducts, drugs, chemicals, alcohol, and other poisons are all detoxified in the liver. When detoxification is suppressed due to liver dysfunction, toxic substances accumulate in the bloodstream and may trigger multiple organ failure.

Another major function of the liver is the continuous production of bile by hepatocytes. Bile passes through the bile duct made up of cholangiocytes and is concentrated and stored in the gallbladder. After eating, this stored bile is discharged into the duodenum to aid in digestion. Bile acids are essential for the formation of the mixed micelles in the small intestine that facilitate solubilization, digestion, and absorption of dietary lipids and fat-soluble vitamins [6]. Bile acids are reabsorbed by the ileum and transported back to the liver via the portal vein, inhibiting further bile acid synthesis. This enterohepatic bile circulation is highly efficient and critical for maintaining metabolic homeostasis.

Lastly, the liver is the major organ of hematopoiesis in the mammalian fetus [7,8]. Hematopoietic stem cells (HSCs) with diverse developmental potential are first found in the embryonic yolk sac, the aorta-gonad-mesonephros region, and the placenta. These HSCs migrate to and colonize the fetal liver, where they undergo expansion and maturation. HSCs from the fetal liver then colonize the fetal bone marrow and spleen, sustaining the embryo until the postnatal bone marrow takes over the function of core hematopoiesis, and the postnatal liver turns to the maintenance of metabolic homeostasis.

Despite the importance of the liver, our knowledge of its mechanisms of development has lagged compared to our understanding of the mechanisms underlying hematopoiesis and cardiac and nervous system development [2]. A major reason for this lag has been a dearth of molecular markers to identify distinct hepatic cell lineages. More than 300 "cluster of differentiation" (CD) molecules, which are cell surface markers of hematopoietic cells, have been defined and monoclonal antibodies raised against them. These tools have greatly accelerated the detailed delineation and characterization of hematopoietic cell lineages and functions. It is only relatively recently that markers capable of identifying and tracking liver cells have been isolated, and that unique mouse and small fish animal models crucial for the analysis of liver development have been devised. This chapter summarizes the use of these tools to increase our comprehension of how this vital organ develops and functions.

2 LIVER DEVELOPMENT

In higher organisms, the liver, hematopoietic system, and adipose tissue represent specialized and distinct organs or functional units. However, there is an intimate relationship among these organs that reflects their close evolutionary underpinnings. Indeed, the systems that control key metabolic and immune functions in higher organisms have evolved from common ancestral structures. One such structure is the *Drosophila* fat body. This structure incorporates the vertebrate homologues of the liver and the hematopoietic and immune systems [9,10]. Studies of Drosophila development have therefore pointed the way toward dissecting mechanisms governing liver development in vertebrates. We will now summarize key aspects of embryonic liver development in vertebrates.

2.1 Chronology of Liver Bud Formation

Within a fertilized mouse egg, the cells of the epiblast begin to differentiate into the three fundamental germ layers, namely the endoderm, mesoderm, and ectoderm, at around embryonic day 6 (E6) (Fig. 1.1) [11,12]. The endoderm forms the thyroid gland, lung, stomach, liver, pancreas, intestine, colon, and urinary bladder. The mesoderm forms muscle, bone, and adipose tissue, while the ectoderm forms the brain, peripheral nervous system, and skin. The murine fetal liver appears in a recognizable shape at about E10, but its development actually begins at E7.5 as an area of differentiation of the foregut endoderm tube, which is derived from medial and lateral domains of developing ventral foregut (Fig. 1.2) [13,14]. At E8.5–E9, this region of the endoderm tube receives several signals that cause some of its component cells to differentiate into hepatoblasts, which are multipotent hepatic progenitor cells marked by the expression of alpha-fetoprotein (AFP). Hepatoblasts proliferate and migrate into the surrounding septum transversum mesenchyme, resulting in liver bud formation at E9.5. Liver bud outgrowth then occurs at E9.5–E14.0.

2.2 Genes Involved in Liver Bud Formation

To dissect the extracellular and intracellular signaling networks involved in liver bud formation, Zaret and colleagues established and analyzed mouse embryo

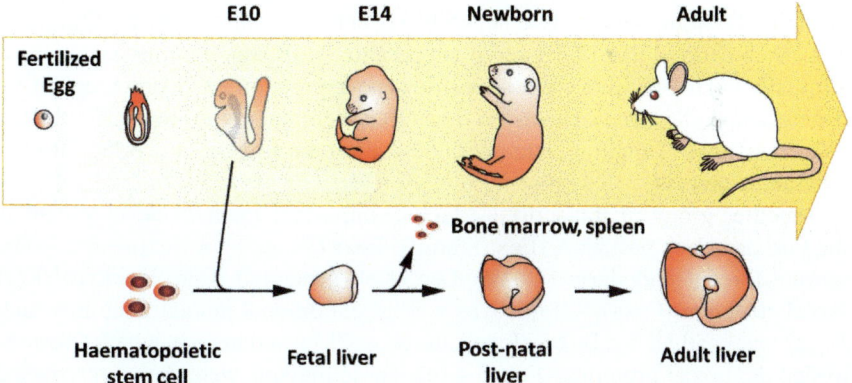

FIGURE 1.1 Overview of mouse liver development. The average gestation period in the mouse is 20 days. Hematopoietic stem cells arising from fertilized eggs flow into the fetal liver via the blood at around E10. HSCs generated in the fetal liver colonize the fetal bone marrow and spleen. After birth, the postnatal liver loses its hematopoietic function and acquires the adult liver functions of detoxification and the maintenance of metabolic homeostasis.

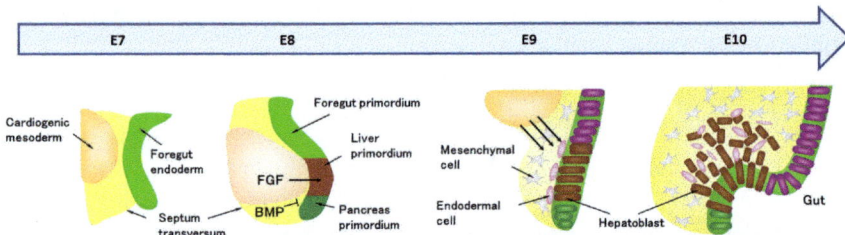

FIGURE 1.2 Overview of mouse liver bud formation. At E8 in the murine embryo, FGF signaling from the cardiac mesoderm, as well as BMP signaling from the septum transversum mesenchyme, are transmitted to the foregut endoderm. Where these signals interact determines the region that becomes the liver primordium. Continued signaling emanating from the cardiogenic mesoderm, mesenchymal cells, and endodermal cells induces hepatoblasts to appear at E9, with hepatic buds forming at E10.

tissue explants and whole embryo cultures [13,14]. In brief, mouse embryos at the 4–6 somite stage were excised from uteri and the yolk sac, optic lobes, and hindgut were carefully removed using electrolytically etched tungsten needles. To isolate the presumptive ventral endoderm, all head, upper cardiac lobe, dorsal, and posterior tissues were dissected away from the ventral foregut, and the lower half of the cardiac lobe was removed. The isolated tissues were placed physically adjacent to one another in culture dishes in various combinations and monitored for morphological changes. This approach revealed that it is the cardiac mesoderm that induces hepatoblast differentiation in the foregut. To identify the signaling pathway inducing hepatoblast differentiation, Zaret et al. tested a variety of ligands and signaling inhibitors using this in vitro system. They found that fibroblast growth factor (FGF) 1 or FGF2 was sufficient to generate hepatoblasts, and that an FGF2 inhibitor suppressed hepatoblast induction [15,16]. FGF8 did not generate hepatoblasts but did promote their proliferation. They then showed that FGFs act in cooperation with bone morphogenetic proteins (BMPs) to upregulate liver-specific gene expression while downregulating pancreas-specific gene expression during endodermal patterning [17]. By the balance of these signaling activities, the embryonic foregut is divided into the liver and pancreas.

Another group of transcription factors important for liver development is the Forkhead box protein A (Foxa) family. Foxa1, 2, and 3 are expressed in the murine foregut endoderm [18,19]. Lee et al. generated gene knockout (KO) Foxa1-deficient (Foxa1$^{-/-}$) mice, as well as conditional mutant mice in which Foxa2 was sandwiched by two loxp sites (Foxa2$^{fl/fl}$), and mice in which Cre controlled the Foxa3 promoter (Foxa3-Cre). These mutants were then intercrossed to produce Foxa1$^{-/-}$; Foxa2$^{fl/fl}$; Foxa3-Cre mice lacking expression of Foxa1 and Foxa2 in the foregut endoderm. The cardiac mesoderm, foregut, and notochord formed properly in these mutants but they lost AFP expression and failed to form the hepatic bud at E9.5. These results indicate that Foxa1 and Foxa2 are required for the establishment of competence within the foregut endoderm and for the onset of hepatogenesis (Fig. 1.3A).

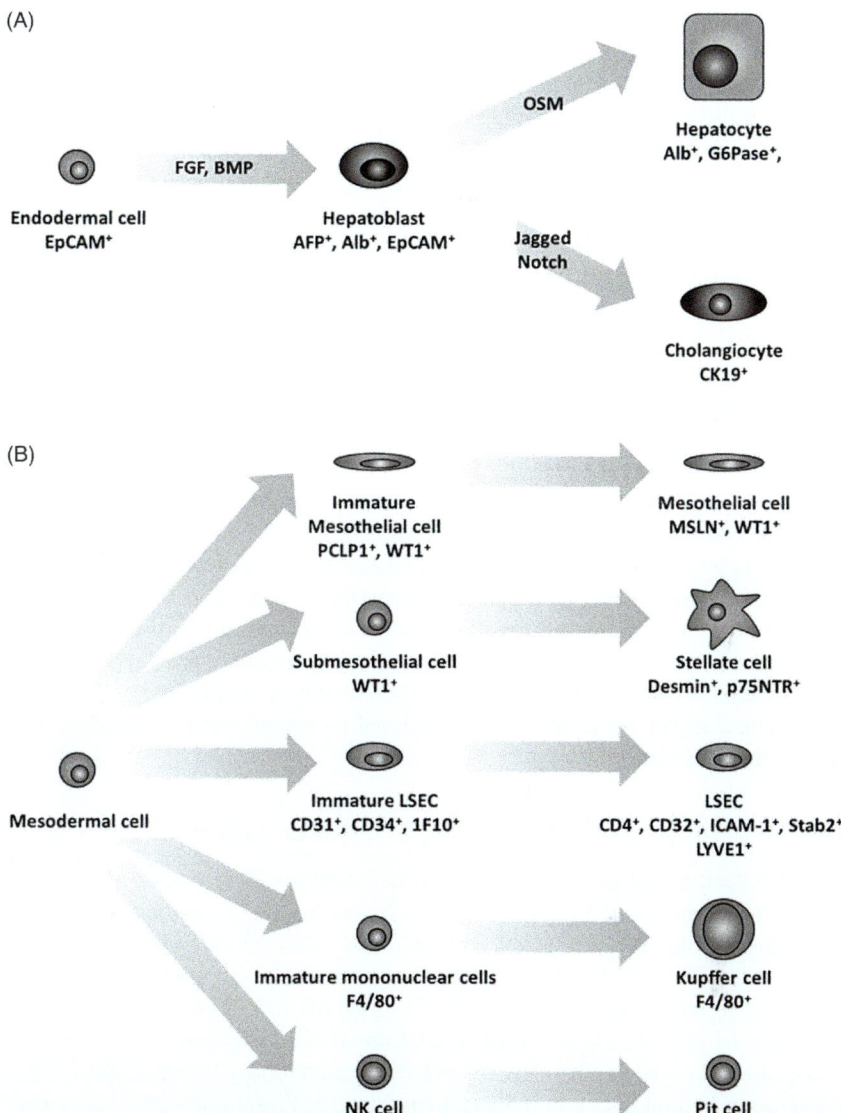

FIGURE 1.3 **Mouse liver cell lineages.** (A) Endodermal cells give rise to hepatoblasts that can differentiate into hepatocytes under the influence of OSM, or into cholangiocytes under the influence of Jagged/Notch signaling. (B) Mesodermal cells differentiate into the indicated immature lineages identified by the indicated markers. These cells in turn differentiate into the mature cell types shown on the right. The molecular markers identifying a given cell type are shown underneath.

The transcription factor Tbx3 is expressed in hepatoblasts at E9.5 [20]. Hepatoblasts of Tbx3$^{-/-}$ mice exhibited severe defects in proliferation as well as dysregulated hepatobiliary lineage segregation. Cholangiocyte (biliary epithelial cell) differentiation was promoted at the expense of hepatocyte differentiation, causing abnormal liver development. At the molecular level, Tbx3 deletion increased expression of the tumor suppressor p19ARF (Cdkn2a), which in turn induced the growth arrest of hepatoblasts and activated a gene expression program driving cholangiocyte differentiation. Thus, Tbx3 plays a crucial role in controlling hepatoblast proliferation and cell-fate determination by suppressing p19ARF expression, and thereby promotes liver organogenesis.

Mouse embryos lacking either of the stress-responsive kinases MKK4 and MKK7 displayed severe anemia and died between E10.5 and E12.5 [21–25]. Although hematopoiesis from yolk sac precursors, vasculogenesis, and hepatic bud formation were normal in these mutants, they showed a greatly reduced number of hepatocytes at E11.5–E12.5. MKK4-deficient hepatocytes subsequently underwent massive apoptosis at E12.5, and MKK7-deficient cells experienced G2/M cell cycle arrest due to impaired CDC2 expression. These results indicate that MKK4 and MKK7 are also crucial for normal hepatoblast proliferation and survival.

The transcription factor Prox1 is expressed in the hepatic primordium and the dorsal pancreatic bud [26,27]. To analyze Prox1's function in these tissues, Oliver and coworkers achieved functional inactivation of Prox1 by utilizing an in-frame insertion of the beta-galactosidase (lacZ) gene. Examination of the resulting Prox1-deficient embryos showed that the proliferation of LacZ$^+$ hepatocytes and their migration into the septum transversum were suppressed at E10.5. These Prox1-deficient embryos formed only a small liver and died at E14.5. Thus, Prox1 contributes to the control of hepatocyte proliferation and migration in the liver bud.

2.3 Hepatocyte and Cholangiocyte Differentiation

At E13.5 during murine embryogenesis, hepatoblasts differentiate into either hepatocytes expressing tyrosine aminotransferase (TAT), or cholangiocytes expressing cytokeratin 19 (CK19) [1,28]. The roles of various cytokines in hepatocyte differentiation have been analyzed using isolated hepatoblasts. Treatment of hepatoblasts in vitro with the interleukin-6 (IL-6) family member oncostatin M (OSM) triggered the induction of TAT expression followed by glycogen accumulation [29]. Because OSM is secreted by hematopoietic cells proliferating in fetal liver and induces hepatocyte differentiation, it is likely that OSM assists in the coordination of liver development and hematopoiesis.

Cholangiocyte precursors are generated at the ductal plate from a monolayer of hepatoblasts surrounding the portal veins, whereas hepatoblasts located away from portal vein areas differentiate into hepatocytes [30]. The evolutionarily conserved Jagged-Notch signaling pathway has been shown to be necessary for normal ductal plate development. When the ligand Jagged binds to the transmembrane Notch receptor, Notch is activated, its intracellular domain (ICD) is cleaved, and this Notch ICD translocates to the cell's nucleus. The ICD interacts with the DNA-binding protein Rbpj (also known as RBPJκ) to mediate changes in gene expression.

To study the role of Notch2 (N2) signaling during intrahepatic bile duct development, N2ICD mice harboring a Notch2ICD construct downstream of a floxed STOP cassette were crossed with Alb-Cre mice to induce conditional expression of Notch2ICD in hepatoblasts [31]. Overexpression of Notch2ICD in hepatoblasts promoted excessive cholangiocyte differentiation, and these cells formed additional bile ducts in periportal regions of the liver as well as ectopic ducts in its lobular regions. The additional periportal ducts were maintained into adulthood and became connected to the biliary tight junction network, resulting in an increased number of bile ducts per portal tract. Remarkably, Notch2ICD-expressing ductal plate remnants were not eliminated during postnatal development, implicating Notch2 signaling in cholangiocyte survival. In contrast, the lobular ectopic ducts did not persist into adulthood, indicating that local signals in the portal environment are important for maintaining bile ducts. In a related study, another group crossed Foxa3-Cre mice with Rbpj$^{loxP/\Delta}$ mice to suppress Notch2 signaling in hepatoblasts [32]. These mutants exhibited a decrease in ductal plate cells at E16.5 and a significant reduction in the number of bile ducts at postnatal day (P) 0. Thus, Notch2 signaling regulates ductal plate development, cholangiocyte differentiation, the induction of tubulogenesis during intrahepatic bile duct development, and chorangiocyte survival.

Mature hepatocytes are characterized by expression of the metabolism- and detoxification-related genes encoding glucose 6-phosphatase (G6Pase), phosphoenolpyruvate carboxykinase (PEPCK), glycogen synthase (GS), cytochrome P450 (Cyp) family enzymes, and tryptophan 2,3-dioxygenase (TDO) [33–36]. The expression of these genes is controlled by the transcription factor C/EBPα. Wang et al. showed that C/EBPα$^{-/-}$ hepatocytes failed to accumulate glycogen and lipids at E18.5 and showed reduced G6Pase, PEPCK, and GS expression at 2 h postbirth [37]. These mutants died from hypoglycemia within 8 h of birth. Thus, C/EBPα is critical for the establishment and maintenance of energy homeostasis, at least in neonatal mice.

2.4 Mesothelial Cell Differentiation

The adult mouse liver is covered by a single layer of mesothelial cells that provides a nonadhesive and protective surface, and also covers underlying capsular fibroblasts, termed submesothelial cells [38]. Mesothelial cells and submesothelial cells are separated by the basal lamina, which is composed of type IV collagen [39]. In contrast, the embryonic foregut that gives rise to the liver is not protected by mesothelial cells, at least up to E9.5 [40]. At E10.5, mesothelial cells start to cover the mesentery, coating the entire gut by E11.5. Immature mesothelial cells show high expression of the cell surface molecule podocalyxin-like protein 1 (PCLP1), which is then downregulated in mature mesothelial cells [41] (Fig. 1.3B). Conversely, the cell surface molecule mesothelin (MSLN) is minimally expressed by immature mesothelial cells but sharply upregulated in mature mesothelial cells. The Wilms Tumor 1 (WT1) protein is stably expressed in both immature and mature mesothelial cells as well as in submesothelial cells. The expression levels of WT1, PCLP1, and MSLN have been used to distinguish among subtypes of mesothelial and submesothelial cells and their developmental stages. For example, Onitsuka et al. studied $WT1^{-/-}$ mice and showed that these mutants display a 75% reduction in PCLP1$^+$ mesothelial cells at E13.5 [41]. Another group later demonstrated that impaired development of hepatic mesothelial cells resulted in decreased hepatocyte proliferation and abnormal liver morphogenesis [42]. Thus, WT1 is essential for mesothelial cell development and liver morphogenesis.

2.5 Hepatic Stellate Cell Differentiation

Hepatic stellate cells in adult mammalian liver are perisinusoidal cells that store vitamin A and produce growth factors, cytokines, prostaglandins, and other bioactive substances [3]. Upon activation, these stellate cells acquire a myofibroblast-like phenotype and show increases in proliferation, mobility, contractility, and synthesis of collagen and other extracellular matrix components. A subpopulation of submesothelial cells that migrates from the basement membrane into the liver is believed to contain the precursors of hepatic stellate cells [38]. WT1$^+$ mesothelial cells isolated from fetal liver have the capacity to store vitamin A [4], but mature hepatic stellate cells that do store vitamin A do not express WT1. Cell lineage analyses using WT1-CreERT2 mice carrying ROSA LacZ have revealed that hepatic stellate cells become labeled with LacZ at E11.5 [43], indicating that WT1$^+$ submesothelial cells develop into WT1$^-$ hepatic stellate cells.

2.6 LSEC Differentiation

The endothelium in the liver has a discontinuous architecture, meaning that fusion of the luminal and abluminal plasma membranes of LSECs occurs at sites other than cell junctions, in areas called "fenestrae." LSECs are derived from hemangioblasts, which are mesodermal progenitors common to endothelial cells and erythrocytes [4]. Immature LSECs gradually lose expression of markers expressed by cells in continuous endothelial layers, including CD31, CD34, and 1F10 antigen, and acquire expression of markers of adult liver sinusoidal cells, including CD4, CD32, ICAM-1, stabilin-2, and LYVE1. Other studies of liver vasculatures have suggested that the portal vein is derived from the umbilical mesenteric vein (omphalomesenteric vein), while the central vein develops from the posterior cardinal vein [44,45].

2.7 Differentiation of Kupffer Cells and Pit Cells

Kupffer cells, which are ameboid in shape, are macrophages that are resident in the liver and adhere to the surface of LSECs [5]. Macrophages first develop in the murine yolk sac at E8.5 [5]. These cells then migrate from the yolk sac via the bloodstream to colonize the fetal liver at E10. Pit cells are liver-associated natural killer cells [46]. Pit cells likely originate in the bone marrow and develop into large granular lymphocytes containing cytotoxic granules. Pit cells also associate with LSECs.

3 STUDIES OF LIVER FORMATION USING ZEBRAFISH

As useful as genetically modified mice are for examining gene functions, this species is not appropriate for forward-genetics studies. Zebrafish, which are small aquarium fish, can be bred in large numbers and require little cost and effort to maintain. Zebrafish eggs are fertilized and develop outside the mother's body. The embryos are transparent, so that organ formation can be observed in the living organism. Conveniently, zebrafish have many of the same organs as mammals, including a liver, gallbladder, pancreas, and heart. Moreover, the zebrafish genome is half the size of the mouse genome. All these attributes make the zebrafish ideal for detailed explorations of developmental genes. In the early 1990s, screening for zebrafish with altered early development was conducted in Tubingen, Germany and Boston, USA [47,48]. Hundreds of mutants were isolated, indicating that large-scale screening using zebrafish was eminently suitable. Since these studies, more than 10 zebrafish mutants showing abnormalities in liver morphogenesis have been characterized [49,50].

To aid in studies of zebrafish development, Stainier's group generated a transgenic zebrafish line in which green fluorescent protein (GFP) was expressed in endoderm and established an experimental system capable of analyzing liver bud formation and liver morphology in detail [51]. This team used this system to screen for mutations affecting endodermal organ morphogenesis and identified the *Prometheus* mutation as causing profound, but transient, defects in liver specification. Positional cloning revealed that *Prometheus* encodes a mutated Wnt2bb protein. Wnt2bb is normally expressed in restricted bilateral domains in the lateral plate mesoderm directly adjacent to the liver-forming endoderm. Two downstream effectors of Wnt2bb are Fgf and Bmp, whose signaling pathways play important roles in zebrafish liver specification [52]. Thus, normal Wnt, Fgf, and Bmp signaling are important for zebrafish liver specification.

Stainer's group then established another transgenic zebrafish line, called "basic helix-loop-helix transcription factor (hand2):enhanced green fluorescent protein (EGFP)", which expressed EGFP in lateral plate mesoderm [53]. Examination of this transgenic fish showed that hepatic stellate cells were labeled with EGFP, confirming that stellate cells are derived from lateral plate mesoderm. To analyze the dynamics of hepatic stellate cells and LSECs during liver development, hand2:EGFP zebrafish were mated to another transgenic zebrafish strain whose LSECs were labeled with mCherry. Monitoring of development in the progeny revealed that LSECs were the first cell type to invade the liver bud, followed by hepatic stellate cells. To analyze the contribution of LSECs to hepatic stellate cell localization, LSECs were removed by mating hand2:EGFP zebrafish to the *cloche* zebrafish mutant, which lacks most endothelial cells. In normal zebrafish, hepatic stellate cells are separated from biliary cells by hepatocytes. In hand2:EGFP;$cloche^{-/-}$ mutants, the hepatic stellate cells were instead closely associated with biliary cells. These results indicated that hepatic stellate cells are derived from lateral plate mesoderm and that stellate cell localization is dependent on LSECs.

4 STUDIES OF LIVER FORMATION USING MEDAKA

The medaka is another small fish species used as an animal model primarily in Japan [54–56]. There are currently around 20 inbred strains of medaka mutants. Medaka embryos are even more transparent than those of zebrafish, making it very easy to observe organs even in the late stages of embryogenesis (Fig. 1.4A). In addition, medaka can breed at a wider range of temperatures than zebrafish (14–34°C), making it possible to isolate useful temperature-sensitive mutations. The medaka genome size is 800 Mb, half the size of the zebrafish genome. Because the complementation rate in medaka is low compared to that in zebrafish, abnormal medaka phenotypes are exposed more readily than in zebrafish. A medaka screening effort spanning 1998–2003 identified 19 mutants

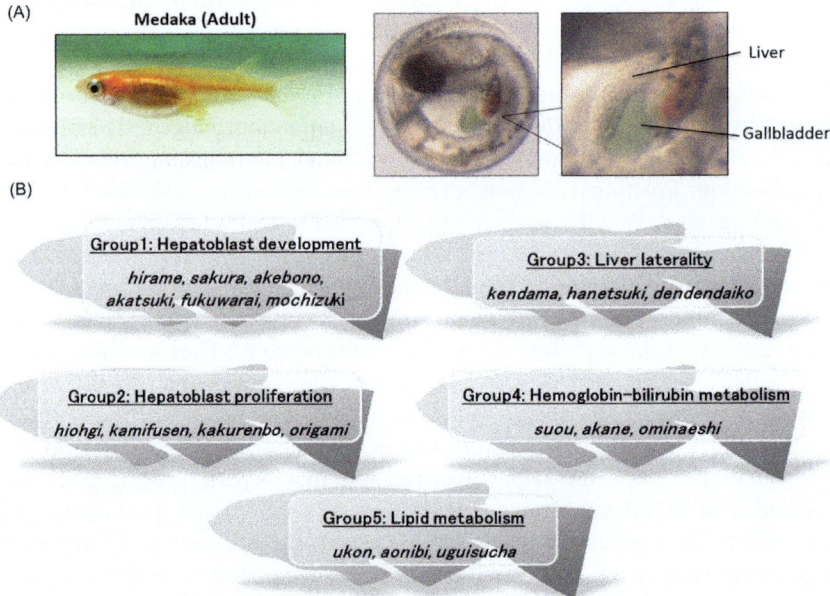

FIGURE 1.4 **Medaka mutations affecting liver formation and function.** (A) *Left*: Gross side view of an adult medaka fish. *Middle*: Gross view of a medaka embryo. *Right*: Higher magnification view of the region in the middle panel containing the liver and gallbladder. (B) The mutations listed were classified into five phenotypic groups based on their impairment of the indicated aspects of liver development or function.

showing hepatic dysfunction and/or defects in liver formation [55,57]. These 19 medaka mutants were then classified into five groups as follows (Fig. 1.4B).

4.1 Group 1: Mutations Affecting Hepatoblast Development

The endodermal rod does not develop in this group of mutants. Their patterning phenotypes are similar to those of *one-eyed-pinhead* zebrafish mutants, in which the endoderm is absent or strongly reduced [58–60]. In *sakura* mutant medaka embryos, the anterior part of the endodermal rod appears to be missing or misspecified, curtailing the growth of the hepatic region.

4.2 Group 2: Mutations Affecting Hepatoblast Proliferation

In *kamifusen* mutant medaka embryos, the liver and gallbladder are malformed, the heart is small, and blood accumulates near the liver and gut. This blood accumulation may reflect a defect either in vasculature formation in the liver or in endothelial cell development.

4.3 Group 3: Mutations Affecting Liver Laterality

Kendama mutant medaka shows a defect in the laterality of the viscera. In zebrafish *flh* and *boz* mutants, the laterality of the liver and heart is randomized and uncoupled [61]. In addition, both the *flh* and *boz* mutations affect early specification of the zebrafish notochord. In contrast, *kendama* mutants do not show any morphological deficits other than the laterality defect. These results suggest that the normal *kendama* gene product is a component of the signaling pathway downstream of *boz* and *flh*.

4.4 Group 4: Mutations Affecting Hemoglobin-Bilirubin Metabolism

Compromised liver function often leads to jaundice in humans. Accordingly, hemoglobin metabolism producing bilirubin in the liver is one of this organ's most important functions. The *suou* mutation in medaka (and its equivalent in zebrafish) were the first mutations in any species identified as affecting hemoglobin–bilirubin metabolism [57]. The *suou* mutant of either species may be useful as a model of human jaundice.

4.5 Group 5: Mutations Affecting Lipid Metabolism

Medaka embryos bearing either the *ukon* or *aonibi* mutations exhibit normal morphologies of the liver, gallbladder, and intestine [57]. However, these mutants show decreased accumulation of a fluorescent metabolite of a phospholipase A2 substrate, PED6, in the gallbladder. These mutations may therefore affect steps in lipid metabolism, such as ingestion and cleavage of lipids within the intestine and/or their subsequent hepatobiliary transport to the gallbladder.

4.6 Analysis of Medaka Mutants

The Group 2 *hiohgi* (*hio*) mutant exhibits a profound (but transient) defect in liver specification that resembles the liver formation defect found in zebrafish *prometheus* mutants [62]. However, in addition to this liver abnormality, *hio* mutants lack pectoral fins and die soon after hatching. Positional cloning showed that the *hio* mutation affects the *raldh2* gene encoding retinaldehyde dehydrogenase type-2 (RALDH2), the enzyme principally responsible for retinoic acid (RA) biosynthesis. Interestingly, in *hio* mutants, expression of Wnt2bb in the lateral plate mesoderm directly adjacent to the liver-forming endoderm is completely lost. Thus, this study revealed the unexpected finding that RA signaling positively regulates the Wnt2bb gene expression required for liver specification in medaka.

The Group 1 *hirame* (*hir*) mutant undergoes a collapse of all tissues, including the foregut, in the direction of gravity [63]. As a result, *hir* embryos exhibit body flattening that reflects their inability to withstand gravity, suggesting reduced tissue tension. Positional cloning revealed that the gene affected by the *hir* mutation encoded Yes-associated protein (YAP), a transcriptional coactivator. Actomyosin-mediated tissue tension was reduced in *hir* embryos, leading to tissue flattening and misalignment, both of which contribute to body flattening. These findings exposed a novel function for YAP in regulating 3D body shape in medaka. Naturally, these findings raised the question of whether YAP also functions in mammalian liver organogenesis.

5 STUDIES OF LIVER SIZE REGULATION IN MICE

It has been reported that the weight of a mammalian liver equals $0.033 \times$ the body weight of the animal$^{0.87}$ [64]. An important regulator of body weight and size in Drosophila is the Hippo pathway [65–67]. In mammals, the orthologue of the Drosophila Hippo pathway is involved in organ size control and cancer formation because it modulates cell proliferation via regulation of YAP activation. Central to the mammalian Hippo pathway is a kinase cascade wherein Mst (the mammalian orthologue of Drosophila Hippo) phosphorylates and activates the adaptor protein Mob and the protein kinase LATS. Activated LATS then phosphorylates YAP, promoting its cytoplasmic retention and thereby inhibiting its activation. Unphosphorylated (activated) YAP translocates into the nucleus, interacts with transcription factors, and induces target gene expression regulating organ size, including that of the liver.

$YAP^{-/-}$ mice die around E8.5, reflecting the early importance of the Hippo pathway [68]. Analysis of $YAP^{fl/fl}$ Alb-Cre mice at E18.5 showed that there was a profound defect in the formation of the ductal plate adjacent to the portal vein

[69]. In addition, the YAP-deficient cholangiocytes in these mutants regressed from P1 to P14 without forming mature bile ducts in the portal mesenchyme. Thus, YAP is critical for bile duct development. Others have demonstrated that MST1$^{fl/fl}$;MST2$^{fl/fl}$;Alb-Cre mice and Mob1a$^{fl/fl}$;Mob1b$^{-/-}$;Alb-Cre mice display hepatomegaly and cancer formation [70–73]. Similarly, liver-specific KO of other Hippo pathway components in mice increases liver size and promotes tumorigenesis [69,70]. Liver-specific YAP over-expression induced by the apolipoprotein E (ApoE) promoter also results in hepatomegaly and liver cancer [66]. Conversely, depletion of YAP suppresses liver cancer formation in Mob1a$^{fl/fl}$;Mob1b$^{-/-}$;Alb-Cre mice [73]. Thus, the liver phenotypes caused by impairments of the Hippo pathway are strongly dependent on YAP.

6 HUMAN LIVER DISEASES

6.1 Hepatocellular Carcinoma

The involvement of Hippo signaling defects in liver cancer in mice has its parallel in humans. Hepatocellular carcinoma (HCC) is one of the most common cancers in humans and is responsible for nearly 600,000 deaths each year worldwide [74]. Although Hippo pathway mutations are extremely rare in human HCC [67,75], YAP hyperactivation is common, being observed in approximately 60% of human HCC [76]. Other work has shown that YAP is essential for HCC initiation, progression, and/or metastasis in humans.

6.2 Metabolic Disorders

Metabolism and detoxification in the mammalian fetus rely on the mother's liver. In the newborn, ammonia accumulating in the neonatal liver is quickly converted to urea through the urea cycle. The urea cycle consists of six steps that are sequentially catalyzed by carbamoyl phosphate synthetase I (CPS1), ornithine transcarbamylase (OTC), argininosuccinate synthetase, argininosuccinase (ASS1), arginase (ARG), argininosuccinate lyase (ASL), and N-acetylglutamate synthase (NAGS) [77,78]. Mutations of the genes encoding these six enzymes can cause urea cycle disorders in human infants with an incidence of 50–150 cases/100,000 live births [78]. An affected baby is irritable or may refuse feedings, or may feed but then vomit repeatedly. The most dramatic symptoms occur in neonates at 24–48 h after birth.

Wilson's disease is a disorder of copper metabolism in the liver and occurs in about 1 in 30,000 people [79]. Wilson's disease is caused by an autosomal recessive mutation in the gene encoding ATPase Copper Transporting Beta (ATP7b), which is involved in the ATP-driven transport of copper in and out of cells. In the absence of functional ATP7b, an overload of copper damages the liver, brain, and other organs. Symptoms of Wilson's disease, which usually appear at age 5–10 years, are the vomiting of blood, loss of muscle control, slowness of movement, loss of memory, and mental confusion.

6.3 Alagille Syndrome

Alagille syndrome is an autosomal dominant disorder characterized by liver disease in combination with heart, skeletal, ocular, facial, renal, and pancreatic abnormalities [80]. JAGGED1 mutations have been identified in 94% of Alagille syndrome patients, and NOTCH2 mutations in another 2% [81]. The estimated prevalence of this syndrome is 1 in every 100,000 live births, with symptoms first observed by age 3 months. The liver of an Alagille syndrome patient shows very few or no intrahepatic bile ducts. Patients suffer from deep jaundice and severe pruritus as a consequence of cholestasis caused by congenital intrahepatic ductopenia. The infant mortality rate is about 10%.

7 CONCLUSION

About 20 years ago, Zaret's group established a beautiful experimental system based on the culture of whole mouse embryos. Many interesting tissue–tissue interactions during liver development were revealed by this approach, and a cogent framework of liver organogenesis was constructed. Subsequent use of genetically modified mice and molecular markers has clarified hepatoblast, hepatocyte, and cholangiocyte cell lineages and the mechanisms of their differentiation. The origins of mesothelial cells, LSECs, hepatic stellate cells, and Kupffer cells have also been revealed by these means. However, the origins and differentiation mechanisms establishing Pit cells and the endothelial cells of the portal vein and central vein are still largely unclear.

Small fish species are useful for forward-genetics analyses and bring distinct advantages in cell visualization technology compared to the mouse. Analysis of zebrafish mutants has revealed liver specification mechanisms that operate through Wnt, Fgf, and Bmp signaling and could not be established by mouse studies. Similarly, examination of the *hio* medaka mutant showed that RA signaling plays a critical role in liver specification upstream of Wnt signaling, and study of the *hir* mutant demonstrated the importance of tissue tension in 3D liver formation. Thus, several different animal models, each with their own advantages, have made different but equally valuable contributions to the progress of liver development research. That being said, some mysteries remain. Most studies of liver development have been traditionally focused on cell proliferation, differentiation, and/or lineage but not cell number. The concept of organ size regulation was introduced to the liver development field by studies of Hippo signaling in Drosophila, which led to the recognition of the importance of the Hippo-YAP pathway in mammalian liver. However, how this pathway assesses the total number of cells in a mammalian liver is still unknown [67].

Liver development research in mice and small fish has opened the door to progress in our understanding of human liver diseases. Although the JAGGED mutations associated with Alagille syndrome were identified in 1998, the underlying mechanism of the liver anomalies in this disease was not solved until

2009, when analysis of genetically modified mice with altered Notch signaling revealed the presence of bile duct disarrangement [82]. Thus, study of these mutants established the likely cause of Alagille syndrome in humans to be bile duct failure due to impaired JAG-Notch signaling. Observations of this type can be tied together to bring new hope to patients with liver cirrhosis or liver cancer, both groups of which typically need liver transplants. Because liver donor shortage is a serious problem, regenerative medicine using liver cells and tissues prepared in vitro is expected to fill the gap. Sekiya et al. have shown that mouse fibroblasts can be directly converted to hepatocyte-like cells (induced hepatocyte-like cells; iHep) by forced expression of Foxa and Hnf4 [83]. Others have demonstrated that culturing endodermal cells derived from induced pluripotent stem cells (iPSCs) with endothelial and mesenchymal cells can give rise to a liver bud with a functional 3D structure [84,85]. These significant advances strongly support the progress of regenerative medicine and illustrate the vital importance of basic research into liver development using a range of animal models.

LIST OF ACRONYMS AND ABBREVIATIONS

3D	Three-dimensional
AFP	Alpha-fetoprotein
Alb	Albumin
CD	Cluster of differentiation
E	Embryonic day
FGF	Fibroblast growth factors
Foxa	Forkhead box protein A
hio	*hiohgi*
hir	*hirame*
KO	Knockout
LSECs	Liver sinusoidal endothelial cells
P	Postnatal day
WT1	Wilms tumor 1
YAP	Yes-associated protein

REFERENCES

[1] Tanaka M, Itoh T, Tanimizu N, Miyajima A. Liver stem/progenitor cells: their characteristics and regulatory mechanisms. J Biochem 2011;149(3):231–9.

[2] Nakamura T, Nishina H. Liver development: lessons from knockout mice and mutant fish. Hepatol Res: Off J Jpn Soc Hepatol 2009;39(7):633–44.

[3] Geerts A. History, heterogeneity, developmental biology, and functions of quiescent hepatic stellate cells. Semin Liver Dis 2001;21(3):311–35.

[4] Poisson J, Lemoinne S, Boulanger C, Durand F, Moreau R, Valla D, et al. Liver sinusoidal endothelial cells: physiology and role in liver diseases. J Hepatol 2017;66(1):212–27.

[5] Naito M, Hasegawa G, Ebe Y, Yamamoto T. Differentiation and function of Kupffer cells. Medical Electron Microscopy: Off J Clin Elect Micro Soc Jpn 2004;37(1):16–28.

[6] de Aguiar Vallim Thomas Q, Tarling Elizabeth J, Edwards Peter A. Pleiotropic roles of bile acids in metabolism. Cell Metabol 2013;17(5):657–69.
[7] Golub R, Cumano A. Embryonic hematopoiesis. Blood Cells, Molec Dis 2013;51(4):226–31.
[8] Migliaccio G, Migliaccio AR, Petti S, Mavilio F, Russo G, Lazzaro D, et al. Human embryonic hemopoiesis: kinetics of progenitors and precursors underlying the yolk sac–liver transition. J Clin Investigat 1986;78(1):51–60.
[9] Hotamisligil GS. Inflammation and metabolic disorders. Nature 2006;444(7121):860–7.
[10] Sondergaard L. Homology between the mammalian liver and the Drosophila fat body. Trends Gen: TIG 1993;9(6):193.
[11] Arnold SJ, Robertson EJ. Making a commitment: cell lineage allocation and axis patterning in the early mouse embryo. Nat Rev Mol Cell Biol 2009;10(2):91–103.
[12] Gordillo M, Evans T, Gouon-Evans V. Orchestrating liver development. Development 2015;142(12):2094–108. Cambridge, England.
[13] Gualdi R, Bossard P, Zheng M, Hamada Y, Coleman JR, Zaret KS. Hepatic specification of the gut endoderm in vitro: cell signaling and transcriptional control. Genes Dev 1996;10(13):1670–82.
[14] Wandzioch E, Zaret KS. Dynamic signaling network for the specification of embryonic pancreas and liver progenitors. Science 2009;324(5935):1707–10. New York, NY.
[15] Jung J, Zheng M, Goldfarb M, Zaret KS. Initiation of mammalian liver development from endoderm by fibroblast growth factors. Science 1999;284(5422):1998. New York, NY.
[16] Calmont A, Wandzioch E, Tremblay KD, Minowada G, Kaestner KH, Martin GR, et al. An FGF response pathway that mediates hepatic gene induction in embryonic endoderm cells. Dev Cell 2006;11(3):339–48.
[17] Rossi JM, Dunn NR, Hogan BL, Zaret KS. Distinct mesodermal signals, including BMPs from the septum transversum mesenchyme, are required in combination for hepatogenesis from the endoderm. Genes Dev. 2001;15(15):1998–2009.
[18] Lee CS, Friedman JR, Fulmer JT, Kaestner KH. The initiation of liver development is dependent on Foxa transcription factors. Nature 2005;435(7044):944–7.
[19] Kaestner KH. The FoxA factors in organogenesis and differentiation. Cur Opin Gen Dev 2010;20(5):527–32.
[20] Suzuki A, Sekiya S, Buscher D, Izpisua Belmonte JC, Taniguchi H. Tbx3 controls the fate of hepatic progenitor cells in liver development by suppressing p19ARF expression. Development 2008;135(9):1589–95. Cambridge, England.
[21] Nishina H, Fischer KD, Radvanyi L, Shahinian A, Hakem R, Rubie EA, et al. Stress-signalling kinase Sek1 protects thymocytes from apoptosis mediated by CD95 and CD3. Nature 1997;385(6614):350–3.
[22] Watanabe T, Nakagawa K, Ohata S, Kitagawa D, Nishitai G, Seo J, et al. SEK1/MKK4-mediated SAPK/JNK signaling participates in embryonic hepatoblast proliferation via a pathway different from NF-kappaB-induced anti-apoptosis. Dev Biol 2002;250(2):332–47.
[23] Wada T, Joza N, Cheng HY, Sasaki T, Kozieradzki I, Bachmaier K, et al. MKK7 couples stress signalling to G2/M cell-cycle progression and cellular senescence. Nature Cell Biol 2004;6(3):215–26.
[24] Ganiatsas S, Kwee L, Fujiwara Y, Perkins A, Ikeda T, Labow MA, et al. SEK1 deficiency reveals mitogen-activated protein kinase cascade crossregulation and leads to abnormal hepatogenesis. Proc Natl Acad Sci U.S.A. 1998;95(12):6881–6.
[25] Nishina H, Vaz C, Billia P, Nghiem M, Sasaki T, De la Pompa JL, et al. Defective liver formation and liver cell apoptosis in mice lacking the stress signaling kinase SEK1/MKK4. Development 1999;126(3):505–16. Cambridge England.

[26] Wigle JT, Oliver G. Prox1 function is required for the development of the murine lymphatic system. Cell 1999;98(6):769–78.
[27] Sosa-Pineda B, Wigle JT, Oliver G. Hepatocyte migration during liver development requires Prox1. Nat Genet. 2000;25(3):254–5.
[28] Kamiya A, Inagaki Y. Stem and progenitor cell systems in liver development and regeneration. Hepatol Res: Off J Jpn Soc Hepatol 2015;45(1):29–37.
[29] Kamiya A, Kinoshita T, Ito Y, Matsui T, Morikawa Y, Senba E, et al. Fetal liver development requires a paracrine action of oncostatin M through the gp130 signal transducer. EMBO J 1999;18(8):2127–36.
[30] Ruebner BH, Blankenberg TA, Burrows DA, SooHoo W, Lund JK. Development and transformation of the ductal plate in the developing human liver. Pedia Pathol 1990;10(1-2):55–68.
[31] Tchorz JS, Kinter J, Muller M, Tornillo L, Heim MH, Bettler B. Notch2 signaling promotes biliary epithelial cell fate specification and tubulogenesis during bile duct development in mice. Hepatology 2009;50(3):871–9.
[32] Zong Y, Panikkar A, Xu J, Antoniou A, Raynaud P, Lemaigre F, et al. Notch signaling controls liver development by regulating biliary differentiation. Development 2009;136(10):1727–39. Cambridge, England.
[33] Adeva-Andany MM, Perez-Felpete N, Fernandez-Fernandez C, Donapetry-Garcia C, Pazos-Garcia C. Liver glucose metabolism in humans. Biosci Rep 2016;36(6).
[34] Adeva-Andany MM, González-Lucán M, Donapetry-García C, Fernández-Fernández C, Ameneiros-Rodríguez E. Glycogen metabolism in humans. BBA Clinical 2016;5:85–100.
[35] Badawy AA. Tryptophan metabolism, disposition and utilization in pregnancy. Biosci Rep 2015;35(5).
[36] Shahabi P, Siest G, Meyer UA, Visvikis-Siest S. Human cytochrome P450 epoxygenases: variability in expression and role in inflammation-related disorders. Pharmacol Therap 2014;144(2):134–61.
[37] Wang ND, Finegold MJ, Bradley A, Ou CN, Abdelsayed SV, Wilde MD, et al. Impaired energy homeostasis in C/EBP alpha knockout mice. Science 1995;269(5227):1108–12. New York, NY.
[38] Lua I, Asahina K. The role of mesothelial cells in liver development, injury, and regeneration. Gut Liver 2016;10(2):166–76.
[39] Asahina K, Tsai SY, Li P, Ishii M, Maxson RE, Sucov HM, et al. Mesenchymal origin of hepatic stellate cells, submesothelial cells, and perivascular mesenchymal cells during mouse liver development. Hepatology 2009;49(3):998–1011. Baltimore, Md.
[40] Wilm B, Ipenberg A, Hastie ND, Burch JBE, Bader DM. The serosal mesothelium is a major source of smooth muscle cells of the gut vasculature. Development 2005;132(23):5317. Cambridge, England.
[41] Onitsuka I, Tanaka M, Miyajima A. Characterization and functional analyses of hepatic mesothelial cells in mouse liver development. Gastroenterology 2010;138(4). 1525- 35.e6.
[42] Ijpenberg A, Pérez-Pomares JM, Guadix JA, Carmona R, Portillo-Sánchez V, Macías D, et al. Wt1 and retinoic acid signaling are essential for stellate cell development and liver morphogenesis. Dev Biol 2007;312(1):157–70.
[43] Asahina K, Zhou B, Pu WT, Tsukamoto H. Septum transversum-derived mesothelium gives rise to hepatic stellate cells and perivascular mesenchymal cells in developing mouse liver. Hepatology 2011;53(3):983–95. Baltimore, Md.
[44] Shiojiri N, Niwa T, Sugiyama Y, Koike T. Preferential expression of connexin37 and connexin40 in the endothelium of the portal veins during mouse liver development. Cell Tissue Res 2006;324(3):547.

[45] Goldman O, Han S, Hamou W, Jodon de Villeroche V, Uzan G, Lickert H, et al. Endoderm generates endothelial cells during liver development. Stem Cell Rep 2014;3(4): 556–65.
[46] Wisse E, Luo D, Vermijlen D, Kanellopoulou C, De Zanger R, Braet F. On the function of pit cells, the liver-specific natural killer cells. Semin Liver Dis 1997;17(4):265–86.
[47] Haffter P, Granato M, Brand M, Mullins MC, Hammerschmidt M, Kane DA, et al. The identification of genes with unique and essential functions in the development of the zebrafish, Danio rerio. Development 1996;123:1–36. Cambridge, England.
[48] Nüsslein-Volhard C. The zebrafish issue of development. Development 2012;139(22):4099. Cambridge England.
[49] Chen JN, Haffter P, Odenthal J, Vogelsang E, Brand M, van Eeden FJ, et al. Mutations affecting the cardiovascular system and other internal organs in zebrafish. Development 1996;123(1):293. Cambridge, England.
[50] Pack M, Solnica-Krezel L, Malicki J, Neuhauss SC, Schier AF, Stemple DL, et al. Mutations affecting development of zebrafish digestive organs. Development 1996;123:321–8. Cambridge, England.
[51] Ober EA, Verkade H, Field HA, Stainier DYR. Mesodermal Wnt2b signalling positively regulates liver specification. Nature 2006;442(7103):688–91.
[52] Shin D, Shin CH, Tucker J, Ober EA, Rentzsch F, Poss KD, et al. Bmp and FGF signaling are essential for liver specification in zebrafish. Development 2007;134(11):2041–50. Cambridge, England.
[53] Yin C, Evason KJ, Maher JJ, Stainier DY. The basic helix-loop-helix transcription factor, heart and neural crest derivatives expressed transcript 2, marks hepatic stellate cells in zebrafish: analysis of stellate cell entry into the developing liver. Hepatology 2012;56(5):1958–70. Baltimore, Md.
[54] Shima A, Mitani H. Medaka as a research organism: past, present and future. Mechan Dev 2004;121(7-8):599–604.
[55] Naruse K, Hori H, Shimizu N, Kohara Y, Takeda H. Medaka genomics: a bridge between mutant phenotype and gene function. Mechan Dev 2004;121(7-8):619–28.
[56] Furutani-Seiki M, Wittbrodt J. Medaka and zebrafish, an evolutionary twin study. Mechan Dev 2004;121(7-8):629–37.
[57] Watanabe T, Asaka S, Kitagawa D, Saito K, Kurashige R, Sasado T, et al. Mutations affecting liver development and function in Medaka, Oryzias latipes, screened by multiple criteria. Mechan Dev 2004;121(7-8):791–802.
[58] Schier AF, Neuhauss SC, Harvey M, Malicki J, Solnica-Krezel L, Stainier DY, et al. Mutations affecting the development of the embryonic zebrafish brain. Development 1996;123:165–78. Cambridge, England.
[59] Schier AF, Neuhauss SC, Helde KA, Talbot WS, Driever W. The one-eyed pinhead gene functions in mesoderm and endoderm formation in zebrafish and interacts with no tail. Development 1997;124(2):327–42. Cambridge, England.
[60] Zhang J, Talbot WS, Schier AF. Positional cloning identifies zebrafish one-eyed pinhead as a permissive EGF-related ligand required during gastrulation. Cell 1998;92(2):241–51.
[61] Chin AJ, Tsang M, Weinberg ES. Heart and gut chiralities are controlled independently from initial heart position in the developing zebrafish. Dev Biol 2000;227(2):403–21.
[62] Negishi T, Nagai Y, Asaoka Y, Ohno M, Namae M, Mitani H, et al. Retinoic acid signaling positively regulates liver specification by inducing wnt2bb gene expression in medaka. Hepatology 2010;51(3):1037–45. Baltimore, Md.

[63] Porazinski S, Wang H, Asaoka Y, Behrndt M, Miyamoto T, Morita H, et al. YAP is essential for tissue tension to ensure vertebrate 3D body shape. Nature 2015;521(7551): 217–21.
[64] Calder WA. Size, function and life history. Dover Publications; 1984.
[65] Zhao B, Wei X, Li W, Udan RS, Yang Q, Kim J, et al. Inactivation of YAP oncoprotein by the Hippo pathway is involved in cell contact inhibition and tissue growth control. Genes Dev 2007;21(21):2747–61.
[66] Dong J, Feldmann G, Huang J, Wu S, Zhang N, Comerford SA, et al. Elucidation of a universal size-control mechanism in drosophila and mammals. Cell 2007;130(6):1120–33.
[67] Yu FX, Zhao B, Guan KL. Hippo pathway in organ size control, tissue homeostasis, and cancer. Cell 2015;163(4):811–28.
[68] Morin-Kensicki EM, Boone BN, Howell M, Stonebraker JR, Teed J, Alb JG, et al. Defects in yolk sac vasculogenesis, chorioallantoic fusion, and embryonic axis elongation in mice with targeted disruption of Yap65. Molec Cell Biol 2006;26(1):77–87.
[69] Zhang N, Bai H, David KK, Dong J, Zheng Y, Cai J, et al. The Merlin/NF2 tumor suppressor functions through the YAP oncoprotein to regulate tissue homeostasis in mammals. Dev Cell 2010;19(1):27–38.
[70] Lu L, Li Y, Kim SM, Bossuyt W, Liu P, Qiu Q, et al. Hippo signaling is a potent in vivo growth and tumor suppressor pathway in the mammalian liver. Proc Nat Acad Sci 2010;107(4):1437–42.
[71] Song H, Mak KK, Topol L, Yun K, Hu J, Garrett L, et al. Mammalian Mst1 and Mst2 kinases play essential roles in organ size control and tumor suppression. Proc Nat Acad Sci 2010;107(4):1431–6.
[72] Nishio M, Hamada K, Kawahara K, Sasaki M, Noguchi F, Chiba S, et al. Cancer susceptibility and embryonic lethality in Mob1a/1b double-mutant mice. J Clin Investig 2012;122(12):4505–18.
[73] Nishio M, Sugimachi K, Goto H, Wang J, Morikawa T, Miyachi Y, et al. Dysregulated YAP1/TAZ and TGF-beta signaling mediate hepatocarcinogenesis in Mob1a/1b-deficient mice. Proc Nat Acad Sci U.S.A. 2016;113(1):E71–80.
[74] McGlynn KA, London WT. The global epidemiology of hepatocellular carcinoma. Pres Fut: Clin Liver Dis 2011;15(2):223–43.
[75] Zanconato F, Cordenonsi M, Piccolo S. YAP/TAZ at the roots of cancer. Cancer Cell 2016;29(6):783–803.
[76] Harvey KF, Zhang X, Thomas DM. The Hippo pathway and human cancer. Nat Rev Cancer 2013;13(4):246–57.
[77] Meijer AJ, Lamers WH, Chamuleau RA. Nitrogen metabolism and ornithine cycle function. Physiol Rev 1990;70(3):701.
[78] Natesan V, Mani R, Arumugam R. Clinical aspects of urea cycle dysfunction and altered brain energy metabolism on modulation of glutamate receptors and transporters in acute and chronic hyperammonemia. Biomed Pharmaco 2016;81:192–202.
[79] Hedera P. Update on the clinical management of Wilson's disease. Applic Clin Genet 2017;10:9–19.
[80] Saleh M, Kamath BM, Chitayat D. Alagille syndrome: clinical perspectives. Applic Clin Genet 2016;9:75–82.
[81] Boyer-Di Ponio J, Wright-Crosnier C, Groyer-Picard MT, Driancourt C, Beau I, Hadchouel M, et al. Biological function of mutant forms of JAGGED1 proteins in Alagille syndrome: inhibitory effect on Notch signaling. Human Molec Genet 2007;16(22):2683–92.

[82] Krantz ID, Colliton RP, Genin A, Rand EB, Li L, Piccoli DA, et al. Spectrum and frequency of jagged1 (JAG1) mutations in Alagille syndrome patients and their families. Am J Human Genet 1998;62(6):1361–9.
[83] Sekiya S, Suzuki A. Direct conversion of mouse fibroblasts to hepatocyte-like cells by defined factors. Nature 2011;475(7356):390–3.
[84] Takebe T, Sekine K, Enomura M, Koike H, Kimura M, Ogaeri T, et al. Vascularized and functional human liver from an iPSC-derived organ bud transplant. Nature 2013;499(7459): 481–4.
[85] Takebe T, Zhang RR, Koike H, Kimura M, Yoshizawa E, Enomura M, et al. Generation of a vascularized and functional human liver from an iPSC-derived organ bud transplant. Nature Proto 2014;9(2):396–409.

Chapter 2

Molecular Mechanisms Regulating the Proliferation and Maturation of Hepatic Progenitor Cells During Liver Development

Hiromi Chikada, Akihide Kamiya
Tokai University School of Medicine, Kanagawa, Japan

1 INTRODUCTION

The liver is the largest organ in the body and has several important metabolic functions such as detoxification, amino acid and lipid metabolism, and expression of serum proteins. The liver contains different types of cells such as hepatocytes, cholangiocytes, sinusoidal endothelial cells, Kupffer cells, and stellate cells. In particular, hepatocytes, the liver parenchymal cells, are the most important cells for liver functions and express many metabolic genes. Cholangiocytes, another epithelial cell type in the liver, form intrahepatic bile ducts and are important for bile acid transport. Hepatocytes and cholangiocytes are differentiated from the same origin, hepatic stem cells, during embryonic liver development.

In mouse liver development, the foregut endoderm develops into the hepatic organ by stimulation of soluble factors derived from the heart and septum transversum. Fibroblast growth factor from the cardiac mesoderm and bone morphogenic protein from the septum transversum are important for differentiation into the fetal liver bud [1,2]. In the next step of liver development, hepatic progenitor cells called hepatoblasts migrate into the septum transversum and expand [3]. At the same time, hematopoietic stem cells, originating in the aorta-gonad-mesonephros region, migrate into the fetal liver and expand. During this mid- to late-fetal liver development stage, fetal hepatocytes start producing several metabolic enzymes for the adult liver functions. In a study on the molecular basis of fetal liver maturation, oncostatin M (OSM), an interleukin 6 family cytokine, was revealed to promote hepatic maturation, as evidenced by the induction of metabolic enzymes, accumulation of glycogen and lipids, and detoxification of ammonia. OSM was expressed in CD45+ hematopoietic cells in the mid-fetal livers, whereas the OSM receptor was mainly detected in hepatic cells [4]. In addition, the interaction between the extracellular matrices and integrins is important for the terminal maturation of fetal hepatocytes [5]. These results suggest that OSM produced from hematopoietic cells and the extracellular matrices produced from the nonparenchymal liver cells play pivotal roles in fetal liver development. In contrast, immature fetal hepatocytes express many soluble factors and cytokines. In addition, proliferation of hematopoietic cells derived from the fetal liver is induced by coculture with fetal hepatocytes [6]. Maturation of fetal hepatocytes significantly decreases their ability to support the proliferation of hematopoietic cells. Thus, the interaction between fetal hepatocytes and hematopoietic cells is important for regulating fetal liver cell proliferation and differentiation [7].

Bile ducts, which contribute to the transport of the bile acids synthesized by mature hepatocytes, are differentiated from hepatoblasts. The hepatoblasts around the portal veins are stimulated by soluble factors and differentiate into ductal plates during the mid-fetal liver stage. These ductal plates form mature bile ducts connecting the bile canaliculi at the apical membrane of hepatocytes and the extrahepatic bile ducts. Mouse model analyses revealed the importance of soluble factors and cell–cell interactions in this step. A gradient of transforming growth factor (TGF)-β signaling is important for the cholangiocytic differentiation of hepatoblasts [8,9]. TGF-β concentration is higher around the portal vein than that near the parenchyma. The Notch-Jagged interaction is also important for biliary system development [10,11]. In particular, Alagille syndrome is an autosomal dominant disorder associated with bile duct paucity and was found to be caused by mutations in Jagged1 [12]. The liver also has several other nonparenchymal cells. Sinusoidal endothelial cells are the specific endothelial cells in the liver and form liver sinusoids. Kupffer cells are the liver-specific macrophages, lining the sinusoid walls. Hepatic stellate cells, called as Ito cells, are the mesenchymal cells in the perisinusoidal region of the liver. This region is called as the space of Disse and is a small area between the sinusoidal endothelial cells and hepatocytes. The interactions of these cells are important for hepatocytic function. For instance, hepatic stellate cells are activated during liver injury and express several extracellular matrix proteins such as collagen. This step is involved in the progression of liver fibrosis.

In this review, we concentrate on the molecular mechanisms regulating the proliferation and differentiation of hepatic stem/progenitor cells. We established an in vitro hepatic maturation and gene transfer culture system using mouse embryonic liver cells and retroviral vectors. Using this culture system, we found several new transcription factors regulating the hepatic progenitor cells. In addition, we also established a human hepatic progenitor cell culture system derived from human pluripotent stem cells. We discuss the possibility of using these progenitor cells for research and therapy concerning severe liver diseases in the future.

2 LIVER REGENERATION AND NORMAL LIVER-DERIVED STEM/PROGENITOR CELLS

One of important characteristics of the liver is its regeneration potential [13]. Loss or damage of liver tissues induces the proliferation of remaining hepatocytes to regenerate the organ mass and liver functions. In this regeneration step, mature hepatocytes temporally acquire a high-proliferative ability and compensate for the loss of parenchymal tissue. However, at times, massive damage of parenchymal cells cannot be repaired by mature-hepatocyte mediated proliferation. In these situations, the stem/progenitor cell population is considered to contribute to liver regeneration [14]. Liver parenchymal epithelial cells, hepatocytes, have a long half-life. Thus, the mature hepatocyte turnover in normal conditions is slower than that of other digestive organ tissue cells such as the intestine epithelial cells. However, the source for the maintenance of mature hepatocytes in noninjured livers remains controversial. A mouse model using Cre-LoxP system-mediated cell-labeling methods can reveal the heritable cells derived from stem and progenitor cells under normal and injured conditions. In these studies, a major theory is that hepatocytes are mainly maintained from preexisting hepatocytes in the normal liver [15]. In addition, a recent study revealed that self-renewing hepatocytes adjacent to the central vein are important for maintaining mature hepatocytes [16]. Hepatocytes positive for the Wnt target gene, Axin2, exist around the central vein and function as stem-like cells. Under severe liver injury conditions, in addition to hepatocyte-mediated regeneration, a putative stem/progenitor cell population has been experimentally characterized by several cell surface markers. For example, we purified $CD13^+CD133^+$ cells from noninjured adult livers and established an efficient single-cell culture system using fetal liver cell-derived conditioned medium and Y-27632, a Rho-associated kinase (ROCK) inhibitor [17]. Colonies derived from these $CD13^+CD133^+$ cells can be expanded for a long time and differentiate into hepatocyte-like and cholangiocyte-like cells under appropriate culture conditions. These results demonstrate the presence of stem and progenitor cells in the $CD13^+CD133^+$ subpopulation of nonhematopoietic cells derived from noninjured postnatal livers (detailed methods were written in [18]).

3 REGULATION OF TRANSCRIPTION FACTORS REGULATING THE PROLIFERATION AND MATURATION OF HEPATIC PROGENITOR CELLS

Hepatic progenitor cells have the bipotency to differentiate into mature hepatocytes and cholangiocytes with high proliferative ability [19]. During fetal liver development, the liver bud derived from the foregut endoderm contains many hepatic progenitor cells, which expand and differentiate into mature cells. As described earlier, we showed that OSM derived from hematopoietic cells is important for hepatic progenitor cell maturation. However, the molecular mechanism regulating the proliferation and maturation of hepatic progenitor cells remains unknown. Particularly, the in vitro characteristics of hepatic progenitor cells at the onset of liver development (embryonic day [E] 9.5–10.5 in mice) have not been elucidated. We therefore established a novel culture system using feeder cells and a ROCK inhibitor for in vitro clonal expansion of candidate hepatic progenitor cells (CD13$^+$Dlk$^+$ cells) from early-fetal livers and found that these cells showed bipotency and could proliferate to form large colonies. In this culture system, we used mouse embryonic fibroblasts (MEFs) as feeder cells. Hepatic progenitor cells derived from mid-fetal livers (mouse E13–14 livers) can expand without feeder cells. In contrast, hepatic progenitor cells derived from early-fetal livers require interaction with feeder cells for in vitro expansion. This is the first investigation indicating that purified hepatic progenitor cells in early-fetal livers have properties distinct from those in mid-fetal livers [20]. In addition, these results suggested that interaction between hepatic progenitor cells and mesenchymal cells is important for the regulation of liver development. The fetal liver has several mesenchymal cells, which are known to be the origin of stellate cells and portal myofibroblasts [21,22]. Thus, we next tried to purify mesenchymal cells from fetal livers [23]. We stained mouse mid-fetal livers with antibodies against several mesenchymal cell surface proteins and detected platelet-derived growth factor receptor (PDGFR) α-positive cells in the parenchymal region of E13.5–14.5 fetal livers. Flow cytometric purification revealed that Dlk1^{mid}PDGFRα$^+$ cells have a mesenchymal morphology and mesenchymal stem/progenitor cell-like properties such as the potential to differentiate into adipocytes, chondrocytes, and osteocytes. Mesenchymal stem/progenitor cells have been found in multiple tissues including the adult bone marrow, adipose tissue, skeletal muscle, skin, umbilical cord, dental pulp, and amniotic fluid [24]. These mesenchymal stem/progenitor cells are also found in fetal liver tissues. We next analyzed the effect of the interaction between hepatic cells and mesenchymal cells using a coculture system. Dlk1^{mid}PDGFRα$^+$ cells possessed the potential to support the proliferation of hepatic progenitor cells through production of soluble paracrine factors. In the bone marrow, mesenchymal stem cells are important for the function of the hematopoietic stem cell niche [25,26]. Thus, it is a common observation that mesenchymal stem-like cells support organ development and maintenance.

Next, we analyzed transcription factors that might be involved in the regulation of hepatic progenitor cells. To find candidate transcription factors, we compared the expression of transcription factors in fetal hepatic progenitor cells to that in mature hepatocytes. One of the genes mainly expressed in hepatic progenitor cells is prospero-related homeobox 1 (prox1) and its binding partner is liver receptor homolog 1 (lrh1) [27]. In order to analyze whether these candidate transcription factors regulate the proliferation and maturation of hepatic progenitor cells, we established a novel primary and gene transfer system using retroviral vectors (Fig. 2.1). CD133$^+$ and Dlk1$^+$ hepatic progenitor cells were purified from fetal livers using flow cytometry or magnetic beads. These

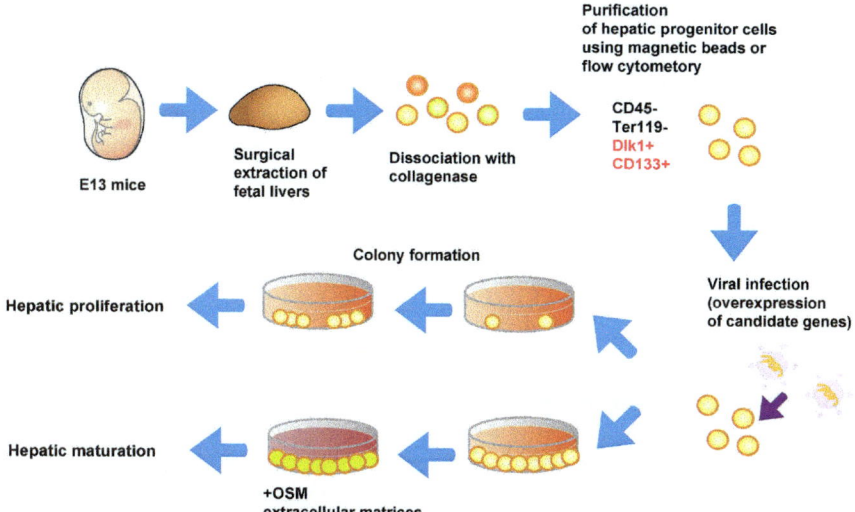

FIGURE 2.1 **Schema of fetal hepatic progenitor cell culture.** E13 mouse livers were minced and digested with collagenase buffer. CD45+Ter119+ Cells were eliminated using magnetic negative selection. Next Dlk1+ hepatic progenitor cells were purified using MACS flow cytometry beads. Purified Dlk1+ hepatic progenitor cells were cultured for the proliferation and differentiation analyses. For the colony culture in the proliferation analyses, progenitor cells were cultured on mitomycin C-treated MEF feeder cells at a low density. The number of colonies was then counted. For hepatic maturation culture in the differentiation assay, progenitor cells were cultured on gelatin-coated dishes with hepatic maturation factors (OSM and EHS matrices). During these cultures, suitable retroviruses expressing several transcription factors were used for gene transfection.

hepatic progenitor cells were cultured and infected with the transcription factor-expressing retroviral vectors. The proliferation of hepatic progenitor cells was then analyzed using a single cell-derived colony culture system. Maturation of the hepatic progenitor cells was analyzed using an in vitro hepatic maturation culture system with OSM and extracellular matrices. Overexpression of Prox1 induced the migration and proliferation of hepatic stem/progenitor cells. In contrast, Lrh1 overexpression inhibited hepatic stem/progenitor cell colony formation. Prox1 supported the long-term growth of hepatic stem/progenitor cells. Expression of $p16^{ink4a}$, the cyclin-dependent kinase inhibitor (CDKi), was suppressed by Prox1 through the regulation of $p16^{ink4a}$ promoter activity. These results suggest that Prox1 and Lrh1 coordinately regulate the development of hepatic stem/progenitor cells [27].

Differentiation of hepatic progenitor cells is also regulated by several transcription factors. Sal-like protein 4 (Sall4) has been shown to regulate organogenesis, embryogenesis, and maintenance of pluripotency. We found that Sall4 plays a crucial role in controlling lineage commitment of hepatoblasts, not only inhibiting their differentiation into hepatocytes, but also driving their differentiation toward cholangiocytes [28]. Overexpression of Sall4 in fetal hepatic progenitor cells induced the expression of cholangiocytic marker genes but not mature hepatocytic marker genes. In addition, elevated SALL4 expression in tumors is associated with poor survival in hepatocellular carcinoma patients [29]. Experimental manipulation of SALL4 expression using virus-mediated gene transfer revealed that the proliferation and differentiation of human hepatocellular carcinoma cell lines are regulated by the amount of SALL4 protein in vitro and in vivo. Particularly, significant growth inhibition accompanied by increased differentiation occurred with SALL4 downregulation. In contrast, overexpression of SALL4 resulted in increased cell proliferation in vitro. Other research also showed that SALL4 could be an oncofetal marker and an attractive therapeutic target in hepatocellular carcinoma [30]. In this study, it was found to regulate phosphatase and tensin homologue (PTEN) and phosphatidylinositol 3-kinase (PI3K)–AKT signaling pathway through interaction with the NuRD (nucleosome remodeling and histone deacetylase) complex. A specific inhibitory 12-AA peptide against SALL4 suppressed this signaling pathway and is a potential therapeutic method for SALL4-induced hepatic tumors. These results suggest that SALL4 is important for cell fate decision and differentiation of hepatic progenitor cells, and that it regulates the stemness of hepatic tumor cells.

We also identified a novel regulator to induce hepatic differentiation using this culture system with the maturation factor OSM and extracellular matrices. Overexpression of Mist1 in hepatic progenitor cell culture induced the expression of mature hepatocytic markers such as carbamoyl-phosphate synthetase1 and several cytochrome P450 (CYP) genes [31]. In contrast, Mist1 suppressed the expression of cholangiocytic markers such as Sox9, Sox17, cytokelatin 19, and Grhl2. In addition, Mist1 induced liver-enriched transcription factors, CCAAT/enhancer-binding protein α and hepatocyte nuclear factor 1α, which are known to be involved in liver functions [32, 33]. Thus, Mist1 is involved in mature hepatocytic function accompanied by the downregulation of cholangiocytic markers during fetal liver development.

4 SIGNALING PATHWAYS REGULATING HEPATOBLAST PROLIFERATION AND DIFFERENTIATION

Hepatic stem/progenitor cells are somatic progenitor cells in the fetal liver, which retain a high proliferative capacity in vivo. In contrast, efficient expansion of hepatoblasts in vitro is difficult without genetic modification. Primary cells cultured in vitro start to express several cell CDKi and stop proliferating [34]. Thus, cell-cycle regulation is important for the long-term expansion of progenitor cells. As shown earlier, individual primary hepatic stem/progenitor cells derived from mouse fetal livers formed large colonies when cocultured with mesenchymal feeder cells [20]. We purified $CD13^+Dlk1^+$ hepatic progenitor cells derived from mid-fetal livers and cultured them with MEFs at a low density. The colony formation activity in this coculture system increased more than that in a feeder-free colony culture. However, these colonies on feeder cells could not proliferate for a long time because the expression of p16/19^{cdkn2a} and p21^{cdkn1a} was upregulated in 4–6 days of culture. These CDKi expressions are important for suppressing hepatic progenitor cell proliferation in vitro because fetal hepatic progenitor cells derived from p16/19^{cdkn2a} knockout mice showed long-term proliferation. We analyzed several cell signaling pathways using signal molecule inhibitors and found that PD0325901, the MEK inhibitor, suppressed the induction of CDKi expression in hepatic progenitor cell culture. A low concentration of PD0325901 partially suppressed MEK activation and supported long-term proliferation of hepatic progenitor cells. In addition, we found that the MEK activation level in hepatic progenitor cells in vivo fetal livers was downregulated. These results indicate that the high activation of MEK signal induces the expression of CDKi and that hepatic progenitor cells proliferate using an MEK-ERK independent pathway during fetal liver development [35].

Cell proliferation is regulated by cell cycle-related molecules and is related to cell fate decision. As described earlier, fetal hepatic progenitor cells have a bipotent differentiation ability and expand highly during fetal liver development. In contrast, in vivo mature hepatocytes barely proliferate under normal conditions. Thus, the correlation between cell-cycle regulation and hepatic maturation is important but not unknown. In analyses of expression for cell cycle molecules during liver development, we found that p57Kip2 (one of the CDKi) was specifically expressed in hepatoblasts and mesenchymal cells of the fetal liver in a spatiotemporal manner. Purification of the hepatic progenitor cells derived from p57$^{Kip2-/-}$ mice fetal livers revealed higher proliferation in these cells than that in cells derived from p57$^{Kip2+/+}$ mice. In addition, p57$^{Kip2-/-}$ knockout mouse livers matured slower than those of wild type mice. To determine whether p57^{Kip2} regulates hepatic progenitor cell maturation by changes in intrinsic molecules, p57$^{Kip2-/-}$ progenitor cell differentiation was analyzed using chimeric mouse technologies. We generated induced pluripotent stem (iPS) cells from p57$^{Kip2+/+}$ or p57$^{Kip2-/-}$ MEFs. These p57 $^{Kip2+/+}$ or p57 $^{Kip2-/-}$ iPS cells were injected into normal mouse blastocysts. In the fetal livers derived from these chimeric blastocysts, the hepatic progenitor cells differentiated from donor p57 $^{Kip2+/+}$ or p57$^{Kip2-/-}$ iPS cells matured with p57$^{Kip2+/+}$ nonparenchymal cells derived from wild-type recipient eggs. Interestingly, the p57 $^{Kip2-/-}$ hepatic progenitor cells could normally differentiate into mature hepatocytes in chimeric mice, which possessed both p57 $^{Kip2+/+}$ and p57 $^{Kip2-/-}$ mesenchymal cells [36]. Thus, the intrinsic activity of p57^{Kip2} is not involved in the maturation of hepatic progenitor cells. It was previously described that mesenchymal cells in fetal livers are important for liver maturation [37]. The homeobox transcription factor Hlx is specifically expressed in mesenchymal cells during early-fetal liver development. Hlx knockout mice have severe hypoplasia of the liver on embryonic day 15. These findings suggested that the extrinsic factors derived (soluble factors and cell–cell interaction) from mesenchymal cells are involved in p57^{Kip2}-mediated maturation of the hepatic progenitor cells in fetal liver development.

In addition to hepatic progenitor cells, some mature hepatocytes also can proliferate and expand in vitro. We previously found that a combination of chemical compounds can induce proliferation in adult mouse hepatocytes [38]. Mature hepatocytes in adult livers in vivo can proliferate after liver injury. Hepatocyte growth factors, epidermal growth factors, insulin, and nicotinamide are known to be important for the expansion of hepatocytes derived from adult livers in vitro [39–41]. However, in vitro expansion of mature hepatocytes is difficult because these cells have gradually lost their proliferative ability and specific functions, particularly those of drug metabolism. Thus, we explored a novel culture condition for mature hepatocyte expansion. We purified mature hepatocytes from adult mouse livers and cultured these cells with several cell-signaling inhibitors. The addition of inhibitors against TGF-β and glycogen synthase kinase (GSK) 3β efficiently induced the proliferation of mature hepatocytes. These hepatocytes could expand for more than 1 week in monolayer culture on extracellular matrix-coated dishes but lost the expression of several hepatocellular functional genes. Expanded hepatocytes recovered these mature hepatic functions such as the cytochrome P450 activity after the maturation culture step using spheroid formation. Three-dimensional (3D) biological structures such as spheroid formation are useful to maintain and induce the functions of mature hepatocytes in vitro [42, 43]. Furthermore, when hepatocytes were cocultured with MEF, addition of an MEK inhibitor at the spheroid formation step enhanced drug-metabolism-related gene expression. These results suggested that a combination of the MEF coculture system along with addition of the inhibitors of TGFβ and GSK3β, induced the expansion of mature hepatocytes. Moreover, the expression of mature hepatic genes and the activity of drug-metabolic enzymes of in vitro expanded hepatocytes were reinduced after spheroid culture. Recently, it was described that the combination of these cell signal inhibitors is involved in the conversion of mature hepatocytes to bipotent progenitor cells [44]. Lineage-tracing studies have recently revealed that the conversion of mature hepatocytes into immature proliferative progenitor cells occurs in vivo during liver regeneration [16]. The addition of Y-27632 (ROCK inhibitor), A-83-01 (TGFβ inhibitor), and CHIR99021 (GSK3β inhibitor) induced the proliferation of mature rat hepatocytes. These cells can expand clonally in vitro and differentiate into both mature hepatocyte-like as well as cholangiocyte-like cells. In addition, these cells can repopulate and expand in liver-injured mice after transplantation. These results suggest that some mature hepatocytes have the potency to dedifferentiate into an immature state and that this phenomenon is important for regeneration from severe liver injury.

5 HEPATIC PROGENITOR CELLS DERIVED FROM HUMAN PLURIPOTENT STEM CELLS

The mechanisms underlying the proliferation and differentiation of human hepatic progenitor cells remain largely unknown because of the difficulty in analyzing the cellular and molecular events in vivo. Pluripotent stem cells, embryonic stem (ES) cells, and inducible pluripotent stem (iPS) cells are multipotent and can differentiate into specialized cells of the three primary germ layers. To establish an in vitro assay system for human liver development, we established a method for differentiating human iPS cells into fetal hepatic progenitors. We found that highly proliferative cells existed in the $CD13^+CD133^+$ fraction of human iPS cells stimulated by hepatic differentiation factors. As described earlier, CD13 and CD133 are mouse hepatic progenitor marker-specific cell surface molecules during the early and intermediate (E9.5–14) stages of fetal development [20,45]. When the purified $CD13^+CD133^+$ cells were cultured at a low density with feeder cells (MEFs) in the presence of suitable growth factors and signaling inhibitors (TGFβ inhibitor A-83-01 and ROCK inhibitor Y-27632), individual cells gave rise to relatively large colonies (detail methods were written in [46]). Individual $CD13^+CD133^+$ cells formed large colonies (contained more than 100 cells) expressing the marker genes of both hepatocytic cells (α-feto protein and hepatocyte nuclear factor 4α) and cholangiocytic cells (cytokeratin 19 and 7). In addition, the colony cells continued to proliferate over long durations. Next, we analyzed the bipotency to differentiate into hepatocytic and cholangiocytic lineages. In a spheroid formation assay using human iPS-derived hepatic progenitor cells, the cells were found to express genes required for mature liver function, such as CYP enzymes. In addition, secretion of albumin, one of the most important liver functions, was also induced in these cells. In an extracellular matrix gel culture supplemented with cytokines, cholangiocytic cells can form cysts with an epithelial polarity, thus demonstrating in vitro tubulogenesis [47]. When hepatic progenitor cells derived from human iPS cells were cultured in a suitable extracellular matrix gel containing collagen and laminin, they eventually formed a cholangiocytic cyst-like structure with epithelial polarity. These results suggest that human iPS cell-derived hepatic progenitor cells have a bipotent tendency to differentiate into mature hepatocytic and cholangiocytic cells. This procedure using an in vitro expansion system might be useful for liver regeneration as well as for determining the molecular mechanisms that regulate liver development.

Cell surface molecules are useful for the identification and characterization of rare cells such as stem/progenitor cells. Hepatic progenitor cells are enriched in the $CD13^+CD133^+$ cell fraction differentiated from human iPS cells. These cells have a high proliferation potential, the ability to differentiate into both hepatocytes and cholangiocytes, and expand in long-term culture; however the characteristics of these cells have remained unknown. We therefore analyzed the correlation between cell surface molecules and the characteristics of human iPS cell-derived hepatic progenitor cells [48]. In addition to hepatic progenitor markers (CD13 and CD133), the cells were coimmunostained for various cell surface markers (116 types). The cells were analyzed and purified by flow cytometry. Twenty types of cell surface molecules were highly expressed on the $CD13^+CD133^+$ cells derived from human iPS cells, suggesting that these molecules are new surface markers for human hepatic progenitor cells. Particularly,

CD221 (insulin-like growth factor receptor), which was highly expressed on CD13$^+$CD133$^+$ cells, was quickly downregulated after in vitro clonal expansion. The proliferative ability of human hepatic progenitor cells was suppressed by a neutralizing antibody and specific inhibitor to CD221, suggesting that the insulin growth factor signal is important for the expansion of hepatic progenitor cells in vitro during colony formation culture with feeder cells. In addition, overexpression of CD221 increased the colony-forming ability of these cells, indicating that downregulation of CD211 on hepatic progenitor cells suppresses the expansion of these cells in vitro. CD340 (erbB2) and CD266 (fibroblast growth factor inducible 14) signals are also involved in the expansion of these progenitor cells. Addition of specific inhibitors against these signals also downregulated the colony forming activity of human hepatic progenitor cells. Interaction with mesenchymal cells is important for regulating the proliferation and differentiation of hepatic progenitor cells. We used MEFs as feeder cells in this in vitro expansion culture. We found that both insulin-like growth factor (a ligand of CD221) and TWEAK (a ligand of CD266) were provided by the MEFs in our culture system. This study revealed the expression profile of cell surface molecules on human iPS cell-derived hepatic progenitor-like cells, and showed that the paracrine interactions between hepatic progenitor-like cells and other cells through specific receptors are important for proliferation.

6 CONCLUSION

In the review, we summarized the molecular mechanisms regulating hepatic proliferation and function in a somatic stem/progenitor cell system. Transcription factors such as the hepatocyte nuclear factor family and the C/EBP family are important for hepatocytic specification and differentiation. In addition, we found several new transcription factors regulating hepatocytic proliferation and function. We also established a new expansion system for human hepatic progenitor cells derived from pluripotent stem cells. These findings may contribute toward novel therapeutic strategies for severe liver diseases. These systems and molecules are useful to analyze the molecular mechanisms regulating the proliferation and differentiation of human hepatic progenitor cells. For example, the mechanism regulating self-renewal activity induction in human hepatic progenitor cells was analyzed using gene transfection methods. The expression analyses of genes and miRNAs revealed that the expression of several specific molecules was changed during liver development. Overexpression or knockdown of these candidate molecules may induce self-renewal activity in human iPS cell-derived hepatic progenitor cells. In addition, induction of maturation in human iPS cell-derived hepatic progenitor cells in vitro is also an important issue. Expanded human iPS cell-derived hepatic progenitor cells expressed immature hepatic genes (i.e., albumin and α-feto protein) but not mature functional genes. Maturation of immature hepatic cells is regulated by several factors such as soluble cytokines, cell-matrix interactions, and cell–cell interactions. As shown earlier, we recently identified several transcription factors, which can induce the expression of functional hepatic genes. These genes might be useful for the differentiation of hepatic progenitor cells derived from human iPS cells to mature hepatocytes.

Because of the existing shortage of donated livers, human ES and iPS cells are expected to be suitable sources of human hepatocytes for cell-transplant therapies and for drug metabolism analyses. One of our future goals is therefore generating large number of functional hepatocytes derived from human pluripotent stem cells using our culture system and analyzing the safety of these hepatocytes for cell therapy (i.e., carcinogenic risks of these cells).

FINANCIAL SUPPORT AND SPONSORSHIP

Our studies described in this review chapter were supported in part by the Grants-in-Aid for Scientific Research from the Ministry of Education, Culture, Sports, Science and Technology of Japan and by the Research and Study Project of Tokai University Educational System General Research Organization.

ACKNOWLEDGMENTS

We thank the Education and Research Support Center of Tokai University for assistance on some of the analyses in our studies described in this review chapter.

There are no conflicts of interest to disclose.

LIST OF ABBREVIATIONS

CDKi cyclin-dependent kinase inhibitor
CYP cytochrome P450
E embryonic day
ES cells embryonic stem cells
iPS cells inducible pluripotent stem cells
lrh1 liver receptor homolog 1
MEF mouse embryonic fibroblast
OSM oncostatin M
PDGFR platelet-derived growth factor receptor
PI3K phosphatidylinositol 3-kinase
Prox1 prospero-related homeobox 1
PTEN phosphatase and tensin homologue
ROCK Rho-associated kinase
TGF-β transforming growth factor-β
Sall4 Sal-like protein 4

REFERENCES

[1] Douarin NM. An experimental analysis of liver development. Med Biol 1975;53(6):427–55.
[2] Houssaint E. Differentiation of the mouse hepatic primordium. I. An analysis of tissue interactions in hepatocyte differentiation. Cell Differ 1980;9(5):269–79.

[3] Watanabe T, Nakagawa K, Ohata S, Kitagawa D, Nishitai G, Seo J, et al. SEK1/MKK4-mediated SAPK/JNK signaling participates in embryonic hepatoblast proliferation via a pathway different from NF-kappaB-induced anti-apoptosis. Dev Biol 2002;250(2):332–47.

[4] Kamiya A, Kinoshita T, Ito Y, Matsui T, Morikawa Y, Senba E, et al. Fetal liver development requires a paracrine action of oncostatin M through the gp130 signal transducer. EMBO J 1999;18(8):2127–36.

[5] Kamiya A, Kojima N, Kinoshita T, Sakai Y, Miyaijma A. Maturation of fetal hepatocytes in vitro by extracellular matrices and oncostatin M: induction of tryptophan oxygenase. Hepatology 2002;35(6):1351–9.

[6] Kinoshita T, Sekiguchi T, Xu MJ, Ito Y, Kamiya A, Tsuji K, et al. Hepatic differentiation induced by oncostatin M attenuates fetal liver hematopoiesis. Proc Natl Acad Sci U.SA. 1999;96(13):7265–70.

[7] Miyajima A, Kinoshita T, Tanaka M, Kamiya A, Mukouyama Y, Hara T. Role of Oncostatin M in hematopoiesis and liver development. Cytokine Growth Factor Rev 2000;11(3):177–83.

[8] Clotman F, Jacquemin P, Plumb-Rudewiez N, Pierreux CE, Van der Smissen P, Dietz HC, et al. Control of liver cell fate decision by a gradient of TGF beta signaling modulated by Onecut transcription factors. Genes Dev 2005;19(16):1849–54.

[9] Antoniou A, Raynaud P, Cordi S, Zong Y, Tronche F, Stanger BZ, et al. Intrahepatic bile ducts develop according to a new mode of tubulogenesis regulated by the transcription factor SOX9. Gastroenterology 2009;136(7):2325–33.

[10] Kodama Y, Hijikata M, Kageyama R, Shimotohno K, Chiba T. The role of notch signaling in the development of intrahepatic bile ducts. Gastroenterology 2004;127(6):1775–86.

[11] Zong Y, Panikkar A, Xu J, Antoniou A, Raynaud P, Lemaigre F, et al. Notch signaling controls liver development by regulating biliary differentiation. Development 2009;136(10):1727–39.

[12] Hofmann JJ, Zovein AC, Koh H, Radtke F, Weinmaster G, Iruela-Arispe ML. Jagged1 in the portal vein mesenchyme regulates intrahepatic bile duct development: insights into Alagille syndrome. Development 2010;137(23):4061–72.

[13] Fausto N, Campbell JS, Riehle KJ. Liver regeneration. Hepatology 2006;43(2 Suppl 1):S45–53.

[14] Lanzoni G, Cardinale V, Carpino G. The hepatic, biliary, and pancreatic network of stem/progenitor cell niches in humans: A new reference frame for disease and regeneration. Hepatology 2016;64(1):277–86.

[15] Yanger K, Knigin D, Zong Y, Maggs L, Gu G, Akiyama H, et al. Adult hepatocytes are generated by self-duplication rather than stem cell differentiation. Cell Stem Cell 2014;15(3):340–9.

[16] Wang B, Zhao L, Fish M, Logan CY, Nusse R. Self-renewing diploid Axin2(+) cells fuel homeostatic renewal of the liver. Nature 2015;524(7564):180–5.

[17] Kamiya A, Kakinuma S, Yamazaki Y, Nakauchi H. Enrichment and clonal culture of progenitor cells during mouse postnatal liver development in mice. Gastroenterology 2009;137(3):1114–26. e1-14.

[18] Kamiya A, Nakauchi H. Enrichment and clonal culture of hepatic stem/progenitor cells during mouse liver development. Methods Mol Biol 2013;945:273–86.

[19] Miyajima A, Tanaka M, Itoh T. Stem/progenitor cells in liver development, homeostasis, regeneration, and reprogramming. Cell Stem Cell 2014;14(5):561–74.

[20] Okada K, Kamiya A, Ito K, Yanagida A, Ito H, Kondou H, et al. Prospective isolation and characterization of bipotent progenitor cells in early mouse liver development. Stem Cells Dev 2012;21(7):1124–33.

[21] Asahina K, Tsai SY, Li P, Ishii M, Maxson RE Jr, Sucov HM, et al. Mesenchymal origin of hepatic stellate cells, submesothelial cells, and perivascular mesenchymal cells during mouse liver development. Hepatology 2009;49(3):998–1011.

[22] Asahina K, Zhou B, Pu WT, Tsukamoto H. Septum transversum-derived mesothelium gives rise to hepatic stellate cells and perivascular mesenchymal cells in developing mouse liver. Hepatology 2011;53(3):983–95.
[23] Ito K, Yanagida A, Okada K, Yamazaki Y, Nakauchi H, Kamiya A. Mesenchymal progenitor cells in mouse foetal liver regulate differentiation and proliferation of hepatoblasts. Liver International: Off J Int Assoc Study Liver 2014;34(9):1378–90.
[24] da Silva Meirelles L, Chagastelles PC, Nardi NB. Mesenchymal stem cells reside in virtually all post-natal organs and tissues. J Cell Sci 2006;119(Pt 11):2204–13.
[25] Sacchetti B, Funari A, Michienzi S, Di Cesare S, Piersanti S, Saggio I, et al. Self-renewing osteoprogenitors in bone marrow sinusoids can organize a hematopoietic microenvironment. Cell 2007;131(2):324–36.
[26] Greenbaum A, Hsu YM, Day RB, Schuettpelz LG, Christopher MJ, Borgerding JN, et al. CXCL12 in early mesenchymal progenitors is required for haematopoietic stem-cell maintenance. Nature 2013;495(7440):227–30.
[27] Kamiya A, Kakinuma S, Onodera M, Miyajima A, Nakauchi H. Prospero-related homeobox 1 and liver receptor homolog 1 coordinately regulate long-term proliferation of murine fetal hepatoblasts. Hepatology 2008;48(1):252–64.
[28] Oikawa T, Kamiya A, Kakinuma S, Zeniya M, Nishinakamura R, Tajiri H, et al. Sall4 regulates cell fate decision in fetal hepatic stem/progenitor cells. Gastroenterology 2009;136(3):1000–11.
[29] Oikawa T, Kamiya A, Zeniya M, Chikada H, Hyuck AD, Yamazaki Y, et al. Sal-like protein 4 (SALL4), a stem cell biomarker in liver cancers. Hepatology 2013;57(4):1469–83.
[30] Yong KJ, Gao C, Lim JS, Yan B, Yang H, Dimitrov T, et al. Oncofetal gene SALL4 in aggressive hepatocellular carcinoma. N Engl J Med 2013;368(24):2266–76.
[31] Chikada H, Ito K, Yanagida A, Nakauchi H, Kamiya A. The basic helix-loop-helix transcription factor, Mist1, induces maturation of mouse fetal hepatoblasts. Scient Rep 2015;514989.
[32] Flodby P, Barlow C, Kylefjord H, Ahrlund-Richter L, Xanthopoulos KG. Increased hepatic cell proliferation and lung abnormalities in mice deficient in CCAAT/enhancer binding protein alpha. J Biol Chem 1996;271(40):24753–60.
[33] Pontoglio M, Barra J, Hadchouel M, Doyen A, Kress C, Bach JP, et al. Hepatocyte nuclear factor 1 inactivation results in hepatic dysfunction, phenylketonuria, and renal Fanconi syndrome. Cell 1996;84(4):575–85.
[34] Chiba T, Zheng YW, Kita K, Yokosuka O, Saisho H, Onodera M, et al. Enhanced self-renewal capability in hepatic stem/progenitor cells drives cancer initiation. Gastroenterology 2007;133(3):937–50.
[35] Kamiya A, Ito K, Yanagida A, Chikada H, Iwama A, Nakauchi H. MEK-ERK activity regulates the proliferative activity of fetal hepatoblasts through accumulation of p16/19(cdkn2a). Stem Cells Dev 2015;24(21):2525–35.
[36] Yanagida A, Chikada H, Ito K, Umino A, Kato-Itoh M, Yamazaki Y, et al. Liver maturation deficiency in p57(Kip2)-/- mice occurs in a hepatocytic p57(Kip2) expression-independent manner. Dev Biol 2015;407(2):331–43.
[37] Hentsch B, Lyons I, Li R, Hartley L, Lints TJ, Adams JM, et al. Hlx homeo box gene is essential for an inductive tissue interaction that drives expansion of embryonic liver and gut. Genes Dev 1996;10(1):70–9.
[38] Ito H, Kamiya A, Ito K, Yanagida A, Okada K, Nakauchi H. In vitro expansion and functional recovery of mature hepatocytes from mouse adult liver. Liver Int. 2012;32(4):592–601.
[39] Richman RA, Claus TH, Pilkis SJ, Friedman DL. Hormonal stimulation of DNA synthesis in primary cultures of adult rat hepatocytes. Proc Natl Acad Sci U.S.A. 1976;73(10):3589–93.

[40] Block GD, Locker J, Bowen WC, Petersen BE, Katyal S, Strom SC, et al. Population expansion, clonal growth, and specific differentiation patterns in primary cultures of hepatocytes induced by HGF/SF, EGF and TGF alpha in a chemically defined (HGM) medium. J Cell Biol 1996;132(6):1133–49.

[41] Inoue C, Yamamoto H, Nakamura T, Ichihara A, Okamoto H. Nicotinamide prolongs survival of primary cultured hepatocytes without involving loss of hepatocyte-specific functions. J Biol Chem 1989;264(9):4747–50.

[42] Koide N, Shinji T, Tanabe T, Asano K, Kawaguchi M, Sakaguchi K, et al. Continued high albumin production by multicellular spheroids of adult rat hepatocytes formed in the presence of liver-derived proteoglycans. Biochem Biophys Res Commun 1989;161(1):385–91.

[43] Takezawa T, Yamazaki M, Mori Y, Yonaha T, Yoshizato K. Morphological and immunocytochemical characterization of a hetero-spheroid composed of fibroblasts and hepatocytes. J Cell Sci 1992;101(Pt 3):495–501.

[44] Katsuda T, Kawamata M, Hagiwara K, Takahashi RU, Yamamoto Y, Camargo FD, et al. Conversion of terminally committed hepatocytes to culturable bipotent progenitor cells with regenerative capacity. Cell Stem Cell 2017;20(1):41–55.

[45] Kakinuma S, Ohta H, Kamiya A, Yamazaki Y, Oikawa T, Okada K, et al. Analyses of cell surface molecules on hepatic stem/progenitor cells in mouse fetal liver. J Hepatol 2009;.

[46] Yanagida A, Nakauchi H, Kamiya A. Generation and In Vitro Expansion of Hepatic Progenitor Cells from Human iPS Cells. Methods Mol Biol 2016;1357:295–310.

[47] Tanimizu N, Miyajima A, Mostov KE. Liver progenitor cells develop cholangiocyte-type epithelial polarity in three-dimensional culture. Molec Biol Cell 2007;18(4):1472–9.

[48] Tsuruya K, Chikada H, Ida K, Anzai K, Kagawa T, Inagaki Y, et al. A paracrine mechanism accelerating expansion of human induced pluripotent stem cell-derived hepatic progenitor-like cells. Stem Cells Dev 2015;24(14):1691–702.

Chapter 3

Plasticity of Liver Epithelial Cells in Healthy and Injured Livers

Naoki Tanimizu, Toshihiro Mitaka
Research Institute for Frontier Medicine, Sapporo Medical University School of Medicine, Sapporo, Japan

1 INTRODUCTION

An epithelial organ consists of multiple types of epithelial cells, which are derived from immature progenitors (or stem cells) existing in an organ bud. On the other hand, it is considered that tissue-specific stem/progenitor cells supply all types of epithelial cells for tissue homeostasis and regeneration in an adult organ. Typically, stem cells locating in the crypt continuously supply secretary and absorptive epithelial cells in the intestine [1]. However, except the gastrointestinal tracts and skin, it remains ambiguous whether tissue stem/progenitor cells generally contribute to homeostasis and regeneration of epithelial organs [2].

It was demonstrated that the combination of transcription factors confers pluripotency onto fibroblasts [3]. Following the sensational achievement, various types of multipotent stem cells have been generated from somatic cells, which have been considered to be terminally differentiated, via "reprogramming" by introducing transcription factors or by using small compounds in vitro and in vivo [4]. These results indicate that differentiated cells potentially have lineage plasticity. In addition to attempts generating stem/progenitor cells, it has been demonstrated that mature epithelial cells show lineage plasticity and dedifferentiate or change their fate upon severe injuries in epithelial organs. In severely injured trachea, secretory cells converted to basal-like pulmonary stem cells [5]. In stomach, Troy$^+$ chief cells dedifferentiated and behaved as gastric stem cells, when the stem cell compartment was experimentally depleted [6]. Even in the intestine, Alpi$^+$ enterocytes dedifferentiated to progenitors when Lgr5$^+$ stem cells were depleted [7]. Dedifferentiation of mature epithelial cells and their conversion into stem cells could be a backup system for tissue stem cell system in these organs. However, it remains unknown whether dedifferentiation of mature epithelial cells results in producing lineage-committed progenitors, multipotent stem cells, or different lineage cells.

The liver contains two types of epithelial cells, namely hepatocytes and cholangiocytes, originated from hepatoblasts during embryonic development [8]. It is well known that the liver has a tremendous regenerative capability: after 70% partial hepatectomy (PHx) or acute liver injuries, the liver can restore nearly the original tissue mass within a week. During this type of regeneration, remaining hepatocytes and cholangiocytes self-duplicate and regenerate the liver tissue. On the other hand, in chronically damaged liver when many hepatocytes lost regenerative capability, liver stem/progenitors (LPCs) are supposed to be activated and supply new hepatocytes for restoring hepatic tissue. Indeed, LPCs have been prospectively isolated based on expression of surface antigens including CD133, epithelial adhesion molecule (EpCAM), CD24, and Thy1 from healthy and injured livers and shown to differentiate into hepatocytes and cholangiocytes in vitro and in vivo [9–11]. Although markers used to identify LPCs are expressed not only in LPCs but also in cholangiocytes, those works indicate that healthy and injured livers contain a population of cells possessing the capability of clonal proliferation and bi-directional differentiation. However, recent studies using genetic lineage tracing have demonstrated that cellular fraction marked by LPC markers do not majorly supply hepatocytes during cellular turnover in healthy liver and during regeneration. On the other hand, it has been demonstrated that mature hepatocytes (MHs) or part of MHs show progenitor-like characteristics after dedifferentiation or lineage conversion in various types of injury models [12–15]. Importantly, those dedifferentiated cells acquire proliferative capability and efficiently redifferentiate into functional MHs. These results suggest that hepatocytes possess capability to alter their differentiation status responding to injuries. In addition to remarkable plasticity of hepatocytes, it has been recognized that hepatocytes as well as cholangiocytes contain heterogeneous cell populations. However, it remains unknown that those subpopulations can be distinguishable for differentiation status and lineage plasticity.

In this chapter, we summarize recent experimental results showing the lineage plasticity of liver epithelial cells and their heterogeneity in developing and adult healthy livers as well as in regenerating livers. We also discuss about possible molecular mechanisms governing dedifferentiation and lineage conversion of hepatocytes and cholangiocytes.

2 LINEAGE SPECIFICATION DURING DEVELOPMENT

The foregut endoderm develops into the liver bud by fibroblast growth factors (FGFs) and bone morphogenetic proteins (BMPs) secreted from the cardiac mesoderm and septum transversum mesenchyme, respectively [16,17]. The liver bud contains hepatoblasts that express EpCAM and delta-like 1 (Dlk-1) at embryonic day 11 (E11) and then lose EpCAM expression around E13 during mouse liver organogenesis [18]. In addition to these surface antigens, hepatoblasts express albumin and hepatocyte nuclear factor 4α (HNF4α), which are expressed in any types of hepatocytes and, thereby, can be defined as the hepatocyte lineage markers, as well as an immature hepatocyte marker, α-fetoprotein. Hepatoblasts isolated between E11 and E14 can clonally proliferate and differentiate into hepatocytes and cholangiocytes [18–21]. Around E15, hepatoblasts near the portal vein (PV) are committed to cholangiocytes depending on Notch and transforming growth factor β (TGFβ) signals, whereas they differentiate into hepatocytes in the parenchymal region [22–24]. These experimental results and observation definitely indicate that hepatoblasts are bipotential fetal LPCs. Interestingly, the lineage specification of cholangiocytes proceeds progressively and radially from the liver hilum to the periphery [25,26]. Beyond this stage, two epithelial lineage cells can be recognized within the liver tissue as well as are isolated as two different cellular populations. In following sections, the lineage plasticity of cholangiocytes and hepatocytes will be discussed.

3 LINEAGE PLASTICITY AND HETEROGENEITY OF CHOLANGIOCYTES

3.1 Lineage Plasticity of Cholangiocytes

3.1.1 Fetal and Neonatal Cholangiocytes

Cholangiocytes express cholangiocyte markers including Sry box containing gene 9 (Sox9), osteopontin (OPN), and EpCAM just after their commitment from hepatoblasts [25]. Taking advantage of the specific expression of one of these markers, cholangiocytes were labeled in Sox9-CreERT2:ROSA-YFP mice at E15.5 by injecting tamoxifen intraperitoneally to pregnant mice [27]. YFP-positive cells differentiated to ductular cholangiocytes, the canal of Hering, and periportal hepatocytes at postnatal day 40 (P40), indicating that cholangiocytes definitely possess bidirectional differentiation potential when they are just committed from hepatoblasts.

To know whether cholangiocyte keep or change their lineage plasticity during development, we isolated $EpCAM^+$ cholangiocytes from livers between E17.5 and 11W by FACS and examined their clonal proliferation and differentiation potential using a colony assay. In the period between E17.5 and 2W after birth, about 1%–2% of $EpCAM^+$ cells could clonally proliferate and show bidirectional differentiation capability, which were demonstrated by the formation of colonies that contained albumin $(ALB)^+$ hepatocytes and cytokeratin 19 $(CK19)^+$ cholangiocytes [28]. We further established clones derived from 1W $EpCAM^+$ cells. Those clones can differentiate to functional hepatocytes in the presence of oncostatin M (OSM) and Matrigel and to cholangiocytes-like cells in the 3 dimensional (3D) culture, in which bipotential LPCs and cholangiocytes establish the apico-basal polarity and form cysts, spheroids with the central lumen [29,30]. Furthermore, when the clones were transplanted through the spleen into livers of nude mice that were treated with retrorsine and 70% PHx (Ret/PHx model), they were engrafted as hepatocytes and cholangiocytes in recipient livers [28,30]. Even beyond 4W, EpCAM$^+$ cells clonally proliferated and form colonies, but the ratio of ALB$^+$ hepatocytes in colony-forming cells became very low. When they were transplanted to nude mice of the Ret/PHx model, the repopulation was inefficient and the engrafted donor cells mostly showed ductular-like structures. Similarly, Kamiya et al. reported that the number of LPCs enriched in CD13$^+$CD133$^+$ cells, which exist in bile ducts, gradually decreased during postnatal development [10]. These results indicate that a subpopulation of cholangiocytes can function as bipotential LPCs by the neonatal period.

3.1.2 Adult Cholangiocytes

The plasticity of adult mature cholangiocytes has been recently examined in healthy and injured livers by genetic lineage tracing in vivo. Furuyama et al. genetically labeled cholangiocytes as SOX9$^+$ cells and demonstrated that those cells supply MHs during cellular turnover in healthy liver [31]. Remarkably, MHs derived from SOX9$^+$ cells mostly replenished the liver tissue within 6 months. However, the subsequent works did not support this result [32–34].

The canal of Hering is the joint between hepatic cord and bile ducts, where resident LPCs are possibly located. This hypothesis is based on the observation that small cells, which are called oval cells named from their ovoid-shaped nucleus, are apparently expanded from the canal of Hering upon chronic liver injuries. Indeed, even though cholangiocytes or LPCs that reside in bile ducts or the canal of Hering do not likely supply new hepatocytes in healthy condition, their contribution in liver regeneration has been further examined [32,35]. Espanol-Suner et al. demonstrated that OPN$^+$ cells supplied a significant number of MHs during the recovery from CDE-diet induced liver injury. In most liver injury models used for mice, MHs are damaged, but a significant number of MHs that maintain proliferative capability may still remain and work for regeneration. Therefore, extreme condition disrupting almost completely MHs may be necessary to activate the plasticity of cholangiocytes. In a transgenic fish expressing nitroreductase under the hepatocyte-specific *fabp10a* promoter, hepatocytes were ablated after administration of metronidazole (MTZ). In this zebrafish line, cholangiocytes converted to hepatocytes and regenerated liver after very severe loss of hepatocytes [36,37]. Lu et al. established a mouse model, in which most of hepatocytes were removed by depleting Mdm2, an E3 ubiquitin ligase, essential for cell survival [38]. When EpCAM$^+$CD24$^+$CD133$^+$ LPCs were transplanted to this model, they differentiated to MHs and repopulated the liver tissue. Collectively, cholangiocytes or LPCs reside in bile ducts or the canal of Hering could convert to hepatocytes after very severe loss of MHs.

3.2 Heterogeneity of Cholangiocytes

The biliary system consists of intrahepatic (IHBD) and extrahepatic bile ducts (EHBD) and gallbladder. Recent works directly analyzing 3D structures of liver tissues clearly showed that IHBDs contain two distinctive structural units: large ducts, which are conventionally called as interlobular bile duct (BDs), run along the PVs, and small ductules, which are branched from large ducts and form mesh-like structures, surround the PV [39,40]. Cholangiocytes in these two structural units may have different physiological functions and proliferative capability. Morphological and functional heterogeneity of IHBDs were correlated with cell size in rats and mice [41–43]. Large and small cholangiocytes were isolated and examined their gene expression profiles. Large cells express cystic fibrosis transmembrane conductance regulator and anion exchanger 2 and respond against secretin more than small ones do. We isolated EpCAM$^+$ cholangiocytes in two different cellular fractions after two-step collagenase perfusion. Collagenase perfusion liberates cells from the parenchyma with nonparenchymal cells (NPCs) and remains undigested connective tissue. EpCAM$^+$ cells liberated from liver tissue by collagenase perfusion were named as "NPC-EpCAM$^+$ cells." On the other hand, from undigested tissue that contained the biliary trees, "biliary tree EpCAM$^+$ cells (BT-EpCAM$^+$ cells)" were isolated. Comparing two populations, the NPC-EpCAM$^+$ fraction contained colony-forming cells more abundantly than the BT-EpCAM$^+$ one (our unpublished data). It can be assumed that "large cholangiocytes" and "BT-cholangiocytes" are isolated from large ducts, whereas "small cholangiocytes" and "NPC-cholangiocytes" are from small ductules. These results support the idea that cholangiocytes contain heterogeneous cell populations in terms of physiological function and proliferative capability.

In chronically injured mouse livers, IHBDs are expanded and the tubular network is rearranged, which is generally called ductular reaction (DR) [40]. The DR is important to protect liver tissue from severe damages [44]. Although it is considered that LPCs are activated and form expanded ductular structures, recent works suggest that DR likely depends on expansion and adaptive rearrangement of preexisting bile ducts [40,45]. Damages on the periportal tissue result in expansion of BDs around the PV, whereas injuries of pericentral hepatocytes induce extension of BDs toward the CV area. Kamimoto et al. randomly labeled cholangiocytes with tdTomato expression by injecting low dose of tamoxifen into CD133-CreERT2:ROSA-tdTomato mice and followed tdTomato$^+$ cells at the single cell level during thioacetamide (TAA)-induced chronic injury [45]. They demonstrated that cholangiocytes in small ductules rather than those in large ducts clonally proliferated and majorly contributed to expansion of duct structures. Moreover, their data indicate that these proliferating cells were not derived from specific parental cells but cholangiocytes stochastically started proliferation and expanded. Interestingly, proliferating cholangiocytes were not necessarily in the vicinity of injured hepatocytes that are near the central vein (CV) in TAA-injury model. In contrast to these results, we observed that in bile duct ligation (BDL)- and 4,4′-diaminodiphenylmethane (DAPM)-injured livers, cholangiocytes in large ducts were positive for Ki67. In combination of the former in vitro works with the recent report, cholangiocytes in small ductules possess potential acquiring strong proliferative capability. However, under a specific physiological demand, cholangiocytes in large ducts may also start proliferation and contribute to protection of liver tissue and/or regeneration. In this flexible system, cholangiocytes can respond to physiological demand for expansion of biliary structure to accommodate or drain excess amount of the bile juice, though molecular mechanisms determining when and how cholangiocytes enter cell cycle remains unknown.

3.3 Molecular Mechanisms Regulating the Plasticity of Cholangiocytes

Comparing properties of colonies derived from different developmental stages, cholangiocytes gradually lose potential differentiating into hepatocytes. By comparing neonatal and adult cholangiocytes, we found that transcription factors related to cholangiocyte differentiation, including hairy enhancer of slit 1 (Hes1), Sox9, hes-related family bHLH transcription factor with YRPW motif 1 (Hey1), and grainyhead-like 2 (Grhl2), are expressed at higher levels in adult cholangiocytes than in neonatal ones. Among those transcription factors, Grhl2 was significantly downregulated in neonatal cells but not in adult ones during hepatocyte differentiation in vitro in the presence of OSM and Matrigel. Furthermore, Grhl2 inhibited expression of HNF4α, CCAAT enhancer binding protein (C/EBP) α, and microRNA 122 (miR122), which are essential factors for hepatocyte differentiation. In addition to its role in regulating lineage plasticity of epithelial cells, it has been demonstrated that Grhl2 regulates formation of intercellular junctions by regulating junctional molecules. In cholangiocyte differentiation, Grhl2 regulates the integrity of tight junctions (TJs) by upregulating expression of Claudin 3 (*CLDN3*) and *CLDN4* and by promoting TJ localization of CLDN4 via upregulation of Rab25, a small GTPase involved in apical protein sorting [46]. These results indicate that EpCAM$^+$ cells lose lineage plasticity while they undergo epithelial maturation.

Examination of the cholangiocyte-driven regeneration model of zebrafish has identified several molecules involved in cholangiocyte-to-hepatocyte conversion. Administration of DAPT (N-[N-(3,5-difluorophenacetyl)-1-alanyl]-S-phenylglycine t-butyl ester), a γ-secretase inhibitor, that has been used to suppress the Notch signal, before hepatocyte depletion by MTZ decreased hepatocyte regeneration, suggesting that the Notch signaling pathway is involved in dedifferentiation of cholangiocytes. In contrast, deficiency of *Wnt2bb* reduced proliferation of hepatocytes derived from dedifferentiated cholangiocytes [36]. Ko et al. examined effects of small compounds on cholangiocyte-to-hepatocyte conversion [47]. They found that inhibitors against bromodomain and extraterminal domain (BET) suppressed dedifferentiation of cholangiocytes into hepatoblast-like cells and their proliferation. *Myc* was a target of BET in regulating proliferation of hepatoblast-like cells. A BET inhibitor also suppressed LPC expansion in mice fed with choline-deficient diet supplemented with ethionine (CDE)-diet. This cholangiocyte-driven regeneralyses of human hepatic diseation model of zebrafish could be a powerful system for screening small compounds and genes to examine their possible roles in the regulation of lineage plasticity of liver epithelial cells.

4 LINEAGE PLASTICITY AND HETEROGENEITY OF HEPATOCYTES

4.1 Lineage Plasticity of Hepatocytes

In healthy condition, hepatocytes show the gene expression profile and the cellular morphology distinctive to cholangiocytes' ones, which are suitable for performing their physiological functions. However, several experimental results suggest that such clear discrimination between hepatocytes and cholangiocytes becomes ambiguous in chronically injured livers. Analyses of human hepatic diseases associated with cholestasis and biliary damages have shown that hepatocytes around the PVs express genes that are normally specific to cholangiocytes [48,49]. When dipeptidyl peptidase IV (DPPIV)$^-$-rats reconstituted with DPPIV$^+$ hepatocytes, in which only donor-derived hepatocytes are DPPIV$^+$, were exposed to DAPM followed by BDL, 45% of the new biliary ductules were DPPIV$^+$, indicating that they were derived from MHs. Nishikawa et al. showed that MHs converted to ductular cholangiocytes in collagen gel in the presence of TNFα [50]. These results suggest that MHs have lineage plasticity and may be able to convert to the cholangiocyte lineage.

Yanger et al. labeled MHs in adeno-associated virus (AAV) 8-Cre:ROSA-YFP mice by tamoxifen injection before inducing liver injuries and found expression of cholangiocyte markers, such as SOX9 and OPN, in hepatocytes around the PVs in 3,5-diethoxycarbonyl-1,4-dihydrocollidine (DDC)-diet, CDE-diet, and BDL-injured livers [12]. Given that SOX9 and OPN are exclusively expressed in bile ducts in healthy liver, MHs may change their fate and convert to cholangiocytes. Consistently, Sekiya et al. demonstrated that MHs labeled in Alb-CreERT2:ROSA-YFP mice administrated with tamoxifen formed ductular structures in DDC-injured livers [14].

Nevertheless, the efficiency of the conversion needs to be elucidated. Yanger et al. provided a quantitative study demonstrating that approximately 4% of the ductular cells were derived from hepatocytes [12], whereas Malato et al. did not identify cholangiocytes derived from hepatocytes [51]. We also found cholangiocyte-like cells derived from hepatocytes in Mx1-Cre:ROSA mice fed with the DDC-diet. In our experiments, approximately 2% of the CK19$^+$ ductular cells were derived from hepatocytes [13]. Nagahama et al. quantitatively analyzed hepatocyte-to-cholangiocyte conversion in different types of chronic injury models and demonstrated that the hepatocyte-to-cholangiocyte conversion in chronic injuries induced by repeated administration of carbon tetrachloride (CCl$_4$) or DAPM. About 5% and 10% of CK19$^+$ duct cells were derived from

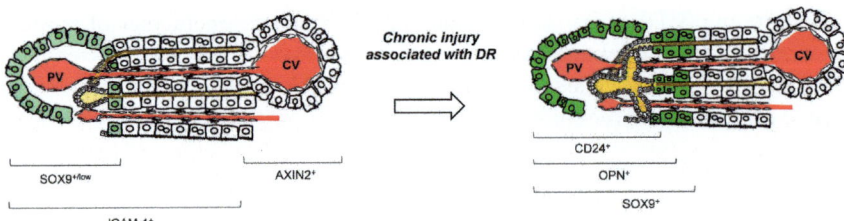

FIGURE 3.1 **Dedifferentiation and lineage conversion of hepatocytes.** The Notch signal induces dedifferentiation of hepatocytes. SOX9+CD24− biphenotypic hepatocytes further become SOX9+CD24+ cells, which have potential of strong proliferative capability. Yes-associated protein (YAP) is suggested to be an upstream of the Notch pathway. Ectopic activation of YAP results in hepatocyte-to-cholangiocyte conversion and eventually formation of intrahepatic cholangiocyte carcinoma. Since the simultaneous activation of the Notch pathway and Akt induces intrahepatic cholangiocarcinoma (ICC) formation, Akt is also involved in this transformation. Actually, hepatocytes robustly dedifferentiate to biphenotypic cells, whereas they barely convert to cholangiocytes forming ductular structures.

hepatocytes in DAPM and CCl_4 injury model, respectively. Taken together, although hepatocytes or part of them likely can convert into cholangiocytes forming the duct structure, the conversion is not an efficient event and its physiological significance remains unknown (Fig. 3.1).

Cellular characteristics of dedifferentiated hepatocytes have been investigated by examining their proliferative and differentiation potential in vitro and their structural properties in vivo. SOX9+ biphenotypic hepatocytes were isolated as GFP+EpCAM− cells from SOX9-EGFP mice fed with DDC-diet or after BDL. They clonally proliferated and re-differentiated to MHs in vitro and in vivo [13,15]. SOX9+ hepatocytes gradually changed from CD24− cells to CD24+ ones during DDC-feeding and the latter one possessed strong proliferative potential [52]. As a result of dedifferentiation, biphenotypic hepatocytes expressed some cholangiocyte markers and partly altered the mode of epithelial polarity. During development, MHs develop the apical domain among adjacent cells, which are called bile canaliculi, whereas cholangiocytes form the typical epithelial polarity; the apical domain surrounds the luminal space with neighboring cells and the basal domain is supported by extracellular matrix [53]. Near the expanded ductular structures, biphenotypic hepatocytes partly established the cholangiocyte type polarity and surrounded the luminal space [12,13], which could be called as "metaplastic" change [15]. However, even in CD24+ status, biphenotypic hepatocytes neither strongly expressed CK19 nor formed the large cyst structure in 3D culture [52]. Given that mature cholangiocytes and bipotential LPCs isolated from neonatal liver establish the cholangiocyte-type epithelial polarity and develop the large cyst structure in the same culture condition, SOX9+ CD24+ biphenotypic hepatocytes did not completely convert to cholangiocytes or bipotential LPCs. Collectively, hepatocytes robustly dedifferentiate and stay at the intermediate status and then redifferentiate into MHs for tissue repair.

In addition to chronic injury models, SOX9$^+$ hepatocytes emerged after acute liver injuries induced by acetaminophen and DAPM. In both models, SOX9$^+$ hepatocytes were near the CVs and PVs in acetaminophen and DAPM injured livers, respectively ([13] and unpublished data). Considering that hepatocytes downregulate Cyps in SOX9$^+$ cells, the temporal conversion into SOX9$^+$ cells in these injury models could be a protective reaction against acute but severe liver damages.

4.2 Heterogeneity of Hepatocytes

Hepatocytes are the liver parenchymal cells performing many metabolic functions. The turnover of hepatocytes is very slow in healthy liver, while they show strong regenerative capacity after acute liver injury; remaining hepatocytes proliferate or grow to compensate the lost tissue [54,55]. In contrast to in vivo regenerative capability, MHs rapidly lose their functions and eventually die in vitro. However, when MHs are cultured in the presence of nicotinamide, a subpopulation of MHs, namely small hepatocytes (SHs) proliferate and form colonies [56]. SHs are highly proliferative and importantly can redifferentiate to MHs both functionally and structurally [57]. This in vitro result suggests that hepatocytes consist of heterogeneous cell populations that may have different proliferative capability and differentiation status.

Recently, the heterogeneity of hepatocytes has been demonstrated in vivo by using genetic lineage tracing (Fig. 3.2). Wang et al. permanently labeled pericentral hepatocytes in Axin-2-CreERT2:ROSA26-mTmG mice by tamoxifen injection and followed their fates [58]. GFP$^+$ hepatocytes gradually expanded and replenished about 40% of the liver mass within a year, suggesting that Axin-2$^+$ hepatocytes have progenitor properties. Planas-Paz et al. also demonstrated that Axin-2$^+$ as well as Lgr5$^+$ hepatocytes exit around the CV [59]. However, in the latter work, Lgr4$^+$ hepatocytes locating throughout the liver lobule rather than pericentral Axin2$^+$Lgr5$^+$ hepatocytes majorly contributed to homeostasis and regeneration. On the other hand, Font-Burgada et al. demonstrated that periportal hepatocytes express a cholangiocyte marker SOX9. They specifically labeled SOX9$^+$ hybrid hepatocytes in Sox9-CreERT:NZG mice and examined their fate in chronic injury [60]. In the transgenic mice, tamoxifen–injection induces nuclear LacZ expression in SOX9$^+$ hybrid hepatocytes as well as cholangiocytes, whereas following infection of AAV expressing optimized FLP recombinase (FLPo) recombinase depletes LacZ and induces green fluorescence protein (GFP) specifically in hybrid hepatocytes. Although SOX9$^+$ periportal "hybrid hepatocytes" did not likely contribute to cellular turnover in healthy liver, they supplied MHs after repeated CCl$_4$ injection and partly converted to CK19$^+$HNF4α^- cholangiocytes, which were incorporated to ductular structures, in DDC-injured liver.

According to the protocol for isolating rat and human SHs, we worked on isolating the cells corresponding to rat SHs in mice. After two-step collagenase

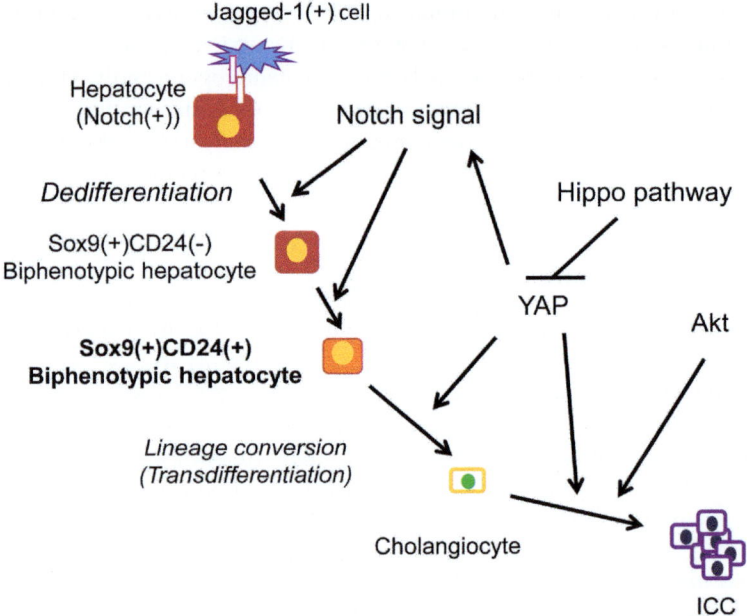

FIGURE 3.2 Heterogeneity of hepatocyte in healthy and injured livers. ICC, Intrahepatic cholangiocarcinoma. In normal liver, SOX9$^+$ hybrid hepatocytes is localized next to the portal vein (PV), whereas AXIN2$^+$ hepatocytes are next to the central vein (CV). SOX9 expression in hybrid hepatocytes is quite low. In addition, ICAM-1$^+$ hepatocytes may exist in liver tissue other than pericentral area. In chronically injured livers in which ductular reaction (DR) is observed, SOX9$^+$ biphenotypic hepatocytes expand around the PV. SOX9$^+$ hepatocytes contain OPN(−) and (+), and CD24(−) and (+) cells. CD24$^+$SOX9$^+$ hepatocytes isolated by FACS can clonally proliferate and efficiently differentiate to MHs in vitro and in vivo.

perfusion, hepatocytes were separated into MH and SH fractions by centrifugation at two different gravities. EpCAM$^-$ICAM-1$^+$ cells in CD31$^-$CD45$^-$ nonhematopoietic/nonendothelial cell fraction were isolated from the SH fraction. ICAM-1$^+$ hepatocytes clonally proliferated on laminin 111-coated dishes to form colonies containing ALB$^+$ cells. Furthermore, part of ICAM-1$^+$ hepatocytes continuously proliferated, while they maintained the potential differentiating into functional hepatocytes in vitro in the presence of OSM and Matrigel. Those ICAM-1$^+$ hepatocytes were engrafted in the liver of the recipient nude mice of Ret/PHx model [61]. These results further indicate that hepatocytes consist of heterogeneous cell populations.

Gene expression profile of SOX9$^+$ hybrid hepatocytes was compared with that of MHs by RNA sequence analysis. Hybrid hepatocytes and MHs equally expressed basic hepatocyte markers including *Hnf1a* and *Hnf4a*. Functional markers including *Cytochrome P450s* (*Cyps*) were downregulated in hybrid hepatocytes, whereas cholangiocyte markers including *Hnf1b* and *Epcam* were hardly detected [60]. Similar to hybrid hepatocytes, ICAM-1$^+$ hepatocytes expressed *Alb* and *Hnf4a* at similar level as MHs, whereas the expression of some *Cyps* were lower than that in MHs. Interestingly, ICAM-1$^+$ hepatocytes expressed *Axin2* and *Lgr5* at lower level as compared with MHs [61]. The gene expression profiles suggest that ICAM-1$^+$ hepatocytes are excluded from the pericentral area and possibly include SOX9$^+$ hybrid hepatocytes (Fig. 3.2).

Significant part of MHs are polyploidy, which is correlated with hepatocyte maturation. Miyaoka et al. demonstrated that binucleated hepatocytes, that include tetraploid and octaploid cells, proliferated through unconventional cell division during regeneration after PHx [55]. However, the long-term proliferative capability could be affected by ploidy. Axin2$^+$ pericentral hepatocytes predominantly contained diploid cells. ICAM-1$^+$ hepatocytes were mostly mononucleated and consisted of diploid cells, whereas their progenies showing long-term proliferation were diploid (unpublished data). Recently, Katsuda et al. reprogrammed MHs into bipotential LPCs in the presence of Y-27632 (Rock inhibitor), A-83-01 (ALK inhibitor), and CHIR99021 (GSK3β inhibitor, potential activator for the Wnt/β-catenin pathway) [62]. They showed that diploid hepatocytes could be the source of induced LPCs.

In future, it is necessary to compare ICAM-1$^+$ hepatocytes with AXIN2$^+$ pericentral hepatocytes and SOX9$^+$ periportal hybrid hepatocytes to reveal the nature of hepatocyte heterogeneity.

4.3 Molecular Mechanisms Regulating Lineage Plasticity of Hepatocytes

4.3.1 Notch Pathway

Given that the Notch signaling pathway induces the cholangiocyte lineage from hepatoblasts, MHs may respond to the signal and change their fate. Indeed, Zong et al. demonstrated that ectopic activation of the Notch signal by expressing the intracellular domain of Notch (NICD), the constitutive active form, in neonatal liver induced hepatocyte-to-cholangiocyte conversion resulting in expansion of ductular structures [63]. Sekiya et al. labeled hepatocytes in Alb-CreERT2:ROSA mice and induced chronic liver injuries. They found that MHs converted to cholangiocytes and increase of ductular structures in a Notch-Hes1 pathway-dependent manner [14]. However, the former study performed at neonatal stage and the latter one contains a problem of the specificity of Alb-CreERT2 labeling [64]. Thus, it should be further examined whether the activation of Notch signal alone can convert adult MHs into cholangiocytes. In addition to regulating epithelial lineages, the Notch signaling pathway is likely involved in pathogenesis. Mice develop intrahepatic cholangiocarcinoma (ICC) by a long-term administration of TAA. Sekiya et al. demonstrated that ICC in this model was derived from MHs and ICC formation depended on the Notch-Hes1 pathway [14,65]. Fan et al. also demonstrated that ICCs were derived from MHs, which NICD and the constitutive active forms of Akt were introduced [66]. These results suggest a possibility that ICCs are derived from hepatocytes.

It should be noted that direct interaction between ligand- and receptor-expressing cells is crucial to activate the Notch signaling pathway. Similar to developing livers, periportal fibroblasts may be the main source of Jagged-1 and activate the signal in periportal hepatocytes to induce their dedifferentiation. Recently, Terada et al. demonstrated that macrophages transiently accumulated in the pericentral area, where hepatocytes are damaged by TAA administration, expressed Jagged-1. They further demonstrated that those macrophages played an important role in hepatocyte-to-cholangiocyte conversion [67].

4.3.2 Hippo Pathway

The Hippo-signaling pathway regulates cell proliferation and organ size. The activation of Hippo pathway results in phosphorylation of the transcriptional coactivator YAP, which is exported from the nucleus to the cytoplasm and then eventually degraded [68]. It is well documented that abnormal activity of the Hippo pathway disturbs proliferative status of liver epithelial cells [69,70]. Overexpression of YAP in hepatocytes increased proliferation and resulted in hepatomegaly. Consistently, depletion of Mst1/2, Nf2, or Ww45, which are involved in phosphorylation of YAP, resulted in hepatomegaly and hepatocellular carcinoma [71–73]. Yimlamai et al. demonstrated that inactivation of the Hippo pathway by introducing a constitutive active form of YAP in hepatocytes resulted in dedifferentiation of hepatocytes [74]. In YAP activated liver, dedifferentiated hepatocytes further lost HNF4α and became cholangiocyte-like cells forming ectopic ductular structures. They further showed that the Notch pathway is a major downstream signal mediating hepatocyte-to-cholangiocyte conversion. In addition, they mentioned that YAP could activate *Sox9* expression in the absence of the Notch signal, suggesting the Hippo pathway regulates the hepatocyte lineage at multiple points. Although it remains unknown whether YAP activation is definitely induced in chronic liver injuries, *connective tissue growth factor (Ctgf)*, a major target of YAP, is upregulated in SOX9$^+$ hepatocytes [15]. It is intriguing how the Hippo pathway is inactivated in chronically injured livers.

4.3.3 FGF Signal

FGFs regulate liver specification in the foregut endoderm and later promote expansion of hepatoblasts [16,75]. To examine roles of the Fgf signals in regenerating liver, Utley et al. generated CMVcre:Rosa26rtTA:tet(O)-Fgf10 and CMVcre:Rosa26rtTA:tet(O)-sFGFR2-IIIb mice to induce ectopic expression of Fgf10 and soluble dominant-negative FGF Receptor-2 IIIb, respectively, by administration of doxycycline. When Fgf10 transgenic mice were fed with DDC-diet, A6$^+$HNF4α$^-$ ductular cells were expanded in consistent with the previous work in which Fgf7-overexpression induced expansion of ductular structure [44]. In addition, the number of A6$^+$HNF4α$^+$ hepatocytes increased in Fgf10-transgenic mice via activation of β-catenin depending on Akt. On the contrary,

A6⁺HNF4α⁺ cells were reduced in dnFGFR mice fed with DDC-diet [76]. The FGF signal likely promotes proliferation of A6⁺ HNF4α⁺ hepatocytes. It is intriguing whether the Fgf signal is also involved in dedifferentiation of MHs.

4.3.4 TGFβ Signal

During DDC-injury, biphenotypic hepatocytes co-expressing SOX9 and HNF4α expand from the periportal area to the parenchymal region. MHs near PV may be activated by the Notch pathway via interaction with periportal mesenchymal cells, whereas it is not clear how MHs receive signals promoting dedifferentiation in the area dissociating from the periportal mesenchymal layer. This raises a possibility that soluble factors are involved in this "partial lineage conversion" in addition to the Notch pathway.

TGFβ signal and Notch signal majorly regulate the commitment of hepatoblasts into cholangiocytes in fetal liver [77]. Accordingly, this signal is expected to be involved in regulating the lineage plasticity of hepatocytes. Interestingly, in the rat model with the combination of BDL and repeated DAPM administration, a significant number of hepatocytes converted into cholangiocytes, where expression of TGFβ was remarkably increased [78]. *Sox9* was upregulated in mouse liver progenitor cell line by TGFβ in vitro [25,79], suggesting that TGFβ is a candidate to induce such a transition in chronically injured livers.

4.3.5 Wnt-β-Catenin Pathway

The Wnt-β-catenin pathway is known to regulate generation of hepatic zonation [80]. In addition, Wnt-β-catenin pathway is involved in ductular expansion [81] and lineage specification of LPCs [82], whereas noncanonical Wnt signal regulates bile duct development [83]. Thompson et al. administrated DDC-diet to transgenic mice overexpressing β-catenin. In addition to expanded tubular structures similarly in the control mice, "atypical hepatocytes" were expanded in the transgenic mice after DDC-administration for 5 months. Those atypical hepatocytes were A6⁺CK19⁻Trop2⁻HNF1β⁻ [84]. Although they did not confirm the origin of "a typical hepatocyte," given that LPCs barely differentiate to hepatocytes in this injury model and that biphenotypic hepatocytes emerged in the transgenic mice were A6⁺CK19⁻, they were probably derived from hepatocytes. They further demonstrated that the transgenic mice recovered from 4 weeks of DDC-injury more quickly than the wild type mice [84]. These results suggest that the Wnt-β-catenin signal is involved in dedifferentiation and/or proliferation of hepatocytes and thereby the signal regulates liver regeneration depending on the lineage plasticity of MHs. The intensive work done by Planas-Paz et al. further demonstrated that liver zonation is strictly controlled by the Wnt-β-catenin single that is modulated by positive regulators including R-spondin (RSPO) and its receptor LGR4/5 as well as negative regulators including transmembrane E3 ubiquitin ligase ZNRF3 and RNF43 [59]. In addition, they showed that the RSPO-LGR4/5 signaling promoted hepatocyte proliferation after PHx [59]. It is intriguing whether these molecules modifying the Wnt-β-catenin single affect dedifferentiation of MHs and their proliferation during liver regeneration after chronic liver injuries.

5 CONCLUDING REMARKS

The concept of the tissue stem cell system is that self-renewable stem cells supply multiple types of cells for tissue homeostasis and regeneration. However, growing evidences suggest that a unidirectional cell supply is not necessarily the universal mechanism for maintaining tissue homeostasis and for regeneration. Even in the intestine, a prototype organ in which Lgr5$^+$ crypt base columnar (CBCs) cells continuously supply all types of intestinal epithelial cells for securing fast cell turnover, epithelial cells de-differentiate and behave facultative stem cells when CBCs are ablated [7]. Likewise, under very severe injuries, mature epithelial cells that are conventionally considered as terminally and irreversibly differentiated could be dedifferentiated or converted into cells possessing properties as stem/progenitor cells [5,6].

In the liver, both hepatocytes and cholangiocytes show lineage plasticity responding to liver injuries. In particular, dedifferentiation of MHs is frequently observed in mouse chronic injury models. Dedifferentiated hepatocytes efficiently redifferentiate into functional hepatocytes in vivo and in vitro. Part of dedifferentiated hepatocytes further convert to cholangiocyte-like cells. The degree of their dedifferentiation is affected by types of injury, though the hepatocyte-to-cholangiocyte conversion is usually limited. However, signals have not been identified to determine whether hepatocytes stay dedifferentiated status or further convert to cholangiocytes. The Notch signal is necessary but not likely sufficient for the lineage conversion. Inactivation of the Hippo pathway resulting in YAP activation induces the lineage conversion but it remains unknown how this pathway is regulated during chronic liver injuries.

In addition to the zonal difference of MHs, the heterogeneity of MHs has been recognized; subpopulations of MHs likely maintain immature status as compared with other hepatocytes. Although direct evidence has not been provided, it is plausible that subpopulations of MHs showing progenitor properties such as SOX9$^+$ hybrid hepatocytes and ICAM-1$^+$ hepatocytes could be more plastic than other hepatocytes. In particular, SOX9$^+$ hybrid hepatocytes are a good candidate for the source of biphenotypic hepatocytes emerged in the chronic liver injury models, in which biphenotypic hepatocytes appear in the periportal region. However, biphenotypic hepatocytes emerging in injured livers show apparently different in gene expression profile and in functions from SOX9$^+$ and ICAM-1$^+$ hepatocytes identified in the healthy liver. Therefore, to prove that a subpopulation of hepatocytes is equipped with lineage plasticity, it is necessary to analyze epigenetic modifications on the promoters of crucial gene correlated with hepatocyte lineage and functions in subpopulations of hepatocytes and MHs.

It remains unclear whether and how the plasticity of MHs contributes to tissue homeostasis, regeneration, and pathogenesis. Dedifferentiated hepatocytes or expanded ductular cells often appear in chronically injured livers, but the appearance of those cells is not correlated with recovery from injured status. Fibrogenic and inflammatory environment may overwhelm expansion of those progenitors and/or suppress their hepatocytic differentiation. Therefore, to protect liver tissue and promote regeneration using dedifferentiated hepatocytes or LPCs, it is a need to find a way to further promote expansion of induced progenitors and their differentiation into MHs. It is important to understand how different ways of epithelial cell supply are coordinated for maintenance of liver functions and for cellular homeostasis as a response to various types of liver injuries. Such information is crucial to find new therapeutic approaches for chronic liver diseases.

In addition to physiological importance, hepatocyte plasticity and heterogeneity are important issues for ex vivo studies. Since metabolic and xenobiotic functions of MHs cannot be maintained in vitro, researchers have made efforts to generate functional hepatocytes from hepatic or nonhepatic stem/progenitor cells usable for not only for hepatocyte transplantation but for generation of artificial livers and establishment of ex vivo drug screening systems. As compared with complicated culture conditions and overexpression of transcription factors for inducing hepatocyte-like cells from multipotent stem cells including ES and iPS cells or from somatic cells, expansion of intrinsic hepatocyte progenitors is simple and safe. Although only 0.1% of ICAM-1^+ cells can be expanded while maintaining redifferentiation potential, the optimization of culture conditions may increase the efficiency for producing functional hepatocytes [62]. Considering that a similar cell population exists in human liver cells, expansion intrinsic hepatocyte progenitors may be one of hopeful methodologies to generate functional human hepatocytes ex vivo in future.

ABBREVIATIONS

BDL	Bile duct ligation
BMP	Bone morphogenetic protein
CCl$_4$	Carbon tetrachloride
C/EBPα	CCAAT enhancer binding protein
CDE	Choline-deficient diet supplemented with ethionine
Cldn	Claudin
CV	Central vein
Cyp	Cytochrome P450s
DAPM	4,4′-diaminodiphenylmethane
DDC	3,5-diethoxycarbonyl-1,4-dihydrocollidine
Dlk1	Delta-like 1
DPPIV	Dipeptidyl peptidase IV
EpCAM	Epithelial adhesion molecule
Fgf	Fibroblast growth factor

Grhl2	Grainyhead-like 2
Hes1	Hairy enhancer of slit 1
Hey1	Hes-related family bHLH transcription factor with YRPW motif 1
HNF4α	Hepatocyte nuclear factor 4α
IHBD	Intrahepatic bile duct
LPC	Liver stem/progenitor cell
MH	Mature hepatocyte
NICD	Intracellular domain of Notch
OSM	Oncostatin M
PHx	Partial hepatectomy
PV	Portal vein
Ret	Retrorsine
RSPO	R-spondin
Sox9	Sry-HMG box protein 9
TAA	Thioacetamide
TJ	Tight junction
TGFβ	Transforming growth factor β

REFERENCES

[1] Clevers H. The intestinal crypt, a prototype stem cell compartment. Cell 2013;154:274–84.
[2] Kopp JL, Grompe M, Sander M. Stem cells versus plasticity in liver and pancreas regeneration. Nat Cell Biol 2016;18:238–45.
[3] Takahashi K, Yamanaka S. Induction of pluripotent stem cells from mouse embryonic and adult fibroblast cultures by defined factors. Cell 2006;126:663–76.
[4] Xu J, Du Y, Deng H. Direct lineage reprogramming: strategies, mechanisms, and applications. Cell Stem Cell 2015;16:119–34.
[5] Tata PR, Mou H, Pardo-Saganta A, Zhao R, Prabhu M, Law BM, et al. Dedifferentiation of committed epithelial cells into stem cells in vivo. Nature 2013;503:218–23.
[6] Stange DE, Koo BK, Huch M, Sibbel G, Basak O, Lyubimova A, et al. Differentiated troy(+) chief cells act as reserve stem cells to generate all lineages of the stomach epithelium. Cell 2013;155:357–68.
[7] Tetteh PW, Basak O, Farin HF, Wiebrands K, Kretzschmar K, Begthel H, et al. Replacement of lost Lgr5-positive stem cells through plasticity of their enterocyte-lineage daughters. Cell Stem Cell 2016;18:203–13.
[8] Zaret KS. Regulatory phases of early liver development: paradigms of organogenesis. Nat Rev Genet 2002;3:499–512.
[9] Suzuki A, Sekiya S, Onishi M, Oshima N, Kiyonari H, Nakauchi H, et al. Flow cytometric isolation and clonal identification of self-renewing bipotent hepatic progenitor cells in adult mouse liver. Hepatology 2008;48:1964–78.
[10] Kamiya A, Kakinuma S, Yamazaki Y, Nakauchi H. Enrichment and clonal culture of progenitor cells during mouse postnatal liver development in mice. Gastroenterology 2009;137 (1114-1126):e1111–1114.
[11] Okabe M, Tsukahara Y, Tanaka M, Suzuki K, Saito S, Kamiya Y, et al. Potential hepatic stem cells reside in EpCAM+ cells of normal and injured mouse liver. Development 2009;136: 1951–60.
[12] Yanger K, Zong Y, Maggs LR, Shapira SN, Maddipati R, Aiello NM, et al. Robust cellular reprogramming occurs spontaneously during liver regeneration. Genes Dev 2013;27:719–24.

[13] Tanimizu N, Nishikawa Y, Ichinohe N, Akiyama H, Mitaka T. Sry HMGbox protein 9-positive (Sox9 +) epithelial cell adhesion molecule-negative (EpCAM-) biphenotypic cells derived from hepatocytes are involved in mouse liver regeneration. J Biol Chem 2014;289:7589–98.

[14] Sekiya S, Suzuki A. Hepatocytes, rather than cholangiocytes, can be the major source of primitive ductules in the chronically injured mouse liver. Am J Pathol 2014;184:1468–78.

[15] Tarlow BD, Pelz C, Naugler WE, Wakefield L, Wilson EM, Finegold MJ, et al. Bipotential adult liver progenitors are derived from chronically injured mature hepatocytes. Cell Stem Cell 2014;15:605–18.

[16] Jung J, Zheng M, Goldfarb M, Zaret KS. Initiation of mammalian liver development from endoderm by fibroblast growth factors. Science 1999;284:1998–2003.

[17] Rossi JM, Dunn NR, Hogan BL, Zaret KS. Distinct mesodermal signals, including BMPs from the septum transversum mesenchyme, are required in combination for hepatogenesis from the endoderm. Genes Dev 2001;15:1998–2009.

[18] Tanaka M, Okabe M, Suzuki K, Kamiya Y, Tsukahara Y, Saito S, et al. Mouse hepatoblasts at distinct developmental stages are characterized by expression of EpCAM and DLK1: drastic change of EpCAM expression during liver development. Mech Dev 2009;126:665–76.

[19] Tanimizu N, Nishikawa M, Saito H, Tsujimura T, Miyajima A. Isolation of hepatoblasts based on the expression of Dlk/Pref-1. J Cell Sci 2003;116:1775–86.

[20] Watanabe T, Nakagawa K, Ohata S, Kitagawa D, Nishitai G, Seo J, et al. SEK1/MKK4-mediated SAPK/JNK signaling participates in embryonic hepatoblast proliferation via a pathway different from NF-kappaB-induced anti-apoptosis. Dev Biol 2002;250:332–47.

[21] Suzuki A, Zheng YW, Kaneko S, Onodera M, Fukao K, Nakauchi H, et al. Clonal identification and characterization of self-renewing pluripotent stem cells in the developing liver. J Cell Biol 2002;156:173–84.

[22] McCright B, Lozier J, Gridley T. A mouse model of Alagille syndrome: Notch2 as a genetic modifier of Jag1 haploinsufficiency. Development 2002;129:1075–82.

[23] Tanimizu N, Miyajima A. Notch signaling controls hepatoblast differentiation by altering the expression of liver-enriched transcription factors. J Cell Sci 2004;117:3165–74.

[24] Clotman F, Jacquemin P, Plumb-Rudewiez N, Pierreux CE, Van der Smissen P, Dietz HC, et al. Control of liver cell fate decision by a gradient of TGF beta signaling modulated by Onecut transcription factors. Genes Dev 2005;19:1849–54.

[25] Antoniou A, Raynaud P, Cordi S, Zong Y, Tronche F, Stanger BZ, et al. Intrahepatic bile ducts develop according to a new mode of tubulogenesis regulated by the transcription factor SOX9. Gastroenterology 2009;136:2325–33.

[26] Tanimizu N, Kaneko K, Itoh T, Ichinohe N, Ishii M, Mizuguchi T, et al. Intrahepatic bile ducts are developed through formation of homogeneous continuous luminal network and its dynamic rearrangement in mice. Hepatology 2016;64:175–88.

[27] Carpentier R, Suner RE, van Hul N, Kopp JL, Beaudry JB, Cordi S, et al. Embryonic ductal plate cells give rise to cholangiocytes, periportal hepatocytes, and adult liver progenitor cells. Gastroenterology 2011;141:1432–8. 1438.e1–4.

[28] Tanimizu N, Kobayashi S, Ichinohe N, Mitaka T. Downregulation of miR122 by grainyhead-like 2 restricts the hepatocytic differentiation potential of adult liver progenitor cells. Development 2014;141:4448–56.

[29] Tanimizu N, Miyajima A, Mostov KE. Liver progenitor cells develop cholangiocyte-type epithelial polarity in three-dimensional culture. Mol Biol Cell 2007;18:1472–9.

[30] Tanimizu N, Nakamura Y, Ichinohe N, Mizuguchi T, Hirata K, Mitaka T. Hepatic biliary epithelial cells acquire epithelial integrity but lose plasticity to differentiate into hepatocytes in vitro during development. J Cell Sci 2013;126:5239–46.

[31] Furuyama K, Kawaguchi Y, Akiyama H, Horiguchi M, Kodama S, Kuhara T, et al. Continuous cell supply from a Sox9-expressing progenitor zone in adult liver, exocrine pancreas and intestine. Nat Genet 2011;43:34–41.
[32] Espanol-Suner R, Carpentier R, Van Hul N, Legry V, Achouri Y, Cordi S, et al. Liver progenitor cells yield functional hepatocytes in response to chronic liver injury in mice. Gastroenterology 2012;143(1564-1575):e1567.
[33] Tarlow BD, Finegold MJ, Grompe M. Clonal tracing of Sox9+ liver progenitors in mouse oval cell injury. Hepatology 2014;60:278–89.
[34] Yanger K, Knigin D, Zong Y, Maggs L, Gu G, Akiyama H, et al. Adult hepatocytes are generated by self-duplication rather than stem cell differentiation. Cell Stem Cell 2014;15:340–9.
[35] Rodrigo-Torres D, Affo S, Coll M, Morales-Ibanez O, Millan C, Blaya D, et al. The biliary epithelium gives rise to liver progenitor cells. Hepatology 2014;60:1367–77.
[36] Choi TY, Khaliq M, Ko S, So J, Shin D. Hepatocyte-specific ablation in zebrafish to study biliary-driven liver regeneration. J Vis Exp 2015;e52785.
[37] He J, Lu H, Zou Q, Luo L. Regeneration of liver after extreme hepatocyte loss occurs mainly via biliary transdifferentiation in zebrafish. Gastroenterology 2014;146:789–800. e788.
[38] Lu WY, Bird TG, Boulter L, Tsuchiya A, Cole AM, Hay T, et al. Hepatic progenitor cells of biliary origin with liver repopulation capacity. Nat Cell Biol 2015;17:971–83.
[39] Takashima Y, Terada M, Kawabata M, Suzuki A. Dynamic three-dimensional morphogenesis of intrahepatic bile ducts in mouse liver development. Hepatology 2015;61:1003–11.
[40] Kaneko K, Kamimoto K, Miyajima A, Itoh T. Adaptive remodeling of the biliary architecture underlies liver homeostasis. Hepatology 2015;61:2056–66.
[41] Glaser SS, Gaudio E, Rao A, Pierce LM, Onori P, Franchitto A, et al. Morphological and functional heterogeneity of the mouse intrahepatic biliary epithelium. Lab Invest 2009;89: 456–69.
[42] Marzioni M, Glaser SS, Francis H, Phinizy JL, LeSage G, Alpini G. Functional heterogeneity of cholangiocytes. Semin Liver Dis 2002;22:227–40.
[43] Ueno Y, Alpini G, Yahagi K, Kanno N, Moritoki Y, Fukushima K, et al. Evaluation of differential gene expression by microarray analysis in small and large cholangiocytes isolated from normal mice. Liver Int 2003;23:449–59.
[44] Takase HM, Itoh T, Ino S, Wang T, Koji T, Akira S, et al. FGF7 is a functional niche signal required for stimulation of adult liver progenitor cells that support liver regeneration. Genes Dev 2013;27:169–81.
[45] Kamimoto K, Kaneko K, Kok CY, Okada H, Miyajima A, Itoh T. Heterogeneity and stochastic growth regulation of biliary epithelial cells dictate dynamic epithelial tissue remodeling. Elife 2016;5:e15034–145.
[46] Dozynkiewicz MA, Jamieson NB, Macpherson I, Grindlay J, van den Berghe PV, von Thun A, et al. Rab25 and CLIC3 collaborate to promote integrin recycling from late endosomes/lysosomes and drive cancer progression. Dev Cell 2012;22:131–45.
[47] Ko S, Choi TY, Russell JO, So J, Monga SP, Shin D. Bromodomain and extraterminal (BET) proteins regulate biliary-driven liver regeneration. J Hepatol 2016;64:316–25.
[48] Crosby HA, Hubscher S, Fabris L, Joplin R, Sell S, Kelly D, et al. Immunolocalization of putative human liver progenitor cells in livers from patients with end-stage primary biliary cirrhosis and sclerosing cholangitis using the monoclonal antibody OV-6. Am J Pathol 1998;152:771–9.
[49] Limaye PB, Bowen WC, Orr AV, Luo J, Tseng GC, Michalopoulos GK. Mechanisms of hepatocyte growth factor-mediated and epidermal growth factor-mediated signaling in transdifferentiation of rat hepatocytes to biliary epithelium. Hepatology 2008;47:1702–13.

[50] Nishikawa Y, Doi Y, Watanabe H, Tokairin T, Omori Y, Su M, et al. Transdifferentiation of mature rat hepatocytes into bile duct-like cells in vitro. Am J Pathol 2005;166:1077–88.

[51] Malato Y, Naqvi S, Schurmann N, Ng R, Wang B, Zape J, et al. Fate tracing of mature hepatocytes in mouse liver homeostasis and regeneration. J Clin Invest 2011;121:4850–60.

[52] Tanimizu N, Ichinohe N, Yamamoto M, Akiyama H, Nishikawa Y, Mitaka T. Progressive induction of hepatocyte progenitor cells in chronically injured liver. Sci Rep 2017;7:39990.

[53] Tanimizu N, Mitaka T. Epithelial Morphogenesis during Liver Development. Cold Spring Harb Perspect Biol 2017;9:a027862.

[54] Michalopoulos GK, DeFrances MC. Liver regeneration. Science 1997;276:60–6.

[55] Miyaoka Y, Ebato K, Kato H, Arakawa S, Shimizu S, Miyajima A. Hypertrophy and unconventional cell division of hepatocytes underlie liver regeneration. Curr Biol 2012;22: 1166–75.

[56] Mitaka T, Mikami M, Sattler GL, Pitot HC, Mochizuki Y. Small cell colonies appear in the primary culture of adult rat hepatocytes in the presence of nicotinamide and epidermal growth factor. Hepatology 1992;16:440–7.

[57] Mitaka T, Sato F, Mizuguchi T, Yokono T, Mochizuki Y. Reconstruction of hepatic organoid by rat small hepatocytes and hepatic nonparenchymal cells. Hepatology 1999;29:111–25.

[58] Wang B, Zhao L, Fish M, Logan CY, Nusse R. Self-renewing diploid Axin2(+) cells fuel homeostatic renewal of the liver. Nature 2015;524:180–5.

[59] Planas-Paz L, Orsini V, Boulter L, Calabrese D, Pikiolek M, Nigsch F, et al. The RSPO-LGR4/5-ZNRF3/RNF43 module controls liver zonation and size. Nat Cell Biol 2016;18: 467–79.

[60] Font-Burgada J, Shalapour S, Ramaswamy S, Hsueh B, Rossell D, Umemura A, et al. Hybrid periportal hepatocytes regenerate the injured liver without giving rise to cancer. Cell 2015;162:766–79.

[61] Tanimizu N, Ichinohe N, Ishii M, Kino J, Mizuguchi T, Hirata K, et al. Liver progenitors isolated from adult healthy mouse liver efficiently differentiate to functional hepatocytes in vitro and repopulate liver tissue. Stem Cells 2016;34:2889–901.

[62] Katsuda T, Kawamata M, Hagiwara K, Takahashi RU, Yamamoto Y, Camargo FD, et al. Conversion of terminally committed hepatocytes to culturable bipotent progenitor cells with regenerative capacity. Cell Stem Cell 2017;20:41–55.

[63] Zong Y, Panikkar A, Xu J, Antoniou A, Raynaud P, Lemaigre F, et al. Notch signaling controls liver development by regulating biliary differentiation. Development 2009;136:1727–39.

[64] Lemaigre FP. Determining the fate of hepatic cells by lineage tracing: facts and pitfalls. Hepatology 2015;61:2100–3.

[65] Sekiya S, Suzuki A. Intrahepatic cholangiocarcinoma can arise from Notch-mediated conversion of hepatocytes. J Clin Invest 2012;122:3914–8.

[66] Fan B, Malato Y, Calvisi DF, Naqvi S, Razumilava N, Ribback S, et al. Cholangiocarcinomas can originate from hepatocytes in mice. J Clin Invest 2012;122:2911–5.

[67] Terada M, Horisawa K, Miura S, Takashima Y, Ohkawa Y, Sekiya S, et al. Kupffer cells induce Notch-mediated hepatocyte conversion in a common mouse model of intrahepatic cholangiocarcinoma. Sci Rep 2016;6:34691.

[68] Ramos A, Camargo FD. The Hippo signaling pathway and stem cell biology. Trends Cell Biol 2012;22:339–46.

[69] Camargo FD, Gokhale S, Johnnidis JB, Fu D, Bell GW, Jaenisch R, et al. YAP1 increases organ size and expands undifferentiated progenitor cells. Curr Biol 2007;17:2054–60.

[70] Dong J, Feldmann G, Huang J, Wu S, Zhang N, Comerford SA, et al. Elucidation of a universal size-control mechanism in Drosophila and mammals. Cell 2007;130:1120–33.

[71] Zhou D, Conrad C, Xia F, Park JS, Payer B, Yin Y, et al. Mst1 and Mst2 maintain hepatocyte quiescence and suppress hepatocellular carcinoma development through inactivation of the Yap1 oncogene. Cancer Cell 2009;16:425–38.
[72] Benhamouche S, Curto M, Saotome I, Gladden AB, Liu CH, Giovannini M, et al. Nf2/Merlin controls progenitor homeostasis and tumorigenesis in the liver. Genes Dev 2010;24:1718–30.
[73] Lee KP, Lee JH, Kim TS, Kim TH, Park HD, Byun JS, et al. The Hippo-Salvador pathway restrains hepatic oval cell proliferation, liver size, and liver tumorigenesis. Proc Natl Acad Sci U S A 2010;107:8248–53.
[74] Yimlamai D, Christodoulou C, Galli GG, Yanger K, Pepe-Mooney B, Gurung B, et al. Hippo pathway activity influences liver cell fate. Cell 2014;157:1324–38.
[75] Berg T, Rountree CB, Lee L, Estrada J, Sala FG, Choe A, et al. Fibroblast growth factor 10 is critical for liver growth during embryogenesis and controls hepatoblast survival via beta-catenin activation. Hepatology 2007;46:1187–97.
[76] Utley S, James D, Mavila N, Nguyen MV, Vendryes C, Salisbury SM, et al. Fibroblast growth factor signaling regulates the expansion of A6-expressing hepatocytes in association with AKT-dependent beta-catenin activation. J Hepatol 2014;60:1002–9.
[77] Lemaigre FP. Development of the biliary tract. Mech Dev 2003;120:81–7.
[78] Limaye PB, Bowen WC, Orr A, Apte UM, Michalopoulos GK. Expression of hepatocytic- and biliary-specific transcription factors in regenerating bile ducts during hepatocyte-to-biliary epithelial cell transdifferentiation. Comp Hepatol 2010;9:9.
[79] Tanimizu N, Kikkawa Y, Mitaka T, Miyajima A. α1- and α5-containing laminins regulate the development of bile ducts via β1 integrin signals. J Biol Chem 2012;287:28586–97.
[80] Benhamouche S, Decaens T, Godard C, Chambrey R, Rickman DS, Moinard C, et al. Apc tumor suppressor gene is the "zonation-keeper" of mouse liver. Dev Cell 2006;10:759–70.
[81] Itoh T, Kamiya Y, Okabe M, Tanaka M, Miyajima A. Inducible expression of Wnt genes during adult hepatic stem/progenitor cell response. FEBS Lett 2009;583:777–81.
[82] Boulter L, Govaere O, Bird TG, Radulescu S, Ramachandran P, Pellicoro A, et al. Macrophage-derived Wnt opposes Notch signaling to specify hepatic progenitor cell fate in chronic liver disease. Nat Med 2012;18:572–9.
[83] Kiyohashi K, Kakinuma S, Kamiya A, Sakamoto N, Nitta S, Yamanaka H, et al. Wnt5a signaling mediates biliary differentiation of fetal hepatic stem/progenitor cells. Hepatology 2013;.
[84] Thompson MD, Awuah P, Singh S, Monga SP. Disparate cellular basis of improved liver repair in beta-catenin-overexpressing mice after long-term exposure to 3,5-diethoxycarbonyl-1,4-dihydrocollidine. Am J Pathol 2010;177:1812–22.

Chapter 4

Stem Cells in the Liver Cancer and the Organ Size Control

Amita Tiyaboonchai*, Yulong Su**, Bin Li*
*Oregon Stem Cell Center, Oregon Health and Science University, Portland, OR, United States;
**Oregon Health and Science University, Portland, OR, United States

1 INTRODUCTION

The liver is an organ involved in food and drug metabolism, detoxification, and alcohol metabolism. Owing to these reasons, the liver is vulnerable to damage by toxins and viruses when performing these defensive duties. Physical exposure to irradiation and heavy metals can lead to breakage of the DNA double chains thus causing damage to the chromosomes, which can lead to cancer. High exposure to toxicants can trigger direct liver injuries, genetic, and epigenetic mutations. Other drugs such as paracetamol/acetaminophen, often induce hepatotoxicity and contribute to 39% of acute liver failures in the USA [1]. Abuse of alcohol can also lead to liver injury, such as alcoholic cirrhosis [2,3]. Hepatitis B (HBV) and C (HCV) viruses could also induce chronic injuries. Around 2% of patients with alcoholic liver cirrhosis develop hepatocellular carcinoma (HCC) per year [4]. Because the cirrhosis in liver cancer is usually not reversible and currently the only curative method is transplantation, it is important to find more efficient therapeutic ways to treat the cancer. Cell therapy such as stem cell transplantation is an alternative way for liver transplantation. Moreover, understanding the role of stem cells in liver cancer could help researchers to find therapeutic target for liver cancer. In this chapter, we will discuss the stem cells, organ size, and cancer in the liver.

2 STEM CELLS IN THE LIVER

With about 1500 of the 17,000 US patients dying each year due to lack of available transplantable livers, according to the American Liver Foundation, there is an increasing demand for a source of alternative donor livers or liver cell therapies to treat severe liver diseases. Transplantation of healthy liver cells has emerged as a promising strategy both for treating patients waiting for a transplant and as an alternative to liver transplant. Adult hepatocytes serve as indispensable resource for research on liver function in both clinical and pharmaceutical settings. Hepatocytes are a type of liver parenchymal cell that are important for liver function in nutrient metabolism, defense against toxins, alcohol processing, and pharmaceutical metabolism. As a result, hepatocytes are vulnerable during the metabolic and detoxification processes, causing liver failure, which is irreversible and a major health concern worldwide. Identifying progenitors/stem hepatocytes cell populations in the liver as well as tumor initiating cells (TIC) in liver cancer has become a major topic of intensive research. Although some murine hepatic stem cells have been documented in the lab, there are still many important features that are hard to find or observe in human hepatocytes. In contrast, hepatic stem cells have provided a platform for research resulting in benefits that have allowed the rescue of human livers from failure [5]. Some of these benefits include the use of hepatocytes as an ex vivo drug screening agent by the pharmaceutical industry and repopulating an injured liver with hepatocyte cells.

Stem cells are cells that have the ability to self-renewal and differentiate, dividing into one daughter cell which retains stem cell identity and one progenitor cell which differentiates into somatic tissues [6]. There are three types of stem cells in mammalian systems: totipotent stem cells, pluripotent stem cells, and multipotent stem cells. Totipotent stem cells are cells such as germ line stem cells that can form an embryo. Pluripotent stem cells, so called embryonic stem cells and induced pluripotent stem cells are able to differentiate into three germ layers [7] and then terminally differentiated tissues such as liver from definitive endoderm. Multipotent stem cells, so called adult stem cells, are capable of regeneration of tissue from the same lineage upon tissue injury caused by mechanical damage or toxicity. Adult stem cells are a rare population inside the body and the stem cells are usually quiescent unless activated by injury. Somatic stem cells require a certain environment called the "niche" to maintain homeostasis. Signaling from the niche regulates stem cell proliferation and differentiation. Regulation of the niche can cause stem cell proliferation, differentiation, senescence, and carcinogenesis.

To address the shortage of functional liver cells available for transplantation, medical focus has been made to use liver stem cells to make hepatocytes. Liver stem cells are easy to handle, can expend in vitro and can differentiate into hepatocytes both in vitro and in vivo. Unfortunately, it was still unknown whether hepatic stem cells can be isolated from a normal liver. Furthermore, despite numerous experiments, little is known about the stem cell inside the liver itself, and questions remain as to how hepatic stem cells differentiate into functional hepatocytes and cholangiocytes (bile duct cells). Thus, a better source for hepatocytes is needed for transplantations. However, the hepatocyte supply is severely limited due to a shortage of donor liver stem cells, which can proliferate and differentiate into hepatocytes both in vitro and in vivo.

One debate in the field of liver regeneration is about the sources of liver stem cells. Hepatocytes themselves are reported to be stem cells. Although hepatocytes are quiescent in the normal liver, it is believed that human and mouse hepatocytes can proliferate and repopulate the injured immune compliment mouse liver [8]. The Grompe lab has established a transgenic mouse model for liver transplantation. The Fumarylacetoacetate hydrolase (Fah) deficient mouse, (Fah$^{-/-}$), which lacks the Fah gene is a model of type I tyrosinemia an inborn error of metabolism which results from the accumulation of fumarylacetoacetate. This type of tyrosinemia can be treated by administration of 2-(2-nitro-4-trifluoro-methylbenzyol)-1,3 cyclohexanedione (NTBC). NTBC is a drug, which targets parahydroxyphenylpyruvic acid dioxygenase, a step in the tyrosine catabolic pathway, preventing the accumulation of fumarylacetoacetate and cure the disease. The group used NTBC water to control the liver injury and did a hepatocyte transplantation so that the donor hepatocytes could survive in the host liver under NTBC withdraw. Azuma et al. used FRG/N mice, which are Fah$^{-/-}$, Rag2$^{-/-}$, and Il2rg$^{-/-}$ deficient animals on the NOD-strain background (FRGN) as an host model, succeeded in generating a humanized mouse liver, thus confirming that hepatocytes could be a source of liver stem cells in vivo [9]. Serial transplantation of donor derived hepatocytes could repopulate host liver [9], demonstrating the stem cell capability of hepatocytes.

In the developing liver, stem/progenitor cells can be found. Using fluorescence activated cell sorting, Suzuki et al. [10,11] and Zheng et al. [12] found hepatic stem cells in fetal rodents. The CD49f$^+$CD29$^+$TER119$^-$CD45$^-$ cells are clonalgenic and bipotential: they express weak hepatocyte marker Albumin and cholangiocytes marker Cytokeratin19, but after differentiation in vitro, strong Albumin was detected. These cells could repopulate an injured liver, thus confirming stem/progenitor capability of this population.

The Canal of Hering, the liver stem cell niche [13], is located in the bile duct area. Bile ducts transduce bile for food digestion and connect the intestines. Studies demonstrated that Lgr5$^+$ bile duct cells are stem cells in both mouse and human liver, these cells could form organoid as those in small intestine, stomach and could repopulate injured mouse liver upon transplantation into host [14,15]. Moreover, Dorrell et al. found that progenitor cells are also harbored in the duct area. They demonstrated in the normal murine liver, hepatic progenitor cells are identified and characterized by the cell surface antibody CD133 and MIC1-1C3 [16,17]. Additionally, they also found MIC1-1C3$^+$CD133$^+$CD26$^-$CD45/31/11b$^-$ cells formed colonies in vitro and most are Sox9$^+$ cells, indicating they are progenitor cells. Moreover, this group found that the MIC1-1C3$^+$CD133$^+$ cells formed organoids and could repopulate an injured mouse liver in vivo [18]. A recent work demonstrate that the Sox9$^+$ cells inside the liver duct cells are not the progenitor cells [19], while in contrast, hepatocytes themselves are the stem cell in murine liver upon injury [20,21]. The question remained though, is every single hepatocyte able to take the duty as a hepatic stem cell or do hepatocytes have heterogeneity? Further study from Font-Burgada and colleague claimed that Sox9$^+$ hepatocytes, which are located near the periportal area, are capable of regenerating the liver under longer term chronic liver injury [22]. Conversely, by lineage tracing, another study group reported the hepatocytes near pericentral vein are the hepatic "stem cells" [23]. To answer the question of whether hepatocytes are stem cells, single cell analyses should be applied to analyze the transcriptome of individual cells. Besides hepatocytes and cholangiocytes, other cell types such as CD133$^+$ stellate cells, which store vitamin A and secrete hepatocyte growth factor (HGF), have also been demonstrated to be a source of hepatic stem/progenitor cells [24]. However, it must be taken into consideration that the stellate cells are also a source of liver fibrosis. Study on stellate cell as stem cells in liver cancer would help control the fibrosis and over growth of tumor. Bone marrow cells, in contrast, which have hematopoietic stem cells, and myelomonocytic cells are found to regenerate mouse liver in vivo [25,26].

3 LIVER CANCER AND CANCER STEM CELLS

Cancer stem cells have many properties in common with adult stem cells including self-renewal, proliferation, migration, and differentiation. There are two major types of liver cancer: HCC and intrahepatic cholangiocellular carcinoma. However, it is under debating that HCC can not only from hepatocytes but also from cholangiocytes [27] in mouse model. Stem cells have been identified as the source of certain types of cancer [28,29], thus tumor initiating cells are thought as cancer stem cells too. Blocking nonstem cells in the tumor would not help inhibit the growth of the tumor. To understand the mechanism of stem cells and find the therapeutic target for TIC cells thus cure the cancer, it is important to find the factors or land markers involved in tumor formation. Many cancer stem cell markers such as ALDH1 [30], CD13 [31], CD24 [32], CD44 [33], CD90 (Thy-1) [34], CD133 [35,36], CD146 [37], DLK1 [38], EpCAM [39,40], and OV6 [41] are reported (Table 4.1). Specifically, in one study, with flow cytometry, Chiba et al. reported side population (SP) cells from hepatic cancer cell lines are TIC because they have high proliferative rate and self-renewal capability in vitro [42], moreover, as low as 1000 SP cells could form tumor in the immune deficient mice, showing these cells are TIC in vivo. Besides, this group also found the oncogene *Bmi1* overexpression in hepatic stem cells could make TIC [43].

It is important to understand the relationship between normal and cancer stem cell. Epithelial-mesenchymal transition (EMT) and Mesenchymal-epithelial transition (MET) are the phenomenons involved with epithelial cancer stem cell migration and relocalization [44]. EMT is that the epithelial cell gaining mesenchymal characters, losing cell polarity, become migrate, and move to metastasis site. MET is a reversible biological process that involves the transition from motile, multipolar or spindle-shaped mesenchymal cells to planar arrays of polarized cells called epithelia. EMT and MET occur in metastasized cancer stem cells and during embryonic development. Understanding how EMT and MET work is important for finding an inhibitor for it to block

TABLE 4.1 Cancer Stem Cell Markers in Liver

Markers	References
ALDH1	[30]
CD13	[31]
CD24	[32]
CD44	[33]
CD90(Thy-1)	[34]
CD133	[35,36]
CD146	[37]
DLK1	[38]
EpCAM	[39,40]
OV6	[41]

liver cancer invasion, and subsequently blocking of other cancer types. Many experiments suggested that hepatocytes experienced EMT in the chronic liver injury and liver cancer [45–48]. However, it was still undetermined if MET found in hepatic stem cells, or liver stem cells. MET was found in liver cancer metastasis [49]. In one study, Li et al. discovered that hepatic stem cells in a normal mouse model's fetal liver have particularly mesenchymal characters, which result in MET [50].

Stem cells having mesenchymal characters mean these cells lose cell connection and polarity. Cell connection and polarity must be retained because they help cells migrate through the body, making them critical for stem cell related regeneration and cancer metastasis. To mimic the liver cancer in the lab, Zheng et al. used 2-Acetylaminofluorene and partial hepatectomy to induce the liver tumor in rat [12], they developed and confirm TIC from adult rat livers with cell surface markers CD133 and CD44. Targeting the CD133$^+$CD44$^+$ cells in rat model could efficiently reduce the reoccurrence of liver cancer.

4 LIVER ORGAN SIZE

The organ size is controlled by many factors. Murine liver could recover itself 7 days after 70% partial hepatectomy. After partial hepatectomy, the extracellular matrix (ECM) is destructing thus making liver tissue soft, which secretes cytokines and induces induce hepatocyte proliferation. Our study showed that after 24 h hepatectomy, mouse liver loose blood and turning pink, easier to perfuse with collagenase than wild type mice (data not shown). The destruction of ECM release growth factors such as HGF thus induces hepatocytes proliferation at a rate that peaks at 24 h [51]. Subsequently, nonparenchymal cells such as hepatic stellate cells, cholangiocytes and Kupffer cells will start to proliferate. HGF is secreted by stellate cells and is important for hepatocyte culture in vitro [52]. Not only HGF, but also epidermal growth factor (EGF) and transforming growth factor (TGF)-α induce hepatocytes to exit quiescent state (G0 phase) and enter G1-S phase [53,54]. However, direct injection of HGF and EGF hormone didn't give evidence of inducing hepatocyte proliferation [55]. These growth factors significantly increase the hepatocyte DNA synthesis. The murine liver restores the liver mass as well as serum albumin, Alanine transaminase, and Aspartate transaminase levels, back to the normal state after partial hepatectomy in 1 week, followed by the termination of liver regeneration. In the end, signaling such as TGF-β inducing hepatocyte apoptosis to ensure the liver does not overgrow. Interruption of liver regeneration such as continuously expressing HGF or TGF-β will cause abnormal liver growth such as cirrhosis or tumor. Ding reported in the hepatectomy, signal from endothelial cells trigger the hepatocyte proliferation and terminate liver regeneration [56]. Besides that, study on liver sinusoidal endothelium cells also demonstrated angiocrine and autocrine signal between endothelium and hepatocytes in hepatectomy [57]. However, the mechanism terminating liver regeneration is still unknown, and it is the critical factor to understand the regulator for liver organ size.

The hippo pathway, which is involved in tumorigenecity, is found in drosophila [58] and murine [59,60] to control liver size. Blocking hippo pathway with verteporfin and can also inhibit the tissue overgrowth by inhibiting YAP/TAZ-TEAD interation [58]. Similar to Hippo pathway, Naugler et al. found fibroblast growth factor 19 (FGF19) is important for organ size regulation in humanized mouse. Those are FRGN mice which were transplanted with human hepatocytes and they found after engraftment, the chimera liver was bigger than that transplanted with mouse derived hepatocytes [61]. Moreover, FGF19$^{-/-}$ FRGN mice are more efficient for receiving human hepatocytes [61]. Besides, there are still many questions under remained, for example, how could the liver regeneration program terminated in the partial hepatectomy? Other than hepatocytes, is there any other cell types in the liver participate in regulating hepatocyte mitosis? How could hepatocyte reconstruct the liver without other cells in the liver, such as endothelium cells and hematopoietic stem cells? Earlier time point such as 12 h after partial hepatectomy would help answer these questions.

5 EPIGENETIC DEREGULATION IN HCC

It is well appreciated that stem cell maintenance and differentiation are controlled by epigenetic regulation, whereas the deregulation of which would sometime lead to carcinogenesis, which in the case of liver, HCC. Although it is well known that the most prominent cause of HCC is chronic liver injury associated with HBV and HCV infection, the progression from chronic damage to cirrhosis and cancer is highly variable between patients. For example, a study from 108 patients with nonalcoholic fatty liver disease showed 45 of them with fibrosis progression, 43 with no progression, whereas 20 patients had fibrosis regression [62]. The molecular mechanisms of hepatocellular carcinogenesis and disease heterogeneity are poorly defined. The common genetic alterations associated with HCC are chromosomal aberrations including gain of 1q, 5, 6p, 7, 8q, 17q, 20, and loss of 1p, 4q, 6q, 8p, 13q, 16, 17p, 21 [63], as well as gene mutations, such as point mutations and small deletions in *P53*, *CTNNB1*, *AXIN1*, and homozygous deletions in *CDKN2A* [64]. Despite genetic factors, it is well established that patient sex, age, diet, environmental exposure, smoking history, and so on also affect the progression of liver diseases. The study of epigenetics is to address how these nongenetic factors influence the genome to a heritable level and stably affect the phenotype of organisms. From a molecular perspective, epigenetic regulation of gene expression involves DNA methylation, posttranslational histone modification, chromatin structural changes, and noncoding RNAs (ncRNAs).

5.1 DNA Methylation

In eukaryote cells, the best-known modification of DNA is covalent attachment of a methyl group to cytosine at its 5th carbon ring. The methylation of cytosine is found mainly within CpG dinucleotides, which are generally associated transcriptional repression when occur in high density of gene promoter regions. The function of gene-body or intergenic DNA methylation is poorly defined. Notably, the me-CpG mark can be recognized and bound by proteins containing methyl binding domain (MBD). The MBD family member, such as MBD1, MBD2, MeCP2 and MBD4, once bound to me-CpG islands, mediate transcriptional repression by recruiting other histone-modifying enzyme to model local chromatin structure [65]. DNA methylation pattern is the most stable epigenetic marker, the active change of which only occurs during early stages of embryogenesis. The integrity of DNA methylation landscape, however, is compromised in cancers, within which hypermethylation of tumor suppressors and hypomethylation of oncogenes have been reported [66,67]. Enzymes affecting DNA methylation involve DNA methyltransferases (e.g., DNMT1, DNMT3a, DNMT3b), and TET family (TET1, TET2, and TET3). DNMT1 maintains DNA methylation between parent and daughter cells by converting the hemimethylated sites on newly synthesized DNA to bimethylation. De novo methylation, however, does exist in nondividing cells, and is potentially mediated by DNMT3a and 3b [68]. Active DNA demethylation process, via replication-independent way, was controversial and inconclusive for many years. Recent studies, however, have provided compelling experimental evidence this process is coupled with the DNA repair machinery [69–71]. The mechanisms underlying TET-mediated demethylation of DNA involve step-wise oxidation of methyl-cytosine resulting in eventual restoration to unmodified state [72,73].

Loss of DNA methylation on genomic scale, or so called DNA hypomethylation, has been a typical feature of human cancers [74]. HCC is not exceptional: many studies have reported DNA methylation changes specific to HCC

[75–77]. Global DNA hypomethylation in HCC exhibit primarily in otherwise stably methylated regions of the genome, mainly intergenic, and body regions, whereas hypermethylation is associated with promoter regions [78]. There are several molecular mechanisms of how global DNA hypomethylation may contribute to the progression of liver carcinogenesis and tumor progression. Genome-wide hypomethylation may result in deregulation of cis-regulatory elements, loss of imprinting, transposition of repetitive DNA elements leading to chromosomal instability, which further promotes mutagenesis for individual genes [79].

Despite the profound global DNA hypomethylation, the changes of CpG islands of individual gene promoter are of particular interest to researchers as they are more predicable for transcriptional outcome and gene functions. A group of genes have been identified in HCC showing association between hypermethylated promoter and diminished mRNA expression, among them many are tumor suppressors, including *NORE1B*, *FBLN1*, *PAX5*, *p15INK4B*, *RELN*, *RB1*, *SOCS3*, *SYC*, *GSTP1*, *NQO1*, *SOCS1*, *PROX1*, *RASSF1A*, *RIZ1*, and *p16INK4A* [80–97]. Conversely, a number of hypomethylated tumor-promoting genes, whose expression levels are up-regulated have also been characterized in human HCC, including *CD147*, *HPA*, *TFF3*, *MAT2A*, *SNCG*, *uPA*, *HKII*, and *VIM* [98–104]. These genes, no matter tumor suppressive or oncogenic, play important roles in controlling cell cycle, proliferation, apoptosis, metabolism, and tumor microenvironment.

The cause of abnormal patterns of DNA methylation in HCC can be attributed to the disruption in the function of DNA methylation machinery. Various studies reported altered gene expression of DNA methyltransferase DNMT1, DNMT3A, and DNMT3B, and MBD family proteins in the tumor genesis and progression of HCC [105–107]. This is demonstrated by a progressive up-regulation of DNMT1, DNMT3A, and DNMT3B from premalignant noncancerous liver tissues to fully-developed HCC [108] and by the significant correlation between overexpression of these DNMTs and CpG-island hypermethylation of tumor-related genes [76].

5.2 Histone Modification and Chromatin Structure

Eukaryotic DNA is packed with specialized proteins called histones at several levels to form tightly packed chromatin. In developmental process or in response to environmental changes, a wide spectrum of gene expression changes need to fine tuned, which require the chromatin to unpacked and repacked accordingly. The most basic level of chromatin packing is called nucleosome, with each core particle consisting of 147 base pairs of DNA wrapped around an octamer of histone proteins (two copies each of H2A, H2B, H3, and H4). Each of the core histones has an unstructured N-terminal amino acid tail extension that is subject to a wide range of covalent, posttranslational modifications, including acetylation, methylation on lysine and arginine, phosphorylation on serine, ubiquitination, sumoylation, and ADP-ribosylation. These modifications either directly influence chromatin structure or comprise signals to be recognized by protein effectors, so called "reader" [109]. Same type modifications but on different residues can have drastically distinct molecular consequences, thus are referred by the field as "histone codes" [110]. Histone acetylation tends to loosen chromatin to be transcriptionally active, whereas trimethylation of lysine 9 of histone 3 (H3K9me3) and H3K27me3 are generally associated with condensed, transcriptionally inactive heterochromatin. However, histone lysine methylation can also promote transcription, as H3K4me1/3 and H3K36me2/3 are generally associated with euchromatin or decondensed regions. Nucleosome structure can be remodeled through insertion of ATP-dependent chromatin remodeling complexes into variants of the core histones. For instance, exchange of H2 for H2A.Z is associated with gene activation [111], while exchange for macroH2A is important for gene repression [112]. Histone modifications are highly dynamic compared to DNA methylations. "Writer" and "eraser" enzymes that respectively add or remove specific modifications can actively change the marks for recruitment of chromatin remodeling complexes such as the SWI/SNF complexes to affect local chromatin context. Mammalian SWI/SNF can slide nucleosomes on DNA, exchange or extrude histones, turning on gene expression [113]. By contrast, repressive chromatin remodelers compress nucleosomes, restricting access to transcription factors and transcription machinery. The most established repressing complex is Polycomb group proteins including polycomb repressor complexes 1 and 2 (PRC1 and PRC2). PRC2 facilitates H3K27 trimethylation via its enzymatic subunit EZH2. The PRC1 complex utilizes the ubiquitin ligases RING1A and RING1B to monoubiquitylates H2AK119. PRC1 can also bind to H3K27me3 caused by PRC2 function to cooperatively repress gene expression [114].

HCC is characterized by a prominent dysregulation of a group of histone-modifying enzymes, including histone deacetylase (HDAC): HDAC1, HDAC2, HDAC3, and SIRT1, as well as histone methyltransferases EZH2, SMYD3, and RIZ1 [115–118]. The function of HDAC involves removal of acetyl groups from lysine residues of a wide range target proteins from coordinating intracellular signaling pathways to chromatin repression. Several recent studies have reviewed an elevated expression of HDAC family HDAC1, HDAC2, HDAC3 [116,117], and SIRT1 in human HCC [119,120], which correlated with pathological features and recurrence of HCC. EZH2 is arguably the best-studied deregulated histone-modifying enzymes in HCC, the overexpression of which in HCC is associated with malignant progression [121]. Increased expression of EZH2 protein can serve as a biomarker distinguishing premalignant lesions from HCC, and EZH2 positivity correlates with poor patient outcome [122]. Mouse model studies revealed a critical role of EZH2 in HCC tumorigenesis, as evidenced by that intratumoral EZH2 depletion resulted in significant tumor regression [123]. The role of EZH2 on liver cell proliferation has been suggested to repress the expressions of multiple negative regulators of the Wnt pathway

[124]. A recent study found that EZH2 together with its structural partners EED, SUZ12, and RBP7, promotes tumorigenesis via silencing several tumor suppressor miRNAs [125]. EZH2 also seems to play a role in immune regulation of HCC: it up-regulates expression of the chemokine receptor CXCR4 by repressing its regulator miR-622; high expression of CXCR4 is usually associated with worse patient overall survival [126].

Additionally, mutational screening revealed 15% and 5% recurrent inactivating alterations in ARID1A and ARID2 respectively in viral-related and alcohol-related liver tumors [63,127], both of which are within the SWI/SNF chromatin-remodeling complex, associated with tumor suppressor functions. Another exon sequencing in viral-related HCCs and matched nontumour tissue revealed missense mutations in H3K4 methyltransferases MLL1, MLL2, MLL3, and MLL4, although the functional consequences of these mutations on histone methylation are yet to be validated [128].

5.3 Noncoding RNAs

It is estimated more than 90% of the transcribed RNAs are not encoded into proteins but carries critical regulatory functions as ncRNAs, including ribosomal RNAs, microRNAs (miRNAs), Piwi-associated RNAs, and long ncRNAs [129]. Among them, the sequence-specific 22-nucleotide miRNAs, are by far the most intensively studied and functionally characterized ncRNAs. MiRNAs are generated primarily by RNA polymerase II as primary miRNAs (pri-miRNA). Following transcription, the stem-loop structure of pri-miRNAs form is recognized and processed by Drosha, creating precursor miRNAs (pre-miRNA). These pre-miRNA are then transported from the nucleus to the cytoplasm by Exportin-5. In the cytoplasm, the pre-miRNAs are further processed by Dicer, generating short miRNA hybrids. After unwinding, one strand of the duplex becomes a mature miRNA. The mechanism underlying miRNA mediated gene regulation is to modulate the expression and/or translation of their target mRNAs usually leading to decreased protein expression [130].

Up to now, there are more than 800 mammalian miRNAs that can potentially target up to one-third of protein-coding genes involved in virtually every cellular behavior. Not surprisingly, deregulation of miRNA expression has been found in cancer [131–133], including HCC [134,135]. Recently, a number of comprehensive studies have documented both up-regulation and down-regulation of numerous miRNAs are observed in HCC. For instance, up-regulated miR-21, miR-17-92, miR-155, miR-191, and miR-221/miR-222 and down-regulated miR-122 are associated with tumor progression; up-regulated miR-21 and miR-151 and down-regulated miR-200 family are involved in cell invasion and metastasis; up-regulated miR-21 and down-regulated miR-122/miR-199a-3p are responsible for chemo-resistance [136–144].

Increasing evidence points to a direct and interdependent link between miRNA regulation and epigenetic alterations, revealing the complexity of epigenetic abnormalities in HCC. For example, overexpression of miR-191 in HCC is attributed to hypomethylation of the mir-191 gene [145], whereas reduced expression of miR-1, miR-124, miR-125b, and miR-203 is associated with DNA hypermethylation [87,146,147]. Conversely, differential expressions of miR-29, miR-152, and miR-200a lead to significant alterations in DNA methylation and histone modifications, potentially through affecting the expression of related epigenetic regulating enzymes [148–150].

6 MANIPULATE LIVER CANCER WITH OVER EXPRESSING ONCOGENE

To gain a greater understanding of how cancer stem cells in the liver may arise and the process of oncogenesis of such stem cells, a model of murine and human hepatocellular (HCC) carcinoma must be generated. One way by which HCC may arise is through random integration of adeno-associated viruses (AAVs). AAVs are members of the parvovirus group. Their genome consists of a single strand of linear DNA with inverted terminal repeats at the termini [151]. AAVs alone are replication incompetent and must be co-infected with adenovirus for efficient replication. Wild type AAVs have been shown to integrate at high frequency at a specific locus in human chromosome 19 known as the Adeno Associated Virus Integration Site 1 (AAVS1) [152]. Recombinant adeno associated virus (rAAV) are promising for use in gene therapy as they have long term stable expression and have been developed to be replication incompetent [153]. rAAVs typically persist as an episome in cells and rarely integrate into the genomic DNA [151,154]. The site-specific integration at the AAVS1 locus has been a loss [155]. rAAVs have been utilized in a number of clinical trials for various diseases including the blood disorder, hemophilia B and the eye disorder, Leber's congenital amaurosis [156].

A number of studies have been carried out to address whether liver targeted rAAV vector integrations into the genome during delivery of vectors for gene therapy can result in insertional mutagenesis and promote oncogenesis. As hepatocytes may act as stem/progenitor cells of the liver when there is liver injury, a greater understanding of the role of rAAVs in HCC is important as rAAV vectors typically target hepatocytes.

In a study on rAAV treatment for mice with the lysosomal storage disease, mucoploysaccharidosis type VII (MPS VII), neonatal MPS VII mice were treated with recombinant AAV2 (rAAV2) vectors expressing β-glucuronidase. Five of the 18 rAAV2 treated animals were found to have HCC and one animal was found to have metastatic angiocarcinoma [157]. Closer examination revealed that the tumors predominantly had an integration of the 5' inverted tandem repeat (ITR) capsid region of the rAAV2 vector. All rAAV2 insertions were located within a 6 kb region on chromosome 12 which contains the two ncRNAs *Rian* and *Mirg*. These genes were discovered to contain a large number of small nucleolar RNA and miRNAs which may regulate a large number of genes and result in profound effects if misregulated [158]. An age matched, control cohort of untreated MPS VII mice did not show any macroscopic lesions when examined. Additionally, of the 25 MPS VII mice that had received the alternative treatment of a bone marrow transplant only one was shown to have HCC, strongly suggesting that the HCC may be a result of the rAAV2 treatment [157]. These studies on MPS VII were carried out in a diseased mouse model which may predispose the mice to tumor formation. To verify these results and determine whether random integration leads to liver oncogenesis, neonatal wild type mice were treated with a rAAV containing the cytomegalovirus enhancer and a chicken β-actin promoter targeted to the murine *Rian* locus. All treated animals developed HCC and terminal liver failure. These animals were confirmed to have the integration of the promoter in the *Rian* locus suggesting that insertion of a promoter/enhancer is sufficient for the transformation of a normal hepatocyte [159].

Similar insertional mutagenesis was discovered during a study of 193 human patient HCC tumor samples, 11 of the tumor samples were found to contain an AAV2 integration. Through deep sequencing, 7 of the tumor samples displayed clonal integration of AAV2 was found in genes known to be HCC driver genes including cyclin A2, cyclin E1, lysine-specific methyltransferase 2B and tumor necrosis super family member 10 leading to overexpression of these genes. Additionally, AAV2 insertions were found in the promoter of telomerase reverse transcriptase (TERT). The majority of the insertions included the 3′ ITR capsid region of the AAV2 vector and were found to be overexpressed. In vitro studies where the respective inserted fragments were reproduced in a reporter vector containing the TERT promoter or the 3′ untranslated region (UTR) of TNSFSF10 confirmed that the insertions resulted in increased gene expression [160]. These findings support the earlier mouse studies which show that insertion and integration of AAV can lead to HCC.

While these studies in mice and human HCC tumors may raise caution for the use of rAAV as a gene therapy vector, gene therapy for hemophilia B patients are underway. Ten patients with severe hemophilia B that received treatment with rAAV8 expressing factor IX did not report any tumors or toxic effects from the vectors for 3 years following the treatment [161].

In addition, a number of long-term studies in adult murine models have suggested that liver targeted rAAV may have little genotoxicity despite a low but detectable rate of integration. A large cohort of adult wild type mice treated with rAAV2 expressing human factor IX mini gene and assessed for HCC after 18 months. The rate of HCC occurrence in the rAAV treated as compared to an untreated control group were similar suggesting that rAAV did not lead to increased HCC incidence. While integration of the AAV vectors was detected, the rate of integration was not higher in the tumors as compared to the neighboring normal tissue [162]. Another study on a large cohort of adult mice that were treated with rAAV vectors at 6–8 weeks of age had similar results. Of the 695 rAAV treated mice, only one mouse developed a tumor that was visible upon macroscopic inspection [163].

rAAVs are a promising gene therapy vector, however, their risk must continue to be assessed in light of both the murine and human HCC tumor studies that show increased incidence of HCC may result from the integration of rAAV leading to an increased expression of proto oncogenes. Expression of these proto oncogenes in liver stem/progenitor cells can be highly consequential.

7 CONCLUSIONS AND FUTURE PERSPECTIVE

Taken together, the study of liver stem cells in normal and injured liver would allow the field to gain an increase understanding of the control of liver regeneration during mechanical and chemical injury, and thus find a therapeutic cure for liver cancer. Although the origin of liver cancer stem cell is illusive, it is important to make animal model for liver cancer such as HCC with AAV. Finding signaling pathways controlling liver organ size would help to define the mechanism of tumor overgrowth. To isolate different cell types from normal and regenerating livers, single cell transcriptome analyses in normal and regenerative liver would benefit this work. Further studies will not only be focusing on hepatocytes themselves, but also other cell types in the liver. These studies may lead to important discoveries including the identification of the niche and paracrine network in liver stem cells.

REFERENCES

[1] Ostapowicz G, Fontana RJ, Schiodt FV, Larson A, Davern TJ, Han SH, et al. Results of a prospective study of acute liver failure at 17 tertiary care centers in the United States. Ann Intern Med 2002;137(12):947–54.

[2] O'Shea RS, Dasarathy S, McCullough AJ. Alcoholic liver disease. Am J Gastroenterol 2010;105(1):14–32. quiz 3.

[3] O'Shea RS, Dasarathy S, McCullough AJ. Practice Guideline Committee of the American Association for the Study of Liver D, Practice Parameters Committee of the American College of G. Alcoholic liver disease. Hepatology 2010;51(1):307–28.

[4] Verma S, Kaplowitz N. Diagnosis, management and prevention of drug-induced liver injury. Gut 2009;58(11):1555–64.

[5] Sakiyama R, Blau BJ, Miki T. Clinical translation of bioartificial liver support systems with human pluripotent stem cell-derived hepatic cells. World J Gastroenterol 2017;23(11):1974–9.

[6] Miyajima A, Tanaka M, Itoh T. Stem/progenitor cells in liver development, homeostasis, regeneration, and reprogramming. Cell Stem Cell 2014;14(5):561–74.

[7] Yamanaka S. Pluripotency and nuclear reprogramming. Philos Trans R Soc Lond B Biol Sci 2008;363(1500):2079–87.

[8] Duncan AW, Dorrell C, Grompe M. Stem cells and liver regeneration. Gastroenterology 2009;137(2):466–81.

[9] Azuma H, Paulk N, Ranade A, Dorrell C, Al-Dhalimy M, Ellis E, et al. Robust expansion of human hepatocytes in Fah-/-/Rag2-/-/Il2rg-/- mice. Nat Biotechnol 2007;25(8):903–10.

[10] Suzuki A, Zheng Y, Kondo R, Kusakabe M, Takada Y, Fukao K, et al. Flow-cytometric separation and enrichment of hepatic progenitor cells in the developing mouse liver. Hepatology 2000;32(6):1230–9.

[11] Suzuki A, Zheng YW, Kaneko S, Onodera M, Fukao K, Nakauchi H, et al. Clonal identification and characterization of self-renewing pluripotent stem cells in the developing liver. J Cell Biol 2002;156(1):173–84.
[12] Zheng YW, Tsuchida T, Taniguchi H. A novel concept of identifying precancerous cells to enhance anti-cancer therapies. J Hepatobiliary Pancreat Sci 2012;19(6):621–5.
[13] Khan FM, Komarla AR, Mendoza PG, Bodenheimer HC Jr, Theise ND. Keratin 19 demonstration of canal of Hering loss in primary biliary cirrhosis: "minimal change PBC"? Hepatology 2013;57(2):700–7.
[14] Huch M, Dorrell C, Boj SF, van Es JH, Li VS, van de Wetering M, et al. In vitro expansion of single Lgr5+ liver stem cells induced by Wnt-driven regeneration. Nature 2013;494(7436):247–50.
[15] Huch M, Gehart H, van Boxtel R, Hamer K, Blokzijl F, Verstegen MM, et al. Long-term culture of genome-stable bipotent stem cells from adult human liver. Cell 2015;160(1–2):299–312.
[16] Dorrell C, Erker L, Lanxon-Cookson KM, Abraham SL, Victoroff T, Ro S, et al. Surface markers for the murine oval cell response. Hepatology 2008;48(4):1282–91.
[17] Dorrell C, Erker L, Schug J, Kopp JL, Canaday PS, Fox AJ, et al. Prospective isolation of a bipotential clonogenic liver progenitor cell in adult mice. Genes Dev 2011;25(11):1193–203.
[18] Dorrell C, Tarlow B, Wang Y, Canaday PS, Haft A, Schug J, et al. The organoid-initiating cells in mouse pancreas and liver are phenotypically and functionally similar. Stem Cell Res 2014;13(2):275–83.
[19] Tarlow BD, Finegold MJ, Grompe M. Clonal tracing of Sox9(+) liver progenitors in mouse oval cell injury. Hepatology 2014;60(1):278–89.
[20] Tarlow BD, Pelz C, Naugler WE, Wakefield L, Wilson EM, Finegold MJ, et al. Bipotential adult liver progenitors are derived from chronically injured mature hepatocytes. Cell Stem Cell 2014;15(5):605–18.
[21] Yanger K, Zong Y, Maggs LR, Shapira SN, Maddipati R, Aiello NM, et al. Robust cellular reprogramming occurs spontaneously during liver regeneration. Genes Dev 2013;27(7):719–24.
[22] Font-Burgada J, Shalapour S, Ramaswamy S, Hsueh B, Rossell D, Umemura A, et al. Hybrid periportal hepatocytes regenerate the injured liver without giving rise to cancer. Cell 2015;162(4):766–79.
[23] Wang B, Zhao L, Fish M, Logan CY, Nusse R. Self-renewing diploid Axin2(+) cells fuel homeostatic renewal of the liver. Nature 2015;524(7564):180–5.
[24] Kordes C, Sawitza I, Muller-Marbach A, Ale-Agha N, Keitel V, Klonowski-Stumpe H, et al. CD133+ hepatic stellate cells are progenitor cells. Biochem Biophys Res Commun 2007;352(2):410–7.
[25] Wang X, Willenbring H, Akkari Y, Torimaru Y, Foster M, Al-Dhalimy M, et al. Cell fusion is the principal source of bone-marrow-derived hepatocytes. Nature 2003;422(6934):897–901.
[26] Willenbring H, Bailey AS, Foster M, Akkari Y, Dorrell C, Olson S, et al. Myelomonocytic cells are sufficient for therapeutic cell fusion in liver. Nat Med 2004;10(7):744–8.
[27] Fan B, Malato Y, Calvisi DF, Naqvi S, Razumilava N, Ribback S, et al. Cholangiocarcinomas can originate from hepatocytes in mice. J Clin Investig 2012;122(8):2911.
[28] Murillo-Garzon V, Kypta R. WNT signalling in prostate cancer. Nat Rev Urol 2017;14(11):683–96.
[29] Cho RW, Clarke MF. Recent advances in cancer stem cells. Curr Opin Genet Dev 2008;18(1):48–53.
[30] Yang CK, Wang XK, Liao XW, Han CY, Yu TD, Qin W, et al. Aldehyde dehydrogenase 1 (ALDH1) isoform expression and potential clinical implications in hepatocellular carcinoma. PLoS One 2017;12(8):e0182208.

[31] Haraguchi N, Ishii H, Mimori K, Tanaka F, Ohkuma M, Kim HM, et al. CD13 is a therapeutic target in human liver cancer stem cells. J Clin Investig 2010;120(9):3326.
[32] Lee TKW, Castilho A, Cheung VCH, Tang KH, Ma S, Ng IOL. CD24+ liver tumor-initiating cells drive self-renewal and tumor initiation through STAT3-mediated NANOG regulation. Cell Stem Cell 2011;9(1):50–63.
[33] Mima K, Okabe H, Ishimoto T, Hayashi H, Nakagawa S, Kuroki H, et al. CD44s regulates the TGF-β–mediated mesenchymal phenotype and is associated with poor prognosis in patients with hepatocellular carcinoma. Cancer Res 2012;72(13):3414–23.
[34] Yang ZF, Ho DW, Ng MN, Lau CK, Yu WC, Ngai P, et al. Significance of CD90+ cancer stem cells in human liver cancer. Cancer Cell 2008;13(2):153–66.
[35] Ma S, Chan KW, Lee TK-W, Tang KH, Wo JY-H, Zheng B-J, et al. Aldehyde dehydrogenase discriminates the CD133 liver cancer stem cell populations. Mol Cancer Res 2008;6(7):1146–53.
[36] Piao LS, Hur W, Kim T-K, Hong SW, Kim SW, Choi JE, et al. CD133+ liver cancer stem cells modulate radioresistance in human hepatocellular carcinoma. Cancer Lett 2012;315(2):129–37.
[37] Chen K, Ding A, Ding Y, Ghanekar A. High-throughput flow cytometry screening of human hepatocellular carcinoma reveals CD146 to be a novel marker of tumor-initiating cells. Biochem Biophys Rep 2016;8:107–13.
[38] Xu X, Liu R-F, Zhang X, Huang L-Y, Chen F, Fei Q-L, et al. DLK1 as a potential target against cancer stem/progenitor cells of hepatocellular carcinoma. Mol Cancer Ther 2012;11(3):629–38.
[39] Yamashita T, Honda M, Nakamoto Y, Baba M, Nio K, Hara Y, et al. Discrete nature of EpCAM+ and CD90+ cancer stem cells in human hepatocellular carcinoma. Hepatology 2013;57(4):1484–97.
[40] Matsumoto T, Takai A, Eso Y, Kinoshita K, Manabe T, Seno H, et al. Proliferating EpCAM-positive ductal cells in the inflamed liver give rise to hepatocellular carcinoma. Cancer Res 2017;77(22):6131–43. canres. 1800.2017.
[41] Yang W, Wang C, Lin Y, Liu Q, Yu L-X, Tang L, et al. OV6+ tumor-initiating cells contribute to tumor progression and invasion in human hepatocellular carcinoma. J Hepatol 2012;57(3):613–20.
[42] Chiba T, Kita K, Zheng YW, Yokosuka O, Saisho H, Iwama A, et al. Side population purified from hepatocellular carcinoma cells harbors cancer stem cell-like properties. Hepatology 2006;44(1):240–51.
[43] Chiba T, Zheng YW, Kita K, Yokosuka O, Saisho H, Onodera M, et al. Enhanced self-renewal capability in hepatic stem/progenitor cells drives cancer initiation. Gastroenterology 2007;133(3):937–50.
[44] Hanahan D, Weinberg RA. Hallmarks of cancer: the next generation. Cell 2011;144(5):646–74.
[45] Ding W, You H, Dang H, LeBlanc F, Galicia V, Lu SC, et al. Epithelial-to-mesenchymal transition of murine liver tumor cells promotes invasion. Hepatology 2010;52(3):945–53.
[46] Jayachandran A, Shrestha R, Dhungel B, Huang I-T, Vasconcelos MYK, Morrison BJ, et al. Murine hepatocellular carcinoma derived stem cells reveal epithelial-to-mesenchymal plasticity. World J Stem Cells 2017;9(9):159–68.
[47] Wang L, Chiou S, Chai C, Hsi E, Chiang C, Huang S, et al. Transcription factor SPZ1 promotes TWIST-mediated epithelial–mesenchymal transition and oncogenesis in human liver cancer. Oncogene 2017;36(31):4405–14.
[48] Yoshida GJ. Emerging role of epithelial-mesenchymal transition in hepatic cancer. J Exp Clin Cancer Res 2016;35(1):141.

[49] Kalluri R, Weinberg RA. The basics of epithelial-mesenchymal transition. J Clin Investig 2009;119(6):1420–8.
[50] Li B, Zheng YW, Sano Y, Taniguchi H. Evidence for mesenchymal-epithelial transition associated with mouse hepatic stem cell differentiation. PLoS One 2011;6(2):e17092.
[51] Michalopoulos GK, DeFrances MC. Liver regeneration. Science 1997;276(5309):60–6.
[52] Zarnegar R, Petersen B, DeFrances MC, Michalopoulos G. Localization of hepatocyte growth factor (HGF) gene on human chromosome 7. Genomics 1992;12(1):147–50.
[53] Skouteris GG, Ord MG, Stocken LA. Regulation of the proliferation of primary rat hepatocytes by eicosanoids. J Cell Physiol 1988;135(3):516–20.
[54] Skouteris GG, Kaser MR. Prostaglandins E2 and F2a mediate the increase in c-myc expression induced by EGF in primary rat hepatocyte cultures. Biochem Biophys Res Commun 1991;178(3):1240–6.
[55] Webber EM, Wu JC, Wang L, Merlino G, Fausto N. Overexpression of transforming growth factor-alpha causes liver enlargement and increased hepatocyte proliferation in transgenic mice. Am J Pathol 1994;145(2):398–408.
[56] Ding BS, Nolan DJ, Butler JM, James D, Babazadeh AO, Rosenwaks Z, et al. Inductive angiocrine signals from sinusoidal endothelium are required for liver regeneration. Nature 2010;468(7321):310–5.
[57] Hu J, Srivastava K, Wieland M, Runge A, Mogler C, Besemfelder E, et al. Endothelial cell-derived angiopoietin-2 controls liver regeneration as a spatiotemporal rheostat. Science 2014;343(6169):416–9.
[58] Yu FX, Zhao B, Guan KL. Hippo pathway in organ size control, tissue homeostasis, and cancer. Cell 2015;163(4):811–28.
[59] Hong L, Cai Y, Jiang M, Zhou D, Chen L. The Hippo signaling pathway in liver regeneration and tumorigenesis. Acta Biochim Biophys Sin (Shanghai) 2015;47(1):46–52.
[60] Reis H, Bertram S, Pott L, Canbay A, Gallinat A, Baba HA. Markers of Hippo-pathway activity in tumor forming liver lesions. Pathol Oncol Res 2017;23(1):33–9.
[61] Naugler WE, Tarlow BD, Fedorov LM, Taylor M, Pelz C, Li B, et al. Fibroblast growth factor signaling controls liver size in mice with humanized livers. Gastroenterology 2015;149(3):728–740.e15.
[62] McPherson S, Hardy T, Henderson E, Burt AD, Day CP, Anstee QM. Evidence of NAFLD progression from steatosis to fibrosing-steatohepatitis using paired biopsies: implications for prognosis and clinical management. J Hepatol 2015;62(5):1148–55.
[63] Guichard C, Amaddeo G, Imbeaud S, Ladeiro Y, Pelletier L, Maad IB, et al. Integrated analysis of somatic mutations and focal copy-number changes identifies key genes and pathways in hepatocellular carcinoma. Nat Genet 2012;44(6):694–8.
[64] Ozturk M. Genetic aspects of hepatocellular carcinogenesis. Semin Liver Dis 1999;19(3):235–42.
[65] Cedar H, Bergman Y. Programming of DNA methylation patterns. Annu Rev Biochem 2012;81:97–117.
[66] Baylin SB. DNA methylation and gene silencing in cancer. Nat Clin Pract Oncol 2005;2(Suppl. 1):S4–S11.
[67] Kulis M, Esteller M. DNA methylation and cancer. Adv Genet 2010;70:27–56.
[68] Law JA, Jacobsen SE. Establishing, maintaining and modifying DNA methylation patterns in plants and animals. Nat Rev Genet 2010;11(3):204–20.
[69] Ito S, Shen L, Dai Q, Wu SC, Collins LB, Swenberg JA, et al. Tet proteins can convert 5-methylcytosine to 5-formylcytosine and 5-carboxylcytosine. Science 2011;333(6047):1300–3.
[70] He YF, Li BZ, Li Z, Liu P, Wang Y, Tang Q, et al. Tet-mediated formation of 5-carboxylcytosine and its excision by TDG in mammalian DNA. Science 2011;333(6047):1303–7.

[71] Cortellino S, Xu J, Sannai M, Moore R, Caretti E, Cigliano A, et al. Thymine DNA glycosylase is essential for active DNA demethylation by linked deamination-base excision repair. Cell 2011;146(1):67–79.
[72] Tahiliani M, Koh KP, Shen Y, Pastor WA, Bandukwala H, Brudno Y, et al. Conversion of 5-methylcytosine to 5-hydroxymethylcytosine in mammalian DNA by MLL partner TET1. Science 2009;324(5929):930–5.
[73] Rasmussen KD, Helin K. Role of TET enzymes in DNA methylation, development, and cancer. Genes Dev 2016;30(7):733–50.
[74] Feinberg AP, Tycko B. The history of cancer epigenetics. Nat Rev Cancer 2004;4(2):143–53.
[75] Stefanska B, Huang J, Bhattacharyya B, Suderman M, Hallett M, Han ZG, et al. Definition of the landscape of promoter DNA hypomethylation in liver cancer. Cancer Res 2011;71(17):5891–903.
[76] Revill K, Wang T, Lachenmayer A, Kojima K, Harrington A, Li J, et al. Genome-wide methylation analysis and epigenetic unmasking identify tumor suppressor genes in hepatocellular carcinoma. Gastroenterology 2013;145(6):1424–1435.e1-25.
[77] Nishida N, Kudo M, Nagasaka T, Ikai I, Goel A. Characteristic patterns of altered DNA methylation predict emergence of human hepatocellular carcinoma. Hepatology 2012;56(3):994–1003.
[78] Villanueva A, Portela A, Sayols S, Battiston C, Hoshida Y, Mendez-Gonzalez J, et al. DNA methylation-based prognosis and epidrivers in hepatocellular carcinoma. Hepatology 2015;61(6):1945–56.
[79] Saito Y, Kanai Y, Sakamoto M, Saito H, Ishii H, Hirohashi S. Overexpression of a splice variant of DNA methyltransferase 3b, DNMT3b4, associated with DNA hypomethylation on pericentromeric satellite regions during human hepatocarcinogenesis. Proc Natl Acad Sci U S A 2002;99(15):10060–5.
[80] Schagdarsurengin U, Wilkens L, Steinemann D, Flemming P, Kreipe HH, Pfeifer GP, et al. Frequent epigenetic inactivation of the RASSF1A gene in hepatocellular carcinoma. Oncogene 2003;22(12):1866–71.
[81] Li X, Hui AM, Sun L, Hasegawa K, Torzilli G, Minagawa M, et al. p16INK4A hypermethylation is associated with hepatitis virus infection, age, and gender in hepatocellular carcinoma. Clin Cancer Res 2004;10(22):7484–9.
[82] Harder J, Opitz OG, Brabender J, Olschewski M, Blum HE, Nomoto S, et al. Quantitative promoter methylation analysis of hepatocellular carcinoma, cirrhotic and normal liver. Int J Cancer 2008;122(12):2800–4.
[83] Roncalli M, Bianchi P, Bruni B, Laghi L, Destro A, Di Gioia S, et al. Methylation framework of cell cycle gene inhibitors in cirrhosis and associated hepatocellular carcinoma. Hepatology 2002;36(2):427–32.
[84] Edamoto Y, Hara A, Biernat W, Terracciano L, Cathomas G, Riehle HM, et al. Alterations of RB1, p53 and Wnt pathways in hepatocellular carcinomas associated with hepatitis C, hepatitis B and alcoholic liver cirrhosis. Int J Cancer 2003;106(3):334–41.
[85] Yuan Y, Wang J, Li J, Wang L, Li M, Yang Z, et al. Frequent epigenetic inactivation of spleen tyrosine kinase gene in human hepatocellular carcinoma. Clin Cancer Res 2006;12(22):6687–95.
[86] Tada M, Yokosuka O, Fukai K, Chiba T, Imazeki F, Tokuhisa T, et al. Hypermethylation of NAD(P)H: quinone oxidoreductase 1 (NQO1) gene in human hepatocellular carcinoma. J Hepatol 2005;42(4):511–9.
[87] Datta J, Kutay H, Nasser MW, Nuovo GJ, Wang B, Majumder S, et al. Methylation mediated silencing of MicroRNA-1 gene and its role in hepatocellular carcinogenesis. Cancer Res 2008;68(13):5049–58.

[88] Miyoshi H, Fujie H, Moriya K, Shintani Y, Tsutsumi T, Makuuchi M, et al. Methylation status of suppressor of cytokine signaling-1 gene in hepatocellular carcinoma. J Gastroenterol 2004;39(6):563–9.
[89] Okochi O, Hibi K, Sakai M, Inoue S, Takeda S, Kaneko T, et al. Methylation-mediated silencing of SOCS-1 gene in hepatocellular carcinoma derived from cirrhosis. Clin Cancer Res 2003;9(14):5295–8.
[90] Zhong S, Tang MW, Yeo W, Liu C, Lo YM, Johnson PJ. Silencing of GSTP1 gene by CpG island DNA hypermethylation in HBV-associated hepatocellular carcinomas. Clin Cancer Res 2002;8(4):1087–92.
[91] Shimoda M, Takahashi M, Yoshimoto T, Kono T, Ikai I, Kubo H. A homeobox protein, prox1, is involved in the differentiation, proliferation, and prognosis in hepatocellular carcinoma. Clin Cancer Res 2006;12(20 Pt. 1):6005–11.
[92] Laerm A, Helmbold P, Goldberg M, Dammann R, Holzhausen HJ, Ballhausen WG. Prospero-related homeobox 1 (PROX1) is frequently inactivated by genomic deletions and epigenetic silencing in carcinomas of the bilary system. J Hepatol 2007;46(1):89–97.
[93] Zhang C, Li H, Wang Y, Liu W, Zhang Q, Zhang T, et al. Epigenetic inactivation of the tumor suppressor gene RIZ1 in hepatocellular carcinoma involves both DNA methylation and histone modifications. J Hepatol 2010;53(5):889–95.
[94] Macheiner D, Heller G, Kappel S, Bichler C, Stattner S, Ziegler B, et al. NORE1B, a candidate tumor suppressor, is epigenetically silenced in human hepatocellular carcinoma. J Hepatol 2006;45(1):81–9.
[95] Okamura Y, Nomoto S, Kanda M, Hayashi M, Nishikawa Y, Fujii T, et al. Reduced expression of reelin (RELN) gene is associated with high recurrence rate of hepatocellular carcinoma. Ann Surg Oncol 2011;18(2):572–9.
[96] Liu W, Li X, Chu ES, Go MY, Xu L, Zhao G, et al. Paired box gene 5 is a novel tumor suppressor in hepatocellular carcinoma through interaction with p53 signaling pathway. Hepatology 2011;53(3):843–53.
[97] Kanda M, Nomoto S, Okamura Y, Hayashi M, Hishida M, Fujii T, et al. Promoter hypermethylation of fibulin 1 gene is associated with tumor progression in hepatocellular carcinoma. Mol Carcinog 2011;50(8):571–9.
[98] Chan CF, Yau TO, Jin DY, Wong CM, Fan ST, Ng IO. Evaluation of nuclear factor-kappaB, urokinase-type plasminogen activator, and HBx and their clinicopathological significance in hepatocellular carcinoma. Clin Cancer Res 2004;10(12 Pt. 1):4140–9.
[99] Xiao Y, Kleeff J, Shi X, Buchler MW, Friess H. Heparanase expression in hepatocellular carcinoma and the cirrhotic liver. Hepatol Res 2003;26(3):192–8.
[100] Zhao W, Liu H, Liu W, Wu Y, Chen W, Jiang B, et al. Abnormal activation of the synuclein-gamma gene in hepatocellular carcinomas by epigenetic alteration. Int J Oncol 2006;28(5):1081–8.
[101] Kitamura Y, Shirahata A, Sakuraba K, Goto T, Mizukami H, Saito M, et al. Aberrant methylation of the vimentin gene in hepatocellular carcinoma. Anticancer Res 2011;31(4):1289–91.
[102] Okada H, Kimura MT, Tan D, Fujiwara K, Igarashi J, Makuuchi M, et al. Frequent trefoil factor 3 (TFF3) overexpression and promoter hypomethylation in mouse and human hepatocellular carcinomas. Int J Oncol 2005;26(2):369–77.
[103] Kong LM, Liao CG, Chen L, Yang HS, Zhang SH, Zhang Z, et al. Promoter hypomethylation up-regulates CD147 expression through increasing Sp1 binding and associates with poor prognosis in human hepatocellular carcinoma. J Cell Mol Med 2011;15(6):1415–28.

[104] Yang H, Huang ZZ, Zeng Z, Chen C, Selby RR, Lu SC. Role of promoter methylation in increased methionine adenosyltransferase 2A expression in human liver cancer. Am J Physiol Gastrointest Liver Physiol 2001;280(2):G184–90.

[105] Saito Y, Kanai Y, Sakamoto M, Saito H, Ishii H, Hirohashi S. Expression of mRNA for DNA methyltransferases and methyl-CpG-binding proteins and DNA methylation status on CpG islands and pericentromeric satellite regions during human hepatocarcinogenesis. Hepatology 2001;33(3):561–8.

[106] Oh BK, Kim H, Park HJ, Shim YH, Choi J, Park C, et al. DNA methyltransferase expression and DNA methylation in human hepatocellular carcinoma and their clinicopathological correlation. Int J Mol Med 2007;20(1):65–73.

[107] Fan H, Zhao ZJ, Cheng J, Su XW, Wu QX, Shan YF. Overexpression of DNA methyltransferase 1 and its biological significance in primary hepatocellular carcinoma. World J Gastroenterol 2009;15(16):2020–6.

[108] Saito Y, Kanai Y, Nakagawa T, Sakamoto M, Saito H, Ishii H, et al. Increased protein expression of DNA methyltransferase (DNMT) 1 is significantly correlated with the malignant potential and poor prognosis of human hepatocellular carcinomas. Int J Cancer 2003;105(4):527–32.

[109] Clapier CR, Cairns BR. The biology of chromatin remodeling complexes. Ann Rev Biochem 2009;78:273–304.

[110] Bannister AJ, Kouzarides T. Regulation of chromatin by histone modifications. Cell Res 2011;21(3):381–95.

[111] Raisner RM, Hartley PD, Meneghini MD, Bao MZ, Liu CL, Schreiber SL, et al. Histone variant H2A.Z marks the 5′ ends of both active and inactive genes in euchromatin. Cell 2005;123(2):233–48.

[112] Doyen CM, An W, Angelov D, Bondarenko V, Mietton F, Studitsky VM, et al. Mechanism of polymerase II transcription repression by the histone variant macroH2A. Mol Cell Biol 2006;26(3):1156–64.

[113] Suganuma T, Workman JL. Signals and combinatorial functions of histone modifications. Ann Rev Biochem 2011;80:473–99.

[114] Blackledge NP, Rose NR, Klose RJ. Targeting polycomb systems to regulate gene expression: modifications to a complex story. Nat Rev Mol Cell Biol 2015;16(11):643–9.

[115] Jiang G, Liu L, Buyse IM, Simon D, Huang S. Decreased RIZ1 expression but not RIZ2 in hepatoma and suppression of hepatoma tumorigenicity by RIZ1. Int J Cancer 1999;83(4):541–6.

[116] Hamamoto R, Furukawa Y, Morita M, Iimura Y, Silva FP, Li M, et al. SMYD3 encodes a histone methyltransferase involved in the proliferation of cancer cells. Nat Cell Biol 2004;6(8):731–40.

[117] Sudo T, Utsunomiya T, Mimori K, Nagahara H, Ogawa K, Inoue H, et al. Clinicopathological significance of EZH2 mRNA expression in patients with hepatocellular carcinoma. Br J Cancer 2005;92(9):1754–8.

[118] Chen J, Zhang B, Wong N, Lo AW, To KF, Chan AW, et al. Sirtuin 1 is upregulated in a subset of hepatocellular carcinomas where it is essential for telomere maintenance and tumor cell growth. Cancer Res 2011;71(12):4138–49.

[119] Wu LM, Yang Z, Zhou L, Zhang F, Xie HY, Feng XW, et al. Identification of histone deacetylase 3 as a biomarker for tumor recurrence following liver transplantation in HBV-associated hepatocellular carcinoma. PLoS One 2010;5(12):e14460.

[120] Choi HN, Bae JS, Jamiyandorj U, Noh SJ, Park HS, Jang KY, et al. Expression and role of SIRT1 in hepatocellular carcinoma. Oncol Rep 2011;26(2):503–10.

[121] Sasaki M, Ikeda H, Itatsu K, Yamaguchi J, Sawada S, Minato H, et al. The overexpression of polycomb group proteins Bmi1 and EZH2 is associated with the progression and aggressive biological behavior of hepatocellular carcinoma. Lab Invest 2008;88(8):873–82.

[122] Cai MY, Tong ZT, Zheng F, Liao YJ, Wang Y, Rao HL, et al. EZH2 protein: a promising immunomarker for the detection of hepatocellular carcinomas in liver needle biopsies. Gut 2011;60(7):967–76.

[123] Chen Y, Lin MC, Yao H, Wang H, Zhang AQ, Yu J, et al. Lentivirus-mediated RNA interference targeting enhancer of zeste homolog 2 inhibits hepatocellular carcinoma growth through down-regulation of stathmin. Hepatology 2007;46(1):200–8.

[124] Cheng AS, Lau SS, Chen Y, Kondo Y, Li MS, Feng H, et al. EZH2-mediated concordant repression of Wnt antagonists promotes beta-catenin-dependent hepatocarcinogenesis. Cancer Res 2011;71(11):4028–39.

[125] Au SL, Wong CC, Lee JM, Fan DN, Tsang FH, Ng IO, et al. Enhancer of zeste homolog 2 epigenetically silences multiple tumor suppressor microRNAs to promote liver cancer metastasis. Hepatology 2012;56(2):622–31.

[126] Wang L, Zhang X, Jia LT, Hu SJ, Zhao J, Yang JD, et al. c-Myc-mediated epigenetic silencing of MicroRNA-101 contributes to dysregulation of multiple pathways in hepatocellular carcinoma. Hepatology 2014;59(5):1850–63.

[127] Li M, Zhao H, Zhang X, Wood LD, Anders RA, Choti MA, et al. Inactivating mutations of the chromatin remodeling gene ARID2 in hepatocellular carcinoma. Nat Genet 2011;43(9): 828–9.

[128] Cleary SP, Jeck WR, Zhao X, Chen K, Selitsky SR, Savich GL, et al. Identification of driver genes in hepatocellular carcinoma by exome sequencing. Hepatology 2013;58(5):1693–702.

[129] Rinn JL, Chang HY. Genome regulation by long noncoding RNAs. Ann Rev Biochem 2012;81:145–66.

[130] Bartel DP. MicroRNAs: target recognition and regulatory functions. Cell 2009;136(2): 215–33.

[131] Garzon R, Calin GA, Croce CM. MicroRNAs in Cancer. Annu Rev Med 2009;60:167–79.

[132] Ventura A, Jacks T. MicroRNAs and cancer: short RNAs go a long way. Cell 2009;136(4): 586–91.

[133] Di Leva G, Croce CM. Roles of small RNAs in tumor formation. Trends Mol Med 2010;16(6):257–67.

[134] Gramantieri L, Fornari F, Callegari E, Sabbioni S, Lanza G, Croce CM, et al. MicroRNA involvement in hepatocellular carcinoma. J Cell Mol Med 2008;12(6A):2189–204.

[135] Mott JL. MicroRNAs involved in tumor suppressor and oncogene pathways: implications for hepatobiliary neoplasia. Hepatology 2009;50(2):630–7.

[136] Meng F, Henson R, Wehbe-Janek H, Ghoshal K, Jacob ST, Patel T. MicroRNA-21 regulates expression of the PTEN tumor suppressor gene in human hepatocellular cancer. Gastroenterology 2007;133(2):647–58.

[137] Connolly E, Melegari M, Landgraf P, Tchaikovskaya T, Tennant BC, Slagle BL, et al. Elevated expression of the miR-17-92 polycistron and miR-21 in hepadnavirus-associated hepatocellular carcinoma contributes to the malignant phenotype. Am J Pathol 2008;173(3):856–64.

[138] Elyakim E, Sitbon E, Faerman A, Tabak S, Montia E, Belanis L, et al. hsa-miR-191 is a candidate oncogene target for hepatocellular carcinoma therapy. Cancer Res 2010;70(20): 8077–87.

[139] Fornari F, Gramantieri L, Ferracin M, Veronese A, Sabbioni S, Calin GA, et al. MiR-221 controls CDKN1C/p57 and CDKN1B/p27 expression in human hepatocellular carcinoma. Oncogene 2008;27(43):5651–61.

[140] Pineau P, Volinia S, McJunkin K, Marchio A, Battiston C, Terris B, et al. miR-221 overexpression contributes to liver tumorigenesis. Proc Natl Acad Sci U S A 2010;107(1):264–9.
[141] Luedde T. MicroRNA-151 and its hosting gene FAK (focal adhesion kinase) regulate tumor cell migration and spreading of hepatocellular carcinoma. Hepatology 2010;52(3):1164–6.
[142] Tomimaru Y, Eguchi H, Nagano H, Wada H, Tomokuni A, Kobayashi S, et al. MicroRNA-21 induces resistance to the anti-tumour effect of interferon-alpha/5-fluorouracil in hepatocellular carcinoma cells. Br J Cancer 2010;103(10):1617–26.
[143] Wong QW, Ching AK, Chan AW, Choy KW, To KF, Lai PB, et al. MiR-222 overexpression confers cell migratory advantages in hepatocellular carcinoma through enhancing AKT signaling. Clin Cancer Res 2010;16(3):867–75.
[144] Fornari F, Gramantieri L, Giovannini C, Veronese A, Ferracin M, Sabbioni S, et al. MiR-122/cyclin G1 interaction modulates p53 activity and affects doxorubicin sensitivity of human hepatocarcinoma cells. Cancer Res 2009;69(14):5761–7.
[145] He Y, Cui Y, Wang W, Gu J, Guo S, Ma K, et al. Hypomethylation of the hsa-miR-191 locus causes high expression of hsa-mir-191 and promotes the epithelial-to-mesenchymal transition in hepatocellular carcinoma. Neoplasia 2011;13(9):841–53.
[146] Furuta M, Kozaki KI, Tanaka S, Arii S, Imoto I, Inazawa J. miR-124 and miR-203 are epigenetically silenced tumor-suppressive microRNAs in hepatocellular carcinoma. Carcinogenesis 2010;31(5):766–76.
[147] Alpini G, Glaser SS, Zhang JP, Francis H, Han Y, Gong J, et al. Regulation of placenta growth factor by microRNA-125b in hepatocellular cancer. J Hepatol 2011;55(6):1339–45.
[148] Braconi C, Kogure T, Valeri N, Huang N, Nuovo G, Costinean S, et al. microRNA-29 can regulate expression of the long non-coding RNA gene MEG3 in hepatocellular cancer. Oncogene 2011;30(47):4750–6.
[149] Huang J, Wang Y, Guo Y, Sun S. Down-regulated microRNA-152 induces aberrant DNA methylation in hepatitis B virus-related hepatocellular carcinoma by targeting DNA methyltransferase 1. Hepatology 2010;52(1):60–70.
[150] Yuan JH, Yang F, Chen BF, Lu Z, Huo XS, Zhou WP, et al. The histone deacetylase 4/SP1/microrna-200a regulatory network contributes to aberrant histone acetylation in hepatocellular carcinoma. Hepatology 2011;54(6):2025–35.
[151] Goncalves MA. Adeno-associated virus: from defective virus to effective vector. Virol J 2005;2:43.
[152] Smith RH. Adeno-associated virus integration: virus versus vector. Gene Ther 2008;15(11):817–22.
[153] Nakai H, Montini E, Fuess S, Storm TA, Grompe M, Kay MA. AAV serotype 2 vectors preferentially integrate into active genes in mice. Nat Genet 2003;34(3):297–302.
[154] Afione SA, Conrad CK, Kearns WG, Chunduru S, Adams R, Reynolds TC, et al. In vivo model of adeno-associated virus vector persistence and rescue. J Virol 1996;70(5):3235–41.
[155] Kearns WG, Afione SA, Fulmer SB, Pang MC, Erikson D, Egan M, et al. Recombinant adeno-associated virus (AAV-CFTR) vectors do not integrate in a site-specific fashion in an immortalized epithelial cell line. Gene Ther 1996;3(9):748–55.
[156] Naldini L. Gene therapy returns to centre stage. Nature 2015;526(7573):351–60.
[157] Donsante A, Vogler C, Muzyczka N, Crawford JM, Barker J, Flotte T, et al. Observed incidence of tumorigenesis in long-term rodent studies of rAAV vectors. Gene Ther 2001;8(17):1343–6.
[158] Donsante A, Miller DG, Li Y, Vogler C, Brunt EM, Russell DW, et al. AAV vector integration sites in mouse hepatocellular carcinoma. Science 2007;317(5837):477.

[159] Wang PR, Xu M, Toffanin S, Li Y, Llovet JM, Russell DW. Induction of hepatocellular carcinoma by in vivo gene targeting. Proc Natl Acad Sci U S A 2012;109(28):11264–9.
[160] Nault JC, Datta S, Imbeaud S, Franconi A, Mallet M, Couchy G, et al. Recurrent AAV2-related insertional mutagenesis in human hepatocellular carcinomas. Nat Genet 2015;47(10):1187–93.
[161] Nathwani AC, Reiss UM, Tuddenham EG, Rosales C, Chowdary P, McIntosh J, et al. Long-term safety and efficacy of factor IX gene therapy in hemophilia B. N Engl J Med 2014;371(21):1994–2004.
[162] Li H, Malani N, Hamilton SR, Schlachterman A, Bussadori G, Edmonson SE, et al. Assessing the potential for AAV vector genotoxicity in a murine model. Blood 2011;117(12):3311–9.
[163] Bell P, Wang L, Lebherz C, Flieder DB, Bove MS, Wu D, et al. No evidence for tumorigenesis of AAV vectors in a large-scale study in mice. Mol Ther 2005;12(2):299–306.

Chapter 5

Exploration for Cell Sources for Liver Regenerative Medicine: "CLiP" as a Dawn of Cell Transplantation Therapy

Kazunori Hosaka*,**, Takeshi Katsuda*, Shuji Terai**, Takahiro Ochiya*
*National Cancer Center Research Institute, Tokyo, Japan; **Niigata University Medical and Dental Hospital, Niigata, Japan

1 INTRODUCTION

Novel therapeutics for serious liver diseases, such as liver cirrhosis and fulminant liver diseases, are in high demand. Liver cirrhosis results from continuous inflammation caused by Hepatitis B or C viral infection, drug use, autoimmunity (e.g., autoimmune hepatitis and primary biliary cirrhosis), alcoholism, and metabolic disease. The causes leading to fatal liver diseases are wide ranging, but the only current therapy is liver transplantation. However, the application of liver transplantation is limited due to a shortage of donor livers. Liver transplantation is classified into brain-dead liver transplantation and living-donor liver transplantation. In brain-dead liver transplantation, obtaining donor livers when they are needed is difficult, especially as treatment for fulminant liver diseases. In contrast, living liver transplantation requires an established human relationship. In addition, because the donors are required to have surgery to provide the liver for transplant, certain risks are inevitable for the donors as well. In addition, the donor's quality of life after surgery should also be considered [1–5]. If the worst-case scenario occurs during or after surgery, the donor will be dead. To provide a safer transplant treatment, alternative treatments for end-stage liver disease should be immediately developed.

Cell transplantation has been proposed as an alternative treatment for end-stage liver disease but has not yet become a clinically effective cure. Liver cell transplantation treatment will contribute to the temporary metabolic support of patients awaiting liver transplantation or spontaneous reversion of their liver disease. However, liver cell transplantation therapy is hampered due to the shortage of fully functional cell sources. The difficulty of stable long-term hepatocyte culture in vitro is the major cause of this shortage. It is well known that mature hepatocytes (MHs) rapidly lose their proliferation, differentiation, and metabolism abilities in vitro [6]. Consequently, hepatocytes useful for cell transplantation therapy can only be obtained from donor livers, which are substantially lacking as mentioned above. Therefore a methodology to expand hepatocytes or their equivalents in vitro is necessary for obtaining adequate numbers of transplantable liver cells. In addition, retaining the mature hepatic functionality of the expanded cells is also important after the liver cell transplantation therapy. Researchers have intensively studied novel methods to generate transplantable hepatic cells in vitro using several candidate cell sources.

In this review, we classify the candidate cell sources into three groups as follows. The first group is native hepatocytes or their derivative cells. The second group is hepatocyte-like cells (HLCs) induced from pluripotent stem cells, including embryonic stem cells (ESCs), induced pluripotent stem cells (iPSDs), induced hepatocyte-like cells (iHeps), and mesenchymal stem cells (MSCs). Finally, the third group is resident liver progenitor cells (LPCs), such as fetal liver cells (FLCs) and adult LPCs. This review presents an overview of the situations and issues surrounding each group while focusing on the possibilities of liver cell transplantation (Fig. 5.1).

In the following sections, we introduce a new candidate cell source, known as chemically induced liver progenitors (CLiPs), which we have recently characterized [7]. We describe how we developed CLiPs and the potential uses and limitations of CLiP technology for cell transplantation therapy.

2 HEPATOCYTES AS THE SOURCE OF LIVER-REPOPULATING CELLS

Native hepatocytes are the most direct cell source for cell transplantation, and their feasibility has long been discussed [8,9]. Although primary human hepatocytes can be purified from donor livers, mainly from cadaveric livers, the quantity required for cell transplantation therapy (approximately $1-5 \times 10^9$ cells) is an obstacle [10]. To satisfy this demand, new methods for stably expanding hepatocytes in long-term in vitro culture are needed. Development of these approaches has been hindered by the fact that hepatocytes rapidly lose their proliferative capacity and hepatic functionality in vitro. Here we present an overview of the possibility of hepatocytes as the source for liver cell transplantation.

Hepatocytes have the remarkable ability to extensively expand in vivo. After a two-thirds partial hepatectomy (PH), residual adult hepatocytes proliferate to compensate for the resected lobes [11]. However, it is well known that hepatocytes rapidly lose their proliferation ability and hepatic function in vitro.

In recent decades, many researchers have struggled to maintain hepatocyte functionality in vitro. Some coculture techniques have been shown to retain hepatocyte proliferation ability and hepatic function. Among these, coculture with 3T3-J2 cells showed remarkable results [12,13]. In addition, Swiss 3T3

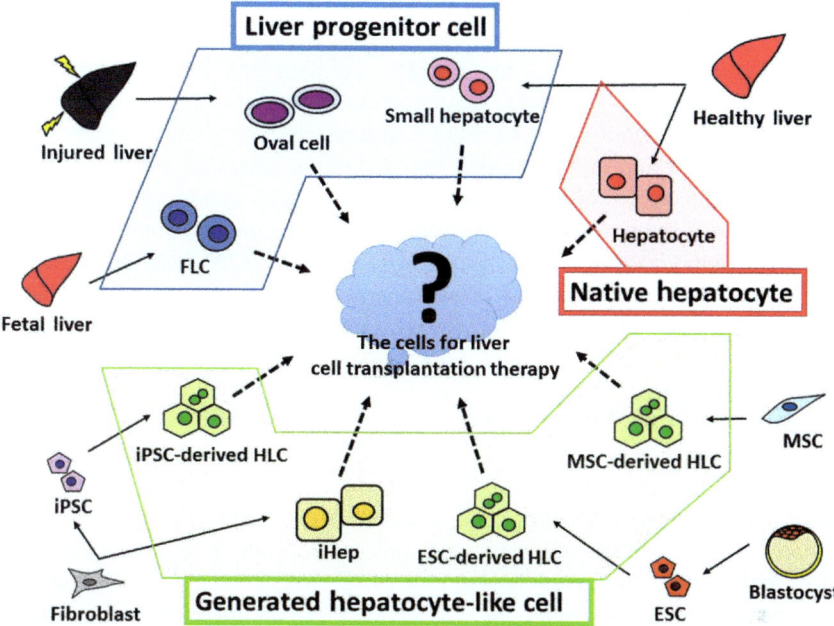

FIGURE 5.1 **Candidate cells for liver cell transplantation therapy.** Researchers approach liver cell transplantation therapy in many ways. Hepatocytes are the most accessible source of cells, although they rapidly lose their abilities in culture conditions. HLCs derived from ESCs and iPSCs are able to provide adequate amounts of cells. However, the clinical applications of these cells are hampered due to their low repopulation capacities and the risk of tumorigenesis. iHeps also have the tumorigenic risks. MSC-derived HLCs lack mature hepatic functions. In addition, there is a possibility that their therapeutic effect would be a part of a paracrine effect. Although FLCs are stable in long-term culture conditions, and their repopulation capacity is substantial, ethical issues are inevitable in the use of fetal liver. Similarly, Adult LPCs, such as oval cells, have the same ethical problems as they are derived from injured livers. Small hepatocytes exhibit a limited engraftment efficiency compared with other hepatocytes, but they can be obtained from healthy liver and undergo multiple cell divisions in vitro to form colonies. These facts led us to investigate small hepatocytes as a leading candidate for cell transplantation therapy.

coculture allowed hepatocytes to proliferate in vitro via secretion of pleiotrophin [14]. Another approach to improve culture conditions involves a 3D culture technique using hydrogels with a liver-specific synthetic extracellular matrix (ECM) [15]. However, these approaches are inadequate for obtaining a sufficient number of transplantable cells, and using feeder cells introduces the risk of feeder cell contamination. Although these modifications are external factors, many attempts have been made to manipulate the hepatocytes themselves to generate a sufficient quantity of transplantable cells.

Immortalization is a major technique used to expand hepatocytes in vitro. Overexpression of immortalizing gene(s) has been achieved using viral transfection methods. NKNT-3 cells are an immortalized hepatocyte cell line that was generated following retroviral transfer of simian virus 40T (SV40T) into primary human hepatocytes. This immortalizing gene can be subsequently excised by Cre/LoxP recombination [16]. Another cell line, cBAL111, was immortalized by the introduction of human telomerase reverse transcriptase (hTERT) [17]. Recently, nonviral gene editing techniques have become available. FoxM1, which is known as a key cell cycle regulator, is delivered into hepatocytes using the nonviral sleeping beauty transposon system [18]. These cell lines have the potential to sufficiently expand to the cell numbers required for transplantation, but hepatic function in these cells is extremely low, and the potential of oncogenic transformation remains due to genetic manipulation. Recently, primary human hepatocytes with low expression of the human papilloma virus (HPV) genes E6 and E7 were expanded in an oncostatin M (OSM)-dependent manner [19]. However, due to the use of the HPV oncogene, this cell line also carries the possibility of tumorigenesis.

Another approach to expand hepatocytes in vitro was attempted using small molecules. Via a high-throughput screen of 12,480 small molecules, Shan et al. identified a few chemical compounds that induce proliferation of human primary hepatocytes in vitro while allowing the cells to maintain their hepatic functions [13]. Although this approach is very promising, no evidence has been presented that the obtained proliferative human hepatocytes can repopulate injured livers or can be stably expanded in long-term culture.

In summary, much progress has been made in refining methods for hepatocyte expansion in vitro. However, at present, it is difficult to maintain hepatocyte proliferation and function in vitro without transduction of exogenous genes. Further studies are required in this field.

3 HEPATOCYTE-LIKE CELLS INDUCED FROM ESCS AS THE SOURCE OF LIVER-REPOPULATING CELLS

ESCs and iPSCs are attractive cell sources for cell transplantation therapeutics due to their intensive proliferation and multi-lineage differentiation abilities. In addition, induced HLCs, known as iHeps, are generated from fibroblasts by direct reprogramming using defined factors and have also attracted investigative attention [20,21]. The use of iHeps is expected to circumvent the difficulties associated with ESCs and iPSCs. However, some concerns regarding the hepatic function of iHeps remain unsolved. Here we discuss the possibility of these cells as a candidate cell source for liver cell transplantation.

Mouse and human ESCs can be induced to differentiate into the hepatic lineage in vitro by sequential administration of defined growth factors. The challenging aspect of this approach is the requirement for mimicking the sequential development cues that occur in vivo. These days, most protocols consist of three steps: directing undifferentiated ESCs into the definitive endoderm (step 1), differentiating the endodermal cells into LPCs (step 2), and inducing the liver progenitors into hepatocytes and subsequent maturation (step 3) [22–31]. In traditional protocols, embryoid bodies (EBs), which mimic the developing embryo, were formed to induce spontaneous ESC differentiation [22–27]. In other protocols, ESCs have been cultured as a monolayer in the presence of exogenous inducers for direct differentiation into the endoderm [28–31]. To improve the functions of ESC-derived hepatocytes, some studies have used small molecules as key factors [29,32]. Hepatic cells obtained in this manner had several liver functions such as albumin secretion, glycogen storage, ICG and LDL uptake, and inducible CYP450 activity, expression of the known fetal liver marker α-fetoprotein (AFP) was also detected. Interestingly, a bio-artificial liver with ESC-derived hepatocytes rescued 90% of hepatectomized mice, proving that ESC-derived hepatocytes are nearly identical to primary hepatocytes [26]. Some studies performed transplantation experiments using human ESC-derived hepatic cells in animals with liver injury, but the repopulation efficiency was much lower than that achieved by native hepatocyte transplantation [28,30,31]. Heo et al. transplanted GFP$^+$ hepatic cells derived from mouse ESCs into the spleens of MUP-uPA/SCID mice [33]. GFP$^+$ foci were observed in the recipient mouse livers, and the repopulation index (RI) was 1.94% ± 5.81% at 82 days after transplantation. Tolosa et al. transplanted human ESC-derived hepatocytes into acetaminophen-treated mice and demonstrated up to 10% liver repopulation [34].

These differentiation protocols are relatively efficient, but the risk of retaining undifferentiated cells causes undesired phenotypic changes, such as teratoma formation. Obtaining a homogeneously differentiated population is important in reducing these risks. Gabriel et al. used a reporter vector containing GFP and puromycin resistance genes driven by an AFP promoter to purify HLCs [35]. Another study used lentiviral vectors encoding GFP driven by the liver-specific apolipoprotein A-II (APOA-II) promoter and purified the GFP$^+$ HLCs by FACS [36]. Fuming et al. reported the most functional ESC-derived induced hepatocytes [37]. They acquired and clonally expanded c-kit$^-$/EpCAM$^+$ cells. The cells could be cultured for 3 months and be passaged over 30 times in vitro without morphological changes. The doubling time was approximately 30 h. The c-kit$^-$/EpCAM$^+$ cells were transplanted into Fah-deficient mice intrasplenically, and at 10 weeks after transplantation, the repopulation efficiency was 24 ± 15%.

In summary, ESC-derived hepatocytes exhibit some promising features. However, considering the relatively low repopulation capacity after transplantation, the feasibility of these cells as a source for therapeutic use is debatable. Moreover, it should be carefully considered that ESC-derived cells still carry the risks of teratoma formation, oncogenic transformation, and immune rejection.

4 HEPATOCYTE-LIKE CELLS INDUCED FROM IPSCS AS THE SOURCE OF LIVER-REPOPULATING CELLS

Within the past decade, iPSCs have been intensively studied as an ideal cell source for cell transplantation. iPSCs have features similar to those observed in ESCs, such as pluripotency and the ability to intensively proliferate. Moreover, iPSCs can circumvent some obstacles involved with the use of ESCs. Here we present an overview of the progress made in therapeutic approaches using iPSC-derived hepatocytes for liver cell transplantation.

iPSCs are ethically acceptable and do not elicit immune responses. iPSCs can be generated by the introduction of four transcription factor genes [38,39], which were later reduced to three genes [40]. iPSCs have a demonstrated ability to differentiate into many cell types, including hepatic cells. Because iPSCs are derived from somatic cells, destruction of embryonic cells can be avoided. In addition, somatic cells can be obtained from patients themselves, thereby minimizing the risk of immune rejection following iPSC-derived transplantation. Indeed, this expectation is supported by the first clinical trial of iPSCs for retina regeneration. The transplanted iPSC-derived retinal pigment epithelium sheet showed no sign of rejection during the 1-year study period [41].

Song et al. developed human HLCs from iPSCs. The hepatic functions of differentiated human iPSC-derived hepatocytes were similar to those observed in human ESC-derived hepatocytes [42]. Other groups have also succeeded in generating human HLCs from iPSCs [10,43]. Karim et al. transplanted human HLCs from iPSCs into the right liver lobe and identified donor cell foci (the RI is unknown) [10]. From these reports, although the use of iPSC-derived hepatocytes is in the early stages, it is obvious that these cells lack mature hepatic function and thus their repopulation capacity is insufficient. Therefore many researchers have struggled to obtain more mature iPSC-derived hepatocytes to improve cell-based therapies. Some methodologies, such as treatment with valproic acid and transduction of hepatic transcription factors (ATF5, c/EBPα, and PROX1), achieved hepatic maturation [44,45]. In addition, some researchers have revised the protocols for efficient differentiation. HNF4α is a key transcription factor for liver development, and the HNF4α-1D isoform is crucial in definitive endoderm differentiation [46]. Other groups have reported growth factor-free protocols that rely on small molecules to differentiate human pluripotent stem cells towards a hepatic lineage [47].

Considering the potential application of iPSC-derived hepatocytes in clinical settings, concerns regarding the use of recombinant factors remain debatable due to cost and safety issues. A growth factor-free protocol that relies on small molecules to differentiate human iPSCs into HLCs would be cost-effective [47]. Another study reported successful generation of mouse iPSCs only with small molecules to address safety concerns [48].

Given that the repopulation capacity of iPSC-derived hepatocytes has improved, these cells are a promising source for cell transplantation therapy. However, their repopulation capacity is still lower than that of primary MHs, which makes the feasibility of iPSC-derived hepatocytes for clinical use questionable. Nagamoto et al. indicated that the low repopulation capacity observed in iPSC-derived hepatocytes is due to apoptosis [49]. Over expression of the antiapoptotic gene FNK in iPSC-derived hepatocytes improved their repopulation efficacy in human liver chimeric mice from 2.5% to 20%. Human primary hepatocytes and human iPSC-derived hepatocytes were transplanted into Gunn rats, which serve as a model of Crigler-Najjar syndrome 1, following X-irradiation of 30% of the liver to promote hepatocyte transplantation [50]. Human primary hepatocytes and human iPSC-derived hepatocytes constituted 14%–20% and 2.5%–7.5% of the preconditioned liver lobe. Although the repopulation efficiency of iPSC-derived hepatocytes has improved, it remains inferior to that of primary hepatocyte transplantation.

iPSCs are a promising source for liver cell transplantation due to their lack of potential immune rejection. To progress the clinical feasibility of iPSCs, a more robust differentiation technique and improved repopulation capacity are needed.

5 INDUCED HEPATOCYTE-LIKE CELLS AS THE SOURCE OF LIVER-REPOPULATING CELLS

Direct reprogramming technology has provided another approach for generating cells with the potential capacity for cell transplantation therapy. This approach allows direct conversion of terminally differentiated somatic cells by defined factors into cells of other lineages without entering a pluripotent state. Skipping the pluripotent cell state reduces risks, such as teratoma formation and tumorigenesis. The concept was demonstrated in the generation of cardiomyocytes (iCMs) [51] and neurons (iNs) [52] via the introduction of three transcription factors. Thus expectations for regenerative medicine using this technique are increasing. Here, we present an overview of the reports concerning direct reprogramming of hepatic lineage cell and the possibilities for liver transplantation.

iHeps were first reported in 2011 by two independent groups [20,21]. Huang et al. created mouse iHeps from tail tip fibroblasts via transduction of Gata4, Hnf1α, and Foxa3 and inactivation of P19Arf. These cells showed hepatic morphology and function. Furthermore, the transplanted iHeps repopulated the recipient Fah-deficient mice at 5%–80% of the total hepatocytes and reduced mouse mortality by approximately half [20]. Sekiya et al. converted mouse fibroblasts into iHeps using two factors, Hnf4α and Foxa1, Foxa2 or Foxa3. These cells possessed hepatic function and rescued Fah-deficient mice (the repopulation efficiency was not shown) [21].

Following these pioneering achievements, researchers revised the direct reprogramming methods by modifying the transduced transcription factors. Bing et al. converted the mouse embryonic fibroblasts (MEFs) into induced hepatic stem cells (iHepSCs) via the addition of Hnf1β and Foxa3 [53]. The doubling time of the modified cells was approximately 20 h, and the cells could be passaged up to 32 times before abnormal karyotypes were observed. The bipotency of these cells to differentiate into both hepatocytes and biliary epithelial cells (BECs) was confirmed in vitro. In addition, transplanted iHepSCs differentiated into hepatocytes in Fah-deficient mice and into BECs in DDC-injured mice in vivo. Du et al. used six transcription factors, HNF1A, HNF4A, HNF6, ATF5,

PROX1, and CEBPA, to convert human fibroblasts into iHeps [54]. The iHeps showed hepatic function, especially CYP activity comparable to that of primary human hepatocytes. iHeps transplanted into Tet-uPA/Rag2 deficient mice repopulated approximately 30% of the total hepatocytes and secreted human albumin at concentrations greater than 300 μg/mL. Although these data represent remarkable progress in the field of iHeps, the number of transcription factors required for iHep generation should be reduced prior to clinical application. Lim et al. succeeded in reducing the number of transcription factors required via the use of small molecules. Using A-83-01, CHIR99021, and BMP4, these authors showed that one transcription factor, Hnf1α, was sufficient to induce direct hepatic reprogramming [55].

In a similar manner, iHeps derived from human fibroblasts through a multipotent progenitor cell (iMPC) state have also been reported [56]. These iHeps, named iMPC-Heps, proliferated extensively and acquired a set of hepatic functions. iMPC-Heps were transplanted into FRG mice and repopulated the liver with 2% of all parenchymal cells.

In vivo direct reprogramming of liver myofibroblasts into iHeps is another way to improve liver fibrosis. Song et al. converted liver myofibroblasts into iHeps in vivo in mouse models of chronic liver disease by adding the appropriate transcription factors (FOXA3, GATA4, HNF1A, and HNF4A) [57]. This conversion attenuated liver fibrosis and provided new insights into treatment for chronic liver disease.

Although iHeps are a promising cell source for liver cell transplantation, they require genetic modifications and therefore retain the potential for unpredictable risks. In addition, the repopulative capacity of iHeps following transplantation into liver-injured animals is still low compared with that of MH transplantation, which calls the therapeutic potential of iHeps into question. Therefore, application of these cells as the source of liver cell transplantation should be further investigated in a stringent manner.

6 INDUCTION OF HEPATIC CELLS FROM MESENCHYMAL STEM CELLS AS THE SOURCE OF LIVER-REPOPULATING CELLS

MSCs are an attractive cell source for regenerative medicine given the concerns surrounding ESCs, iPSCs, and iHeps. MSCs are present in adult mesodermal tissues, such as the bone marrow, adipose tissue and cartilage, and can differentiate into several types of mesodermal cells, including osteoblasts, chondrocytes, adipocytes, and myoblasts, under varying culture conditions [58]. In addition, it has been demonstrated that MSCs have "stem cell plasticity," which is the ability of adult stem cells to cross lineage barriers and adopt the expression profiles and functional phenotypes of cells unique to other tissues, including the liver [59]. MSCs exhibited hepatic differentiation in vitro using protocols similar to those for directing ESCs towards the hepatic lineage [60–64]. MSC-derived HLCs demonstrated some hepatic functions, such as albumin production, glycogen storage, urea secretion, uptake of LDL, and CYP450 activity. Moreover, it was demonstrated that MSCs also differentiate into HLCs in vivo after transplantation [65]. A previous study from our laboratory demonstrated that HLCs generated from adipose tissue-derived mesenchymal stem cells (AT-MSCs) were transplantable [64].

It is controversial as to whether MSCs have the potential for cell transplantation therapy due to their inefficient hepatic function in vivo. Previous reports demonstrated that MSCs could engraft and differentiate after transplantation into injured livers [66,67]. Aurich et al. transplanted human peritoneal and subcutaneous AT-MSCs with or without in vitro differentiation into immunodeficient Pfp/Rag2 mice [68]. Their RIs with or without differentiation were 4%–21% and 0.3%–1.4%, respectively, which indicates that the repopulation capacity of differentiated AT-MSCs is approximately 13–15 times more efficient than that of undifferentiated AT-MSCs. Although, this study reported a highly efficient repopulation capacity, the serum levels of hALB remained relatively low. From this result, the hepatic ability of HLCs derived from MSCs is questionable. Another group reported that transplanted human bone marrow-derived MSCs (BM-MSCs) had the potential to migrate into normal and injured liver, but their differentiation into HLCs was a rare event (ranging from 0.1%–0.23%) [69]. In addition, the possibility of a paracrine effect was indicated as the trigger of the therapeutic benefits of MSCs. BM-MSCs transplanted into CCl4-treated mice reduced oxidative stress in recipient mice and accelerated resident hepatocyte repopulation [70]. From these results, the use of MSCs as a substitute for hepatocytes is controversial. We should consider every possibility for MSC therapeutics, such as factors to suppress inflammation.

7 LIVER PROGENITOR/STEM CELLS AS THE SOURCE OF LIVER-REPOPULATING CELLS

Liver progenitor/stem cells (LPCs) lack mature hepatic functionality but have greater in vitro proliferative capacity compared with MHs and thus are regarded as a candidate cell source for liver transplantation. There are two types of LPCs: those isolated from fetal liver and those from adult liver. FLCs contain bipotent cells known as hepatoblasts that can differentiate into both hepatocytes and BECs. However, adult LPCs are classified into two categories, hepatoblast-like bipotent cells and unipotent cells. In this section, we summarize reports of trials using LPCs for transplantation and discuss the feasibility of these approaches for therapeutic application.

FLCs exhibit a substantial repopulative capacity when transplanted into injured liver. Engraftment upon transplantation has been a stringent criteria to determine the "stemness" of a hepatic stem cell [71]. In mouse studies, several types of nonhematopoietic cell populations isolated from fetal liver have been identified as expressing bipotent stem-like phenotypes. These populations are isolated based on their surface antigens, including Cd49f, Cd29, Dlk, c-Met,

Cd133, Ecad, and CD13 [72–77], and have been shown to engraft into injured liver upon transplantation. Although the repopulation efficiency of these cells is not clear in these reports [72–74,76,77], substantial replacement of injured liver tissue has been confirmed in some studies [76,77]. In addition, Nierhoff et al. demonstrated that unfractionated FLCs as well as Ecad$^+$-sorted cells were able to repopulate up to 80% of recipient mouse liver tissue [75].

Rat studies with FLCs demonstrated a more attractive feature for these cells as a candidate cell source for transplantation therapy. Excellent studies by Oertel and colleagues demonstrated that FLCs were able to repopulate 60%–80% of injured recipient liver tissue [78–81]. Furthermore, these researchers demonstrated that FLCs can repopulate normal livers as well. To effectively repopulate the liver with transplanted cells, most researchers use recipient animals that have sustained a liver injury in which hepatocytes are continuously damaged as a result of genetic modifications (e.g., uPA transgenic and Fah-knockout mice) or hepatocyte proliferation is blocked by irradiation or chemical treatment (e.g., retrorsine). Otherwise, even with severe damage to the liver, such as that induced by PH, resident hepatocytes extensively proliferate and the repopulation contribution by transplanted cells becomes very limited. However, surprisingly, Oertel et al. found that FLCs [82,83], especially the Dlk$^+$ fraction [84], but not MHs, substantially repopulated remnant liver after PH (up to 24%) in the absence of any additional hepatic injury. This repopulation resulted from the greater proliferative activity of transplanted FLCs and reduced apoptosis of their progeny compared with that of host hepatocytes, which underscored the excellent repopulative capacity of FLCs.

LPCs have been thought to reside in the adult liver and contribute to tissue repair when hepatocyte proliferation is compromised. Such cells can be isolated and cultured in vitro, at least to some extent, and thus have been regarded as a candidate cell source for transplantation therapy. Oval cells (OCs) emerge from chronically injured liver tissue in which the proliferation of resident MHs is decreased [85]. OCs express immature hepatic markers, such as AFP, and BEC markers, such as CK19. Cumulative histological analyses have suggested that OCs can differentiate into both hepatocytes and BECs, and thus, these cells have been regarded as bipotent LPCs. OCs can be experimentally induced by administration of 2-acetylaminofluorene (2-AAF) combined with PH or D-galactosamine (D-gal) treatment in rats and administration of a 3,5-diethoxycarbonyl-1,4-dihidro-collidine (DDC)-containing diet or a choline-deficient ethionine-supplemented (CDE) diet combined with PH in mice. Wang et al. investigated the repopulative capacity of mouse OCs induced by a DDC/PH protocol [86]. Transplanted OCs substantially repopulated the injured liver of Fah-KO recipient mice (the RI is not clear), and their repopulative efficiency was comparable with that of MHs as assessed by a competitive transplantation assay. Yovchev et al. also demonstrated significant repopulation in a retrorsine/PH rat model by transplanting OCs isolated from D-gal-treated rats. When OCs were transplanted into retrorsine/PH rat livers, these cells achieved 70%–90%

repopulation of the host liver. The authors further identified Epcam⁺ OCs as the cells responsible for the extensive repopulative capacity.

In rats, another LPC population known as small hepatocytes (SHs) exists in addition to OCs. SHs were identified as having a hepatocyte-like morphology but were clearly small-sized cells with a colony forming capacity in vitro [87]. Importantly, SHs can be isolated from the non-parenchymal fraction of injured [88] or healthy liver [87]. Furthermore, SHs can engraft into the liver of retrorsine/PH rats at equivalent or even higher efficiency than MHs at up to 2–3 weeks after transplantation [89,90], but the long-term engraftment capacity is decreased compared with MHs [90]. Icam-1⁺ SH-like cells have recently been identified in normal mouse liver [91]. Icam-1⁺ cells have been shown to differentiate into MHs in vitro and engraft into retrorsine/PH livers, although the engraftment was not very efficient. Encouragingly, several studies have reported that SH-like cells can be isolated from the human liver as well [92,93]. In summary, the bipotent LPCs known as OCs are generally obtained from injured liver and exhibit efficient engraftment. However, SHs, which are unipotent LPCs, can be obtained from healthy or injured liver tissue, but their engraftment efficiency is limited compared with that of MHs or injury-induced OCs.

Although fetal and adult LPCs may offer therapeutic advantages when considering pluripotent stem cell-based approaches, several limitations must be addressed before they can be used clinically. First, in the case of fetal LPC usage, the ethical issues associated with cells of fetal origin are unavoidable [94]. In particular, there is a concern regarding the increase in induced abortion to secure fetal tissue donation. Second, in the case of both fetal and adult LPCs, the availability of transplantable cells is limited, considering that cells with the most efficient repopulative capacity must be selected using antibodies for specific surface antigens, such as Dlk1. Such sorting procedures require sophisticated laboratory skills, and the subsequent number of transplantable cells is substantially decreased after cell sorting. In addition, although LPCs have been reported to be expandable in vitro, it is unclear whether fetal and adult LPCs retain repopulative capacity during long-term in vitro culture.

8 EMERGENCE OF LPCs FROM MHs USING SMALL MOLECULES

Using small molecule signaling inhibitors, we recently identified a novel type of LPC in rats and mice that can be generated from MHs in vitro [7]. Named as CLiPs, these cells exhibit extensive repopulative capacity upon transplantation into animals with injured livers, thereby offering a novel therapeutic option. In this section, we summarize our recent report on CLiPs and describe the generation of these cells (Fig. 5.2).

The primary motivation of our original study was to develop a method for long-term culture of SHs. As described above, SHs can be obtained not only

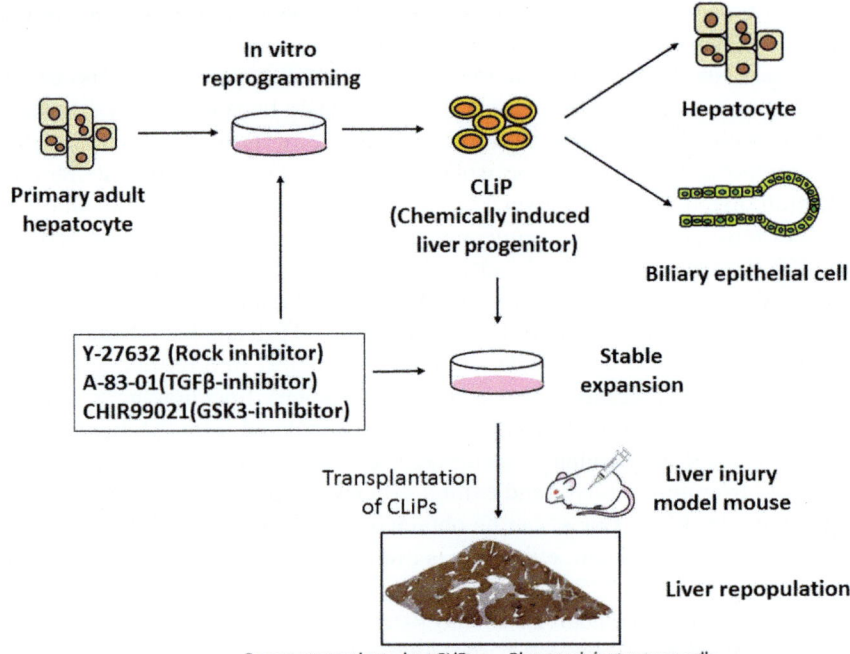

FIGURE 5.2 **Chemical reprogramming generates expandable liver progenitor cells with regenerative capacity.** A chemical cocktail of Y-27632 (ROCK inhibitor), A-83-01 (TGFβ-inhibitor), and CHIR99021 (GSK3-inhibitor) converts primary adult hepatocytes into chemically induced liver progenitors (CLiPs) in vitro. In addition, CLiPs have bipotentiality to differentiate into hepatocytes and biliary epithelial cells and are stably cultured in vitro. Moreover, CLiPs transplanted into a mouse model of liver injury exhibited a high repopulation efficiency of 75%–90% of the recipient liver.

from injured liver but also from healthy liver and undergo multiple cell divisions in vitro to form colonies. In addition, compared with putative resident bipotent LPCs, which are Thy1 (Cd90)-expressing cells found in the nonparenchymal cell fraction, SHs exhibit more efficient engraftment and proliferation capacity in vivo after transplantation into retrorsine/PH rats [90,95]. However, SH proliferation was limited in vitro, and subculturing these cells was difficult. Thus, it was questionable whether we could obtain sufficient amount of SHs for transplantation therapy (a more recent study reported that SHs can be passaged at least four times [96]). This situation led us to speculate whether stable expansion of SHs in vitro would advance cell transplantation therapeutics.

This speculation was prompted by our previous works in which a small molecule cocktail enabled the culture of several types of stem/progenitor cells. Kawamata et al. demonstrated that a cocktail of four small molecules, Y-27632 (ROCK inhibitor), PD0325901 (MAPK inhibitor), A-83-01 (TGF-β receptor inhibitor), and CHIR99021 (GSK3 inhibitor), allowed stable culture of rat ESCs [97,98], p63$^+$/Ck14$^+$ multipotent mammary tumor cells [99], and normal mammary gland progenitor cells (unpublished data). Given the stem/progenitor-like phenotype of SHs, these findings led us to hypothesize that certain combinations of these four molecules would allow stable SH culture with more efficient proliferation. Concomitantly, as mentioned above, recent protocols for inducing HLCs from pluripotent stem cells have used small molecules [29,32,34,47], some of which overlap with the four molecules used by Kawamata et al.

To test our idea, we isolated SHs and investigated the effects of the above-mentioned four small molecules on their proliferation efficiency. We cultured SHs in the presence of all possible combinations of these four molecules. To confirm that proliferative cells were derived from SHs, we cultured MHs as a negative control under the same conditions. In this miniscreening, we found several small molecule combinations that accelerated SH proliferation. Unexpectedly, however, MHs, which have been known to be resistant to in vitro expansion, also expanded extensively in the presence of several combinations of these four molecules. Given that MHs compose approximately 70%–80% of the total liver cell population, MHs would be a more feasible cell source for application to transplantation. Thus we shifted our target from the proliferation of SHs to that of MHs. In addition, of all the possible combinations of the four small molecules, treatment with Y-27632, A-83-01, and CHIR99021 (termed YAC) endowed MHs with the highest proliferative capacity, and thus, we focused on MH-derived proliferative cells induced by YAC.

9 LPC-LIKE CHARACTERISTICS OF YAC-INDUCED PROLIFERATIVE CELLS

We noticed that YAC-induced proliferative cells (YAC-cells) were reprogrammed into an LPC-like state. Morphologically, the YAC-cells resembled fetal LPCs, given their small size, and exhibited an increased nucleus/cytoplasm ratio compared with MHs [100]. Notably, the gene and protein expression levels of multiple LPC markers, including Afp, Sox9, Epcam, and Cd44, were elevated during the 2-week culture under YAC stimulation. Expression of some LPC marker genes, including Sox9, Epcam, and Cd44, was also induced even in the absence of YAC. This observation implied that MHs undergo spontaneous dedifferentiation, probably due to a certain in vitro environmental effect. Of note, Afp, one of the most stringent LPC markers, was upregulated only in the presence of YAC. This led us to speculate that the phenotypic alterations in YAC-cells were representing not just dedifferentiation, but more likely reprogramming into LPC-like state. Thus, we further interrogated the phenotypic characteristics of YAC-cells by assessing their bipotentiality, which is the strict criterion for hepatic stem cells.

10 YAC-CELL BIPOTENTIALITY

We assessed the hepatic differentiation ability of YAC-cells using OSM, dexamethasone, and Matrigel [101]. YAC-cells that underwent hepatic induction exhibited an MH-like morphology, with a lower nucleus/cytoplasm ratio and bile canaliculi-like small intercellular spaces. Some of these cells were binucleated, which is a feature of terminally differentiated hepatocytes. Indeed, hepatic-induced YAC-cells exhibited some hepatic functions, including Alb secretion, glycogen storage, Cyp1a activity, urea production, and bile secretion as assessed by fluorescein diacetate assay. Accordingly, transcriptome analysis demonstrated that hepatic-induced YAC-cells had a gene expression profile similar to that of primary MHs.

Moreover, we found that YAC-cells were able to differentiate into functional BECs. This characteristic was identified by chance while we were searching for a culture condition where YAC-cells would be stably cultured long-term (later we found that Matrigel coating of culture plates allowed for long-term culture of these cells). When the YAC-cells were co-cultured on cell cycle-arrested MEFsin mTeSR1 medium (this medium is typically used to maintain pluripotency in ESCs and iPSCs) supplemented with YAC, they formed ductal/cystic structures. Such structures were observed in earlier studies using primary rat BECs [102–104]. Thus we further characterized the ductal/cystic structures formed by YAC-cells by qRT-PCR. YAC-cells that were cultured in mTeSR1 medium showed higher expression of BEC-related genes, including Ck19, Aqp1, Cftr, and Ae2, and lower expression levels of hepatic marker genes such as Alb and Afp. Furthermore, when cultured in the presence of secretin, which is known to induce osmotic water transport in BECs [105], the cystic/ductal spaces were readily enlarged as observed by phase contrast microscopy. In the presence of fluorescein diacetate, the ductal structures exported and accumulated the metabolized fluorescein into their luminal spaces, thereby strongly suggesting the bile secretion/storage capacity of the cystic/ductal structures formed by YAC-cells. Collectively, it was demonstrated that YAC converted MHs into proliferative and bipotent LPCs. Thus we designated YAC-cells as CLiPs.

We also confirmed the applicability of this method to derive mouse CLiPs. Although we confirmed the bipotentiality of mouse CLiPs, their biliary differentiation capacity was much lower than that of rat CLiPs. By contrast, mouse CLiPs differentiated into the hepatic lineage more efficiently, suggesting that mouse and rat CLiPs have distinct differentiation statuses. Tanimizu et al. also observed a similar tendency in ICam-1[+] mouse adult LPCs, and proposed that the progenies of mouse MHs retained their identities as hepatocytes more tightly than did rat MHs [91,106].

11 IN VITRO REPROGRAMMING OF MHs TO LPCs

In parallel with the above-mentioned observations, we noticed that several earlier studies in rats and mice documented similar observations [107–111], although the adopted conversion methods were distinct from ours. The earliest study by

Block et al. reported that a combination of HGF, EGF, and TGFα converted rat MHs to LPC-like cells [107]. They also showed redifferentiation of these LPC-like cells to MHs, as well as further conversion into bile duct-like structures. Subsequently, several studies also demonstrated that MHs had the potential to dedifferentiate into LPCs [108,110,111], and one of these studies mentioned morphological and immunological similarities between these progenitor cells and OCs [111]. However, it was not clear whether such MH-derived LPC-like cells can be expanded in vitro without changing their phenotypes over the long-term. This was addressed in a study by Fougère-Deschatrette et al., who demonstrated continuous in vitro culture with mouse MH-derived cells [108]. In addition, although three of these studies performed repopulation assays using the MH-derived LPC-like cells [108,110,111], none succeeded in substantial repopulation (the RIs were not described, but seemed to be less than 10% in each study). Thus we were motivated to investigate whether CLiPs could be expanded in vitro long term and whether they could repopulate injured liver with substantial efficiency.

12 STABLE LONG-TERM EXPANSION OF CLiPs

We found that CLiPs were able to undergo a large number of passages and maintain their hepatic differentiation capacity. When cultured on a matrigel-coated plate, although rat CLiPs exhibited a temporarily decreased proliferative capacity between approximately passages 2 and 7, they thereafter stably proliferated up to at least 20 passages (so far, each donor rat-derived CLiP population has not reached senescence in our experiments). Rat CLiPs also proliferated on a collagen-coated plate, but the plating efficiency was decreased, thereby allowing a less efficient subculturing of rat CLiPs. It is unlikely that the enormous proliferation ability of rat CLiPs is the consequence of oncogenic transformation, since these cells immediately cease proliferation upon YAC withdrawal. This highlights the difference between CLiPs and other proliferative hepatic cells that are obtained by transduction of oncogenic genes [112]. Mouse CLiPs were also capable of long-term culture for more than 20 passages. Unlike rat CLiPs, mouse CLiPs did not undergo an evident decrease in proliferation rate during the earlier passages. Importantly, even after long-term culture (more than 10 passages), rat and mouse CLiPs exhibited hepatic differentiation capacity. In particular, some CLiP clones established from independent donor rats maintained their hepatic differentiation capacity at an equivalent efficiency compared with primary CLiPs, as assessed by microarray analysis.

13 IN VIVO REPOPULATION CAPACITY OF CLiPs IN CHRONICALLY INJURED LIVER

Most strikingly, rat CLiPs contributed to substantial repopulation of chronically injured liver. We established three clones from three different donor-derived CLiPs and transplanted each clone into uPA transgenic mice crossed with

SCID mice (cDNA-uPA/SCID mice) [113]. Eight weeks after transplantation, 75%–90% of all recipient mouse livers were repopulated with rat cells. The repopulated areas were positive for a series of hepatic function-related genes and exhibited glycogen storage. Indeed, isolated rat cells showed gene expression profiles resembling that of rat primary hepatocytes. In addition, in some chimeric livers, a few ductal structures stained positively with rat-specific Ck19 antibody. These results verified the bipotentiality of CLiPs in vivo and demonstrated that bipotential CLiPs can almost completely repopulate the chronically injured liver without obvious tumorigenicity.

14 DIPLOID HEPATOCYTES AS THE ORIGIN OF CLiPs

Finally, we identified diploid hepatocytes as the cells from which CLiPs originate. Given the fact that not all the hepatocytes had the ability to be reprogrammed into CLiPs, we speculated that a limited population of MHs has the capability to be converted into CLiPs. During this study, a report was published indicating that hepatocytes with a high proliferation ability were adjacent to the central vein and that these cells were rich in diploid cells [114]. In general, normal somatic cells are diploid, but MHs are composed of heterogeneous ploidy status, and include diploid, tetraploid and octaploid cells. In addition, 85% of hepatocytes are polyploid in rodents [115]. Inspired by Wang's report, we investigated the possible relationship between the ploidy status of MHs and their ability to be reprogrammed into CLiPs. As a result, whereas octaploid MHs had no proliferative capacity, diploid and tetraploid MHs proliferated and formed colonies. Furthermore, we found that the cells with extensive proliferative capacity and the ability to produce colonies with more than 300 cells over 10-day culture were restricted to diploid cells. This fact led to the idea that diploid hepatocytes are the cells from which CLiPs originate. Since diploid hepatocytes compose 2%–5% of the total MH populations in our MH preparation, and approximately 15% of diploid MHs can form large colonies with more than 300 cells, we estimated that 0.3%–0.8% of total MHs have the ability to convert into CLiPs.

15 INSIGHTS FROM CLiPs INTO REGENERATIVE THERAPY AND BASIC LIVER BIOLOGY

CLiPs offer a novel cell source for liver regenerative therapy. As discussed in the previous sections, many approaches for cell transplantation therapy have been proposed by generation of HLCs from stem cells including ESCs, iPSCs, and MSCs, or iHeps. However, the generated hepatic cells usually exhibit a very limited repopulation capacity upon transplantation. Moreover, these approaches often involve gene modification processes, which carry the risk of tumorigenic transformation. By contrast, many animal studies have demonstrated that FLCs as well as LPCs derived from adult livers, especially from injured tissue, have extensive repopulative capacity. Nonetheless, the feasibility regarding the accessibility and the stable in vitro expandability of LPCs have limited their applicability to cell-based therapy. From this point-of-view, CLiPs have the potential to overcome these concerns and possibly serve as an ideal cell source.

For the realization of CLiP-based regenerative liver medicine, it is necessary to develop a method to generate human CLiPs. So far, we have not succeeded in generating human CLiPs from human MHs using the same culture protocol as that used for rat and mouse CLiPs. This implies possible differences between human and rodent MHs. One possible explanation for these differences is that human MHs require different signaling cues to obtain proliferative activity. This possibility might be assessed by comparing human and rodent MHs treated with YAC. Such an analysis might provide information regarding any possible signaling pathways that impede the reprogramming of human MHs. Another possible explanation might be the different telomerase activity levels in human and rodent cells. It is known that in rodents, even normal cells, including hepatocytes, have telomerase activity [116–118]. Furthermore, it has been reported that the telomerase activity of rat MHs spontaneously increases when they are cultured in vitro [117]. Indeed, we confirmed that rat CLiPs in both primary and stable cultures, as well as primary rat MHs, showed telomerase activity. In summary, method optimization is essential to unlock the potential mechanism by which human MHs resist reprogramming into CLiPs.

Rodent CLiP studies can contribute to the basic understanding of liver regeneration and adult LPC origin. There is a longstanding debate regarding resident LPCs in the adult liver. OCs have been identified as facultative LPCs that emerge during chronic liver disease, but their origin has remained unclear [71]. It has long been believed that OCs reside in the intermediate region between the parenchyma and ductal region, which is known as the canals of Hering [119]. This hypothesis was derived from an observation of active OC expansion from the portal area. However, there has not been sufficient evidence to support this hypothesis. By contrast, lineage-tracing studies have demonstrated that MHs are exclusively derived from other MHs after several types of liver injury [120,121]. Accordingly, it has also been suggested that BECs, a putative OC origin cell, minimally contribute to MH generation after liver injury [122–125]. Furthermore, recent studies have strongly suggested that MHs can be reprogrammed into proliferative bipotent LPCs in response to chronic liver injury in mice [126–129] and rats [130]. Our in vitro study might recapitulate such in vivo observations and thus might support a better understanding of this phenomenon.

ACKNOWLEDGMENTS

This research was supported by the Research Program on Hepatitis from the Japan Agency for Medical Research and Development to Shuji Terai (17fk0210101h0001).

LIST OF ACRONYMS AND ABBREVIATIONS

2-AAF	2-Acetylaminofluorene
AFP	α-Fetoprotein
AT-MSCs	Adipose tissue-derived mesenchymal stem cells
BECs	Biliary epithelial cells
BM-MSCs	Bone marrow-derived MSCs
CDE	Choline-deficient ethionine-supplemented diet
CLiPs	Chemically induced liver progenitors
DDC	3,5-Diethoxycarbonyl-1,4-dihidro-collidine
D-gal	D-Galactosamine
ECM	Extracellular matrix
ESCs	Embryonic stem cells
FLCs	Fetal liver cells
HLCs	Hepatocyte-like cells
HPV	Human papilloma virus
hTERT	Human telomerase reverse transcriptase
iHeps	Induced hepatocyte-like cells
iPSCs	Induced pluripotent stem cells
LPCs	Liver progenitor cells
MHs	Mature hepatocytes
MSCs	Mesenchymal stem cells
OSM	Oncostatin M
PH	Partial hepatectomy
SHs	Small hepatocytes
SV40T	Simian virus 40T

REFERENCES

[1] Xu D, Long X, Xia Q. A review of life quality in living donors after liver transplantation. Int J Clin Exp Med 2015;8(1):20–6.
[2] Suh S-W, Lee K-W, Lee J-M, Choi Y, Yi N-J, Suh K-S. Clinical outcomes of and patient satisfaction with different incision methods for donor hepatectomy in living donor liver transplantation. Liver Transplant 2015;21(1):72–8.
[3] Kroencke S, Nashan B, Fischer L, Erim Y, Schulz K-H. Donor quality of life up to two years after living donor liver transplantation. Transplantation 2014;97(5):582–9.
[4] Pascher A. Donor evaluation, donor risks, donor outcome, and donor quality of life in adult-to-adult living donor liver transplantation. Liver Transplant 2002;8(9):829–37.
[5] Olthoff KM, Smith AR, Abecassis M, Baker T, Emond JC, Berg CL, et al. Defining long-term outcomes with living donor liver transplantation in North America. Ann Surg 2015;262(3):465–75.
[6] Mitaka T. The current status of primary hepatocyte culture. Int J Exp Pathol 1998;(79):369–91.
[7] Katsuda T, Kawamata M, Hagiwara K, Takahashi R, Yamamoto Y, Camargo FD, et al. Conversion of terminally committed hepatocytes to culturable bipotent progenitor cells with regenerative capacity. Cell Stem Cell. Elsevier Inc. 2017;20(1):41–55.
[8] Fox IJ, Chowdhury JR. Hepatocyte transplantation. Am J Transplant 2004;4(10):7–13.
[9] Weber A, Groyer-Picard M-T, Franco D, Dagher I. Hepatocyte transplantation in animal models. Liver Transplant 2009;15(1):7–14.

[10] Si-Tayeb K, Noto FK, Nagaoka M, Li J, Battle MA, Duris C, et al. Highly efficient generation of human hepatocyte-like cells from induced pluripotent stem cells. Hepatology 2010;51(1):297–305.

[11] Michalopoulos GK. Liver regeneration after partial hepatectomy. Am J Pathol. American Society for Investigative Pathology 2010;176(1):2–13.

[12] Cho CH, Berthiaume F, Tilles AW, Yarmush ML. A new technique for primary hepatocyte expansion in vitro. Biotechnol Bioeng 2008;101(2):345–56.

[13] Shan J, Schwartz RE, Ross NT, Logan DJ, Thomas D, Duncan Sa, et al. Identification of small molecules for human hepatocyte expansion and iPS differentiation. Nat Chem Biol. Nature Publishing Group 2013;9(8):514–20.

[14] Sato H, Funahashi M, Kristensen DB, Tateno C, Yoshizato K. Pleiotrophin as a Swiss 3T3 cell-derived potent mitogen for adult rat hepatocytes. Exp Cell Res 1999;246(1):152–64.

[15] Skardal A, Smith L, Bharadwaj S, Atala A, Soker S, Zhang Y. Tissue specific synthetic ECM hydrogels for 3-D in vitro maintenance of hepatocyte function. Biomaterials 2012;33(18):4565–75.

[16] Kobayashi N, Fujiwara T, Westerman KA, Inoue Y, Masakiyo S, Hirofumi N, et al. Prevention of acute liver failure in rats with reversibly immortalized human hepatocytes. Science 2000;287(5456):1258–62.

[17] Deurholt T, van Til NP, Chhatta AA, ten Bloemendaal L, Schwartlander R, Payne C, et al. Novel immortalized human fetal liver cell line, cBAL111, has the potential to differentiate into functional hepatocytes. BMC Biotechnol 2009;9(1):89.

[18] Xiang D, Liu C, Wang M-J, Li J, Chen F, Yao H, et al. Non-viral FoxM1 gene delivery to hepatocytes enhances liver repopulation. Cell Death Dis 2014;5(5):e1252.

[19] Levy G, Bomze D, Heinz S, Ramachandran SD, Noerenberg A, Cohen M, et al. Long-term culture and expansion of primary human hepatocytes. Nat Biotechnol 2015;33(12):1264–71.

[20] Huang P, He Z, Ji S, Sun H, Xiang D, Liu C, et al. Induction of functional hepatocyte-like cells from mouse fibroblasts by defined factors. Nature. Nature Publishing Group 2011;475(7356):386–9.

[21] Sekiya S, Suzuki A. Direct conversion of mouse fibroblasts to hepatocyte-like cells by defined factors. Nature. Nature Publishing Group 2011;475(7356):390–3.

[22] Chinzei R. Embryoid-body cells derived from a mouse embryonic stem cell line show differentiation into functional hepatocytes. Hepatology 2002;36(1):22–9.

[23] Gouon-Evans V, Boussemart L, Gadue P, Nierhoff D, Koehler CI, Kubo A, et al. BMP-4 is required for hepatic specification of mouse embryonic stem cell–derived definitive endoderm. Nat Biotechnol 2006;24(11):1402–11.

[24] Hamazaki T, Iiboshi Y, Oka M, Papst PJ, Meacham AM, Zon LI, et al. Hepatic maturation in differentiating embryonic stem cells in vitro. FEBS Lett 2001;497(1):15–9.

[25] Kuai X. Generation of hepatocytes from cultured mouse embryonic stem cells. Liver Transplant 2003;9(10):1094–9.

[26] Soto-Gutiérrez A, Kobayashi N, Rivas-Carrillo JD, Navarro-Álvarez N, Zhao D, Okitsu T, et al. Reversal of mouse hepatic failure using an implanted liver-assist device containing ES cell–derived hepatocytes. Nat Biotechnol 2006;24(11):1412–9.

[27] Yamada T, Yoshikawa M, Kanda S, Kato Y, Nakajima Y, Ishizaka S, et al. In vitro differentiation of embryonic stem cells into hepatocyte-like cells identified by cellular uptake of Indocyanine Green. Stem Cells 2002;20(2):146–54.

[28] Cai J, Zhao Y, Liu Y, Ye F, Song Z, Qin H, et al. Directed differentiation of human embryonic stem cells into functional hepatic cells. Hepatology 2007;45(5):1229–39.

[29] Hannan NRF, Segeritz C-P, Touboul T, Vallier L. Production of hepatocyte-like cells from human pluripotent stem cells. Nat Protoc 2013;8(2):430–7.
[30] Hay DC, Fletcher J, Payne C, Terrace JD, Gallagher RCJ, Snoeys J, et al. Highly efficient differentiation of hESCs to functional hepatic endoderm requires ActivinA and Wnt3a signaling. Proc Natl Acad Sci USA 2008;105(34):12301–6.
[31] Teratani T, Yamamoto H, Aoyagi K, Sasaki H, Asari A, Quinn G, et al. Direct hepatic fate specification from mouse embryonic stem cells. Hepatology 2005;41(4):836–46.
[32] Touboul T, Chen S, To CC, Mora-Castilla S, Sabatini K, Tukey RH, et al. Stage-specific regulation of the WNT/β-catenin pathway enhances differentiation of hESCs into hepatocytes. J Hepatol. European Association for the Study of the Liver 2016;64(6):1315–26.
[33] Heo J, Factor VM, Uren T, Takahama Y, Lee J-S, Major M, et al. Hepatic precursors derived from murine embryonic stem cells contribute to regeneration of injured liver. Hepatology 2006;44(6):1478–86.
[34] Tolosa L, Caron J, Hannoun Z, Antoni M, López S, Burks D, et al. Transplantation of hESC-derived hepatocytes protects mice from liver injury. Stem Cell Res Ther 2015;6(1):246.
[35] Gabriel E, Schievenbusch S, Kolossov E, Hengstler JG, Rotshteyn T, Bohlen H, et al. Differentiation and selection of hepatocyte precursors in suspension spheroid culture of transgenic murine embryonic stem cells. In: Kerkis I, editor. PLoS One 2012;7(9):e44912.
[36] Yang G, Si-Tayeb K, Corbineau S, Vernet R, Gayon R, Dianat N, et al. Integration-deficient lentivectors: an effective strategy to purify and differentiate human embryonic stem cell-derived hepatic progenitors. BMC Biol 2013;11(1):86.
[37] Li F, Liu P, Liu C, Xiang D, Deng L, Li W, et al. Hepatoblast-like progenitor cells derived from embryonic stem cells can repopulate livers of mice. Gastroenterology 2010;139(6). 2158.e8–69.e8.
[38] Takahashi K, Yamanaka S. Induction of pluripotent stem cells from mouse embryonic and adult fibroblast cultures by defined factors. Cell 2006;126(4):663–76.
[39] Yu J, Vodyanik MA, Smuga-Otto K, Antosiewicz-Bourget J, Frane JL, Tian S, et al. Induced pluripotent stem cell lines derived from human somatic cells. Science. 2007;318(5858): 1917–20.
[40] Ichida JK, Blanchard J, Lam K, Son EY, Chung JE, Egli D, et al. A small-molecule inhibitor of Tgf-β signaling replaces Sox2 in reprogramming by inducing nanog. Cell Stem Cell. Elsevier Ltd 2009;5(5):491–503.
[41] Mandai M, Watanabe A, Kurimoto Y, Hirami Y, Morinaga C, Daimon T, et al. Autologous induced stem-cell–derived retinal cells for macular degeneration. N Engl J Med 2017;376(11):1038–46.
[42] Song Z, Cai J, Liu Y, Zhao D, Yong J, Duo S, et al. Efficient generation of hepatocyte-like cells from human induced pluripotent stem cells. Cell Res. Nature Publishing Group 2009;19(11):1233–42.
[43] Sullivan GJ, Hay DC, Park I-H, Fletcher J, Hannoun Z, Payne CM, et al. Generation of functional human hepatic endoderm from human induced pluripotent stem cells. Hepatology 2010;51(1):329–35.
[44] Nakamori D, Takayama K, Nagamoto Y, Mitani S, Sakurai F, Tachibana M, et al. Hepatic maturation of human iPS cell-derived hepatocyte-like cells by ATF5, c/EBPα, and PROX1 transduction. Biochem Biophys Res Commun. Elsevier Ltd 2016;469(3):424–9.
[45] Kondo Y, Iwao T, Yoshihashi S, Mimori K, Ogihara R, Nagata K, et al. Histone deacetylase inhibitor valproic acid promotes the differentiation of human induced pluripotent stem cells into hepatocyte-like cells. In: Johnson R, editor. PLoS One 2014;9(8):e104010.

[46] Hanawa M, Takayama K, Sakurai F, Tachibana M, Mizuguchi H. Hepatocyte nuclear factor 4 alpha promotes definitive endoderm differentiation from human induced pluripotent stem cells. Stem Cell Rev Rep 2016;1–10.
[47] Siller R, Greenhough S, Naumovska E, Sullivan GJ. Small-molecule-driven hepatocyte differentiation of human pluripotent stem cells. Stem Cell Rep 2015;4(5):939–52.
[48] Hou P, Li Y, Zhang X, Liu C, Guan J, Li H, et al. Pluripotent stem cells induced from mouse somatic cells by small-molecule compounds. Science 2013;341(6146):651–4.
[49] Nagamoto Y, Takayama K, Tashiro K, Tateno C, Sakurai F, Tachibana M, et al. Efficient engraftment of human induced pluripotent stem cell-derived hepatocyte-like cells in uPA/SCID mice by overexpression of FNK, a Bcl-xL mutant Gene. Cell Transplant 2015;24(6):1127–38.
[50] Chen Y, Li Y, Wang X, Zhang W, Sauer V, Chang CJ, et al. Amelioration of hyperbilirubinemia in gunn rats after transplantation of human induced pluripotent stem cell-derived hepatocytes. Stem Cell Rep 2015;5(1):22–30.
[51] Ieda M, Fu J-D, Delgado-Olguin P, Vedantham V, Hayashi Y, Bruneau BG, et al. Direct reprogramming of fibroblasts into functional cardiomyocytes by defined factors. Cell. Elsevier Ltd 2010;142(3):375–86.
[52] Vierbuchen T, Ostermeier A, Pang ZP, Kokubu Y, Südhof TC, Wernig M. Direct conversion of fibroblasts to functional neurons by defined factors. Nature 2010;463(7284):1035–41.
[53] Yu B, He ZY, You P, Han QW, Xiang D, Chen F, et al. Reprogramming fibroblasts into bipotential hepatic stem cells by defined factors. Cell Stem Cell. Elsevier Inc. 2013;13(3):328–40.
[54] Du Y, Wang J, Jia J, Song N, Xiang C, Xu J, et al. Human hepatocytes with drug metabolic function induced from fibroblasts by lineage reprogramming. Cell Stem Cell. Elsevier Inc. 2014;14(3):394–403.
[55] Lim K, Lee S, Gao Y, Kim K-P, Song G, An S, et al. Small molecules facilitate single factor-mediated hepatic reprogramming. Cell Rep 2016;15(4):814–29.
[56] Zhu S, Rezvani M, Harbell J, Mattis AN, Wolfe AR, Benet LZ, et al. Mouse liver repopulation with hepatocytes generated from human fibroblasts. Nature. Nature Publishing Group 2014;508(7494):93–7.
[57] Song G, Pacher M, Balakrishnan A, Yuan Q, Tsay H-C, Yang D, et al. Direct reprogramming of hepatic myofibroblasts into hepatocytes in vivo attenuates liver fibrosis. Cell Stem Cell 2016;18(6):797–808.
[58] Chamberlain G, Fox J, Ashton B, Middleton J. Concise review: mesenchymal stem cells: their phenotype, differentiation capacity, immunological features, and potential for homing. Stem Cells 2007;25(11):2739–49.
[59] Banas A, Yamamoto Y, Teratani T, Ochiya T. Stem cell plasticity: learning from hepatogenic differentiation strategies. Dev Dyn 2007;236(12):3228–41.
[60] Schwartz RE, Reyes M, Koodie L, Jiang Y, Blackstad M, Lund T, et al. Multipotent adult progenitor cells from bone marrow differentiate into functional hepatocyte-like cells. J Clin Invest 2002;109(10):1291–302.
[61] Lee K-D, Kuo TK-C, Whang-Peng J, Chung Y-F, Lin C-T, Chou S-H, et al. In vitro hepatic differentiation of human mesenchymal stem cells. Hepatology 2004;40(6):1275–84.
[62] Hong SH, Gang EJ, Jeong JA, Ahn C, Hwang SH, Yang IH, et al. In vitro differentiation of human umbilical cord blood-derived mesenchymal stem cells into hepatocyte-like cells. Biochem Biophys Res Commun 2005;330(4):1153–61.
[63] Seo MJ, Suh SY, Bae YC, Jung JS. Differentiation of human adipose stromal cells into hepatic lineage in vitro and in vivo. Biochem Biophys Res Commun 2005;328(1):258–64.

[64] Banas A, Teratani T, Yamamoto Y, Tokuhara M, Takeshita F, Quinn G, et al. Adipose tissue-derived mesenchymal stem cells as a source of human hepatocytes. Hepatology 2007;46(1):219–28.

[65] Sato Y, Fujiwara H, Zeng BX, Higuchi T, Yoshioka S, Fujii S. Platelet-derived soluble factors induce human extravillous trophoblast migration and differentiation: platelets are a possible regulator of trophoblast infiltration into maternal spiral arteries. Blood 2005;106(2):428–35.

[66] Aurich I, Mueller LP, Aurich H, Luetzkendorf J, Tisljar K, Dollinger MM, et al. Functional integration of hepatocytes derived from human mesenchymal stem cells into mouse livers. Gut 2007;56(3):405–15.

[67] Banas A, Teratani T, Yamamoto Y, Tokuhara M, Takeshita F, Osaki M, et al. IFATS collection: in vivo therapeutic potential of human adipose tissue mesenchymal stem cells after transplantation into mice with liver injury. Stem Cells 2008;26(10):2705–12.

[68] Aurich H, Sgodda M, Kaltwasser P, Vetter M, Weise A, Liehr T, et al. Hepatocyte differentiation of mesenchymal stem cells from human adipose tissue in vitro promotes hepatic integration in vivo. Gut 2009;58(4):570–81.

[69] di Bonzo LV, Ferrero I, Cravanzola C, Mareschi K, Rustichell D, Novo E, et al. Human mesenchymal stem cells as a two-edged sword in hepatic regenerative medicine: engraftment and hepatocyte differentiation versus profibrogenic potential. Gut 2008;57(2):223–31.

[70] Kuo TK, Hung S, Chuang C, Chen C, Shih YV, Fang SY, et al. Stem cell therapy for liver disease: parameters governing the success of using bone marrow mesenchymal stem cells. Gastroenterology 2008;134(7). 2111–21.e3.

[71] Miyajima A, Tanaka M, Itoh T. Stem/progenitor cells in liver development, homeostasis, regeneration, and reprogramming. Cell Stem Cell. Elsevier Inc. 2014;14(5):561–74.

[72] Suzuki A, Zheng Y, Kondo R, Kusakabe M, Takada Y, Fukao K, et al. Flow-cytometric separation and enrichment of hepatic progenitor cells in the developing mouse liver. Hepatology 2000;32(6):1230–9.

[73] Suzuki A, Zheng Y, Kaneko S, Onodera M, Fukao K, Nakauchi H, et al. Clonal identification and characterization of self-renewing pluripotent stem cells in the developing liver. J Cell Biol 2002;156(1):173–84.

[74] Tanimizu N. Isolation of hepatoblasts based on the expression of Dlk/Pref-1. J Cell Sci 2003;116(9):1775–86.

[75] Nierhoff D, Ogawa A, Oertel M, Chen Y-Q, Shafritz Da. Purification and characterization of mouse fetal liver epithelial cells with high in vivo repopulation capacity. Hepatology 2005;42(1):130–9.

[76] Kakinuma S, Ohta H, Kamiya A, Yamazaki Y, Oikawa T, Okada K, et al. Analyses of cell surface molecules on hepatic stem/progenitor cells in mouse fetal liver. J Hepatol. European Association for the Study of the Liver 2009;51(1):127–38.

[77] Kamiya A, Kakinuma S, Yamazaki Y, Nakauchi H. Enrichment and clonal culture of progenitor cells during mouse postnatal liver development in mice. Gastroenterology. Elsevier Inc. 2009;137(3). 1114–26.e14.

[78] Dabeva MD, Petkov PM, Sandhu J, Oren R, Laconi E, Hurston E, et al. Proliferation and differentiation of fetal liver epithelial progenitor cells after transplantation into adult rat liver. Am J Pathol 2000;156(6):2017–31.

[79] Sandhu JS, Petkov PM, Dabeva MD, Shafritz DA. Stem cell properties and repopulation of the rat liver by fetal liver epithelial progenitor cells. Am J Pathol 2001;159(4):1323–34.

[80] Oertel M. Repopulation of rat liver by fetal hepatoblasts and adult hepatocytes transduced ex vivo with lentiviral vectors. Hepatology 2003;37(5):994–1005.

[81] Oertel M, Menthena A, Chen Y-Q, Shafritz DA. Properties of cryopreserved fetal liver stem/progenitor cells that exhibit long-term repopulation of the normal rat liver. Stem Cells 2006;24(10):2244–51.

[82] Oertel M, Menthena A, Dabeva MD, Shafritz DA. Cell competition leads to a high level of normal liver reconstitution by transplanted fetal liver stem/progenitor cells. Gastroenterology 2006;130(2):507–20.

[83] Oertel M, Menthena A, Chen Y-Q, Shafritz DA. Comparison of hepatic properties and transplantation of Thy-1(+) and Thy-1(-) cells isolated from embryonic day 14 rat fetal liver. Hepatology 2007;46(4):1236–45.

[84] Oertel M, Menthena A, Chen Y-Q, Teisner B, Jensen CH, Shafritz D. Purification of fetal liver stem/progenitor cells containing all the repopulation potential for normal adult rat liver. Gastroenterology 2008;134(3):823–32.

[85] Fausto N, Campbell JS. The role of hepatocytes and oval cells in liver regeneration and repopulation. Mech Dev 2003;120(1):117–30.

[86] Wang X, Foster M, Al-Dhalimy M, Lagasse E, Finegold M, Grompe M. The origin and liver repopulating capacity of murine oval cells. Proc Natl Acad Sci 2003;100(Suppl. 1):11881–8.

[87] Mitaka T, Mikami M, Sattler GL, Pitot HC, Mochizuki Y. Small cell colonies appear in the primary culture of adult rat hepatocytes in the presence of nicotinamide and epidermal growth factor. Hepatology 1992;16(2):440–7.

[88] Kon J, Ooe H, Oshima H, Kikkawa Y, Mitaka T. Expression of CD44 in rat hepatic progenitor cells. J Hepatol 2006;45(1):90–8.

[89] Katayama S, Tateno C, Asahara T, Yoshizato K. Size-dependent in vivo growth potential of adult rat hepatocytes. Am J Pathol. American Society for Investigative Pathology 2001;158(1):97–105.

[90] Ichinohe N, Kon J, Sasaki K, Nakamura Y, Ooe H, Tanimizu N, et al. Growth ability and repopulation efficiency of transplanted hepatic stem cells, progenitor cells, and mature hepatocytes in retrorsine-treated rat livers. Cell Transplant 2012;21(1):11–22.

[91] Tanimizu N, Ichinohe N, Ishii M, Kino J, Mizuguchi T, Hirata K, et al. Liver progenitors isolated from adult healthy mouse liver efficiently differentiate to functional hepatocytes in vitro and repopulate liver tissue. Stem Cells 2016;34(12):2889–901.

[92] Sasaki K, Kon J, Mizuguchi T, Chen Q, Ooe H, Oshima H, et al. Proliferation of hepatocyte progenitor cells isolated from adult human livers in serum-free medium. Cell Transplant 2008;17(10–11):1221–30.

[93] Ooe H, Chen Q, Kon J, Sasaki K, Miyoshi H, Ichinohe N, et al. Proliferation of rat small hepatocytes requires follistatin expression. J Cell Physiol 2012;227(6):2363–70.

[94] Ishii T. Fetal stem cell transplantation: past, present, and future. World J Stem Cells 2014;6(4):404.

[95] Ichinohe N, Tanimizu N, Ooe H, Nakamura Y, Mizuguchi T, Kon J, et al. Differentiation capacity of hepatic stem/progenitor cells isolated from D-galactosamine-treated rat livers. Hepatology 2013;57(3):1192–202.

[96] Ishii M, Kino J, Ichinohe N, Tanimizu N, Ninomiya T, Suzuki H, et al. Hepatocytic parental progenitor cells of rat small hepatocytes maintain self-renewal capability after long-term culture. Sci Rep. Nature Publishing Group 2017;7(December 2016):46177.

[97] Kawamata M, Ochiya T. Generation of genetically modified rats from embryonic stem cells. Proc Natl Acad Sci 2010;107(32):14223–8.

[98] Kawamata M, Ochiya T. Two distinct knockout approaches highlight a critical role for p53 in rat development. Sci Rep 2012;2:945.

[99] Kawamata M, Katsuda T, Yamada Y, Ochiya T. In vitro reconstitution of breast cancer heterogeneity with multipotent cancer stem cells using small molecules. Biochem Biophys Res Commun. Elsevier Ltd 2017;482(4):750–7.

[100] Katsuda T, Teratani T, Ochiya T, Sakai Y. Transplantation of a fetal liver cell-loaded hyaluronic acid sponge onto the mesentery recovers a Wilson's disease model rat. J Biochem 2010;148(3):281–8.

[101] Kamiya A, Kojima N, Kinoshita T, Sakai Y, Miyaijma A. Maturation of fetal hepatocytesin vitro by extracellular matrices and oncostatin M: induction of tryptophan oxygenase. Hepatology 2002;35(6):1351–9.

[102] Hashimoto W, Sudo R, Fukasawa K, Ikeda M, Mitaka T, Tanishita K. Ductular network formation by rat biliary epithelial cells in the dynamical culture with collagen gel and dimethylsulfoxide stimulation. Am J Pathol 2008;173(2):494–506.

[103] Sirica a E, Gainey TW. A new rat bile ductular epithelial cell culture model characterized by the appearance of polarized bile ductsin vitro. Hepatology 1997;26(3):537–49.

[104] Katsuda T, Kojima N, Ochiya T, Sakai Y. Biliary epithelial cells play an essential role in the reconstruction of hepatic tissue with a functional bile ductular network. Tissue Eng Part A 2013;19(21–22):2402–11.

[105] Marinelli RA, Pham L, Agre P, LaRusso NF. Secretin promotes osmotic water transport in rat cholangiocytes by increasing aquaporin-1 water channels in plasma membrane. Evidence for a secretin-induced vesicular translocation of aquaporin-1. J Biol Chem 1997;272(20):12984–8.

[106] Tanimizu N, Mitaka T. Which is better source for functional hepatocytes? Stem Cell Investig 2017;4:12.

[107] Block GD. Population expansion, clonal growth, and specific differentiation patterns in primary cultures of hepatocytes induced by HGF/SF, EGF and TGF alpha in a chemically defined (HGM) medium. J Cell Biol 1996;132(6):1133–49.

[108] Fougère-Deschatrette C, Imaizumi-Scherrer T, Strick-Marchand H, Morosan S, Charneau P, Kremsdorf D, et al. Plasticity of hepatic cell differentiation: bipotential adult mouse liver clonal cell lines competent to differentiate in vitro and in vivo. Stem Cells 2006;24(9):2098–109.

[109] Koenig S, Krause P, Hosseini ASA, Dullin C, Rave-Fraenk M, Kimmina S, et al. Noninvasive imaging of liver repopulation following hepatocyte transplantation. Cell Transplant 2009;18(1):69–78.

[110] Krause P, Unthan-Fechner K, Probst I, Koenig S. Cultured hepatocytes adopt progenitor characteristics and display bipotent capacity to repopulate the liver. Cell Transplant 2014;23(7):805–17.

[111] Chen Y, Wong PP, Sjeklocha L, Steer CJ, Sahin MB. Mature hepatocytes exhibit unexpected plasticity by direct dedifferentiation into liver progenitor cells in culture. Hepatology 2012;55(2):563–74.

[112] Ramboer E, De Craene B, De Kock J, Vanhaecke T, Berx G, Rogiers V, et al. Strategies for immortalization of primary hepatocytes. J Hepatol. European Association for the Study of the Liver 2014;61(4):925–43.

[113] Tateno C, Kawase Y, Tobita Y, Hamamura S, Ohshita H, Yokomichi H, et al. Generation of novel chimeric mice with humanized livers by using hemizygous cDNA-uPA/SCID mice. In: Aldabe R, editor. PLoS One 2015;10(11):e0142145.

[114] Wang B, Zhao L, Fish M, Logan CY, Nusse R. Self-renewing diploid Axin2$^+$ cells fuel homeostatic renewal of the liver. Nature 2015;524(7564):180–5.

[115] Gentric G, Desdouets C. Polyploidization in liver tissue. Am J Pathol. American Society for Investigative Pathology 2014;184(2):322–31.
[116] Seluanov A, Chen Z, Hine C, Sasahara THC, Ribeiro AACM, Catania KC, et al. Telomerase activity coevolves with body mass not lifespan. Aging Cell 2007;6(1):45–52.
[117] Nozawa K, Kurumiya Y, Yamamoto A, Isobe Y, Suzuki M, Yoshida S. Up-regulation of telomerase in primary cultured rat hepatocytes. J Biochem 1999;126(2):361–7.
[118] Golubovskaya VM, Presnell SC, Hooth MJ, Smith GJ, Kaufmann WK. Expression of telomerase in normal and malignant rat hepatic epithelia. Oncogene 1997;15(10):1233–40.
[119] Kordes C, Haussinger D. Hepatic stem cell niches. J Clin Invest. 2013;123(5):1874–80.
[120] Malato Y, Naqvi S, Schürmann N, Ng R, Wang B, Zape J, et al. Fate tracing of mature hepatocytes in mouse liver homeostasis and regeneration. J Clin Invest 2011;121(12):4850–60.
[121] Wang Y, Huang X, He L, Pu W, Li Y, Liu Q, et al. Genetic tracing of hepatocytes in liver homeostasis, injury, and regeneration. J Biol Chem 2017;292(21):8594–604.
[122] Carpentier R, Su?er RE, van Hul N, Kopp JL, Beaudry J-B, Cordi S, et al. Embryonic ductal plate cells give rise to cholangiocytes, periportal hepatocytes, and adult liver progenitor cells. Gastroenterology. Elsevier Inc. 2011;141(4). 1432–38.e4.
[123] Español–Suñer R, Carpentier R, Van Hul N, Legry V, Achouri Y, Cordi S, et al. Liver progenitor cells yield functional hepatocytes in response to chronic liver injury in mice. Gastroenterology. Elsevier Inc. 2012;143(6). 1564–75.e7.
[124] Rodrigo-Torres D, Affò S, Coll M, Morales-Ibanez O, Millán C, Blaya D, et al. The biliary epithelium gives rise to liver progenitor cells. Hepatology 2014;60(4):1367–77.
[125] Yanger K, Knigin D, Zong Y, Maggs L, Gu G, Akiyama H, et al. Adult hepatocytes are generated by self-duplication rather than stem cell differentiation. Cell Stem Cell. Elsevier Inc. 2014;15(3):340–9.
[126] Tanimizu N, Nishikawa Y, Ichinohe N, Akiyama H, Mitaka T. Sry HMG box protein 9-positive (Sox9$^+$) epithelial cell adhesion molecule-negative (EpCAM$^-$) biphenotypic cells derived from hepatocytes are involved in mouse liver regeneration. J Biol Chem 2014;289(11):7589–98.
[127] Yanger K, Zong Y, Maggs LR, Shapira SN, Maddipati R, Aiello NM, et al. Robust cellular reprogramming occurs spontaneously during liver regeneration. Genes Dev 2013;27(7):719–24.
[128] Tarlow B, Pelz C, Naugler W, Wakefield L, Wilson E, Finegold M, et al. Bipotential adult liver progenitors are derived from chronically injured mature hepatocytes. Cell Stem Cell. Elsevier Inc. 2014;15(5):605–18.
[129] Yimlamai D, Christodoulou C, Galli GG, Yanger K, Pepe-Mooney B, Gurung B, et al. Hippo pathway activity influences liver cell fate. Cell. Elsevier Inc. 2014;157(6):1324–38.
[130] Yovchev MI, Locker J, Oertel M. Biliary fibrosis drives liver repopulation and phenotype transition of transplanted hepatocytes. J Hepatol. European Association for the Study of the Liver 2016;64(6):1348–57.

Chapter 6

Generation of Hepatocytes by Transdifferentiation

Pengyu Huang*, Qiwen Chen**
*ShanghaiTech University, Shanghai, China; **Shanghai Cancer Center, Shanghai Medical School, Fudan University, Shanghai, China

1 INTRODUCTION

The hepatocyte is an important cell type in parenchymal tissues of the liver and involves in many liver functions, such as detoxification, carbohydrate metabolism, lipid metabolism, secretion of albumin, clotting factors, and complements. Thus the hepatocyte is widely used as an experimental model for studies on liver disease. As hepatocytes serve as the major locus for drug metabolism, primary hepatocytes are also used for drug safety assessments [1]. Importantly still, loss of functional hepatocytes is a primary cause of many liver diseases. Therefore hepatocyte-based therapies, such as cell-replacement therapy, bioartificial liver (BAL) system, and bioengineered liver, hold promises for the treatment of certain liver diseases [2–4].

In the efforts to obtain functional hepatocytes, various technologies have been developed to isolate and purify hepatocytes from liver tissues [5]. However, the maintenance and expansion of hepatocytes are still challenging, making freshly isolated and cryopreserved primary hepatocytes the principle cell sources. For decades, many efforts have been made to generate hepatocytes from alternative cell sources. One possible approach is the differentiation of embryonic stem cells and induced pluripotent stem cells (iPSCs) into hepatocytes [6–10].

In recent years, an increasing number of reports have claimed the direct transdifferentiation of hepatocytes from nonhepatic cells [11–14]. The term "transdifferentiation" was initially employed to describe the conversion of one cell lineage to another as observed in silkmoth [15]. Mammalian cells have been thought to be more resistant to such cell lineage conversion. Nevertheless, conversion of nonhepatic cells to hepatocytes has been studied for decades. In 1981, Scarpelli and Rao found hepatocyte-like cells in the pancreas of Syrian golden hamsters during pancreas regeneration [16]. Since then, many protocols have been developed to induce hepatic conversion from pancreatic cells *in vitro* and *in vivo* [17]. However, considering the lack of pancreatic cell sources, generation of hepatocytes from more easily accessible cell types could be more applicable.

Several investigations claimed that mesenchymal stem cells (MSCs) from bone marrow, hematopoietic stem cells (HSCs), and other adult stem cells, had the plasticity to give rise to a much wider range of cell types than previously thought [18–21]. Under the stimulation of growth factor and cytokine cocktails, hepatocyte-like cells expressing hepatic marker genes were generated from MSCs [20,22], HSCs [23], and even fibroblasts [24]. However, the quality of hepatocytes generated by such strategies is controversial due to lack of definitive evidence of hepatic functions.

In 1987 a pioneering work about the direct conversion of fibroblasts to myoblasts by enforced expression of MyoD was reported, providing a novel strategy of cell lineage conversion by "master regulators" [25]. The success of generating iPSCs further strengthened the notion that certain "master regulators" could change cell fate [26], which leads to the progress of transdifferentiation of nonhepatic cells to hepatocytes by hepatic transcription factors [11–14]. These hepatocytes generated from nonhepatic cells by transdifferentiation were termed induced hepatocytes (iHep cells).

2 GENERATION OF iHEP CELLS

Because cell fate determination requires the expressions of lineage-specific transcription factors, the enforced expression of lineage-specific transcription factors was extensively used to induce cell fate conversion. In pursuing hepatic transdifferentiation, the key issue is to find an appropriate combination of hepatic transcription factors. Others and we independently screened several transcription factors important for liver development and liver functions. We found that Foxa3, Hnf1α, and Gata4 were essential for mouse hepatic transdifferentiation, while Sekiya and Suzuki found that Hnf4α combined with any one of Foxa1, Foxa2, and Foxa3 would induce hepatic conversion in mouse embryonic fibroblasts (MEF) [12,14]. It is not surprising that different combinations of transcription factors could result in a similar cell fate conversion process. Similar cases were reported in generating iPS cells. Several transcription factors, such as GATA3 and Nr5a2, could be used as a substitute of Oct4 [27,28]. To facilitate hepatic transdifferentiation and enable the long-term proliferation of iHep cells, we inactivated p19Arf, a component of the p53 pathway. The p53 pathway has been reported to impede reprogramming [29–31]. Importantly, previous work has also demonstrated that inactivation of p19Arf allowed mouse primary hepatocytes to infinitely proliferate *in vitro* [32]. Using this strategy, we achieved a transdifferentiation efficiency of 23% as indicated by albumin expression. In Sekiya and Suzuki's protocol, spontaneous iHep cell proliferation was found after repeated infection of retrovirus expressing the aforementioned hepatic transcription factors. It is interesting that these iHep cells overcome cell cycle arrest after several passages [14]. Significantly, in addition to studies in culture cells, *in vivo* transdifferentiation also holds potential as a therapeutic strategy for liver diseases. Through the transduction of hepatic transcription factors, liver myofibroblasts can be converted to hepatocytes, which helps attenuate liver fibrosis [33,34].

Human cells were often thought to be more resistant to cell fate conversion. The combination of transcription factors used in mouse hepatic transdifferentiation was not sufficient to convert human fibroblasts to hepatocytes. After a *de-novo* screening, we used FOXA3, HNF1A, and HNF4A to induce human iHep cells (also named hiHep cells) [13]. Unlike in mouse cells, inactivation of neither p53 nor RB was sufficient to support the long-term proliferation of iHep cells. Subsequently, we overexpressed SV40 large T antigen, which suppressed both p53 and RB [35]. Human iHep cells expressing SV40 large T antigen were immortalized and were capable of forming colonies.

In the context of regenerative medicine, using cells introduced with exogenous genes arouses concerns about their uncertain influences. Thus reprogramming without the involvement of exogenous genetic materials is a more appealing approach. In 2013 a group used small-molecule compounds to induce pluripotency in mouse fibroblasts, inspiring efforts for induction of transdifferentiation with compound cocktails [36]. Such small-molecule-compounds-based transdifferentiation has helped achieve generation of neurons [37,38], neural stem cells [39,40], cardiomyocytes [41,42], and multipotent endodermal progenitor cells [43]. The multipotent endodermal progenitor cells induced by small-molecules are capable of efficient hepatic differentiation, providing a promising approach to generate hepatocytes from nonhepatic cells without delivery of genetic materials [43]. Nevertheless, direct hepatic conversion by small compounds has not yet been reported. Hereafter, we will focus on iHep cells generated by enforced expression of transcription factors *in vitro*.

3 DELIVERY SYSTEMS

When inducing transdifferentiation, the approach used for delivery of exogenous genes significantly affects the transdifferentiation efficiency and safety, particularly in introducing the transdifferentiated cells for clinical use. Currently, numerous delivery strategies have been developed.

3.1 Integrative Delivery Approaches

Retroviral and lentiviral expression systems have been used extensively for expression of exogenous transcription factors to induce reprogramming and transdifferentiation. The envelope protein used in virus packaging determines the cell types that could be infected. Among them, VSV-G envelope protein is often used for its broad tropism over a range of species and cell types. The multiplicity of infection (MOI) is an important parameter when using viral expression systems. Excessive MOI often leads to massive donor cell deaths, while low MOI results in inefficient transdifferentiation. The window of optimal MOI is typically relatively narrow. From our experience, MOI ranging between 1 and 2 is usually used for hepatic conversion from mouse and human fibroblasts. The silencing of exogenous genes is another critical issue. The expression levels of hepatic transcription factors under the commonly used promoters, such as EF1-alpha and CMV, are usually much higher than those of endogenous genes in hepatocytes. Sustained excessive expression of lineage-specific transcription factors could be toxic for the cells (unpublished data). Moreover, silence or attenuation of exogenous lineage-specific genes is critical for activation and balance of endogenous gene network [44].

3.2 Nonintegrative Delivery Approaches

Although retroviral and lentiviral systems allow efficient and reproducible delivery of foreign genes, genome integration of the viral DNA brings about concerns on tumorigenicity. Nonintegrating viral systems, such as adenovirus, adeno-associate virus (AAV), and sendai virus, are thought to be safer and permit higher titer after concentration. Adenovirus and sendai virus efficiently infect a broad range of cell types and have been used to generate iPSCs [45–47], possibly providing nonintegration delivery methods for hepatic transdifferentiation. There are several stereotypes of AAV differing in their tropisms among cell types, which is very useful when *in vivo* hepatic transdifferentiation is induced in specific nonhepatic cells [33].

Nonviral systems are also being developed and used for cell fate conversion. Plasmid-based strategies, such as episomal vectors, are easy to implement in various donor cell types, though integration events may still happen in a proportion of donor cells [48–50]. Mouse iHep cells induced by episomal vector-based expressions of *Foxa3*, *Hnf1α*, and *Gata4* have shown similar morphology, marker gene expressions and hepatic functions as compared to primary hepatocytes [51]. Other nonviral delivery approaches, such as RNA delivery and protein delivery, avoid the delivery of genetic materials. Serial transduction of synthetic mRNA encoding Yamanaka factors showed relatively high efficiency in reprogramming [52]. However, direct delivery of recombinant proteins is more challenging, as a large amount of recombinant proteins are needed, and the efficiency is relatively low [53,54]. Thus these innovative works in the generation of iPSCs by direct delivery of synthetic mRNAs and recombinant proteins suggested possible directions in the future development of nonintegrating strategies for hepatic transdifferentiation.

4 DONOR CELL TYPES

4.1 Fibroblasts

When inducing mouse iHep cells, MEF and adult tail-tip fibroblasts are often used as starting cells, because of their easy accessibility [12,14]. For human iHep cells, considering the potential future application of iHep cells in regenerative medicine, it is important to choose appropriate donor cell types.

An advantage of iHep cells for future application in regenerative medicine is the potential of generating patient-specific hepatocytes. Thus the starting cells should be easily accessible and expandable *in vitro*. Currently, the most commonly used cell type is fibroblast, which has been considered as the first choice because of its more established use for cell fate conversion. Adult fibroblasts can be easily isolated from skin or scar tissues. We have successfully generated functional iHep cells from adult fibroblasts isolated from scar tissues, though at a lower efficiency as compared to fetal fibroblasts [13].

4.2 Mesenchymal Stem Cells

MSCs, first isolated from the bone marrow [55] and then from many other tissues, such as fat [56], skeletal muscle [57], spleen, and liver [58], are also an appealing cell source. MSC-related clinical trials have been conducted worldwide [59]. Importantly, many culture methods for MSCs have been developed to meet the requirement for clinical use.

4.3 Urinary Cells

Isolation of fibroblasts and MSCs usually require invasive medical procedures. Recently, urinary cells (possibly renal epithelial cells) were isolated from urine by a noninvasive method [60,61]. To isolate urinary cells, about 50–200 mL urine was collected and subjected to a straightforward procedure. This process contained no operative wound or source of pain for patients, suggesting that urinary cell would be a much more acceptable source for donors. We have compared human iHep cells generated from fetal fibroblasts, adult fibroblasts, MSCs, and urinary cells. It turned out that human iHep cells generated from different donor cell types showed similar hepatic marker gene expression levels and hepatic functions (unpublished data).

5 EVALUATION OF iHEP CELL QUALITY

Several proof of principle works have demonstrated the clinical potential of iHep cells in regenerative medicine. However, there are no standardized criteria currently available to evaluate the quality of iHep cells. Considering that hepatocytes are responsible for numerous functions of the liver, it is impossible to characterize all phenotypes of iHep cells. However, in previous reports, we have suggested some methods that aid in characterizing iHep cells.

Morphological change is the first sign of successive generation of iHep cells. Approximately 5–7 days after the infection of hepatic transcription factors in fibroblasts, epithelial iHep colonies appear in the cell culture. After cells reach 100% confluence, iHep cells display polygonal shape with distinct cell borders.

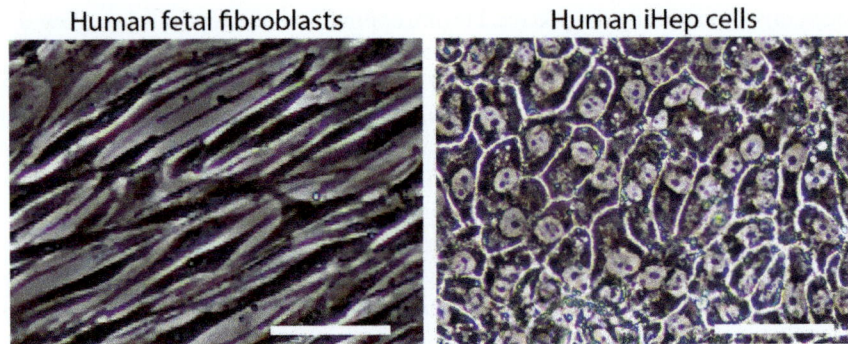

FIGURE 6.1 **Morphology of human fetal fibroblasts and iHep cells.** Human iHep cells were generated by enforced expression of FOXA3, HNF1A, HNF4A, and SV40 large T antigen in human fetal fibroblasts for 30 days.

Further culture of iHep cells will help the cells to acquire morphology with the following features of mature hepatocyte: clear and distinct nuclei, tight cell-to-cell junctions, binucleus, and formation of bile canaliculi (Fig. 6.1).

Currently, gene expression analysis is an essential standard for molecular biology labs. Many hepatic marker genes, such as ALBUMIN, AAT, TJP1, and CDH1, have been extensively used to characterize hepatocytes generated by differentiation and transdifferentiation [6,12–14]. Additionally, fully reprogrammed iHep cells should lose the expression of donor cell marker genes. If iHep cells are generated from fibroblasts, fibrotic genes, such as COL1A1 and FSP1, should be silent. For more comprehensive comparison of the gene expression profiles between iHep cells and primary hepatocytes, mRNA sequencing can be used to evaluate the expressions of genes related to liver functions.

The functional levels are critical parameters in characterizing iHep cells. There are many functional assays to assess hepatic functions. For example, PAS staining is adopted to evaluate glycogen storage, and Albumin ELISA is used to evaluate the ability to produce secretory proteins. Additionally, emphases have been placed on certain functions as relevant to their specific applications. For example, hepatocytes are generally used as an *in vitro* system for drug metabolism studies in drug development. Thus the activities of drug metabolism enzymes, such as cytochromes P450 (P450s) and uridine diphosphate glucuronosyltransferases, are important for this setting. For treatment of liver failure patients, a more comprehensive view should be taken into consideration, such as ammonia clearance, secretion of coagulation factors.

In vivo repopulation and functions are used to comprehensively evaluate the quality of iHep cells. Several liver injury animal models have been developed for this attempt. For example, fumarylacetoacetate hydrolase (Fah)-deficient mouse and albumin-urokinase-type plasminogen activator (Alb-uPA) mouse are widely used for evaluation of hepatic functions *in vivo* [62,63]. Functional iHep cells should be able to efficiently integrate into the liver tissues, repopulate the liver, and rescue the mice from liver injury induced death.

6 APPLICATIONS AND CHALLENGES

Recently the possibility of clinical use of iHep cells is being focused on, and researchers are exploring the potential application of iHep cells in disease treatment. Previous studies have clearly shown the *in vivo* integration ability of iHep cells [12,14]. Cell-replacement therapy is one of the important approaches in regenerative medicine. Especially when the therapeutic cells originate within the patient, immune rejection could be minimized. Another approach for liver regenerative medicine is iHep cell based BAL system. Recently such a device was developed for the treatment of a drug-induced porcine acute liver failure model. Exhibiting the improved liver functions and statistically significant improvement of survival, human iHep cell based BAL system showed promise as a therapeutic strategy for liver failure [64]. Albeit the success of using human iHep cells for the treatment of animal disease models, many challenges must be addressed before moving forward to clinical trials.

The first challenge is the large-scale expansion of iHep cells. Cell-replacement therapy, BAL systems, and bioengineered liver organoids require a significant number of functional cells. Currently, primary hepatocytes, as well as hepatocyte-like cells generated by differentiation and transdifferentiation, are not expandable *in vitro*, unless cell cycle regulation pathways were modified by genetic methods [13,65,66]. Though the p19Arf-deficient mouse iHep cells and human iHep cells expressing SV40 large T antigen did not form tumors after subcutaneous injection into immune-deficient mice, inactivation of such tumor suppressors still arouses concerns on tumorigenicity. Considering that serial transplantation experiments have shown the robust expansion potential of hepatocytes *in vivo* [67], the establishment of *in vitro* niche supporting long-term proliferation of hepatocytes and iHep cells would offer a safer approach. Such culture conditions have been developed for expansion of liver stem cells [68–70]. Nevertheless, rapid progress is being made in the expansion of hepatocytes and iHep cells, which may help obtain enough cells for clinical applications in the near future.

The second major challenge is maturation and maintenance of functional iHep cells. To date, iHep cells are not comparable to primary hepatocytes in many functional assays. Moreover, hepatocytes lining along the porto-central axis of the liver lobule are highly functionally heterogeneous [71]. Wnt, hypoxia, Ras signaling, and pituitary hormones, as well as some unknown factors, regulate the heterogeneity of hepatocytes residing in the liver lobule [71], making it even more challenging to realize all hepatic functions in a homogenous culture condition. However, advancements in 3D culture systems and engineered liver organoid may offer more complex, but more sophisticated systems for further maturation of iHep cells [72,73].

SUMMARY

Accumulating reprogramming strategies are being developed to generate hepatocytes (iHep cells) from nonhepatic cells. These iHep cells are believed to acquire many hepatic functions and are capable of integrating into liver tissues, thus offering a promising cell source for regenerative medicine. Although iHep cells are currently imperfect, these cells will likely impact the future development of regenerative medicine strategies for liver diseases.

BOX 6.1 Generation of iHep cells from human adult fibroblasts

Culture Medium

The culture media used here were described in the previous report [13]. For human fibroblast culture: DMEM/F12 (Hyclone) supplemented with 10% (v/v) fetal bovine serum, 0.1 mM β-mercaptoethanol (Sigma–Aldrich), 1×MEM nonessential amino acids solution (Gibco), and 4 ng/mL bFGF (Peprotech). For iHep cell culture: DMEM/F12 (Hyclone) supplemented with 0.544 mg/L $ZnCl_2$ (Sinopharm), 0.75 mg/L $ZnSO_4$ $7H_2O$ (Sinopharm), 0.2 mg/L $CuSO_4$ $5H_2O$ (Sinopharm), 0.025 mg/L $MnSO_4$ (Sinopharm), 2 g/L bovine serum albumin (Sigma–Aldrich), 2 g/L galactose (Sigma–Aldrich), 0.1 g/L ornithine, 0.8 g/L arginine HCl, 0.03 g/L proline, 0.61 g/L nicotinamide, 1×insulin-transferrin-sodium selenite media supplement (Sigma–Aldrich), 40 ng/mL TGFα (Peprotech), 40 ng/mL EGF (Peprotech), 10 M dexamethasone, and 1% fetal bovine serum (Sigma–Aldrich).

Procedure

1. Human dermal tissues are taken after written consent approved by an Ethics Committee of the respective institution.
2. Mince the dermal tissues into 8 mm^3 pieces with scissors. Place three pieces per 6 cm collagen-coated dish, and add 5 ml fibroblast culture medium.
3. Incubate the dermal tissue pieces for 2–4 weeks until cell culture becomes dense enough for passaging. Replace the medium with fresh fibroblast culture medium every 5 days. Fibroblasts will migrate out of the tissue pieces during this time.
   CRITICAL STEP: The tissue pieces should not be suspended during this time, or fibroblasts will fail to migrate out.
4. Split 2×10^5 fibroblasts per 6 cm collagen-coated dish. Incubate the cells with lentivirus expressing SV40 large T antigen (MOI = 2). Add 4 mg/mL polybrene solution to the final concentration of 4 μg/mL.
5. After 48 h infection, replace the medium with fresh fibroblast culture medium.
6. When fibroblast culture reaches 80%–90% confluence, suspend 2×10^5 fibroblasts expressing SV40 large T in 3 ml fibroblasts culture medium and mix with 2 ml viral supernatant containing FOXA3, HNF1A, and HNF4A lentivirus (MOI = 1.25 for each transcription factor). Add 4 mg/mL polybrene solution to the final concentration of 4 μg/mL. Incubate the mixture in a 6 cm dish.
   CRITICAL STEP: The starting fibroblast number and MOI are absolutely important for iHep cell generation.

7. After 24 h infection of three transcription factors (3TFs), aspirate the medium and add fresh human fibroblast culture medium.
8. After 48 h infection of 3TFs, replace the medium with fresh human iHep cell culture medium supplemented with 10% (v/v) fetal bovine serum.
9. From day 4 after infection of 3TFs, replace the medium daily with fresh human iHep cell culture medium.
10. Initial epithelial iHep colonies should be evident within 5–7 days. Passage the cells before overgrowth takes place.
11. For further maturation of iHep cells, seed 2×10^5 iHep cells in a 6 cm collagen-coated dish and replace fresh iHep cell culture medium every 2 days. Do not passage the cells until distinct cell borders and clear nucleus are formed. It usually takes 6–12 days for iHep cells to acquire mature phenotypes.

REFERENCES

[1] Hengstler JG, Utesch D, Steinberg P, Platt K, Diener B, Ringel M, et al. Cryopreserved primary hepatocytes as a constantly available in vitro model for the evaluation of human and animal drug metabolism and enzyme induction. Drug Metab Rev 2000;32:81–118.
[2] Baptista PM, Siddiqui MM, Lozier G, Rodriguez SR, Atala A, Soker S. The use of whole organ decellularization for the generation of a vascularized liver organoid. Hepatology 2011;53:604–17.
[3] Dhawan A, Puppi J, Hughes RD, Mitry RR. Human hepatocyte transplantation: current experience and future challenges. Nat Rev Gastroenterol Hepatol 2010;7:288–98.
[4] Struecker B, Raschzok N, Sauer IM. Liver support strategies: cutting-edge technologies. Nat Rev Gastroenterol Hepatol 2014;11:166–76.
[5] Berry MN, Edwards AM. The hepatocyte review. Dordrecht, Boston: Kluwer Academic Publishers; 2000.
[6] Cai J, Zhao Y, Liu YX, Ye F, Song ZH, Qin H, et al. Directed differentiation of human embryonic stem cells into functional hepatic cells. Hepatology 2007;45:1229–39.
[7] Kubo A, Shinozaki K, Shannon JM, Kouskoff V, Kennedy M, Woo S, et al. Development of definitive endoderm from embryonic stem cells in culture. Development 2004;131:1651–62.
[8] Si-Tayeb K, Noto FK, Nagaoka M, Li JX, Battle MA, Duris C, et al. Highly efficient generation of human hepatocyte-like cells from induced pluripotent stem cells. Hepatology 2010;51:297–305.
[9] Song ZH, Cai J, Liu YX, Zhao DX, Yong J, Duo SG, et al. Efficient generation of hepatocyte-like cells from human induced pluripotent stem cells. Cell Res 2009;19:1233–42.
[10] Yamada T, Yoshikawa M, Kanda S, Kato Y, Nakajima Y, Ishizaka S, et al. In vitro differentiation of embryonic stem cells into hepatocyte-like cells identified by cellular uptake of indocyanine green. Stem Cells 2002;20:146–54.
[11] Du Y, Wang J, Jia J, Song N, Xiang C, Xu J, et al. Human hepatocytes with drug metabolic function induced from fibroblasts by lineage reprogramming. Cell Stem Cell 2014;14:394–403.
[12] Huang P, He Z, Ji S, Sun H, Xiang D, Liu C, et al. Induction of functional hepatocyte-like cells from mouse fibroblasts by defined factors. Nature 2011;475:386–9.
[13] Huang P, Zhang L, Gao Y, He Z, Yao D, Wu Z, et al. Direct reprogramming of human fibroblasts to functional and expandable hepatocytes. Cell Stem Cell 2014;14:370–84.

[14] Sekiya S, Suzuki A. Direct conversion of mouse fibroblasts to hepatocyte-like cells by defined factors. Nature 2011;475:390-3.
[15] Selman K, Kafatos FC. Transdifferentiation in the labial gland of silk moths: is DNA required for cellular metamorphosis? Cell Differ 1974;3:81-94.
[16] Scarpelli DG, Rao MS. Differentiation of regenerating pancreatic cells into hepatocyte-like cells. Proc Natl Acad Sci USA 1981;78:2577-81.
[17] Shen CN, Horb ME, Slack JM, Tosh D. Transdifferentiation of pancreas to liver. Mech Dev 2003;120:107-16.
[18] Ferrari G, Cusella-De Angelis G, Coletta M, Paolucci E, Stornaiuolo A, Cossu G, et al. Muscle regeneration by bone marrow derived myogenic progenitors. Science 1998;279:1528-30.
[19] Gussoni E, Soneoka Y, Strickland CD, Buzney EA, Khan MK, Flint AF, et al. Dystrophin expression in the mdx mouse restored by stem cell transplantation. Nature 1999;401:390-4.
[20] Jiang YH, Jahagirdar BN, Reinhardt RL, Schwartz RE, Keene CD, Ortiz-Gonzalez XR, et al. Pluripotency of mesenchymal stem cells derived from adult marrow. Nature 2002;418:41-9.
[21] Mezey E, Chandross KJ, Harta G, Maki RA, McKercher SR. Turning blood into brain: cells bearing neuronal antigens generated in vivo from bone marrow. Science 2000;290:1779-82.
[22] Lee KD, Kuo TK, Whang-Peng J, Chung YF, Lin CT, Chou SH, et al. In vitro hepatic differentiation of human mesenchymal stem cells. Hepatology 2004;40:1275-84.
[23] Lagasse E, Connors H, Al-Dhalimy M, Reitsma M, Dohse M, Osborne L, et al. Purified hematopoietic stem cells can differentiate into hepatocytes in vivo. Nat Med 2000;6:1229-34.
[24] Lysy PA, Smets F, Sibille C, Najimi M, Sokal EM. Human skin fibroblasts: from mesodermal to hepatocyte-like differentiation. Hepatology 2007;46:1574-85.
[25] Davis RL, Weintraub H, Lassar AB. Expression of a single transfected cDNA converts fibroblasts to myoblasts. Cell 1987;51:987-1000.
[26] Takahashi K, Yamanaka S. Induction of pluripotent stem cells from mouse embryonic and adult fibroblast cultures by defined factors. Cell 2006;126:663-76.
[27] Heng JCD, Feng B, Han JY, Jiang JM, Kraus P, Ng JH, et al. The nuclear receptor Nr5a2 can replace Oct4 in the reprogramming of murine somatic cells to pluripotent cells. Cell Stem Cell 2010;6:167-74.
[28] Shu J, Wu C, Wu Y, Li Z, Shao S, Zhao W, et al. Induction of pluripotency in mouse somatic cells with lineage specifiers. Cell 2013;153:963-75.
[29] Hong H, Takahashi K, Ichisaka T, Aoi T, Kanagawa O, Nakagawa M, et al. Suppression of induced pluripotent stem cell generation by the p53-p21 pathway. Nature 2009;460:1132-5.
[30] Li H, Collado M, Villasante A, Strati K, Ortega S, Canamero M, et al. The Ink4/Arf locus is a barrier for iPS cell reprogramming. Nature 2009;460. 1136-U1101.
[31] Utikal J, Polo JM, Stadtfeld M, Maherali N, Kulalert W, Walsh RM, et al. Immortalization eliminates a roadblock during cellular reprogramming into iPS cells. Nature 2009;460. 1145-U1112.
[32] Mikula M, Fuchs E, Huber H, Beug H, Schulte-Hermann R, Mikulits W. Immortalized p19(ARF null) hepatocytes restore liver injury and generate hepatic progenitors after transplantation. Hepatology 2004;39:628-34.
[33] Rezvani M, Espanol-Suner R, Malato Y, Dumont L, Grimm AA, Kienle E, et al. In vivo hepatic reprogramming of myofibroblasts with AAV vectors as a therapeutic strategy for liver fibrosis. Cell Stem Cell 2016;18:809-16.
[34] Song GQ, Pacher M, Balakrishnan A, Yuan QG, Tsay HC, Yang DK, et al. Direct reprogramming of hepatic myofibroblasts into hepatocytes in vivo attenuates liver fibrosis. Cell Stem Cell 2016;18:797-808.
[35] Ahuja D, Saenz-Robles MT, Pipas JM. SV40 large T antigen targets multiple cellular pathways to elicit cellular transformation. Oncogene 2005;24:7729-45.

[36] Hou P, Li Y, Zhang X, Liu C, Guan J, Li H, et al. Pluripotent stem cells induced from mouse somatic cells by small-molecule compounds. Science 2013;341:651–4.
[37] Hu WX, Qiu BL, Guan WQ, Wang QY, Wang M, Li W, et al. Direct conversion of normal and Alzheimer's disease human fibroblasts into neuronal cells by small molecules. Cell Stem Cell 2015;17:204–12.
[38] Li X, Zuo XH, Jing JZ, Ma YT, Wang JM, Liu DF, et al. Small-molecule-driven direct reprogramming of mouse fibroblasts into functional neurons. Cell Stem Cell 2015;17:195–203.
[39] Cheng L, Hu W, Qiu B, Zhao J, Yu Y, Guan W, et al. Generation of neural progenitor cells by chemical cocktails and hypoxia. Cell Res 2014;24:665–79.
[40] Pu WJ, Zhang H, Huang XZ, Tian XY, He LJ, Wang Y, et al. Mfsd2a(+) hepatocytes repopulate the liver during injury and regeneration. Nat. Commun. 2016;7.
[41] Cao N, Huang Y, Zheng JS, Spencer CI, Zhang Y, Fu JD, et al. Conversion of human fibroblasts into functional cardiomyocytes by small molecules. Science 2016;352:1216–20.
[42] Fu YB, Huang CW, Xu XX, Gu HF, Ye YQ, Jiang CZ, et al. Direct reprogramming of mouse fibroblasts into cardiomyocytes with chemical cocktails. Cell Res 2015;25:1013–24.
[43] Wang YF, Qin JH, Wang SY, Zhang WC, Duan JL, Zhang J, et al. Conversion of human gastric epithelial cells to multipotent endodermal progenitors using defined small molecules. Cell Stem Cell 2016;19:449–61.
[44] Hotta A, Ellis J. Retroviral vector silencing during iPS Cell Induction: an epigenetic Beacon that signals distinct Pluripotent States. J Cell Biochem 2008;105:940–8.
[45] Fusaki N, Ban H, Nishiyama A, Saeki K, Hasegawa M. Efficient induction of transgene-free human pluripotent stem cells using a vector based on Sendai virus, an RNA virus that does not integrate into the host genome. Proc Jpn Acad B-Phys 2009;85:348–62.
[46] Seki T, Yuasa S, Oda M, Egashira T, Yae K, Kusumoto D, et al. Generation of induced pluripotent stem cells from human terminally differentiated circulating T cells. Cell Stem Cell 2010;7:11–4.
[47] Stadtfeld M, Nagaya M, Utikal J, Weir G, Hochedlinger K. Induced pluripotent stem cells generated without viral integration. Science 2008;322:945–9.
[48] Gonzalez F, Monasterio MB, Tiscornia G, Pulido NM, Vassena R, Morera LB, et al. Generation of mouse-induced pluripotent stem cells by transient expression of a single nonviral polycistronic vector. Proc Natl Acad Sci USA 2009;106:8918–22.
[49] Jia FJ, Wilson KD, Sun N, Gupta DM, Huang M, Li ZJ, et al. A nonviral minicircle vector for deriving human iPS cells. Nat Methods 2010;7. 197-U146.
[50] Okita K, Nakagawa M, Hong HJ, Ichisaka T, Yamanaka S. Generation of mouse induced pluripotent stem cells without viral vectors. Science 2008;322:949–53.
[51] Kim J, Kim KP, Lim KT, Lee SC, Yoon J, Song G, et al. Generation of integration-free induced hepatocyte-like cells from mouse fibroblasts. Sci Rep-UK 2015;5.
[52] Warren L, Manos PD, Ahfeldt T, Loh YH, Li H, Lau F, et al. Highly efficient reprogramming to pluripotency and directed differentiation of human cells with synthetic modified mRNA. Cell Stem Cell 2010;7:618–30.
[53] Kim D, Kim CH, Moon JI, Chung YG, Chang MY, Han BS, et al. Generation of human induced pluripotent stem cells by direct delivery of reprogramming proteins. Cell Stem Cell 2009;4:472–6.
[54] Zhou HY, Wu SL, Joo JY, Zhu SY, Han DW, Lin TX, et al. Generation of induced pluripotent stem cells using recombinant proteins. Cell Stem Cell 2009;4:381–4.
[55] Friedenstein AJ, Petrakova KV, Kurolesova AI, Frolova GP. Heterotopic of bone marrow. Analysis of precursor cells for osteogenic and hematopoietic tissues. Transplantation 1968;6:230–47.
[56] Zuk PA, Zhu M, Mizuno H, Huang J, Futrell JW, Katz AJ, et al. Multilineage cells from human adipose tissue: implications for cell-based therapies. Tissue Eng 2001;7:211–28.

[57] Williams JT, Southerland SS, Souza J, Calcutt AF, Cartledge RG. Cells isolated from adult human skeletal muscle capable of differentiating into multiple mesodermal phenotypes. Am Surgeon 1999;65:22–6.

[58] Anker PSI, Noort WA, Scherjon SA, Kleuburg-van der Keur C, Kruisselbrink AB, van Bezoolien RL, et al. Mesenchymal stem cells in human second-trimester bone marrow, liver, lung, and spleen exhibit a similar immunophenotype but a heterogeneous multilineage differentiation potential. Haematologica 2003;88:845–52.

[59] Squillaro T, Peluso G, Galderisi U. Clinical trials with mesenchymal stem cells: an update. Cell Transplantation 2016;25:829–48.

[60] Zhou T, Benda C, Dunzinger S, Huang YH, Ho JC, Yang JY, et al. Generation of human induced pluripotent stem cells from urine samples. Nat Protoc 2012;7:2080–9.

[61] Zhou T, Benda C, Duzinger S, Huang YH, Li XY, Li YH, et al. Generation of induced pluripotent stem cells from urine. J Am Soc Nephrol 2011;22:1221–8.

[62] Azuma H, Paulk N, Ranade A, Dorrell C, Al-Dhalimy M, Ellis E, et al. Robust expansion of human hepatocytes in Fah-/-/Rag2-/-/Il2rg-/- mice. Nat. Biotechnol. 2007;25:903–10.

[63] Rhim JA, Sandgren EP, Degen JL, Palmiter RD, Brinster RL. Replacement of diseased mouse liver by hepatic cell transplantation. Science 1994;263:1149–52.

[64] Shi XL, Gao YM, Yan YP, Ma HC, Sun LL, Huang PY, et al. Improved survival of porcine acute liver failure by a bioartificial liver device implanted with induced human functional hepatocytes. Cell Res 2016;26:206–16.

[65] Kobayashi N, Fujiwara T, Westerman KA, Inoue Y, Sakaguchi M, Noguchi H, et al. Prevention of acute liver failure in rats with reversibly immortalized human hepatocytes. Science 2000;287:1258–62.

[66] Levy G, Bomze D, Heinz S, Ramachandran SD, Noerenberg A, Cohen M, et al. Long-term culture and expansion of primary human hepatocytes. Nat Biotechnol 2015;33:1264.

[67] Overturf K, al-Dhalimy M, Ou CN, Finegold M, Grompe M. Serial transplantation reveals the stem-cell-like regenerative potential of adult mouse hepatocytes. Am. J. Pathol. 1997;151:1273–80.

[68] Huch M, Dorrell C, Boj SF, van Es JH, Li VSW, van de Wetering M, et al. In vitro expansion of single Lgr5(+) liver stem cells induced by Wnt-driven regeneration. Nature 2013;494:247–50.

[69] Huch M, Gehart H, van Boxtel R, Hamer K, Blokzijl F, Verstegen MMA, et al. Long-term culture of genome-stable bipotent stem cells from adult human liver. Cell 2015;160:299–312.

[70] Katsuda T, Kawamata M, Hagiwara K, Takahashi RU, Yamamoto Y, Camargo FD, et al. Conversion of terminally committed hepatocytes to culturable bipotent progenitor cells with regenerative capacity. Cell Stem Cell 2017;20:41–55.

[71] Halpern KB, Shenhav R, Matcovitch-Natan O, Toth B, Lemze D, Golan M, et al. Single-cell spatial reconstruction reveals global division of labour in the mammalian liver. Nature 2017.

[72] Brophy CM, Luebke-Wheeler JL, Amiot BP, Khan H, Remmel RP, Rinaldo P, et al. Rat hepatocyte spheroids formed by rocked technique maintain differentiated hepatocyte gene expression and function. Hepatology 2009;49:578–86.

[73] Takebe T, Sekine K, Enomura M, Koike H, Kimura M, Ogaeri T, et al. Vascularized and functional human liver from an iPSC-derived organ bud transplant. Nature 2013;499:481.

Chapter 7

Generation of Liver Organoids and Their Potential Applications

Li-Ping Liu*,**, Yu-Mei Li*, Ning-Ning Guo*, Lu-Yuan Wang*, Hiroko Isoda**, Nobuhiro Ohkohchi**, Hideki Taniguchi[†], Yun-Wen Zheng*,**,[†]

*The Affiliated Hospital of Jiangsu University, Zhenjiang, Jiangsu, China; **University of Tsukuba, Ibaraki, Japan; [†]Yokohama City University, Yokohama, Japan

1 INTRODUCTION

Understanding the physiological functions and pathophysiology of the human body begins with studying in vitro tissue. However, the sources of human tissues are subject to various restrictions, such as tissue preservation and the need for prompt acquisition. Animal models could substitute for this shortage. However, it is difficult to compensate for species differences between humans and animals.

The development of stem cell research, especially the generation of human-induced pluripotent stem cells (iPSCs), is expected to address this issue. Much progress has been made in the differentiation of iPSCs to various somatic cell types, including hepatocytes. To study liver functions, most of the conventional two-dimensional (2D) methods are unlikely to recapitulate the complex interactions and functions found in vivo. As tissues and organs in the human body form as a result of an intricate exchange of signals between cell–cell and cell–environment, three-dimensional (3D) organoid models could establish and maintain the intercellular and cell–environment communication and thus mimic the organization and function of organs in vivo. Functional tissue-specific organoids, such as optic cup [1], brain [2], stomach [3], intestine [4], liver [5], pancreas [6], kidney [7], and prostate [8], have been generated using iPSCs and adult stem cells (ASCs). These organoids have shown similar physiogenesis and morphogenesis compared to the normal tissues.

Currently established liver organoid models are summarized in Table 7.1. During the process of organoid formation, a large number of biological parameters are required. These include the spatial organization of heterogeneous tissue-specific cells, cell–cell, and cell–matrix interactions, and certain physiological functions of tissue-specific cells [9]. Liver organoids can be derived directly by liver tissue isolation, and can also be independently established by

TABLE 7.1 Function and Application of Liver-related Organoids From Different Species, Cell Sources, and Culture Conditions

Species	Derivation or cell source		Function and application	Culture condition: surface type (coating); growth factor; and other supplements	Reference
Mouse	Hepatocytes or progenitor cells	Hepatocytes	(1) Tx	Suspended: dexamethasone, EGF	Hickey RD 2015 [75]
		LPCs	(1) Protein expression: ALB, CK19	Scaffold	Minguet S 2003 [76]
			(1) Protein expression: CK18, HNF4α; (2) Tx	Suspended: insulin-like growth factor-II, EGF, insulin	Yap KK 2013 [77]
			(1) Gene expression: ductal progenitor markers; (2) protein expression: SOX9, CK19, HNF1β, HNF4α, ALB; (3) Tx	Embedded (matrigel): N2 supplement, B27, nicotinamide, dexamethasone, Y27632, rmEGF, rmHGF, rmWnt3a, and rhRspo1	Yimlamai D 2014 [78]
		FLCs	(1) Gene expression: *Afp*, *Alb*, α1-antitrypsin, glucose-6 phosphatase, *Ck19*, *Bgp*; (2) protein expression: ALB, AFP, CK7; (3) differentiation potential; (4) Tx	Suspended: B27, ITS-X, EGF, bFGF, HGF	Tsuchiya A 2005 [79]
	Cholangiocytes		(1) Protein expression: integrin b-1 subunit, E-cadherin, ZO-1, cellular tubulin; (2) biliatresone washout experiments	Embedded (collagen-Matrigel mixture)	Waisbourd-Zinman O 2016 [80]
	Ductal cells		(1) Differentiation potential	Embedded (Matrigel): EGF, R-spondin, Noggin, Wnt3a, FGF-10, HGF, [Leu15]-Gastrin, nicotinamide, N-acetylcysteine, ADF, N2, B27.	Flanagan DJ 2016 [81]
	ESCs/iPSCs/MSCs derived hepatocyte-like cells		(1) Gene expression: *Alb*, *Ist-1*, *Tat*, *Cyp7A1*, *Mrp2*; (2) differentiation potential; (3) protein expression: ALB, AAT, CK-18, ECAD, LST-1; (4) glycogen storage; (5) indocyanine green uptake; (6) Tx	Suspended/embedded (collagen Type I)	Gabriel E 2012 [82]
			(1) Ammonia removal activity; (2) albumin concentration (ELISA); (3) ex vivo experiment	Scaffold	Mizumoto H 2012 [83]
		ESCs	(1) Glucose levels; (2) albumin concentration (ELISA); (3) activity of ammonia detoxification; (4) urea synthesis ability	Scaffold: transferrin, hydrocortisone-21-hemisuccinate, ascorbic acid, insulin, niacinamide, dexamethasone	Teratani T 2005 [84]
			(1) Gene expression: *Sox17*, *Hnf3β*, *Afp*, *Alb*, *Aat*, *Glut2*, *Gys2*, *Tat*, *Lst-1*, *CK8*, *CK18*, *CK19*, *CD34*, etc.; (2) protein expression: AAT, AFP, ALB, CK18, CD34, c-KIT, THY-1, DLK; (3) differentiation potential; (4) PAS staining	Suspended/embedded (fibronectin)	Drobinskaya I 2008 [85]
		MSCs	(1) Gene expression: E-cadherin, *Cox 32*; (2) protein expression: pFAK, E-cadherin; (3) Albumin secretion (ELISA); (4) urea secretion	Scaffold: dexamethasone, ITS+ premix, HGF, OSM	Roh H 2015 [86]

Human and mouse	Coculture	HepaRG cells and human HSCs	(1) Gene expression: *Alb, Cy3a4, Gsta1, Slco1b1*; (2) Phase I hepatocyte metabolic capacity; (3) albumin secretion; (4) drug responsiveness	Suspended (Greiner)	Leite SB 2016 [87]
		Hepatocytes and NIH3T3	(1) Albumin secretion (ELISA); (2) ammonia elimination; (3) inducibility of cytochrome P450-CYP1A1 enzyme activity	Scaffold: EGF, dexamethasone, insulin	Seo SJ 2006 [88]
	Liver tissue	Liver organoid unit of liver tissue	(1) Protein expression: ALB, HNF4α, CK8; (2) structure of bile ducts and blood vessels; (3) Tx	Scaffold (type I collagen)	Mavila N 2017 [89]
	Gallbladder cells		(1) Gene expression: *Cd44, Prom1, Sox17, Prom1, Cldn3, Epcam, Itga6*, etc. (2) differentiation potential; (3) PAS staining; (4) Tx	embedded (Matrigel): nicotinamide, EGF, FGF10, HGF, R-spondin 1, noggin	Lugli N 2016 [90]
		Hepatocytes	(1) Gene expression (transcriptomic analyses): endogenous and xenobiotic metabolism, phase I enzymes, transporters; (2) sensitivity to model DILI compounds; (3) compound-specific transcriptional toxicity effects	Not given; insulin, transferrin, dexamethasone	Bell CC 2017 [91]
		Hepatocytes	(1) Albumin secretion (ELISA); (2) Alpha-1-antitrypsin production (ELISA); (3) urea production; (4) protein expression: ALB, alpha-1-antitrypsin, CK18	Scaffold (poly(vinyl alcohol)): insulin, dexamethasone, EGF	Török E 2011 [92]
			(1) Protein expression: HNF4α, CK18, ALB, CYP4503A; (2) CYP450 enzyme inductions; (3) urea secretion; (4) albumin secretion; (5) Phase I and II Enzyme Expression and Activity	Suspended	Tostões RM 2012 [93]
Human	Primary hepatocytes/ progenitor cells/cell lines		(1) Lactate measurements; (2) oxygen concentration; (3) exposure to Antimycin A and Bosentan	Suspended: HGF, EGF	Weltin A 2017 [94]
			(1) Gene expression: *HNF4A, PXR, ALB, GS, CPS1, CYP1A2, YP3A4, CYP2C9*; (2) protein expression: ALB, HNF4α; (3) albumin secretion (ELISA); (4) ammonia detoxification; (5) P450 activity; (6) Efflux transporter activity	Suspended	Rebelo SP 2014 [95]
		HepaRG cells	(1) Gene expression: HNF4α, TAT, PXR, CYP 450s, GSTA1, UGT1A1, UGT1A2, etc; (2) drug toxicity and metabolism evaluation	Scaffold (gelatin)	Wang J 2016 [96]
			(1) Gene expression: HNF4A, ALB, CYP3A4, KRT19, MDR1, MRP2; (2) CYP3A4 enzyme activity; (3) CYP3A4 induction; (4) formation of bile canaliculi; (5) functional polarity	Embedded (nanofibrillar cellulose and hyaluronan–gelatin hydrogels)	Malinen MM 2014 [97]
		FLCs	(1) Protein expression: AFP, CK19, CK18	Embedded (collagen): EGF	Lázaro CA 2003 [98]
	Bile duct cells		(1) Gene expression: *PROM1, LGR5, SOX9, OC2, HNF4α, KRT19, KRT7, EPHB2*; (2) differentiation potential; (3) gene expression after differentiation: *ALB, CYP3A4*, cytochromes, *APOB*, complement factors (C3); (4) protein expression: ALB, MRP4; (5) glycogen accumulation; (6) LDL uptake; (7) albumin production; (8) CYP3A4 activity; (9) Midazolam metabolism; (10) Tx	Embedded (Matrigel or Basement Membrane Extract, Type 2): N2, B27, N-Acetylcysteine, gastrin, EGF, RSPO1 conditioned medium, FGF10, HGF, Nicotinamide, A83.01, Forskolin, Noggin, Wnt, etc	Huch M 2015 [71]

(*Continued*)

TABLE 7.1 Function and Application of Liver-related Organoids From Different Species, Cell Sources, and Culture Conditions (cont.)

Species	Derivation or cell source	Function and application	Culture condition: surface type (coating); growth factor; and other supplements	Reference
	ESCs	(1) Gene expression: *FOXA2, HEX, AFP, HNF4a, C/EBPb, ALB*; (2) protein expression: CK19, CK18, AFP, ALB; (3) differentiation potential; (4) GGT activity; (5) glycogen accumulation; (6) ICG uptake; (7) urea secretion; (8) albumin secretion; (9) cytochrome P450 3A4 activity	Suspended/embedded (collagen type I); B27, Wnt3a, BMP4, insulin, transferrin, ITS, HGF, OSM, DEX	Kim SE 2013 [99]
		(1) Gene expression: *UCT1A1, ALB, ARG1, HNF4a, FVII, CX32*; (2) protein expression: ASGPR, PEPCK, AFP, ALB, CK8, CK18; (3) ALB secretion (ELISA); (4) CYP450 activities; (5) Biliary secretion	Suspended	Subramanian K 2014 [100]
	iPSCs	(1) Protein expression: EpCAM; (2) differentiation potential	Suspended: Activin A, B27, HGF, FGF-4	Tian L 2016 [101]
		(1) Protein expression: CEBPα, AAT, fibrinogen, HNF4α, CK19, DLK1, LGR5, AAT, CK7b, AQP1+, ALB; (2) hematopoietic properties; (3) neuronal niche	Embedded (Matrigel)	Guye P 2016 [55]
ESCs/iPSCs/ MSCs derived hepatocyte-like cells		(1) Gene expression: *AFP, CYP1A1, ALB, CYP3A4, CYP1A2, ASGPR, MRP2, AAT, HNF4α*; (2) albumin production (ELISA); (3) urea production; (4) CYP1A2 and CYP3A4 activity and induction; (5) response to paradigm hepatotoxicants	Scaffold: rh/m/r Activin-A, rmWnt3a, rhF-GF2, rhBMP4, rmFollistatin-288, rmFGF8b, rhFGF4, rhFGF1, rhHGF, rhOSM	Tasnim F 2016 [102]
	iPSCs/ESCs	(1) Gene expression: *ALB*, CYP enzymes, conjugating enzymes, hepatic transporters, hepatic nuclear receptors, and hepatic transcription factors, bile canaliculi transporters; (2) ALB secretion; (3) urea secretion; (4) drug metabolism capacity; (5) CYP induction potency; (6) susceptibility to hepatotoxic drugs	Embedded (Matrigel): dexamethasone, HGF, FGF1, FGF4, FGF10, OSM	Takayama K 2012 [56]
		(1) Gene expression: *HNF4α, AFP, AAT, ALB, NR, GST, CYPs*, etc.; (2) protein expression: ALB, AFP, ASGPR1, CK18, HNF4α; (3) PAS staining; (4) LDL and ICG uptake assay; (5) ALB secretion; (6) urea secretion; (7) bile canaliculi junctional formation; (8) CYP activities; (9) metabolizing activity by exposure to xenobiotics	Dexamethasone, HGF, ITS medium, OSM	Kim JH 2015 [103]
	MSCs	(1) Gene expression: *CK18, AFP, CYP3A4, HNF4A, CYP1A1, CEBPA, ALB*, etc. (2) albumin secretion (ELISA); (3) urea secretion; (4) diclofenac cytotoxicity; (5) PAS staining; (6) CYP activity assays; (7) UGTs activity assay; (8) bupropion and diclofenac conversion	Suspended (with or without collagen): dexamethasone, EGF, FGF-2, HGF, FGF-4, nicotinamide, ITS, OSM	Cipriano M 2017 [104]

	Cells	Assays	Culture conditions	Reference
Coculture	PHH and NPCs	(1) Albumin secretion; (2) drug-metabolizing enzymes activity (induction)	Suspended: insulin, transferrin, sodium selenite, dexamethasone	Bell CC 2016 [105]
	PHH and human ADSCs	(1) Albumin secretion; (2) urea secretion; (3) activity of cytochrome P450 and CYP3A4 enzymatic; (4) gene expression: *CYP3A7, ALB, CYP3A4*, glutamine synthetase, *CYP1B1, CK18*.	Suspended (BSA)	No da Y 2012 [50]
	Hepatocytes and NPC fraction	(1) Protein expression: CK7, CK19; (2) albumin secretion (ELISA)	Scaffold (collagen): nicotinamide, ascorbic acid 2-phosphate, HGF, EGF, insulin, dexamethasone	Sugimoto S 2005 [106]
	Hepatocytes, LSECs, and MSCs	(1) Gene expression: *CK8, CK18, ALB*, etc; (2) drug metabolizing enzymes	Embedded (Matrigel)	Ramachandran SD 2015 [53]
	Hepatocytes, LSECs, and MSCs	(1) Albumin production; (2) glucose concentration; (3) urea secretion	Embedded (Matrigel)	Mattei G 2017 [107]
	Hepatocytes and MSCs	(1) Albumin secretion (ELISA); (2) urea synthesis; (3) CYP450 activity; (4) Efflux transporter activity; (5) repeated-dose toxicity testing	Suspended	Rebelo SP 2015 [108]
	HepaRG cells, HUVECs, HSCs, and primary macrophages	(1) Glucose consumption; (2) lactate formation; (3) albumin synthesis; (4) urea synthesis	Biochip	Gröger M 2016 [51]
	HepaRG cells and HSCs	(1) Gene expression: *ALB, CY3A4, GSTA1, SLCO1B1*; (2) Phase I hepatocyte metabolic capacity; (3) albumin secretion; (4) drug responsiveness	Suspended (Greiner)	Leite SB 2015 [87]
	HepaRG, HUVECs, PBMCs, and LX-2 stellate cells	(1) Protein expression: ZO-1, transferrin, ASGPR-1, MRP-2; (2) formation of hepatocyte microvilli; (3) CYP3A4 metabolite formation	Biochip	Rennert K 2015 [109]
	FLCs and HUVECs	(1) Protein expression: AFP, CYP2A, CYP3A, CK18, CK19, ALB; (2) albumin concentrations; (3) urea secretion	Scaffold (matrigel): dexamethasone, cAMP, hProlactin, hGlucagon, niacinamide, alipoic acid, hEGF, hHGF, hGH, triiodothyronine, hHDL	Baptista PM 2011 [42]
	PHH or iPS-derived hepatocytes and 3T3-J2 murine fibroblasts	(1) Cytochrome P450 enzyme activity; (2) protein expression: HNF4α, ALB, HNF1β; (3) inducible CYP1A1 and CYP2C9 activity	Suspended (pluronic F-127): glucagon, hydrocortisone	Schepers A 2016 [110]
	iPSC-hepatic endoderm cells, MSCs, and HUVECs	(1) Gene expression: *AFP, RBP4, TTR, ALB*, etc. (2) protein expression: AFP, ALB; (3) ALB and AAT production (ELISA); (4) drug metabolism activity; (5) Tx	Embedded (matrigel): dexamethasone, OSM, HGF, singleQuots	Takebe T 2013 [5]

(*Continued*)

TABLE 7.1 Function and Application of Liver-related Organoids From Different Species, Cell Sources, and Culture Conditions (cont.)

Species	Derivation or cell source		Function and application	Culture condition: surface type (coating); growth factor; and other supplements	Reference
Rat	Tissue	Liver tissues	(1) Protein expression: ALB	Scaffold (type I collagen)	Takezawa T 2000 [111]
		Mixed hepatic cells	(1) Protein expression: CK19, GST-P, ALB, transferrin, CK8, CK18, Cx32, glycogen, AFP, CK7, CK19, Cx43, etc. (2) gene expression: *Alb*, *Cx32*, *TO*; (3) bile canalicular structures	Embedded (rat tail collagen): EGF, nicotinamide, ascorbic acid, 2-phosphate, dimethyl sulfoxide	Mitaka T 1999 [112]
			(1) Tx	Suspended: dexamethasone, HGF, EGF	Michalopoulos GK 2002 [113]
	Hepatocytes/small hepatocytes/liver cells	Hepatocytes	(1) Biliary excretion; (2) albumin secretion; (3) urea synthesis; (4) maintenance of drug-metabolizing enzymes; (5) drug transporter expression; (6) drug absorption experiment	Scaffold	Nugraha B 2011 [114]
			(1) Albumin synthesis (ELISA); (2) urea synthesis; (3) glucose concentration measurement; (4) CYP450 measurement	Suspended	Tostões RM 2011 [115]
			(1) Albumin production (ELISA); (2) urea production; (3) excretory function of Mrp2; (4) CYP induction assay	Sandwich (Type I collagen): EGF, insulin, dexamethasone, linoleic acid	Xia L 2012 [116]
			(1) Gene expression: *Cdh1*, *Nags*, *Ass*, *Asl*, *Otc*, *Arg1*, etc; (2) albumin concentration (ELISA); (3) Cytochrome P450 activity	Suspended	Brophy CM 2009 [117]
			(1) Albumin secretion; (2) urea synthesis; (3) cytochrome P450 1A activity	Embedded (collagen I): glucagons, hydrocortisone, insulin	Chen AA 2009 [118]
			(1) Albumin secretion; (2) urea syntheses; (3) activities of CYP3A11, CYP2C9, and phase II enzymes; (4) drug metabolism; (5) drug toxicity tests; (6) induction and inhibition test of hepatocyte metabolism; (7) induction and inhibition tests of drug hepatotoxicity	Scaffold	Yan S 2015 [119]
			(1) Urea secretion; (2) albumin secretion; (3) gene expression: *Alb*, *Cps1*, *Arg1*, *Asl*, *Ass1*	Embedded (Aminated-ELP)	Weeks CA 2016 [120]
			(1) Tx	Suspended	Uchida S 2013 [121]
			(1) Tx	Suspended: EGF, insulin, dexamethasone, linoleic acid	Tong WH 2015 [122]
			(1) Albumin secretion (ELISA); (2) urea secretion; (3) protein expression: ALB; (4) gene expression: *G6Pase*, *Tat*	Suspended: hydrocortisone, EGF, glucagon, insulin	Siltanen C 2017 [123]

(1) Albumin secretion (ELISA); (2) urea production; (3) protein expression: albumin; (4) gene expression: *Alb, G6Pase, Tat*	Suspended: hydrocortisone, EGF, glucagon, insulin	Santoh M 2016 [124]
(1) Gene expression: CYP enzymes, sulfotransferase, UDP-glucuronosyltransferases; (2) hepatotoxicity induced by APAP	Suspended (poly-L-lysine)	Sanoh S 2014 [125]
(1) Protein expression: CK-18, ALB, HNF4-α, OATP-C, MRP2; (2) Phase I and Phase II activities; (3) levels of Phase I and II metabolites	Not given	Pinheiro PF 2017 [126]
(1) Albumin secretion; (2) urea synthesis; (3) 3MC-induced EROD activity; (4) response to APAP-induced hepatotoxicity	Embedded (hybrid RGD/galactose substratum): EGF, dexamethasone, insulin, linoleic acid	Du Y 2006 [127]
(1) Gene expression: *Letf6, Hnf-4, C/Ebp-β, Alb, Cps, Otc, Ass, Asl, Arg*; (2) albumin secretion; (3) ammonia removal	Suspended: insulin, hydrocortisone, EGF, proline, linoleic acid	Fukuda J 2006 [128]
(1) Gene expression: *Alb, Cyp2b2, Cyp8b1, Ck19, Dmbt1, Jagged1, Notch1, Ttr, claudin2, Afp*, etc.; (2) protein expression: ALB, HNF-1, HNF-4α, CK8, CK19, CK20	Embedded (type I collagen): insulin, EGF	Nishikawa Y 2005 [129]
(1) Glucose consumption and lactate production; (2) albumin secretion (ELISA); (3) urea secretion	Scaffold: EGF, dexamethasone, bovine insulin, ascorbic acid	Dvir-Ginzberg M 2004 [130]
(1) Albumin secretion (ELISA); (2) urea synthesis; (3) P450 enzymatic activity	Scaffold: EGF, dexamethasone, insulin, linoleic acid	Chua KN 2005 [131]
(1) Gene expression: *Cyp2b1, Cyp2b2*; (2) protein expression: CYP2B1/2; (3) distribution of PROD activity	Suspended: EGF, dexamethasone, insulin, linoleic acid	Tzanakakis ES 2001 [132]
(1) CYP1A1 activity; (2) measurement of EROD activity	Embedded (Primaria or collagen): insulin, EGF, linoleic acid	Wu FJ 1999 [133]
(1) CYP1A1/2 activity; (2) measurement of EROD activity	Suspended: dexamethasone, EGF, glucagon, tripeptide Gly-His-Lys, apo-transferrin, linoleic acid	Hsiao CC 1999 [134]
(1) Albumin secretion; (2) ammonium metabolism; (3) urea synthesis; (4) gluconeogenesis; (5) Tx	Embedded (PVLA-RPU): EGF, insulin	Sato Y 1994 [135]
(1) Protein expression: ALB, transferrin, P4501A1; (2) Cytochrome P450 induction; (3) Lidocaine metabolism; (4) UDP-glucuronyltransferase activity	Embedded (poly-HEMA): glucagon, dexamethasone, insulin, prolactin, EGF, somatotropin, linoleic acid, selenous acid	Tong JZ 1992 [136]

(Continued)

TABLE 7.1 Function and Application of Liver-related Organoids From Different Species, Cell Sources, and Culture Conditions (cont.)

Species	Derivation or cell source	Function and application	Culture condition: surface type (coating); growth factor; and other supplements	Reference
	Liver cells	(1) Protein expression: ALB, transferrin, AFP	Embedded (poly-HEMA): glucagon, dexamethasone, insulin, prolactin, EGF, somatotropin, linoleic acid, selenous acid	Tong JZ 1990 [137]
		(1) Albumin production; (2) tyrosine aminotransferase activity	Embedded (poly-HEMA)	Landry J 1985 [138]
		(1) Albumin secretion	Embedded (collagen, fibronectin, laminin): insulin, selenium, EGF	Koide N 1990 [139]
		(1) Gene expression: *Pcna*, *CD34*, *Afp*, *Alb*, *Tat*, *Pepck*; (2) protein expression: CK18; (3) albumin secretion; (4) 7-ethoxycoumarin conversion	Scaffold (collagen): EGF, ITS, BSA, insulin, dexamethasone, transferrin, selenious acid, linoleic acid	Dvir-Ginzberg M 2008 [140]
		(1) Albumin secretion; (2) urea synthesis	Scaffold	Liu Z 2017 [141]
		(1) Urea secretion; (2) NO synthesis; (3) amino acid metabolism; (4) GSH and GSSG levels; (5) P450 1A1 activity	Suspended	Xu J 2003 [142]
		(1) Galactose consumption; (2) pyruvate consumption; (3) glucose secretion; (4) albumin secretion; (5) total protein; (6) cellular γ-GT, LDH, GPT, GOT activiy	Suspended	Ma M 2003 [143]
	FLCs	(1) Gene expression: *Alb*; (2) ALB concentration (ELISA); (3) Tx	Scaffold (gelatin)	Ye J 2016 [144]
	Small hepatocytes	(1) Gene expression: *Mrp1*, *Afp*; (2) protein expression: AFP, ALB, NTCP, OATP1B2, BSEPp, MRP2; (3) albumin secretion (ELISA)	Embedded (collagen): galactose, L-proline, insulin, nicotinamide, ascorbic acid, dexamethasone, EGF	Sidler Pfändler MA 2004 [145]
		(1) Expression of *Cyp1a1/2*, *Cyp2b1*, *Cyp3a2*, *Cyp4a1*; (2) induction of activities of CYP1A, CYP2B, CYP3A, and CYP2E	Embedded (collagen, Matrigel): EGF, insulin, dexamethasone	Miyamoto S 2005 [146]
	Duct cells	(1) Gene expression: *Cd44*, *Prom1*, *Sox9*, *Krt19*, *Pparg*, *Tbx3*, *Alb*, *Cyp3a11*; (2) differentiation potential; (3) Tx	Embedded (Matrigel): WNT3A, Noggin, RSPO1, EGF, gastrin, HGF, Fgf10, nicotinamide, Y-27632	Kuijk EW 2016 [147]
Coculture	Rat hepatocytes and mouse fibroblasts	(1) Total protein content; (2) albumin secretion; (3) cytochrome P450 activities	Scaffold: EGF, glucagon, insulin, dexamethasone	Lu HF 2005 [148]
	Rat hepatocytes and human fibroblasts	(1) Protein expression: albumin	Embedded (PNIPAAm/type-I collagen)	Takezawa T 1992 [149]
	Small hepatocytes and NPCs	(1) Albumin concentration (ELISA); (2) protein expression: transferrin, haptoglobin, fibrinogen; (3) gene expression: *Alb*, transferrin, *Cyp1a1*, etc. (4) secretion of urea	Scaffold (collagen)	Harada K 2003 [150]

Cells	Assays	Format	Reference
Primary hepatocyte and HSCs	(1) Protein expression: albumin, cytochrome P450 reductase; (2) albumin secretion; (3) urea secretion	Suspended: EGF, dexamethasone, insulin, nicotinamide, L-ascorbic acid	Lee SA 2013 [151]
Rat hepatocytes and human MSCs	(1) Gene expression: *Hnf4a*, *Hnf6*, cytochrome P450, family 1, subfamily A, *Cyp1a1*, *Cyp1a2*, albumin, *Nags*, *Cps1*, *Otc*, *Ass*, *Asl*, *Arg1*; (2) albumin production (ELISA); (3) ammonia clearance; (4) urea production; (5) diazepam metabolism	Suspended: insulin, transferrin	Bao J 2013 [152]
Rat hepatocytes, and HUVECs/rat MSCs/mouse NIH/3T3 fibroblasts	(1) Urea production; (2) albumin secretion	Scaffold	Chou MJ 2013 [153]
Rat hepatocytes and mouse NIH/3T3 fibroblasts	(1) Albumin secretion (ELISA); (2) CYP activities; (3) CYP induction assay	Suspended	Ikeda Y 2012 [154]
Hepatocytes and pancreatic islet cells	(1) Albumin secretions; (2) urea secretion; (3) activity of cytochrome P450 3A4 (CYP3A4); (4) Tx	Suspended (BSA)	Jun Y 2013 [155]
Hepatocyte and mouse NIH/3T3 fibroblasts/bovine aortic endothelial cells	(1) Albumin secretion (ELISA)	Suspended	Otsuka H 2013 [156]
Rat hepatocytes, and mouse fibroblasts and HUVECs	(1) Albumin secretion (ELISA); (2) urea synthesis; (3) cytochrome P-450 expression; (4) bile canaliculi formation	Scaffold	Liu Y 2014 [157]
Hepatocytes and HSCs	(1) Gene expression: *Alb*, *Cyp450*; 2) albumin secretion (ELISA)	Suspended (BSA): EGF, insulin, dexamethasone, nicotinamide, L-ascorbic acid	Lee G 2016 [158]
Rat hepatocytes and HUVECs	(1) Protein expression: ALB, CYP450; (2) urea secretion; (3) CYP3A4 enzymatic activity; (4) gene expression: *Alb*, *Cyp2e1*, *Pecam*, *Vwf*, *Mmp9*	Scaffold	Jeong GS 2016 [159]
Rat hepatocytes and HUVECs	(1) Albumin secretion (ELISA); (2) urea secretion; (3) protein expression: cytochrome P450 3A4 (CYP3A4); (4) induction of CYP3A4 activity	Suspended (alginate/collagen): EGF, glucagon, insulin, hydrocortisone	Chan HF 2016 [160]
Hepatocytes and MSCs	(1) Albumin synthesis (ELISA); (2) urea secretion	Scaffold: insulin, glucagon, hydrocortisone, L-proline, EGF	Alzebdeh DA 2017 [161]

(*Continued*)

TABLE 7.1 Function and Application of Liver-related Organoids From Different Species, Cell Sources, and Culture Conditions (cont.)

Species	Derivation or cell source		Function and application	Culture condition: surface type (coating); growth factor; and other supplements	Reference
		Hepatocytes and HSCs	(1) LDH leakage; (2) protein expression: SMA	Scaffold (PLA): nicotinamide, insulin	Thomas RJ 2006 [162]
		Hepatocyte and NPCs	(1) Albumin production (ELISA); (2) diazepam metabolism; (3) glycogen content	Not given	Ambrosino G 2005 [163]
		Hepatocytes with/without pancreatic islet cells	(1) Ammonia removal; (2) urea secretion rate; (3) albumin concentration: (ELISA); (4) measurement of insulin secretion	Suspended: dexamethasone, insulin, EGF, prolactin, growth hormone, selenium dioxide	Lee KW 2005 [164]
		Hepatocytes and stellate cells	(1) Cytochrome P-450 isoenzymes induction; (2) metabolic activity: CYP-1A1, CYP-1A2, CYP-2B1 isoenzymes; (3) albumin secretion (ELISA)	Scaffold (PLA)	Riccalton-Banks L 2003 [165]
Porcine			(1) Albumin production; (2) diazepam metabolism; (3) ammonia detoxification and ureagenesis	Suspended: insulin, transferrin, selenium	Nyberg SL 2005 [166]
			(1) Tx	Scaffold	Yamashita Y 2003 [167]
	Hepatocytes		(1) Ammonium removal; (2) urea synthesis; (3) 7-ethoxycoumarin (7EC) metabolism; (4) albumin secretion (ELISA)	Suspended: insulin, dexamethasone, mEGF	Sakai Y 1999 [168]
			(1) Urea production; (2) total protein assays	Suspended: insulin, dexamethasone, glucagon, EGF, liver growth factor, transferrin, linoleic acid	Lazar A 1995 [169]
	Coculture	Primary hepatocytes and MSCs	(1) Albumin secretion (ELISA); (2) protein expressions: ALB, CYP3A1	Not given: dexamethasone, insulin, transferrin, selenium	Gu J 2009 [170]
Dog	Biliary ducts		(1) Gene expression: CD133, LGR5, KRT19, SOX9, FOXA1, HNF4, CYP3A12; (2) protein expression: SOX9, HNF1b, K19, K7, HNF4A, ALB; (3) differentiation potential	Embedded (Matrigel): B27, N2, N-acetylcysteine, gastrin, EGF, R-spondin-1, FGF1, nicotinamide, HGF, Noggin, Wnt3a, Y-27632, A83-01	Nantasanti S 2015 [171]
Oncorhynchus mykiss	Liver tissue		(1) Pharmaceutical metabolism	Suspended (pHEMA)	Baron MG 2017 [172]
Yorkshire swine	Liver tissue		(1) Urea production; (2) albumin secretion	Embedded (collagen I)	Irani K 2010 [173]

ADSCs, Human adipose-derived stem cells; EGF, epidermal growth factor; ESCs, embryonic stem cell; FCF, fibroblast growth factor; FLCs, fetal liver cells; HGF, hepatocyte growth factor; HSCs, hepatic stellate cells; HUVECs, human umbilical vein endothelial cells; iPSCs, induced pluripotent stem cells; LPCs, liver progenitor cells; LSECs, liver sinusoidal endothelial cells; MSCs, mesenchymal stem cells; NPCs, nonparenchymal cells; OSM, oncostatin M; PBMCs, peripheral blood mononuclear cells; PHH, primary human hepatocytes; Tx, transplantation.

self-assembly of HepaRG cells, primary hepatocytes, duct cells, or embryonic stem cells (ESCs), iPSCs, and mesenchymal stem cell (MSC)-derived hepatocytes. Additionally, research is focusing intensively on the coculture of hepatocytes with other supportive cells such as human umbilical vein endothelial cells (HUVECs), MSCs, stellate cells, and fibroblasts, which reportedly improve the function of liver organoids (Table 7.1). Liver functions in these organoids have been evaluated to different extents, such as liver-specific gene expression, albumin and urea secretion, and drug-metabolizing enzyme activities. Notably, in vivo functions of liver organoids after transplantation within animal models have been achieved in some studies.

To mimic the in vivo 3D environment, various essential parameters, such as hydrogel, scaffolds, bioreactor for suspended culture, or particular growth factors, are required for organoid formation. Notably, self-organization is vital for the organoid formation and is crucial in the integration of other supportive lineages. The theory of self-organization, which encompasses self-assembly, self-patterning, and self-driven morphogenesis, is highly reflected in the establishment of liver organoids. For instance, Takebe and colleagues succeeded in generating liver buds that self-organized with hepatic endoderm cells from human iPSCs, HUVECs, and human MSCs [5]. It is assumed that mesenchyme-driven self-condensation enables the inclusion of optimal heterotypic cell lineages for the assessment of self-organization. This study reminds us that cellular interaction and cell-community are critical for organoid generation and function maintenance.

In this chapter, we review the literature describing cell–cell interactions in the multiple-cell society during liver development and the generation of liver organoid by combining primary hepatocytes or stem cell-derived hepatocyte-like cells with nonparenchymal cells in the 3D culture system. Additionally, humanized animal models established by hepatocyte transplantation and their potential applications are discussed.

2 LIVER DEVELOPMENT AND MICROENVIRONMENT

Recently, the concept of a specific stem-cell microenvironment (also called the stem cell niche) has received increasing attention. Hematopoietic stem cells (HSCs) provide an example. They are the best-characterized adult stem cells. Their maintenance, regulation of self-renewal and differentiation is very dependent on the microenvironment [10]. Different types of signals are exchanged between HSCs and the cellular constituents of the stem cell niche, such as osteoblasts, endothelial, mesenchymal cells, and other cell types [11]. Furthermore, cell–cell contact is also responsible for the apparently unlimited proliferative capacity and inhibition of maturation of HSCs. These messages provided by the HSC niche clearly indicate that a 3D reconstruction of the niche with appropriate cells is indispensable for the generation of a stable environment for stem cells [11].

Mature liver is a complex and highly organized architecture incorporating numerous cell types, including hepatocytes, cholangiocytes, endothelial cells, stellate cells, and Kupffer cells. Similar to that required by HSCs, cell–cell communication between different cell types and the microenvironment are also required for the early stage of liver development. The mesenchymal and endothelial cell microenvironment is essential for hepatic differentiation and early liver development [12]. When hepatic bud formation begins in response to the signals released from the cardiac mesoderm and the transverse septum, cells in the ventral endodermal epithelium proliferate and invade the surrounding septum transversum mesenchyme [13]. Several dynamic cellular and molecular mechanisms are participants in this phase, which is an orderly progress for hepatic induction and specification. Besides hepatoblasts which are the major cell type, other cell types such as endothelial cells, hematopoietic elements, and developing stellate cells also contribute to the integration of the liver bud through intercellular reactions and exchange of signals. Endothelial cells play a significant role during liver bud formation and interact with hepatic endodermal cells migrating into the mesenchyme of the septum transversum. Neurturin secreted by endothelial cells can help direct the migration of hepatoblasts [14]. Ablation of endothelial cells leads to the impaired migration and selective defect in later hepatic outgrowth. Hepatic specification is also induced by the endothelial cell niche through the dual repression of Wnt and Notch signaling in the early embryo [15]. These results suggest that endothelial cells are critical for the earliest stages of liver organogenesis [16].

After specification and migration into the septum transversum, hepatoblasts undergo proliferation and differentiate into hepatocytes and cholangiocytes at embryonic day 13.5 in mice, 24 hours postfertilization in zebrafish, and 56–58 days of gestation in humans [17]. Sinusoidal endothelial cells, stellate cells, and hematopoietic elements secrete multiple growth factors and cytokines. They exhibit cross talk with hepatoblasts and participate in the regulation of their expansion and differentiation. Key signaling pathways involved in this process include bone morphogenetic protein (BMP), fibroblast growth factor (FGF), hepatocyte growth factor (HGF), Wnt, Notch, and retinoic acid (RA) signaling. For example, BMP from the septum transversum mesenchyme is important for specification of liver fate and subsequent outgrowth of the liver endoderm [18]. Mesenchyme surrounding the developing liver bud secretes HGF, and its receptor, c-met, is expressed on embryonic hepatocytes [19]. Endothelial cells of hepatic sinusoids act as an instructive niche and secrete Wnt9a or Wnt2 to modulate hepatoblast proliferation [20,21]. Stellate cells located in the space of Disse are the major mesenchymal component of the liver. Like endothelial cells, stellate cells of the embryonic sinusoidal wall secrete Wnt9a to promote the growth of hepatocytes [20]. Besides sinusoidal endothelial cells and stellate cells, hematopoietic cells also play an important role in hepatic morphogenesis. They secrete oncostatin M, which enhances hepatocyte differentiation and maturation [22]. In addition, mesothelial and submesothelial cells are rich sources of multiple factors such as HGF, pleiotrophin, and midkine, which increase the proliferation of hepatoblasts and hepatocytes [23]. Knockout of Wilm's tumor 1 homologue, one of the markers of mesothelial cells, leads to hypoplastic livers because of the decreased numbers of hepatoblasts and hepatocytes [24]. Thus with these important spatiotemporal signals, hepatoblasts are directed to differentiate into either hepatocytes or cholangiocytes.

Therefore during embryonic development, multiple cellular communities work together and contribute to liver organogenesis. Using a similar model, it is feasible to combine primary hepatocytes or stem cell-derived hepatocyte-like cells with other supportive cells, such as vascular endothelial progenitor cells, MSCs, or hematopoietic cells, to simulate the in vivo microenvironment during the process of in vitro liver organoid formation.

3 HEPATIC CELL SOURCES, FUNCTION, AND TOXICITY ASSESSMENT

Hepatocytes constitute the main bulk of the liver. These cells perform several essential functions and the transplantation of primary human hepatocytes (PHHs) is regarded as a safe therapy for end-stage liver disease. PHHs feature high gene expression and function of drug enzymes and transporters. Thus they are considered the gold standard for studying liver function, especially for hepatotoxicity studies. Nevertheless, hepatic functions that include enzyme activities, glucose metabolism, and ammonia detoxification can only be well maintained for about 72 hours during PHH culture [25]. After 72 hours, these functions are rapidly lost due to the poor proliferation of PHHs in vitro. In addition, the scarcity of donors also limits their application. Hepatocyte function varies from batch to batch due to different quality of the tissue, which is a direct source of poor experimental reproducibility.

To overcome these limitations, multiple hepatoma cell lines and genetically engineered cells that have an unlimited life span and which are readily available have been explored. Unfortunately, compared to PHHs, most hepatoma cell lines have shown limited expression of drug-metabolizing enzymes such as cytochromes (CYPs) [26]. Furthermore, in transgenic hepatocytes, only one or two enzymes can be satisfactorily transfected and a mutagenic risk still remains. These problems make them unsuitable for hepatotoxicity studies and cell transplantation.

In recent years, direct reprogramming has been developed to directly generate hepatocytes from other somatic cell types. Huang et al. [27] succeeded in establishing a lentiviral expression system of FOXA3, HNF1A, and HNF4A factors to generate human induced hepatocytes (hiHeps) from fibroblasts. These hiHeps express hepatic gene programs and display characteristic functions of mature hepatocytes, such as CYP P450 activity and biliary drug clearance. They also restore the liver function in mice with induced acute liver failure and fatal metabolic liver disease, and prolong their survival. Furthermore, with the help of a bioartificial liver based on hiHeps, patients with liver failure have experienced significant clinical improvement (personal communication). In general, these studies are a crucial step toward realizing the potential of direct reprogramming in regenerative medicine.

An increasing number of approaches are being explored in stem cell research. Adult stem cells, ESCs, and iPSCs are promising alternative sources for functional hepatocytes. Human liver stem cells (HLSCs) in the adult liver contribute to liver regeneration after injury. They can be easily obtained from a liver biopsy and expanded in vitro. These stem cells can differentiate into mature hepatocytes capable of expressing liver-specific markers and CYPs, and also gain liver-specific metabolic activities, such as detoxification abilities, cleavage of the CYP substrate, and urea production [28]. ESCs, which are another alternative source for hepatocytes, have been extensively studied due to their self-renewing ability and high proliferation rate. Functional hepatocytes derived from ESCs express liver-specific genes at physiological levels and exhibit liver-specific functions comparable to those of PHHs [29]. However, ethical considerations are one of the obstacles for the clinical application of ESCs. There have not been any clinical trials using ESCs to treat liver diseases in human patients. In addition, only allogeneic transplantation can be achieved with ESCs.

Compared to the aforementioned potential hepatocytic sources, iPSCs have attracted more interest and attention due to their donor diversity, unlimited long-term supply, stability, and functional consistency. They also bypass the ethical concerns surrounding ESCs. The successful differentiation of human iPSCs into hepatocytes has been reported [30]. These hepatocytes (iPSC-Heps) have reasonable synthetic and metabolic capacities that are similar to those derived from ESCs [31]. They possess many hepatic functions that include albumin secretion, glycogen storage, drug metabolism, drug transportation, and lipogenesis [32]. The global gene expression pattern of iPSC-Heps resembles PHHs more closely than those of most hepatoma cell lines [33]. These cells are also functional in vivo and so, could offer an alternative for treatment of liver diseases. For example, iPSC-derived hepatocyte-like cells can rescue lethal fulminant hepatic failure in a nonobese diabetic severe combined immunodeficient mouse model [32]. Therefore iPSC-Heps represent a considerable improvement over PHHs and tumor cell lines, and serve as a good model for toxicology studies, drug discovery, and in regenerative medicine.

Besides these potential functions, another significant advantage of iPSC-Heps is that they can be generated from patient-specific iPSCs, which carry individual genetic backgrounds. Thus they can be applied in the evaluation of personalized hepatotoxicity and drug response, which may help to decrease the risk and cost associated with clinical trials. As iPSC-Heps are capable of maintaining the characterization of disease, they can also be used to study inherited metabolic disorders of the liver, which include tyrosinemia, Crigler-Najjar syndrome, alpha1-antitrypsin deficiency, familial hypercholesterolemia, and glycogen storage disease type 1 [34]. Recently, gene-editing technologies have been applied to correct gene mutations in patient-specific iPSC-Heps [35]. Therefore iPSC-Heps might be an ideal source for clinical applications.

Recent clinical studies based on stem cells have already been undertaken to treat liver diseases. For instance, human fetal liver-derived stem cells and hepatic progenitor cells have shown significant improvement of clinical and biochemical parameters in patients with liver cirrhosis caused by different etiologies [36]. Although these stem cells are a promising potential replacement for PHHs in clinical use, much remains to be done. The primary issue is safety. According to the Food and Drug Administration Cellular, Tissue and Gene Therapies Advisory Committee for the clinical application of human ESCs, "optimal study designs for preclinical assessment of inappropriate differentiation, including tumorigenic potential, of investigational cellular products derived from undifferentiated hESCs" must be highlighted. Thus "undifferentiated cells should not be tolerated in hESCs-derived cellular therapies for use in humans" and "it is highly desirable to develop standardized tumorigenicity assays" in which an immunocompromised rodent model is suggested to test for tumorgenicity. The Committee also recommended that "toxicity studies in preclinical models include the same route of administration anticipated for clinical trials." Therefore more efforts are needed to address this issue before the clinical application of these stem cell-derived hepatocytes is achieved [37].

4 3D VS. 2D CULTURE

Hepatocytes are widely used in toxicology screenings, expansion for cell therapy, and to establish bioartificial liver devices. So, optimizing their culture conditions to fully maintain their functions is important. As mentioned earlier, the standard culture system for PHHs is the conventional monolayer culture because in freshly isolated hepatocytes, the hepatic functions and drug-metabolizing enzymes are close to those in vivo. Unfortunately, the cell morphology and liver-specific functions are rapidly lost when the cells are removed from their natural environment because PHHs are highly dependent on cell–cell contact and the surrounding microenvironment. To solve this problem and generate more stable systems for hepatocyte growth, many studies have focused on improving reproducibility and standardization of hepatocytes.

Quality of a cell culture system differs depending on its microenvironment. Sustained cellular contact is crucial for maintaining hepatocyte differentiation and function. The interaction between the extracellular matrix (ECM) and hepatocytes is also critical for normal homeostasis. The ECM can help to improve the cellular adaptation to the environment and maintain a stable hepatocyte function. In 2D culture systems, hepatocytes are usually forced to alter their cytoskeleton to adopt a flattened morphology. This limits cell–cell and cell–matrix interactions, and hinders the maintenance of hepatocyte functions [38]. A 3D system provides hepatocytes with a better microenvironment for the normal expression of liver-specific functions.

A number of techniques are being currently developed to maintain the 3D configuration of cells in vitro during the culture of hepatocytes. Hydrogels are polymer networks that provide a cell-friendly 3D ECM condition similar to that found in native tissues. This environment also allows more intimate cell–cell contacts than conventional 2D conditions. Signaling molecules can be safely incorporated in the hydrogel, and then released to stimulate surrounding cells [39]. Currently, the most commonly used hydrogel product in liver organoid generation is Matrigel™ (Corning). The product is based on ECM extracted from mouse sarcoma. It prolongs aspects of hepatocyte function beyond those observed in collagen sandwich cultures [40]. In addition, collagen and gelatin are also frequently used to mimic the 3D ECM.

Scaffold is another novel approach to 3D culture, which can increase the formation of spheroids. Liver bioengineering has sparked great interest since the introduction of liver acellular scaffolds generated by perfusion decellularization [41]. These decellularized scaffolds preserve their tissue microarchitecture and an intact vascular network that can be used for recellularization by perfusion of different cell populations. For instance, human fetal liver and endothelial cells engrafted in this liver decellularized scaffold and displayed typical endothelial, hepatic, and biliary epithelial markers, thus creating a liver-like tissue in vitro and facilitating the establishment of a functional bioengineered liver [42]. Other bioengineering advances have enabled extensive progress, which could improve behavior and organization of 3D organoids, and help to bridge the gap between in vitro and in vivo environments.

Besides these external elements, self-assembly generation of 3D liver organoids can greatly improve hepatocyte functions because these spheroids can retain most of the cell–cell and cell–ECM contacts, liver-specific cell polarity, and bile canoliculi, which reflect the close contact and self-production of ECM [43]. In 2D culture, cell junctions such as anchoring and gap junctions are expressed in the same pattern as that in vivo. However, they persist for only a short period after seeding of hepatocytes and depend on the cell density to a great extent [25]. In contrast, in the spheroid model, cellular contacts are reestablished during self-assembly and the cell density in a hepatosphere is always the same. Therefore cell–cell contacts could be sustained for a longer period than in 2D culture, which will extend the functionality of hepatocytes. As a result, expression of hepacytic genes and drug metabolism functions in spheroids are enhanced when compared to those in monolayer-cultured cells. For example, rat hepatocyte spheroids display higher expression of genes encoding cell adhesion molecules (integrin 3, cadherin 1, Cx32), transcription factors (HNF4α, CCAAT/enhancer binding protein b), protein and metabolic enzymes (albumin, glucose-6-phosphate, tryptophan 2,3-dioxygenase, arginase 1, and CYP7A1), and transporters (OATPs, MDR2, BSEP) than those in monolayer cultures [44]. These expressions can be maintained for a longer time, while they are rapidly lost in monolayers. Other liver-specific functions such as urea secretion, phase I and phase II enzyme activities, and the capacity to metabolize diphenhydramine and troglitazone are enhanced in rat hepatocyte spheroids [45].

Conclusively, 3D culture systems are helpful for hepatocytes to achieve closer cell–cell interaction, better cellular morphology and structure, longer culture life, higher specific gene expression, and better metabolism functions.

As discussed in Section 7.2, the cells in a multiple cellular community work together and contribute to liver organogenesis during embryonic development. As cell–cell contacts are beneficial for cellular maintenance, incorporating other types of cells in the culture system to recapitulate the microenvironment may enhance hepatocyte viability and liver-specific functions. Coculture of different types of cells, particularly of those present in liver tissue, enables the communication through paracrine factors and aids the formation of more functional cellular structures. In a 3D system, different cells are allowed to recognize and conform in a way that is similar to that in native tissues. Cross talk and the resulting endogenous production of the required biological molecules can significantly recapitulate the signaling molecular environments and is helpful to maintain hepatocyte function. Coculture with primary hepatocytes and nonparenchymal cells (NPCs), such as endothelial, stellate cells and fibroblasts, can facilitate spheroid formation and enhance hepatocyte-specific functions (Table 7.1). For example, a spheroid system that combines rat hepatocytes with hepatic stellate cells was developed. In the system, stellate cells contribute to the generation of ECM supporting hepatocyte stability and help spheroid self-organization [46]. Besides, hepatocyte-specific functionalities such as CYP-450 activity and albumin secretion can also be increased in coculture compared to that in monoculture. Another similar 3D coculture system with hepatocytes and stellate cell lines demonstrated better maintenance of the expression of hepatic markers such as albumin and CYP2B1/2 [47]. Furthermore, a 3D triculture model employing rat hepatocytes, and stellate and endothelial cells was successfully established with a microporous membrane [48]. In this condition, stellate cells intercalated and mediated the communication between hepatocytes and endothelial cells. Interestingly, coculture with different cell types from different species also achieved success in a 3D system. In another study, HUVECs helped to generate rat hepatocyte spheroids and contributed to the generation of a dense vascular network in the spheroids [49]. These coculture models in 3D system should permit the investigations of the heterotypic cellular community in vitro.

Besides rat or mouse models, the characteristics of human liver organoids established by coculture have also been widely explored, especially in recent years (Table 7.1). For example, coculturing PHHs and human adipose-derived stem cells (ADSCs) in concave microwells can successfully generate hepatocytic spheroids having a uniform size and shape [50]. ADSCs contribute to the significantly higher viability and greater functions of hepatocytes in spheroids, such as albumin secretion and enzymatic activity, when compared with monocultured spheroids. The authors assumed that these effects may be due to the transdifferentiation potential and paracrine healing effects of ADSCs on human hepatocytes. Similar studies using coculture of hepatocytes with stellate cells [51], Kupffer cells [52], HUVECs [5], sinusoidal endothelial cells [53], and fibroblasts [54] have also highlighted the importance of a heterotypic cellular community during the organoid formation. Although liver spheroids generated by PHH display multiple functions in the 3D environment, their limited proliferative capability and difficulty in acquisition require the development of other alternative hepatocytes, such as those derived from iPSCs or somatic stem cells [55]. Using a nanopillar plate, Takayama et al. established a 3D spheroid culture and differentiation system that is potentially useful in drug toxicity testing [56]. In this system, hepatocytes derived from human ESCs and human iPSCs displayed higher viability and a higher level of hepatocellular functions than hepatocytes derived from a monoculture system. The enhanced functions include secretion of albumin (Alb) and urea, expression of hepatocyte-related genes, drug metabolism capacity, and potency of CYP induction. However, the in vivo functions remain unclear. To address this issue, another study tried to transplant iPSC-derived liver buds to immunodeficient mice. Successful maturation into tissue resembling the adult liver was achieved and the rescue of drug-induced lethal liver failure was observed. These studies provide more evidence for the clinical application of liver organoids in regenerative medicine [5].

To summarize this section, cell–cell communication and the extracellular microenvironment are necessary to improve hepatocyte functions. The 3D culture system mimics the in vivo environment and is more suitable to maintain hepatic characteristics than monolayer culture. Generation of liver organoids by combining heterotypic cellular coculture and the 3D system is a promising strategy for properly investigating drug hepatotoxicity and for tissue engineering.

5 HUMANIZED LIVER ANIMAL MODELS ESTABLISHED BY HEPATOCYTE TRANSPLANTATION

The liver is central in many human-specific biological processes and animal models are commonly used to study human liver disease and drug discovery. However, even in models that incorporate engineered human genes, there are still some important features of human biology that cannot be replicated. To address this, several humanized mouse models have been successfully established by transplantation of human hepatocytes, stem cells-derived hepatocytes, and even liver organoids. These models are proving to be helpful in exploring the in vivo functions of hepatocytes from different sources, and are germane in studies of human liver metabolism, drug toxicity, and hepatotropic infections.

One of the key factors required for successful establishment of humanized liver by transplantation of hepatocytes in mice is the elimination of the endogenous hepatocytes to prevent competition with the human cells. This is achieved using transgenic immune-deficient mice. The first model developed for the liver was the albumin-uroplasminogen activator (uPA) transgene mouse [57]. In this model, constitutive expression of uPA causes hepatocytes damage and leads to liver injury, which confers a growth advantage for transplanted cells, such as human hepatocytes. Subsequently, Tateno et al. succeeded in generating chimeric uPA+/+/SCID mice. In these mice, the livers could be replaced by more than 90% with human hepatocytes [58]. Measuring human Alb concentration permits the easy estimation of the degree of replacement by human hepatocytes. Thus interindividual variabilities of pharmacokinetics and toxicity can be evaluated using these chimeric mice with hepatocytes from different donors [59]. However, the uPA mouse model has major practical limitations. These include poor breeding efficiency, narrow time window for transplantation and propensity for kidney disease [58]. Azuma et al. developed a new humanized mouse model, Fah−/−/Rag2−/−/Il2rg−/− (FRG), which combines fumarylacetoacetate hydrolase (Fah) knockout and severely immunodeficiency [60]. FRG mice grow well and are fully fertile if continuously given a protective drug, 2-(2-nitro-4-trifluoromethylbenzoyl)-1,3-cyclohexanedione (NTBC), in their drinking water. When NTBC is withdrawn, hepatocellular injury is caused by the accumulation of the toxic metabolite, fumarylacetoacetate, because of a defect in the tyrosine catabolic pathway. As a result, wild-type cells are deficient in this model, which can extensively repopulate the human hepatocytes upon transplantation. Extensive liver humanization can be achieved. Although the FRG liver humanization model described here is superior to the uPA model in several ways and has been widely used, only mature human hepatocytes can be successfully used. To overcome these limitations, Alb-TRECK/SCID mice were used for in vivo differentiation of human hepatic stem cells (HpSCs) and humanized liver generation, which will have functional applications in studies involving drug metabolism and drug–drug interactions and will promote other in vivo and in vitro studies [61]. Another commonly used model is TK-NOG mice [62]. In this model, herpes simplex virus type 1 thymidine kinase (HSVtk) transgene was expressed within the liver of highly immunodeficient NOG mice (TK-NOG). The hepatotoxic transgene can be conditionally activated by gancyclovir, which is a small-molecule drug. Then, mouse liver cells are ablated and hepatic injury occurs, which permit the engraftment of transplanted human cells. In most of these models, the repopulation of the liver with human hepatocytes can reach over 70% [63]. Thus these models can provide an optimized platform for liver regeneration.

Recently, the establishment of a humanized animal model has become feasible by combining liver tissue with other humanized tissues. AFC8-hu HSC/Hep mice was such a system which contain human thymus, blood, and liver [64]. Although many of the properties remain unknown, it is hopeful to apply vascularized human liver organoids and transplant to mice or rats with a humanized immune system. This approach will increase the understanding of the interactive effects among human liver, blood, and immune system, and will hopefully allow modeling of the disease process, drug screening, and cell transplantation.

Some other humanized liver chimeric models have been successfully established based on stem cells. To investigate the potential contribution of human iPSC-Heps in liver regeneration, a liver injury transplantation model was developed based on the NOD/Lt-SCID/IL-2R γ g−/−(NSG) mouse, one of the best rodent models for human cell engraftment analysis [65]. In this liver cirrhosis model, iPSC-Heps at various differentiation stages are engrafted and repopulate the liver tissue of mice, with engraftment efficiency similar to that of ESC-derived hepatocytes. Another mouse model with lethal fulminant hepatic failure was developed by administering diphtheria toxin in Alb-toxin receptor-mediated cell knockout (Alb-TRECK)/SCID mice [61,66]. With this system, HpSCs successfully repopulated the mouse liver and the repopulation rate in some mice was nearly 100%, which rescued the lethal mice. In addition, more human Alb was detected in the humanized mice with hepatic stem cell transplantation compared to the transplantation of primary fetal liver cells.

As mentioned earlier, mouse models with humanized liver are applicable in different research fields. The most common application to date is in drug discovery and development. In recent years, numerous studies have shown that mice with a highly humanized liver provide a good platform to study human liver-mediated metabolism and toxicology. For example, in chimeric mouse models, human CYP P450 genes can be expressed [58] and major human P450 proteins, such as CYP1A2, CYP2C9, and CYP3A4, can be detected [67]. Thus these models allow assessments that are valuable in predicting human drug metabolism and drug–drug interactions. Chimeric mice can correctly identify the predominant human drug metabolite before human testing. One example is Clemizole, a drug that is in clinical development for the treatment of hepatitis C virus (HCV) infection [68].

Mouse models that feature humanized liver are also potentially valuable for studies of the etiology and therapy of human viral liver diseases, and hepatocyte biology. For a long time, there has been a shortage of small animal models of hepatitis B virus (HBV) and HCV infection. Using uPA/recombinant activation gene-2 (RAG-2) mice, Dandri et al. successfully established the model of productive HBV infection and found that human hepatocytes were permissive for HBV infection in a xenogenic liver [69]. Later, another practical system based on a FRG mouse model was generated in which the humanized liver could be infected by HBV and HCV, and responded to antiviral treatment [70]. These models have expanded the experimental possibilities for studying HBV, HCV, and other human hepatotropic pathogen infections, and are proving useful for antiviral drug testing. Recently, 3D liver organoids derived from human iPSCs were successfully simulated ex vivo. This microenvironment models HBV infection and can be used to explore virus–host interactions [in press]. This is the first study that has successfully reproduced the HBV life cycle and liver physiology in vitro. The data suggest that hiPSC-derived liver organoids might be a powerful model for research on viral physiology and pathology, and for the development of new antiviral compounds relevant to transplantation.

6 SUMMARY AND PERSPECTIVES

Liver transplantation is the only existing modality for treating end-stage liver disease. Alternatives are needed owing to the critical shortage of donor livers. Cellular therapy using human fetal liver-derived stem cells has great potential in conservatively managing liver diseases and could serve as a supportive therapy. However, the source and the function of these special stem cells are still limited. Organoids derived from stem cells, especially from iPSCs, retain their organ identity and genome stability. They have become a potential source for regenerative medicine therapy and may overcome the barrier posed by limited donor organs. Moreover, if organoids are derived from healthy tissue of the same individual, allogeneic immune response upon transplantation could also be prevented.

Precision medicine with theory of scientific, rational, and individual treatment requires building an individualized disease model. The initial obstacles in achieving this goal are how to direct patient-derived stem cells to differentiate more precisely and how to produce high level of functional organs. Huch et al. [71] generated liver organoids derived from patients with two inherited disorders, α1-antitrypsin deficiency and Alagille syndrome. These patient-specific organoids expanded in vitro and mimicked the in vivo pathology. Recently, it was shown that functional repair by homologous recombination using CRISPR/Cas9 technology is feasible in intestinal stem cells organoids of cystic fibrosis patients. The corrected allele is expressed and is fully functional, as measured in clonally expanded organoids [72]. These results open experimental avenues for disease modeling and gene therapy. In the near future, gene-editing technology could be combined with patient-specific organoid transplantation in humanized mice models to conduct in vivo gene correction for hereditary defects. Additionally, a rat humanized model with X-SCID [73] and Fah($-/-$) [74] generated by gene editing is being developed in our laboratory. This model will be an excellent means to study human diseases.

Stem or progenitor cells are an exciting prospect for future cell transplantations and may prove to be a sustainable alternative source of cells. Nevertheless, the various strategies developed so far have failed to generate hepatic tissue in vitro that is bioequivalent to native liver tissue. Hence, it is vital to understand the strengths and shortcomings of current strategies for liver regeneration in vitro. Optimization is required before clinical applications become feasible. For instance, the difference in iPS reprogramming and epigenetic cell memory can cause variability in subsequent differentiation procedures. Better understanding of iPSC reprogramming and generation of more consistent iPSC cultures is needed. Future research will hopefully yield improved ways to maintain hepatocyte functions in liver diseases, as well as ways to enhance proliferation, engraftment, survival, and other desirable features for cell transplantation therapy. Moreover, other obstacles are linked with important technical issues, such as creating targeted cells under GMP conditions.

ACKNOWLEDGMENTS

This research was supported in part by grants from National Natural Science Foundation, China (No. 81573053 to Li YM); Science and Technology Project of Zhenjiang (No. SS2015024 to Li YM); by the Jiangsu innovative and entrepreneurial project for the introduction of high-level talent (to Zheng YW) and the Jiangsu science and technology planning project, China (BE2015669 to Zheng YW), and by Ministry of Education, Culture, Sports, Science, and Technology of Japan, KAKENHI, No. 16K15604 to Zheng YW; by Novartis Pharma Research Foundation (to Zheng YW); and Japan Science and Technology Research Partnership for Sustainable Development (SATREPS) project FY2015 (to Isoda H)., by the grants from National Natural Science Foundation, China (No. 81770621 to Zheng YW).

REFERENCES

[1] Eiraku M, et al. Self-organizing optic-cup morphogenesis in three-dimensional culture. Nature 2011;472(7341):51–6.
[2] Lancaster MA, et al. Cerebral organoids model human brain development and microcephaly. Nature 2013;501(7467):373–9.
[3] Barker N, et al. Lgr5(+ve) stem cells drive self-renewal in the stomach and build long-lived gastric units in vitro. Cell Stem Cell 2010;6(1):25–36.
[4] Spence JR, et al. Directed differentiation of human pluripotent stem cells into intestinal tissue in vitro. Nature 2011;470(7332):105–9.
[5] Takebe T, et al. Vascularized and functional human liver from an iPSC-derived organ bud transplant. Nature 2013;499(7459):481–4.
[6] Huch M, et al. Unlimited in vitro expansion of adult bi-potent pancreas progenitors through the Lgr5/R-spondin axis. EMBO J 2013;32(20):2708–21.
[7] Takasato M, et al. Directing human embryonic stem cell differentiation towards a renal lineage generates a self-organizing kidney. Nat Cell Biol 2014;16(1):118–26.
[8] Karthaus WR, et al. Identification of multipotent luminal progenitor cells in human prostate organoid cultures. Cell 2014;159(1):163–75.
[9] Yin X, et al. Engineering stem cell organoids. Cell Stem Cell 2016;18(1):25–38.
[10] Schofield R. The relationship between the spleen colony-forming cell and the haemopoietic stem cell. Blood Cells 1978;4(1–2):7–25.
[11] Wilson A, Trumpp A. Bone-marrow haematopoietic-stem-cell niches. Nat Rev Immunol 2006;6(2):93–106.
[12] Zaret KS. Regulatory phases of early liver development: paradigms of organogenesis. Nat Rev Genet 2002;3(7):499–512.
[13] Nakamura T, Nishina H. Liver development: lessons from knockout mice and mutant fish. Hepatol Res 2009;39(7):633–44.
[14] Tatsumi N, et al. Neurturin-GFRalpha2 signaling controls liver bud migration along the ductus venosus in the chick embryo. Dev Biol 2007;307(1):14–28.
[15] Han S, et al. An endothelial cell niche induces hepatic specification through dual repression of Wnt and Notch signaling. Stem Cells 2011;29(2):217–28.
[16] Matsumoto K, et al. Liver organogenesis promoted by endothelial cells prior to vascular function. Science 2001;294(5542):559–63.

[17] Gordillo M, Evans T, Gouon-Evans V. Orchestrating liver development. Development 2015;142(12):2094–108.
[18] Rossi JM, et al. Distinct mesodermal signals, including BMPs from the septum transversum mesenchyme, are required in combination for hepatogenesis from the endoderm. Genes Dev 2001;15(15):1998–2009.
[19] Schmidt C, et al. Scatter factor/hepatocyte growth factor is essential for liver development. Nature 1995;373(6516):699–702.
[20] Matsumoto K, et al. Wnt9a secreted from the walls of hepatic sinusoids is essential for morphogenesis, proliferation, and glycogen accumulation of chick hepatic epithelium. Dev Biol 2008;319(2):234–47.
[21] Ding BS, et al. Inductive angiocrine signals from sinusoidal endothelium are required for liver regeneration. Nature 2010;468(7321):310–5.
[22] Kamiya A, et al. Fetal liver development requires a paracrine action of oncostatin M through the gp130 signal transducer. EMBO J 1999;18(8):2127–36.
[23] Onitsuka I, Tanaka M, Miyajima A. Characterization and functional analyses of hepatic mesothelial cells in mouse liver development. Gastroenterology 2010;138(4):1525–35. 1535. e1–6.
[24] Ijpenberg A, et al. Wt1 and retinoic acid signaling are essential for stellate cell development and liver morphogenesis. Dev Biol 2007;312(1):157–70.
[25] Godoy P, et al. Recent advances in 2D and 3D in vitro systems using primary hepatocytes, alternative hepatocyte sources and non-parenchymal liver cells and their use in investigating mechanisms of hepatotoxicity, cell signaling and ADME. Arch Toxicol 2013;87(8):1315–530.
[26] Rodriguez-Antona C, et al. Cytochrome P450 expression in human hepatocytes and hepatoma cell lines: molecular mechanisms that determine lower expression in cultured cells. Xenobiotica 2002;32(6):505–20.
[27] Huang P, et al. Direct reprogramming of human fibroblasts to functional and expandable hepatocytes. Cell Stem Cell 2014;14(3):370–84.
[28] Shafritz DA, et al. Liver stem cells and prospects for liver reconstitution by transplanted cells. Hepatology 2006;43(2 Suppl. 1):S89–98.
[29] Duan Y, et al. Differentiation and enrichment of hepatocyte-like cells from human embryonic stem cells in vitro and in vivo. Stem Cells 2007;25(12):3058–68.
[30] Takahashi K, et al. Induction of pluripotent stem cells from adult human fibroblasts by defined factors. Cell 2007;131(5):861–72.
[31] Sullivan GJ, et al. Generation of functional human hepatic endoderm from human induced pluripotent stem cells. Hepatology 2010;51(1):329–35.
[32] Chen YF, et al. Rapid generation of mature hepatocyte-like cells from human induced pluripotent stem cells by an efficient three-step protocol. Hepatology 2012;55(4):1193–203.
[33] Gao X, Liu Y. A transcriptomic study suggesting human iPSC-derived hepatocytes potentially offer a better in vitro model of hepatotoxicity than most hepatoma cell lines. Cell Biol Toxicol 2017;33(4):407–21.
[34] Rashid ST, et al. Modeling inherited metabolic disorders of the liver using human induced pluripotent stem cells. J Clin Invest 2010;120(9):3127–36.
[35] Yusa K, et al. Targeted gene correction of alpha1-antitrypsin deficiency in induced pluripotent stem cells. Nature 2011;478(7369):391–4.
[36] Khan AA, et al. Human fetal liver-derived stem cell transplantation as supportive modality in the management of end-stage decompensated liver cirrhosis. Cell Transplant 2010;19(4):409–18.

[37] FDA Cellular, Tissue and Gene Therapies Advisory Committee. 2008 April 10–11; Available from: https://www.fda.gov/ohrms/dockets/ac/08/minutes/2008-0471M.htm.
[38] Berthiaume F, et al. Effect of extracellular matrix topology on cell structure, function, and physiological responsiveness: hepatocytes cultured in a sandwich configuration. FASEB J 1996;10(13):1471–84.
[39] Drury JL, Mooney DJ. Hydrogels for tissue engineering: scaffold design variables and applications. Biomaterials 2003;24(24):4337–51.
[40] Moghe PV, et al. Cell–cell interactions are essential for maintenance of hepatocyte function in collagen gel but not on matrigel. Biotechnol Bioeng 1997;56(6):706–11.
[41] Baptista PM, et al. Whole organ decellularization – a tool for bioscaffold fabrication and organ bioengineering. Conf Proc IEEE Eng Med Biol Soc 2009;2009:6526–9.
[42] Baptista PM, et al. The use of whole organ decellularization for the generation of a vascularized liver organoid. Hepatology 2011;53(2):604–17.
[43] Peshwa MV, et al. Mechanistics of formation and ultrastructural evaluation of hepatocyte spheroids. In Vitro Cell Dev Biol Anim 1996;32(4):197–203.
[44] Sakai Y, Yamagami S, Nakazawa K. Comparative analysis of gene expression in rat liver tissue and monolayer- and spheroid-cultured hepatocytes. Cells Tissues Organs 2010;191(4):281–8.
[45] Miranda JP, et al. Towards an extended functional hepatocyte in vitro culture. Tissue Eng Part C Methods 2009;15(2):157–67.
[46] Thomas RJ, et al. The effect of three-dimensional co-culture of hepatocytes and hepatic stellate cells on key hepatocyte functions in vitro. Cells Tissues Organs 2005;181(2):67–79.
[47] Abu-Absi SF, Hansen LK, Hu WS. Three-dimensional co-culture of hepatocytes and stellate cells. Cytotechnology 2004;45(3):125–40.
[48] Kasuya J, et al. Hepatic stellate cell-mediated three-dimensional hepatocyte and endothelial cell triculture model. Tissue Eng Part A 2011;17(3–4):361–70.
[49] Inamori M, Mizumoto H, Kajiwara T. An approach for formation of vascularized liver tissue by endothelial cell-covered hepatocyte spheroid integration. Tissue Eng Part A 2009;15(8):2029–37.
[50] No da Y, et al. Functional 3D human primary hepatocyte spheroids made by co-culturing hepatocytes from partial hepatectomy specimens and human adipose-derived stem cells. PLoS ONE 2012;7(12):pe50723.
[51] Groger M, et al. Monocyte-induced recovery of inflammation-associated hepatocellular dysfunction in a biochip-based human liver model. Sci Rep 2016;6:21868.
[52] Skardal A, et al. Bioprinting cellularized constructs using a tissue-specific hydrogel bioink. J Vis Exp 2016;110:pe53606.
[53] Ramachandran SD, et al. In vitro generation of functional liver organoid-like structures using adult human cells. PLOS ONE 2015;10(10):e0139345.
[54] Kobayashi A, et al. Preparation of stripe-patterned heterogeneous hydrogel sheets using microfluidic devices for high-density coculture of hepatocytes and fibroblasts. J Biosci Bioeng 2013;116(6):761–7.
[55] Guye P, et al. Genetically engineering self-organization of human pluripotent stem cells into a liver bud-like tissue using Gata6. Nat Commun 2016;7:10243.
[56] Takayama K, et al. 3D spheroid culture of hESC/hiPSC-derived hepatocyte-like cells for drug toxicity testing. Biomaterials 2013;34(7):1781–9.
[57] Rhim JA, et al. Replacement of diseased mouse liver by hepatic cell transplantation. Science 1994;263(5150):1149–52.

[58] Tateno C, et al. Near completely humanized liver in mice shows human-type metabolic responses to drugs. Am J Pathol 2004;165(3):901–12.
[59] Katoh M, et al. Chimeric mice with humanized liver. Toxicology 2008;246(1):9–17.
[60] Azuma H, et al. Robust expansion of human hepatocytes in Fah−/−/Rag2−/−/Il2rg−/− mice. Nat Biotechnol 2007;25(8):903–10.
[61] Zhang RR, Zheng YW, Taniguchi H. Generation of a humanized mouse liver using human hepatic stem cells. J Vis Exp 2016;29(114):e54167.
[62] Hasegawa M, et al. The reconstituted 'humanized liver' in TK-NOG mice is mature and functional. Biochem Biophys Res Commun 2011;405(3):405–10.
[63] Grompe M, Strom S. Mice with human livers. Gastroenterology 2013;145(6):1209–14.
[64] Washburn ML, et al. A humanized mouse model to study hepatitis C virus infection, immune response, and liver disease. Gastroenterology 2011;140(4):1334–44.
[65] Liu H, et al. In vivo liver regeneration potential of human induced pluripotent stem cells from diverse origins. Sci Transl Med 2011;3(82):82ra39.
[66] Zhang RR, et al. Human hepatic stem cells transplanted into a fulminant hepatic failure Alb-TRECK/SCID mouse model exhibit liver reconstitution and drug metabolism capabilities. Stem Cell Res Ther 2015;6:49.
[67] Katoh M, et al. In vivo induction of human cytochrome P450 enzymes expressed in chimeric mice with humanized liver. Drug Metab Dispos 2005;33(6):754–63.
[68] Nishimura T, et al. Using chimeric mice with humanized livers to predict human drug metabolism and a drug–drug interaction. J Pharmacol Exp Ther 2013;344(2):388–96.
[69] Dandri M, et al. Repopulation of mouse liver with human hepatocytes and in vivo infection with hepatitis B virus. Hepatology 2001;33(4):981–8.
[70] Bissig KD, et al. Human liver chimeric mice provide a model for hepatitis B and C virus infection and treatment. J Clin Invest 2010;120(3):924–30.
[71] Huch M, et al. Long-term culture of genome-stable bipotent stem cells from adult human liver. Cell 2015;160(1–2):299–312.
[72] Schwank G, et al. Functional repair of CFTR by CRISPR/Cas9 in intestinal stem cell organoids of cystic fibrosis patients. Cell Stem Cell 2013;13(6):653–8.
[73] Mashimo T, et al. Generation of knockout rats with X-linked severe combined immunodeficiency (X-SCID) using zinc-finger nucleases. PLoS ONE 2010;5(1):pe8870.
[74] Zhang L, et al. Efficient liver repopulation of transplanted hepatocyte prevents cirrhosis in a rat model of hereditary tyrosinemia type I. Sci Rep 2016;6:31460.
[75] Hickey RD, et al. Noninvasive 3-dimensional imaging of liver regeneration in a mouse model of hereditary tyrosinemia type 1 using the sodium iodide symporter gene. Liver Transpl 2015;21(4):442–53.
[76] Minguet S, et al. A population of c-Kit(low)(CD45/TER119)-hepatic cell progenitors of 11-day postcoitus mouse embryo liver reconstitutes cell-depleted liver organoids. J Clin Invest 2003;112(8):1152–63.
[77] Yap KK, et al. Enhanced liver progenitor cell survival and differentiation in vivo by spheroid implantation in a vascularized tissue engineering chamber. Biomaterials 2013;34(16):3992–4001.
[78] Yimlamai D, et al. Hippo pathway activity influences liver cell fate. Cell 2014;157(6):1324–38.
[79] Tsuchiya A, et al. Long-term extensive expansion of mouse hepatic stem/progenitor cells in a novel serum-free culture system. Gastroenterology 2005;128(7):2089–104.

[80] Waisbourd-Zinman O, et al. The toxin biliatresone causes mouse extrahepatic cholangiocyte damage and fibrosis through decreased glutathione and SOX17. Hepatology 2016;64(3):880–93.
[81] Flanagan DJ, et al. Isolation and culture of adult intestinal, gastric, and liver organoids for cre-recombinase-mediated gene deletion. Humana Press, Methods Mol Biol 2016;1–11.
[82] Gabriel E, et al. Differentiation and selection of hepatocyte precursors in suspension spheroid culture of transgenic murine embryonic stem cells. PLoS ONE 2012;7(9):e44912.
[83] Mizumoto H, et al. Evaluation of a hybrid artificial liver module based on a spheroid culture system of embryonic stem cell-derived hepatic cells. Cell Transplant 2012;21(2–3):421–8.
[84] Teratani T, et al. Long-term maintenance of liver-specific functions in cultured ES cell-derived hepatocytes with hyaluronan sponge. Cell Transplant 2005;14(9):629–35.
[85] Drobinskaya I, et al. Scalable selection of hepatocyte- and hepatocyte precursor-like cells from culture of differentiating transgenically modified murine embryonic stem cells. Stem Cells 2008;26(9):2245–56.
[86] Roh H, et al. Cellular behaviour of hepatocyte-like cells from nude mouse bone marrow-derived mesenchymal stem cells on galactosylated poly(D,L-lactic-co-glycolic acid). J Tissue Eng Regen Med 2015;9(7):819–25.
[87] Leite SB, et al. Novel human hepatic organoid model enables testing of drug-induced liver fibrosis in vitro. Biomaterials 2016;78:1–10.
[88] Seo SJ, et al. Enhanced liver functions of hepatocytes cocultured with NIH 3T3 in the alginate/galactosylated chitosan scaffold. Biomaterials 2006;27(8):1487–95.
[89] Mavila N, et al. Functional human and murine tissue-engineered liver is generated from adult stem/progenitor cells. Stem Cells Transl Med 2017;6(1):238–48.
[90] Lugli N, et al. R-spondin 1 and noggin facilitate expansion of resident stem cells from non-damaged gallbladders. EMBO Rep 2016;17(5):769–79.
[91] Bell CC, et al. Transcriptional, functional, and mechanistic comparisons of stem cell-derived hepatocytes, HepaRG cells, and three-dimensional human hepatocyte spheroids as predictive in vitro systems for drug-induced liver injury. Drug Metab Dispos 2017;45(4):419–29.
[92] Torok E, et al. Primary human hepatocytes on biodegradable poly(L-lactic acid) matrices: a promising model for improving transplantation efficiency with tissue engineering. Liver Transpl 2011;17(2):104–14.
[93] Tostoes RM, et al. Human liver cell spheroids in extended perfusion bioreactor culture for repeated-dose drug testing. Hepatology 2012;55(4):1227–36.
[94] Weltin A, et al. Accessing 3D microtissue metabolism: Lactate and oxygen monitoring in hepatocyte spheroids. Biosens Bioelectron 2017;87:941–8.
[95] Rebelo SP, et al. HepaRG microencapsulated spheroids in DMSO-free culture: novel culturing approaches for enhanced xenobiotic and biosynthetic metabolism. Arch Toxicol 2015;89(8):1347–58.
[96] Wang J, et al. Engineering EMT using 3D micro-scaffold to promote hepatic functions for drug hepatotoxicity evaluation. Biomaterials 2016;91:11–22.
[97] Malinen MM, et al. Differentiation of liver progenitor cell line to functional organotypic cultures in 3D nanofibrillar cellulose and hyaluronan-gelatin hydrogels. Biomaterials 2014;35(19):5110–21.
[98] Lazaro CA, et al. Establishment, characterization, and long-term maintenance of cultures of human fetal hepatocytes. Hepatology 2003;38(5):1095–106.
[99] Kim SE, et al. Engraftment potential of spheroid-forming hepatic endoderm derived from human embryonic stem cells. Stem Cells Dev 2013;22(12):1818–29.
[100] Subramanian K, et al. Spheroid culture for enhanced differentiation of human embryonic stem cells to hepatocyte-like cells. Stem Cells Dev 2014;23(2):124–31.

[101] Tian L, et al. Efficient and controlled generation of 2D and 3D bile duct tissue from human pluripotent stem cell-derived spheroids. Stem Cell Rev 2016;12(4):500–8.
[102] Tasnim F, et al. Functionally enhanced human stem cell derived hepatocytes in galactosylated cellulosic sponges for hepatotoxicity testing. Mol Pharm 2016;13(6):1947–57.
[103] Kim JH, et al. Enhanced metabolizing activity of human ES cell-derived hepatocytes using a 3D culture system with repeated exposures to xenobiotics. Toxicol Sci 2015;147(1):190–206.
[104] Cipriano M, et al. Self-assembled 3D spheroids and hollow-fibre bioreactors improve MSC-derived hepatocyte-like cell maturation in vitro. Arch Toxicol 2017;91(4):1815–32.
[105] Bell CC, et al. Characterization of primary human hepatocyte spheroids as a model system for drug-induced liver injury, liver function and disease. Sci Rep 2016;6:25187.
[106] Sugimoto S, et al. Hepatic organoid formation in collagen sponge of cells isolated from human liver tissues. Tissue Eng 2005;11(3–4):626–33.
[107] Mattei G, et al. On the adhesion-cohesion balance and oxygen consumption characteristics of liver organoids. PLOS ONE 2017;12(3):e0173206.
[108] Rebelo SP, et al. Three-dimensional co-culture of human hepatocytes and mesenchymal stem cells: improved functionality in long-term bioreactor cultures. J Tissue Eng Regen Med 2015;11(7):2034–45.
[109] Rennert K, et al. A microfluidically perfused three dimensional human liver model. Biomaterials 2015;71:119–31.
[110] Schepers A, et al. Engineering a perfusable 3D human liver platform from iPS cells. Lab Chip 2016;16(14):2644–53.
[111] Takezawa T, et al. Concept for organ engineering: a reconstruction method of rat liver for in vitro culture. Tissue Eng 2000;6(6):641–50.
[112] Mitaka T, et al. Reconstruction of hepatic organoid by rat small hepatocytes and hepatic nonparenchymal cells. Hepatology 1999;29(1):111–25.
[113] Michalopoulos GK, et al. Hepatocytes undergo phenotypic transformation to biliary epithelium in organoid cultures. Hepatology 2002;36(2):278–83.
[114] Nugraha B, et al. Galactosylated cellulosic sponge for multi-well drug safety testing. Biomaterials 2011;32(29):6982–94.
[115] Tostoes RM, et al. Perfusion of 3D encapsulated hepatocytes – a synergistic effect enhancing long-term functionality in bioreactors. Biotechnol Bioeng 2011;108(1):41–9.
[116] Xia L, et al. Tethered spheroids as an in vitro hepatocyte model for drug safety screening. Biomaterials 2012;33(7):2165–76.
[117] Brophy CM, et al. Rat hepatocyte spheroids formed by rocked technique maintain differentiated hepatocyte gene expression and function. Hepatology 2009;49(2):578–86.
[118] Chen AA, et al. Modulation of hepatocyte phenotype in vitro via chemomechanical tuning of polyelectrolyte multilayers. Biomaterials 2009;30(6):1113–20.
[119] Yan S, et al. Hepatocyte spheroid culture on fibrous scaffolds with grafted functional ligands as an in vitro model for predicting drug metabolism and hepatotoxicity. Acta Biomater 2015;28:138–48.
[120] Weeks CA, et al. Effect of amine content and chemistry on long-term, three-dimensional hepatocyte spheroid culture atop aminated elastin-like polypeptide coatings. J Biomed Mater Res A 2017;105(2):377–88.
[121] Uchida S, et al. An injectable spheroid system with genetic modification for cell transplantation therapy. Biomaterials 2014;35(8):2499–506.
[122] Tong WH, et al. Constrained spheroids for prolonged hepatocyte culture. Biomaterials 2016;80:106–20.

[123] Siltanen C, et al. One step fabrication of hydrogel microcapsules with hollow core for assembly and cultivation of hepatocyte spheroids. Acta Biomater 2017;50:428–36.
[124] Santoh M, et al. Acetaminophen induces accumulation of functional rat CYP3A via polyubiquitination dysfunction. Sci Rep 2016;6:21373.
[125] Sanoh S, et al. Fluorometric assessment of acetaminophen-induced toxicity in rat hepatocyte spheroids seeded on micro-space cell culture plates. Toxicol In Vitro 2014;28(6):1176–82.
[126] Pinheiro PF, et al. Hepatocyte spheroids as a competent in vitro system for drug biotransformation studies: nevirapine as a bioactivation case study. Arch Toxicol 2017;91(3):1199–211.
[127] Du Y, et al. 3D hepatocyte monolayer on hybrid RGD/galactose substratum. Biomaterials 2006;27(33):5669–80.
[128] Fukuda J, Sakai Y, Nakazawa K. Novel hepatocyte culture system developed using microfabrication and collagen/polyethylene glycol microcontact printing. Biomaterials 2006;27(7):1061–70.
[129] Nishikawa Y, et al. Transdifferentiation of mature rat hepatocytes into bile duct-like cells in vitro. Am J Pathol 2005;166(4):1077–88.
[130] Dvir-Ginzberg M, et al. Ultrastructural and functional investigations of adult hepatocyte spheroids during in vitro cultivation. Tissue Eng 2004;10(11–12):1806–17.
[131] Chua KN, et al. Stable immobilization of rat hepatocyte spheroids on galactosylated nanofiber scaffold. Biomaterials 2005;26(15):2537–47.
[132] Tzanakakis ES, et al. Probing enhanced cytochrome P450 2B1/2 activity in rat hepatocyte spheroids through confocal laser scanning microscopy. Cell Transplant 2001;10(3):329–42.
[133] Wu FJ, et al. Enhanced cytochrome P450 IA1 activity of self-assembled rat hepatocyte spheroids. Cell Transplant 1999;8(3):233–46.
[134] Hsiao CC, et al. Receding cytochrome P450 activity in disassembling hepatocyte spheroids. Tissue Eng 1999;5(3):207–21.
[135] Sato Y, et al. A new three-dimensional culture system for hepatocytes using reticulated polyurethane. Hepatology 1994;19(4):1023–8.
[136] Tong JZ, et al. Long-term culture of adult rat hepatocyte spheroids. Exp Cell Res 1992;200(2):326–32.
[137] Tong JZ, Bernard O, Alvarez F. Long-term culture of rat liver cell spheroids in hormonally defined media. Exp Cell Res 1990;189(1):87–92.
[138] Landry J, et al. Spheroidal aggregate culture of rat liver cells: histotypic reorganization, biomatrix deposition, and maintenance of functional activities. J Cell Biol 1985;101(3):914–23.
[139] Koide N, et al. Formation of multicellular spheroids composed of adult rat hepatocytes in dishes with positively charged surfaces and under other nonadherent environments. Exp Cell Res 1990;186(2):227–35.
[140] Dvir-Ginzberg M, Elkayam T, Cohen S. Induced differentiation and maturation of newborn liver cells into functional hepatic tissue in macroporous alginate scaffolds. FASEB J 2008;22(5):1440–9.
[141] Liu Z, et al. Three-dimensional hepatic lobule-like tissue constructs using cell-microcapsule technology. Acta Biomater 2017;50:178–87.
[142] Xu J, Ma M, Purcell WM. Biochemical and functional changes of rat liver spheroids during spheroid formation and maintenance in culture: II. Nitric oxide synthesis and related changes. J Cell Biochem 2003;90(6):1176–85.
[143] Ma M, Xu J, Purcell WM. Biochemical and functional changes of rat liver spheroids during spheroid formation and maintenance in culture: I. Morphological maturation and kinetic

changes of energy metabolism, albumin synthesis, and activities of some enzymes. J Cell Biochem 2003;90(6):1166–75.
[144] Ye J, Shirakigawa N, Ijima H. Fetal liver cell-containing hybrid organoids improve cell viability and albumin production upon transplantation. J Biosci Bioeng 2016;121(6):701–8.
[145] Sidler Pfandler MA, et al. Small hepatocytes in culture develop polarized transporter expression and differentiation. J Cell Sci 2004;117(Pt 18):4077–87.
[146] Miyamoto S, et al. Expression of cytochrome P450 enzymes in hepatic organoid reconstructed by rat small hepatocytes. J Gastroenterol Hepatol 2005;20(6):865–72.
[147] Kuijk EW, et al. Generation and characterization of rat liver stem cell lines and their engraftment in a rat model of liver failure. Sci Rep 2016;6:22154.
[148] Lu HF, et al. Three-dimensional co-culture of rat hepatocyte spheroids and NIH/3T3 fibroblasts enhances hepatocyte functional maintenance. Acta Biomater 2005;1(4):399–410.
[149] Takezawa T, et al. Morphological and immuno-cytochemical characterization of a heterospheroid composed of fibroblasts and hepatocytes. J Cell Sci 1992;101(Pt 3):495–501.
[150] Harada K, et al. Rapid formation of hepatic organoid in collagen sponge by rat small hepatocytes and hepatic nonparenchymal cells. J Hepatol 2003;39(5):716–23.
[151] Lee SA, et al. Spheroid-based three-dimensional liver-on-a-chip to investigate hepatocyte-hepatic stellate cell interactions and flow effects. Lab Chip 2013;13(18):3529–37.
[152] Bao J, et al. Serum-free medium and mesenchymal stromal cells enhance functionality and stabilize integrity of rat hepatocyte spheroids. Cell Transplant 2013;22(2):299–308.
[153] Chou MJ, et al. Application of open porous poly(D,L-lactide-co-glycolide) microspheres and the strategy of hydrophobic seeding in hepatic tissue cultivation. J Biomed Mater Res A 2013;101(10):2862–9.
[154] Ikeda Y, et al. Long-term survival and functional maintenance of hepatocytes by using a microfabricated cell array. Colloids Surf B Biointerfaces 2012;97:97–100.
[155] Jun Y, et al. 3D co-culturing model of primary pancreatic islets and hepatocytes in hybrid spheroid to overcome pancreatic cell shortage. Biomaterials 2013;34(15):3784–94.
[156] Otsuka H, et al. Micropatterned co-culture of hepatocyte spheroids layered on non-parenchymal cells to understand heterotypic cellular interactions. Sci Technol Adv Mater 2013;14(6):065003.
[157] Liu Y, et al. Hepatocyte cocultures with endothelial cells and fibroblasts on micropatterned fibrous mats to promote liver-specific functions and capillary formation capabilities. Biomacromolecules 2014;15(3):1044–54.
[158] Lee G, et al. Reproducible construction of surface tension-mediated honeycomb concave microwell arrays for engineering of 3D microtissues with minimal cell loss. PLOS ONE 2016;11(8):pe0161026.
[159] Jeong GS, et al. Viscoelastic lithography for fabricating self-organizing soft micro-honeycomb structures with ultra-high aspect ratios. Nat Commun 2016;7:11269.
[160] Chan HF, Zhang Y, Leong KW. Efficient one-step production of microencapsulated hepatocyte spheroids with enhanced functions. Small 2016;12(20):2720–30.
[161] Alzebdeh DA, Matthew HW. Metabolic oscillations in co-cultures of hepatocytes and mesenchymal stem cells: effects of seeding arrangement and culture mixing. J Cell Biochem 2017;118(9):3003–15.
[162] Thomas RJ, et al. Hepatic stellate cells on poly(dl-lactic acid) surfaces control the formation of 3D hepatocyte co-culture aggregates in vitro. Eur Cell Mater 2006;11:16–26. [discussion 26].
[163] Ambrosino G, et al. Isolated hepatocytes versus hepatocyte spheroids: in vitro culture of rat hepatocytes. Cell Transplant 2005;14(6):397–401.
[164] Lee KW, et al. Influence of pancreatic islets on spheroid formation and functions of hepatocytes in hepatocyte-pancreatic islet spheroid culture. Tissue Eng 2004;10(7–8):965–77.

[165] Riccalton-Banks L, et al. Long-term culture of functional liver tissue: three-dimensional co-culture of primary hepatocytes and stellate cells. Tissue Eng 2003;9(3):401–10.
[166] Nyberg SL, et al. Rapid, large-scale formation of porcine hepatocyte spheroids in a novel spheroid reservoir bioartificial liver. Liver Transpl 2005;11(8):901–10.
[167] Yamashita Y, et al. Efficacy of a larger version of the hybrid artificial liver support system using a polyurethane foam/spheroid packed-bed module in a warm ischemic liver failure pig model for preclinical experiments. Cell Transplant 2003;12(2):101–7.
[168] Sakai Y, et al. A new bioartificial liver using porcine hepatocyte spheroids in high-cell-density suspension perfusion culture: in vitro performance in synthesized culture medium and in 100% human plasma. Cell Transplant 1999;8(5):531–41.
[169] Lazar A, et al. Formation of porcine hepatocyte spheroids for use in a bioartificial liver. Cell Transplant 1995;4(3):259–68.
[170] Gu J, et al. Heterotypic interactions in the preservation of morphology and functionality of porcine hepatocytes by bone marrow mesenchymal stem cells in vitro. J Cell Physiol 2009;219(1):100–8.
[171] Nantasanti S, et al. Disease modeling and gene therapy of copper storage disease in canine hepatic organoids. Stem Cell Rep 2015;5(5):895–907.
[172] Baron MG, et al. Pharmaceutical metabolism in fish: using a 3-D hepatic in vitro model to assess clearance. PLOS ONE 2017;12(1):pe0168837.
[173] Irani K, et al. Mechanical dissociation of swine liver to produce organoid units for tissue engineering and in vitro disease modeling. Artif Organs 2010;34(1):75–8.

Chapter 8

Methods for Engineering of Multicellular Spheroids to Reconstitute the Liver Tissue

Nobuhiko Kojima*, Fumiya Tao*, Hirotaka Mihara*, Shigehisa Aoki**
*Yokohama City University, Yokohama, Japan; **Saga University, Saga, Japan

1 INTRODUCTION

Organs and tissues of our body have three-dimensional (3D) microstructures by orderly arranging multiple kinds of cells. In order to develop miniature organs that can be used in industries and medicine, we are working on how to reproduce 3D microstructures by forming spaces and filling extracellular matrices (ECMs) in multicellular spheroids (MCSs). The behavior of stem cells and cancer in the liver is also related to the 3D microstructure, and the reconstitution of the microstructure in vitro can be a powerful tool in the study of those cells.

MCS using hepatocytes is very popular way of 3D culture. To form MCS, cells are seeded on a nontreated plate, and cultured under static or gentle shaking conditions. There are other methods/tools such as a hanging drop method, U- or V-shaped bottom 96-well plates. 3D MCS culture is often used to induce higher functions of hepatocytes. However because of their chaotic internal structures, these MCSs do not exhibit the expected functions. An approach to overcome the limit of MCS functions is to fabricate MCS with internal microstructures similar to the actual liver. In order to create 3D hepatic tissues with microstructures, it is popular to utilize sponge-like scaffolds coated with ECMs. With this method, it is possible to control the microstructures by changing pore size/shape or altering ECMs on the surfaces of scaffolds. In practice, however, it is almost impossible to efficiently adhere cells to whole complex surfaces of the scaffolds. Most of the cells do not adhere to the scaffolds, and in some cases it may be necessary to incubate for a period of weeks to cover the surfaces with proliferated cells. If the cells have low adhesive and/or growth property, it is impossible to perform 3D culture. It means that sponge-like scaffolds are not enough to imitate 3D microstructures in the liver.

Recently, we establish a unique method to gather cells, hydrogel beads and ECMs using a methylcellulose medium. With this method, we can form a

MCS with microchannel spaces and/or thin layers of ECMs in couple of days. The MCS is completely different from conventional MCSs or 3D tissues using sponge-like scaffolds. In this chapter, we introduce the method and discuss the effect of the microstructures in the hepatic MCS.

2 RAPID FORMATION OF MCS USING METHYLCELLULOSE MEDIUM

Methylcellulose is produced from cellulose by partially replacing a part of the hydroxyl group to a methoxy group. Because of the modification, methylcellulose can be dispersed in water. Methylcellulose is generally used as a thickening food additive. It is also popular in the life science field because about 1% methylcellulose medium is used for colony formation assay of hematopoietic stem/progenitor cells. When methylcellulose is dissolved in a culture medium at a concentration of 3%, it becomes a very viscous solution. We found that this kind of high viscous solution can rapidly form the cell aggregation state which helps to form MCS [1].

Fig. 8.1 shows the process of cell aggregation in the methylcellulose medium. We poured the 3% methylcellulose medium into a culture dish using a positive displacement type micropipette, and the medium was left for a while to remove air bubbles. Then, 1 μl of a normal culture medium in which the cells were suspended at a concentration of 1×10^6 to 1×10^7 cells/ml was injected into the 3% methylcellulose medium. Since the methylcellulose medium swelled while absorbing the normal culture medium, the suspended cells were forced to aggregate. When 1 μl of the cell suspension at a concentration of 1×10^6 cells/ml was injected, approximately 1,000 cells were gathered in 10 min. When we used Hep G2 or Huh7 cells, intercellular adhesion was started immediately and they form a MCS in several hours by continuing culture in the methylcellulose medium. It is possible to change the concentration of methylcellulose.

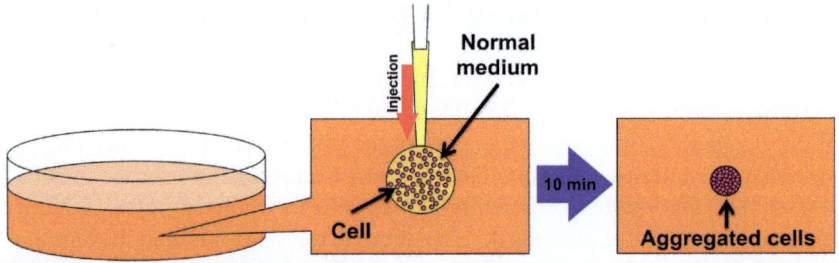

FIGURE 8.1 Rapid aggregation of cells by 3% methylcellulose medium. 3% methylcellulose medium is high viscosity solution and swells very well when they contact normal culture medium. If a little amount normal culture medium is injected into the methylcellulose medium, the normal culture medium is immediately absorbed by the methylcellulose medium. Suspended cells are therefore gathered rapidly and we can culture the aggregated cells in the methylcellulose medium until they biologically adhere each other.

However, as the concentration decreases, the aggregation becomes slow so that the cells are less likely to be aggregated. About 1% is minimum concentration of methylcellulose medium for effective aggregation. Conversely, it is difficult to disperse the methylcellulose in the culture medium at more than 3%. The aggregation can be observed not only in the methylcellulose medium but also in all solutions in which the polymer is dispersed. We prefer to use methylcellulose because it is cheap, it has little influence on pH, and it is well recognized as a safe food additive. It is better to reduce injecting volume of the normal culture medium to shorten the time until completion of the aggregation. We are usually using an ordinary micropipette to inject normal culture medium suspending cells and the range of the injection volume is 0.5–1 µl. This is because the volume is a limit for the manual operation. We can choose the injection volume 0.1 µl or less when using a robotic operation.

The advantage of using the 3% methylcellulose medium is the ability to rapidly bring the cells into aggregated state. In spheroid formation by gyratory culture, growth of clumps takes time since it is only a moment in motion that the cell comes into contact with another cell. On the other hand, cells aggregated in the methylcellulose medium can maintain a stable contact with adjacent cells so that the cells surely form intercellular adhesion by their own biological property. Previously we tired to measure the time that was required for establishment of cell-to-cell adhesion of Hep G2 cells. In that experiment, we trapped two single cells with two optical tweezers, contacted the two single cells for a certain period of time, and then tried to separate the cells. When we contacted the cells for 1 s, almost all pairs were able to be disengaged, indicating that such short time is not enough to form cell-to-cell adhesion. As the contact time increased, the cell adhesion became steady. We concluded that at least 5 min was required to form stable cell adhesion between Hep G2 cells [2]. Of course, the adhesiveness is different between cell types and the results would be different depending on the trapping power of the optical tweezers. However, it is clearly disadvantage to form cell-to-cell adhesion in a dynamic rotation culture that cannot allow cells to contact for a long time. It is necessary to bring cells into contact for at least a few minutes to surely make intercellular adhesion. Our rapid cell aggregation method using 3% methylcellulose medium is therefore suitable to surely form MCS.

High viscosity of the methylcellulose medium provides other merits. As the MCS can be kept floating in the methylcellulose medium, it is possible to prevent the MCS from contacting to the bottom of the dish. In the same way, because the position of the MCS does not change in the methylcellulose medium, it is easy to observe the same MCS continuously for entire culture period. The viscosity becomes a problem when we need to remove the MCS from the methylcellulose medium. In that step, MCS can be recovered easily by lowering the viscosity by addition of a cellulase reagent that can digest cellulose backbone. It needs 30–60 min for enzymatic digestion. There is room for developing a method that can recover MCS in a simpler way in the future.

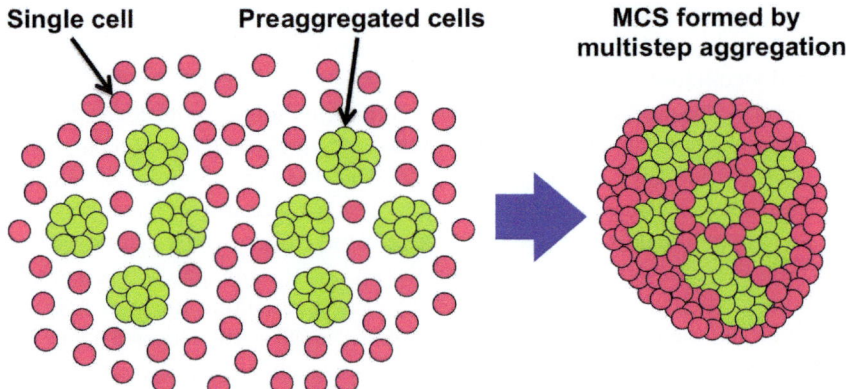

FIGURE 8.2 **Multistep aggregation of cells.** Because the 3% methylcellulose medium can aggregate particles those are from 100 nm to 100 μm in diameter, it is possible to aggregate single cells and pre-aggregated MCS together as shown here. This is an option to make MCS with microstructures.

Using rapid and forced aggregation system by methylcellulose medium, it is very easy to form a MCS consisting of multi types of cells even their adhesive properties are largely different. Immediately after gathering, the positions of cells are completely random in the heterogeneous MCS. Under such "unstable" condition, cells tend to migrate spontaneously and show self-organized tissue patterns depending on the combination of the cell types [1]. Besides, cells with considerably weak adhesiveness can form MCS. In an extreme case, it is possible to reaggregate suspended bone marrow cells, including so many blood cells, into bone marrow-like tissue [3]. As explained in Section 3, materials that do not have biological adhesiveness (e.g., hydrogel beads) can contribute to the formation of aggregated state together with other cells [4]. In addition, as methylcellulose medium can aggregate particulate materials with a diameter from 100 nm to 100 μm, it is possible to structuralize MCS by multistep aggregation as shown in Fig. 8.2 [1]. This will also be mentioned in a later section, macromolecules like collagen and polysaccharides are available to be used in MCS formation in methylcellulose medium [5]. Methylcellulose has low reactivity with cells, that is one of reasons that we choose it for cell aggregation, but it is also attractive to mix biologically active materials like cytokines and/or ECMs into the methylcellulose medium itself to control organization or functions of the MCS. As described above, the method using methylcellulose medium is useful to create various microstructures depending on ideas.

3 MICROCHANNEL FORMATION IN MCS

Three-dimensional culture by MCS formation is attracting attention in recent years as a culture method close to living bodies. For example, albumin secretion by hepatocytes shows higher activity in 3D MCS cultures than in

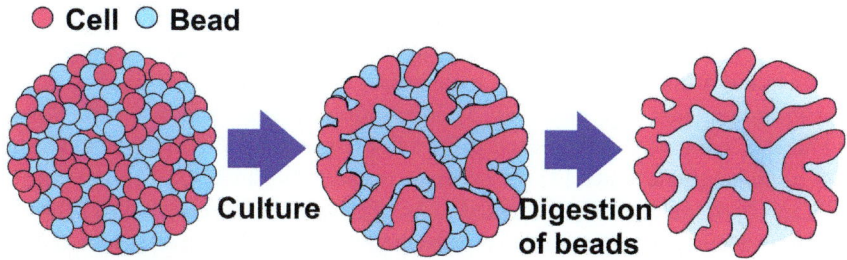

FIGURE 8.3 **Formation of MCS with microchannel-like structures.** MCS with mixture of cells and hydrogel beads make network-like patterns. After aggregation, the cell adhere each other so that the hybrid MCS has microchannel-like structures. The alginate hydrogel beads can be digested by alginate lyase immediately to open the spaces.

two-dimensional plate cultures [6]. Because expression of genes related to drug metabolism such as CYP is also induced, pharmaceutical companies are using MCS for drug screening research. In contrast with the benefits, the problem of MCS culture is limitation of gases/nutrients exchange, and the center part of larger MCS has tendency to become necrotic [7–9]. It is also a problem that the applied drugs work only on cells of MCS surfaces. These are because conventional MCSs are formed by closely packed cells and there are no blood vessels carrying oxygen and nutrients like the living body. Although it would be ideal to create a luminal structure by vascular endothelial cells inside the MCS, it is impossible to reconstitute such functional vessel-like structures by merely mixing the vascular endothelial cells into the MCS. Therefore, we focused on a concept enhancing gases/nutrients exchange by making microchannels as shown in Fig. 8.3 instead of building vasculatures by endothelial cells. By forming "hybrid MCS" comprising of same number of cells and hydrogel beads having about 20 μm in diameter, it is expected that hydrogel beads spontaneously aligned as "branching channel patterns" inside of the MCS. Hydrogel beads have relatively large molecular weight cut off, so that gases, nutrients, and spherical proteins like albumin can easily diffuse in the hydrogel beads [10–13]. If there is an enzyme that can digest the hydrogel, it is possible to make microchannels with actual void spaces. In order to realize such a concept, we had to develop technologies for producing hydrogel beads which have almost same size as cells, and for forming hybrid MCS including half amount of hydrogel beads having no biological adhesiveness.

Alginic acid is a polysaccharide contained in marine algae, and its solution has the feature of instantaneously turn into a gel by contact with another solution containing calcium ion. Alginic acid is a stable material with little toxicity, and is widely used from food additive to immunoprotective membrane. Alginate gel can be solubilized again by not only calcium ion chelating but also enzymatic digestion using alginate lyase. Because of these characteristics, alginic acid is one of the best materials to make hydrogel beads for our concept. In order to prepare alginate hydrogel beads having the same diameter as cells, droplets of a 1.5% sodium alginate solution were discharged by an inkjet nozzle with a 25 μm bore size and let them gelate by receiving these droplets with a 5% calcium chloride solution. As a result, alginate hydrogel beads having 20 μm in diameter were obtained [4]. In addition, it was confirmed that hydrogel beads were able to be digested in 10 min using a culture medium containing 200 μg/ml alginate lyase [4]. In order to prepare hybrid MCS comprising of cells and hydrogel beads, we decided to use the rapid agglutination method with the methylcellulose medium. As already mentioned, the formation of an aggregated state using the swelling phenomenon of the methylcellulose medium does not require biological properties. As the aggregated state was maintained in the methylcellulose medium until the cell adhered each other, we were able to obtain the hybrid MCS containing 50% of hydrogel beads. Observation with laser confocal microscopy revealed that the internal microstructures like microchannels were fabricated as we expected [4].

Using normal MCS consisting of 10,000 cells or hybrid MCS consisting of 5,000 cells and same number of hydrogel beads, albumin secretion activity was measured at after 2 days from MCS formation and normalized by cell number. Diameter of both MCS was about 700–800 μm. Generally, the MCS under such condition have a tendency to show necrosis in the core part at after 3–4 days from MCS formation. It means that there is some restriction of material exchange. The hybrid MCS containing alginate hydrogel beads showed an improvement in albumin secretion activity compared to the normal MCS [4]. It suggests that embedded hydrogel beads worked as a microchannels to enhance material exchange and this approach is effective to regulate microenvironment in the 3D MCS culture. Next, to confirm the connectivity of the microchannels formed by the hydrogel beads, the beads were removed with alginate lyase digestion and the MCS with actual void spaces was immediately immersed in a medium in which fluorescent particles having a diameter of 1 μm were suspended. After 15 min of incubation, the fluorescent particles adhering to the cell was observed with a confocal microscope. In the normal MCS formed with only cells, the fluorescent particles adhered only to the cells existing most outer layer of the MCS, whereas in the MCS with microstructures, the fluorescent signals were able to be detected whole body of the MCS [4]. It was revealed that the void spaces formed by the digestion of hydrogel beads in the hybrid MCS functioned as connected microchannels that allow passing through objects with a diameter of 1 μm.

To exhibit the functions of epithelial cells, it is necessary to reconstruct epithelial structures. For example, hepatocytes construct a hepatic cord structure composed of one or two layers of hepatocytes in liver lobules, and this microstructure is important for their functions. On the other hand, a conventional MCS has disorganized structure with tightly packed cells. Cancer cells such as Hep G2 cells can survive even in multilayered conditions. However, such conditions are abnormal in our body and not suitable for normal epithelial cells. Actually, mouse primary fetal hepatic cells are difficult to cultivate as MCS. When they were cultured in the MCS state for about a week in the presence of oncostatin M, a differentiation inducing factor, cells on the surfaces of MCS continued to express an epithelial marker cytokeratin 8/18, but the cells inside of the MCS did not express the marker. It was thought that epithelial-mesenchymal transition occurred in such abnormal conditions. On the basis of the result, the conventional MCS is not so adequate for epithelial cell culture. In the MCS with microchannels, the cells form sheet-like structures of one to two layers due to the existence of the hydrogel beads. Epithelial-mesenchymal transition is suppressed and primary fetal hepatic cells were able to be safely cultured in such MCS with hydrogel beads [14]. The sheet-like hepatocyte alignments of 1 or 2 layers in the hybrid MCS are similar to the cord-like structure in the liver lobule, and the spaces shaped by the hydrogel beads are similar to the sinusoidal structures. The method to culture the cells with hydrogel beads is not only for increasing material exchange ability, but also for realizing 3D culture reconstituting structures of epithelial hepatic cells in liver lobules.

By using alginate hydrogel beads as a spacer, it is possible to control the microstructures inside of the MCS and enhance the material exchange ability by diffusion. This improves oxygen supply to the interior of the MCS, but if the amount of oxygen in the culture medium is low, the microstructure cannot improve the oxygen deficiency in the MCS. Stirring of the culture medium is one of methods for increasing the dissolved oxygen, but there is a problem that cells are damaged due to mechanical stresses. Particularly, the hybrid MCS containing 50% hydrogel beads are breakable, so that vigorous stirring is not available. We therefore utilized a culture plate which has gas permeable membrane made of polydimethylsiloxane (PDMS) on the bottom of plate. PDMS is a type of silicone and has excellent gas permeability as well as optical transparency. Even under the static culture condition, the PDMS membrane improves the oxygen supply into the culture medium [15, 16]. As mentioned above, the central part of the MCS composed of 10,000 Hep G2 cells undergoes necrosis at 3–4 days of culture. Hybrid spheroids mixed with alginate hydrogel beads can delay the necrosis of the cells somewhat, but necrosis was still observed at 5–6 days of culture. When the culture was carried out using the PDMS membrane plate, it was found that the hybrid MCS was able to be cultured without necrosis for at least 10 days (Fig. 8.4). This long-term culture system is suitable to the repeat-dose studies using anti-cancer drugs and so on.

4 REGULATION OF HEPATIC FUNCTIONS BY ECM LOADING

MCS formation basically uses only cells, but various ECMs exist in actual organs. The use of ECM, therefore, would be an important option to control the functions of the MCS. We found that the rapid cell aggregation method using the methylcellulose medium can agglutinate not only insoluble

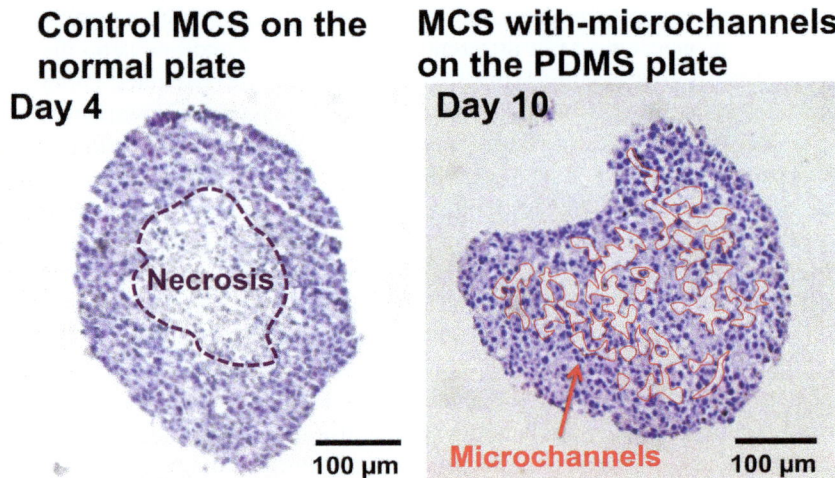

FIGURE 8.4 **Longterm culture system for the repeat-dose studies.** Normal MCS shows necrotic core in 3–4 days when it is formed with 10,000 of Hep G2 cells. The hybrid MCS cultured on the PDMS plate does not show any necrosis for at least 10 days. This culture method is suitable for the repeat-dose studies.

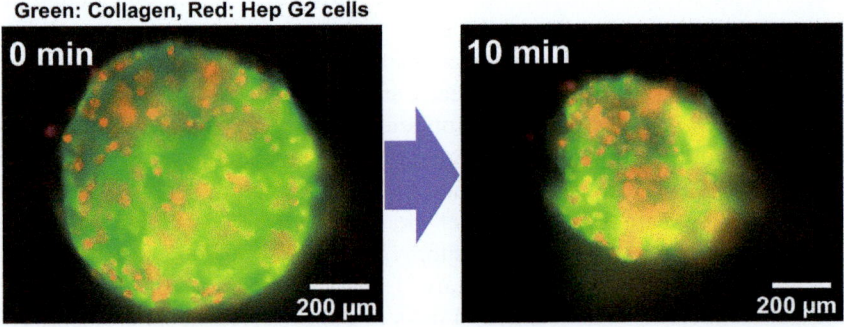

FIGURE 8.5 **ECM condensation in the 3% methylcellulose medium.** Not only cells but also ECMs can be gathered when they are injected into the 3% methylcellulose medium. Collagen was conjugated with FITC, and Hep G2 cells were pre-stained with red-fluorescent dye.

particulate matters such as cells and hydrogel beads but also macromolecular proteins such as collagen and matrigel dispersed in the culture medium. In principle, various kinds of water-dispersible macromolecules, such as polysaccharides, can be concentrated in our method. Fig. 8.5 shows that when we injected culture medium suspending Hep G2 cells and dispersing collagen to a methylcellulose medium, the cells and the collagen were concentrated and formed MCS including ECM. In this experiment, the pH of the collagen solution was adjusted to 7.4, and it was cooled on ice to prevent gelation until injection. The tips of the micropipette and 3% methylcellulose medium were cooled until the injection procedure. After injection, the temperature of the collagen solution gradually increased, and finally the injected collagen solution turned into a gel. By this technique, the cells can be encapsulated in an ECM gel capsule. It is popular using emulsion by mixing water and oil to make ECM capsules. Some cells were damaged due to contacting with oil by the emulsion method. On the other hand, there was no toxicity in our method because it does not need oil.

Do the ECM capsules exert higher functions as expected? We encapsulated Hep G2 cells using matrigel that has potential to enhance hepatic functions. Hep G2 cells were suspended in commercially available matrigel solution to a concentration of 2×10^6 cells/ml, and 1 µl of them was injected into a 3% methylcellulose medium using a micropipette. Matrigel solution containing cells turned into a gel while it was aggregating in a 3% methylcellulose medium. After 2 days culture from injection, the capsules were isolated from the methylcellulose medium, fixed and sectioned to observe inside of the capsules. In the matrigel capsule, the cells were not directly adhered to each other, and the cells were scattered in the capsule. A part of isolated capsules at 2 days were transferred into a fresh culture medium to measure albumin secretion activity for 24 h. The cell in matrigel capsule showed lower albumin secretion activity than those in the control MCS. The reason the cells did show lower function was thought that cells were dispersed inside of matrigel capsule and there is no direct contact between cells. It means that thick ECM layer blocked cell-to-cell communication and inhibited liver functions.

If the thick ECM layer diminishes the functions, how does the thin ECM layer work? In the previous encapsulation, cells were suspended in a commercially available undiluted matrigel solution. Therefore, after injection into the methylcellulose medium, matrigel solution turned into a gel before the cells are completely concentrated. In fact, the diameter of the gel capsule was larger than the normal MCS. We tried to delay or inhibit the gelation by decreasing the concentration of matrigel solution. As the concentration of matrigel was lowered, the diameter of the capsule gradually decreased, and when it was diluted 30 times, the size of capsule became almost the same as the matrigel-free MCS (Fig. 8.6). In 30 times diluted condition, it was not a typical capsule but almost same as normal MCS by observation of paraffin sections. It did not mean that there was no ECM inside of the MCS but there was thin film-like layer of ECM between the cells. In case of using FITC-labeled collagen, 40 times dilution was better to make thin layer (Fig. 8.7). The dilution rate should be considered for each ECM reagent. We named such MCS with thin-layered ECM as "ECM-loaded MCS" to distinguish from the ECM capsule. Interestingly, matrigel-loaded MCSs showed higher albumin secretion activity than the control MCSs (Fig. 8.8). There was a specific condition under which both cell-cell and cell-to-ECM were effective at the same time. These results indicate the fact that we can induce functions by guiding the microenvironment of the cells.

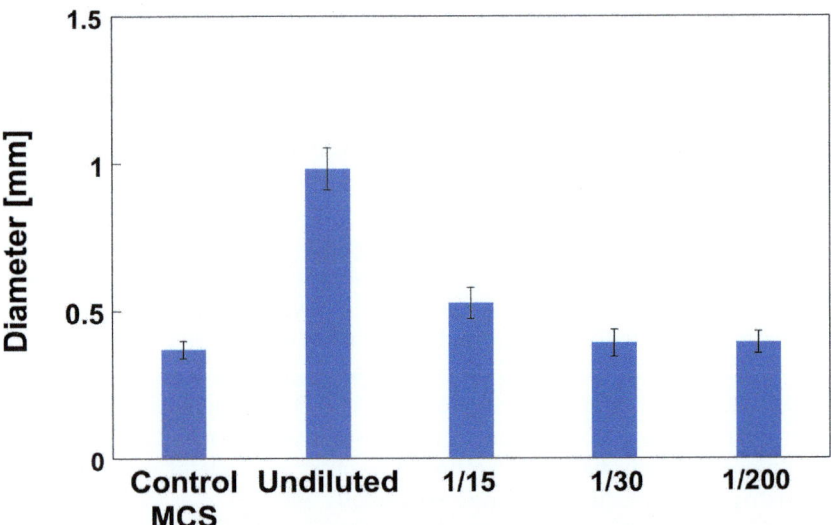

FIGURE 8.6 **Diameters of the MCS comprising various concentrations of matrigel.** When we used undiluted matrigel, they rapidly turned to gel in the 3% methylcellulose medium. The diameter, therefore, became larger. As matrigel solution was diluted, the size of MCS was reduced.

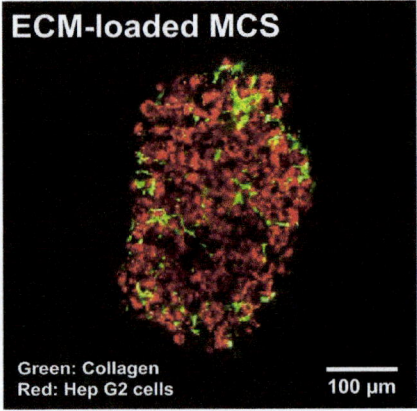

FIGURE 8.7 **ECM-loaded MCS.** If the ECM solution was enough diluted to prevent gelation, we can make the MCS which has thin-layered ECM. In this picture, we used 40 times diluted FITC-labeled type I collagen (green) and pre-stained Hep G2 cells (red).

Matrigel-loaded MCS resulted in induction of the albumin secretion activity. What kind of effects can be obtained by applying other ECMs? To answer such question, same kind of experiments were performed by using type I collagen. MCS containing type I collagen at various concentrations were prepared by stepwise dilution and injected them to the 3% methylcellulose medium together with Hep G2 cells. Type I collagen capsule formed by using undiluted type I collagen showed decreased albumin secretion activity, and type I collagen could

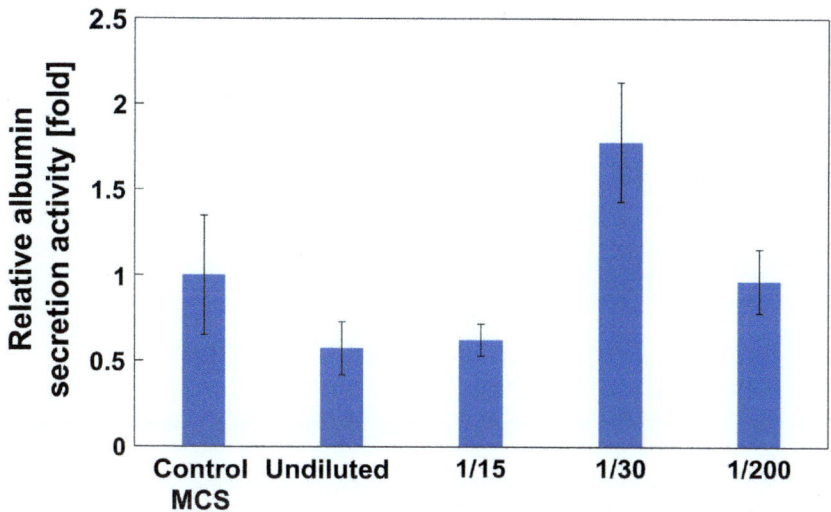

FIGURE 8.8 **Albumin secretion activity of the MCS comprising various concentrations of matrigel.** Cell-to-cell interaction is indispensable to show albumin secretion activity. Therefore, thick layer of matrigel reduced the activity as shown in the undiluted and the 15 times diluted conditions. On the other hand, the MCS having less amount of matrigel (30 times diluted) that existed as thin-layers in the MCS exert higher activity compared with the control MCS.

not induce the activity in any conditions. These results showed that improvement of albumin secretion activity in matrigel-loaded MCS was matrigel specific. Does type I collagen have no effect on hepatocyte functions? Although albumin secretion activity was not improved, type I collagen activated cell proliferation ability under the condition of type I collagen-loaded MCS. Conversely, this growth stimulating effect was not seen in matrigel-loaded MCS, indicating that the effects of EMC-loading on the cellular functions are different depends on the type of ECM. By using methylcellulose medium, various polymers can be loaded in MCS as thin layers, so this system is also useful to investigate 3D microenvironments that can regulate stem or tumor cells behaviors.

5 CELL POLARITY FORMATION BY ECM LOADING

The effect of matrigel-loading on hepatocyte MCS is not limited to improving metabolic activity. It is well known that matrigel is a mixture of ECMs including basement membrane components, laminin, type IV collagen, and so on, which related to cell polarity acquisition. There is almost no basement membrane structure between hepatocytes and sinusoidal endothelial cells in the liver. Researchers have different opinions as to whether induction of cell polarity by matrigel has physiological significance or not. However in order to discuss certain aspects of epithelial hepatocytes, it is important to establish a 3D culture system that promotes the acquisition of cell polarity.

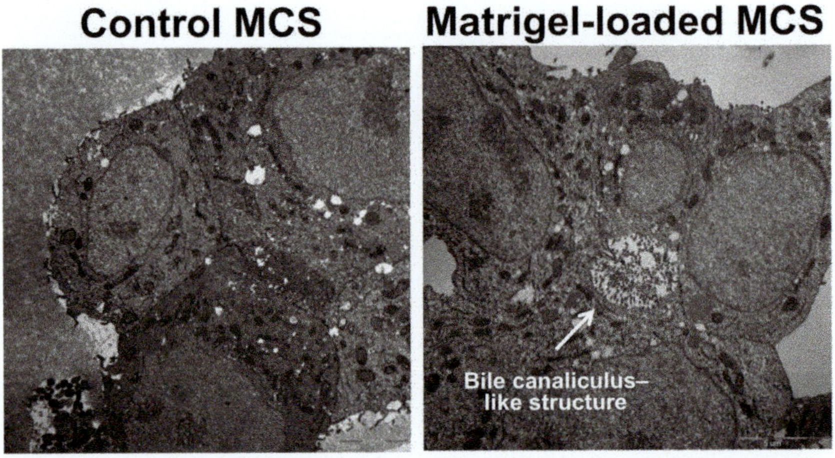

FIGURE 8.9 Formation of bile canaliculus-like structure in the matrigel-loaded MCS. Matrigel-loaded MCS using Hep G2 cells had a tendency to form bile canaliculus-like structures. In those structures, microvilli were detected. They were able to be detected in the MCSs those were fixed after 2 days from injection into the 3% methylcellulose medium.

As described in the previous section, matrigel-loaded MCS was formed by injecting 30 times diluted matrigel solution suspending 2,000 Hep G2 cells into the 3% methylcellulose medium. Matrigel-loaded MCSs were removed form the methylcellulose medium after 2 days from injection, fixed, and embedded to paraffin. Sections were evaluated by immunostaining. There was no difference in the expression and localization of E-cadherin regardless of the presence or absence of matrigel, whereas those of ZO-1 were clearly induced in the matrigel-loaded MCS. In addition, electron microscopic observation revealed that the structures seem to be bile canaliculi were often found in the matrigel-loaded MCS (Fig. 8.9). On the other hand, the expression and localization of MRP2 was still insufficient even in the sample filled with matorigel. ECM loading was partially effective to induce cell polarity, but we need further study to establish precise cell polarity as close as the actual liver. We are now trying to apply this method for aggregating human primary hepatocytes and human iPS cell-derived hepatocyte-like cells. We believe that it would improve their culture condition and enhance the liver functions in vitro.

6 CONCLUSION

As summarized in Table 8.1, the 3% methylcellulose medium is useful for rapid cell aggregation, forming stable cell-cell contact, keeping MCS to away from the bottom of the plate, enabling MCS formation using low adhesive cells or preaggregated smaller MCSs, beads embedding, and ECM loading. On the other hand, this method has a limitation that the 3% methylcellulose medium has relatively higher viscosity. It is important to develop a medium the viscosity

TABLE 8.1 Advantage and Disadvantage of This Method

	Gyratory culture	96 well U-shaped	Hanging drop	This method
Rapid aggregation	−	+	+	+++
Stable cell-cell contact	−	+	+	+++
Without contact from the plate	−	+	+++	+++
Availability for low adhesive cells	−	+	+	+++
Wide range of diameters	−	+	+	+++
Beads embedding	−	−	−	+++
ECM loading	−	−	−	+++
Preparation and Recovery	+++	+	−	−

Comparison between this method and conventional methods. Strong advantage: +++; Advantage: +; disadvantage: −.

is controllable. This unique method gives microstructures to MCS, and is effective on material exchange, epithelial cell culture, and forming cell polarity. Miniaturized organs with realistic microstructures are expected to contribute understanding stem cells and cancer in 3D condition as well as developing medical and pharmaceutical applications.

LIST OF ACRONYMS AND ABBREVIATIONS

MCS multicellular spheroid
ECM extracellular matrix
3D three-dimensional
PDMS polydimethylsiloxane

REFERENCES

[1] Kojima N, Takeuchi S, Sakai Y. Rapid aggregation of heterogeneous cells and multiple-sized microspheres in methylcellulose medium. Biomaterials 2012;33(18):4508–14.
[2] Kojima N, Miura K, Matsuo T, Nakayama H, Komori K, Takeuchi S, et al. Rapid and direct cell-to-cell attachment using avidin-biotin binding system: large aggregate formation in suspension culture and small tissue element formation having a precise microstructure using optical tweezers. J Robot Mechatr 2010;22(5):619–22.
[3] Sayo K, Aoki S, Kojima N. Fabrication of bone marrow-like tissue in vitro from dispersed-state bone marrow cells. Regenerative Therapy 2016;3:32–7.
[4] Kojima N, Takeuchi S, Sakai Y. Fabrication of microchannel networks in multicellular spheroids. Sens Actuators B: Chem 2014;198:249–54.
[5] F. Tao, A. Sato, S. Sakuma, S. Aoki, F. Arai and N. Kojima. A novel method for loading ECM thin layer into multicellular spheroid, submitted for publication.

[6] Matsushita T, Ijima H, Koide N, Funatsu K. High albumin production by multicellular spheroids of adult rat hepatocytes formed in the pores of polyurethane foam. Appl Microbiol Biotechnol 1991;36(3):324–6.

[7] Sutherland RM, Sordat B, Bamat J, Gabbert H, Bourrat B, Mueller-Klieser W. Oxygenation and differentiation in multicellular spheroids of human colon carcinoma. Cancer Res 1986;46(10):5320–9.

[8] Acker H, Carlsson J, Mueller-Klieser W, Sutherland RM. Comparative pO_2 measurements in cell spheroids cultured with different techniques. Br J Cancer 1987;56(3):325–7.

[9] Carlsson J, Acker H. Relations between pH, oxygen partial pressure and growth in cultured cell spheroids. Int J Cancer 1988;42(5):715–20.

[10] Tanaka H, Matsumura M, Veliky IA. Diffusion characteristics of substrates in Ca-alginate gel beads. Biotechnol Bioeng 1983;26(1):53–8.

[11] Lanza RP, Kuhtreiber WM, Ecker D, Staruk JE, Chick WL. Xenotransplantation of porcine and bovine islets without immunosuppression using uncoated alginate microspheres. Transplantation 1995;59(10):1377–84.

[12] DeGroot AR, Neufeld RJ. Encapsulation of urease in alginate beads and protection from α-chymotrypsin with chitosan membranes. Enzyme Microbial Technol 2001;29(6–7):321–7.

[13] Dembczynski R, Jankowski T. Determination of pore diameter and molecular weight cut-off of hydrogel-membrane liquid-core capsules for immunoisolation. J Biomater Sci Polymer Ed 2001;12(9):1051–8.

[14] Motoyama W, Sayo K, Mihara H, Aoki S, Kojima N. Induction of hepatic tissues in multicellular spheroids composed of murine fetal hepatic cells and embedded hydrogel beads. Regenerative Therapy 2016;3:7–10.

[15] Nishikawa M, Yamamoto T, Kojima N, Komori K, Fujii T, Sakai Y. Stable immobilization of rat hepatocytes as hemispheroids onto collagen-conjugated poly-dimethylsiloxane (PDMS) surfaces: importance of direct oxygenation through PDMS for both formation and function. Biotechnol Bioeng 2008;99(6):1472–81.

[16] Nishikawa M, Kojima N, Komori K, Yamamoto T, Fujii T, Sakai Y. Enhanced maintenance and functions of rat hepatocytes induced by combination of on-site oxygenation and coculture with fibroblasts. J Biotechnol 2008;133(2):253–60.

Chapter 9

Role of Platelet, Blood Stem Cell, and Thrombopoietin in Liver Regeneration, Liver Cirrhosis, and Liver Diseases

Tomohiro Kurokawa, Nobuhiro Ohkohchi
University of Tsukuba, Tsukuba, Ibaraki, Japan

1 INTRODUCTION

Platelets are anuclear blood cells derived from thrombopoietin (TPO)-stimulated megakaryocytes; they not only play a critical role in hemostatsis, but also secrete several growth factors, including platelet-derived growth factor (PDGF) and hepatocyte growth factor (HGF) which are required for tissue regeneration or repair [1–6].

Platelet-rich plasma, which is an autologous concentration of platelets in a small volume of plasma, has already been used in the dental implantation, maxillofacial surgery, and plastic surgery for the promotion of regenerating damaged tissue [7].

Platelets contain a large number of secretary granules [8]. Platelets are activated by various types of stimulation and release active substances from the granules [9–11]. Three types of secretary granules are recognized: alpha granules, dense granules, and lysosomal granules. Each granule contains secretory substances, such as HGF, PDGF, insulin-like growth factor-1 (IGF-1), vascular endothelial growth factor (VEGF), epidermal growth factor (EGF), transforming growth factor-β (TGF-β), serotonin, adenosine diphosphate (ADP), adenosine tri-phosphate (ATP), sphingosine 1-phosphate (S1P), among others [8,12–15].

There are several negative effects of platelet degranulation, such as inflammation [16], malignancy [17–19], and immune response [20–23].

Since Trousseau et al. first reported the excessive blood coagulation in cancer patients with elevated platelet counts in 1865, a large number of studies have been conducted on cancer and platelets [24].

Cancer development involves the following progression: (1) separation from the primary tumor and infiltration of blood vessels; (2) transportation through blood flow; (3) adhesion to the capillary walls of distant organs; (4) extravascular migration; and (5) proliferation at the new site to form a metastatic lesion. Of course, platelets are involved in almost every step of this process.

Platelets are involved at each stage of cancer metastasis from separation of cancer cells from the primary tumor to proliferation at the metastatic site. Therefore inhibiting the platelet–cancer interaction by targeting platelets may lead to the development of novel therapeutic agents to treat cancer metastasis. Recently, there have been reports that aspirin significantly reduced the mortality rates of adenocarcinoma but did not inhibit the mortality rate for other types of cancer [25,26]. Furthermore, the mechanism of action regarding this purported antimetastatic effect of aspirin can also be observed in cells other than cancer cells, and the increased bleeding tendency with an aspirin regimen is problematic.

In other words, it is important to target not platelets, which have critical physiological activities, but rather cancer-specific platelet activating factors, including integrins, cathepsin B, Necl-5, MMP2, GPIb, and podoplanin, all of which are expressed on cancer cells.

On the other hand, positive effects have been reported, such as hemostasis [27], wound healing [28–31], and tissue regeneration [32–38].

Platelets are reported to accumulate in the liver under pathological conditions such as cholestasis [39], liver cirrhosis (LC) [40], ischemia/reperfusion [41–44], the residual liver after hepatectomy [45], and viral hepatitis [46].

2 BLOOD STEM CELL

Platelets are produced from megakaryocytes in bone marrow. Cultivation of megakaryocytes became possible in 1994 due to the cloning of TPO, a cytokine required for megakaryocyte hematopoiesis; this advancement enabled progress in the elucidation of megakaryocyte hematopoiesis and the mechanism of thrombopoiesis. In addition, advances in both in vitro and in vivo real-time imaging technology have helped to elucidate the mechanism of thrombopoiesis in detail. The current predominant model is the "proplatelet model," which holds that protrusions and distal ends of long, narrow, string-shaped proplatelets extending from megakaryocytes are released as platelets [47]. Megakaryocyte proliferation/differentiation/maturation and the release of platelets are temporospatially regulated by various humoral factors, transcription factors, and microenvironments. These processes can be broadly divided into the following three stages: (1) differentiation from hematopoietic stem cells into megakaryocytes, (2) megakaryocyte maturation, and (3) release of platelets from megakaryocytes.

Megakaryocytes are believed to differentiate from hematopoietic stem cells via megakaryocyte-erythroid progenitor cells (MEPs). The crucial transcription factor that determines whether MEPs differentiate into erythroid cells or megakaryocytes is c-MYB. In genetically modified mice in which exogenous DNA is inserted into the regulatory region of the c-Myb gene to inhibit its function, the function of c-MYB in MEP is diminished, and the red blood cell count is reduced; however, in this context, differentiation into megakaryocytes is also enhanced, and the platelet count in peripheral blood is increased.

Crucial transcription factors in the megakaryocyte maturation process, which is characterized by multinucleation and enlargement of the nucleus, include stem cell leukemia (SCL)/T-cell acute lymphocytic leukemia protein 1 (TAL1), GATA-binding factor-1 (GATA-1), acute myeloid leukemia 1 protein (AML1), and friend leukemia integration 1 transcription factor (FLI-1). Mice in deficient in these genes demonstrate reduced platelet counts and accumulation of small, undifferentiated megakaryocytes.

Nuclear factor, erythroid 2 (NF-E2) is a transcription factor that regulates the process of platelet production from mature megakaryocytes. While NF-E2-knockout mice exhibit accumulation of multinucleated, enlarged megakaryocytes, they also completely lack proplatelet formation and exhibit a phenotype characterized by a markedly reduced platelet count [48]. Both megakaryocyte proliferation/differentiation/maturation and hematopoietic stem cell maintenance require TPO-mediated signaling.

Transcription factors that regulate the proliferation and differentiation of megakaryocytes include Runt-related transcription factor 1 (RUNX1), GATA-1, FLI-1, and NF-E2; these factors act synergistically during megakaryocyte-specific protein expression. Megakaryocyte maturation and platelet release require interaction with other cells and intercellular matrices. In the bone tissue margin, proplatelet formation is inhibited by contact with collagen. Megakaryocyte progenitor cells differentiated from hematopoietic stem cells migrate from the bone tissue margin to the vicinity of sinusoids. This migration occurs in accordance with oxygen tension and the concentration gradient of stromal-derived factor 1 (SDF-1), a chemokine produced by bone marrow stromal cells that promotes platelet formation via CXC chemokine receptor type 4 (CXCR4) [49].

Mature megakaryocytes adhere to bone marrow sinusoids and elongate proplatelets within these sinusoids. Blood flow fragments the elongated proplatelets in blood vessels and carries them to the periphery. However, the proplatelet fragments released in vessels are 10–100 times larger than circulating platelets [47].

To be converted into normal-size platelets, they must be treated while in circulation.

Platelet production does not end with the release of proplatelets from megakaryocytes in bone marrow; they subsequently mature into normal-size platelets in circulation. The proplatelets released from megakaryocytes are fragmented, shrunk, and increased in number via repeated reciprocal morphological transformations into a disc, figure-8, or barbell shape.

Protrusions then emerge from platelets, and multiple platelets are produced from distal and central bulging [50]. Blood transfusion is a fundamental medical treatment that originated from the discovery of blood types in the early 20th century and remains an essential aspect of regenerative medicine. However, platelet preparations, despite recent prolongation of the preservation period, can only be used for 4 days due to the need to prevent proliferation of bacteria by preservation at room temperature. Furthermore, in patients with hematologic diseases who require repeated blood transfusions, alloantibodies are produced for human leukocyte antigens (HLAs) and human platelet antigens (HPAs) that differ from the patient's own antigens, thereby triggering platelet transfusion refractoriness (PTR). Patients with PTR caused by anti-HLA antibodies or anti-HPA antibodies require HLA/HPA-matched platelet transfusions; however, securing donors is extremely difficult [51].

These factors result in an imbalance between supply and demand, thus generating interest in research that seeks to develop an alternative transfusion source to blood donation. Despite research in which hematopoietic stem cells (which are somatic stem cells) are used as a source to manufacture platelets ex vivo, this research has not been put to practical use due to the absence of an ex vivo human hematopoietic stem cell amplification method in which the cells' functions can be retained.

Using our independently developed human mesenchymal stromal cells, we have succeeded in cultivating larger numbers of megakaryocytes, the sources of platelet production, than reported in previous studies; however, this method has not produced enough platelets to serve as a substitute for donated blood [52].

Similarly, studies in which umbilical cord blood-derived $CD34^+$ cells are used as a source for red blood cell production have been unable to amplify these cells ex vivo; thus, no technique has yet been developed that can produce 10^{12} red blood cells, the number of red blood cells needed for a single transfusion [53].

Alternatively, pluripotent embryonic stem cells (ES cells) and induced pluripotent stem cells (iPS cells) [54] can be grown semipermanently in vitro and can resolve the crucial issue of "yielding the number of original cells." In iPS cell-based regenerative medicine, posttransplantation cancer is a major concern. In response to this concern, safety for platelets and red blood cells can easily be ensured with a combination of anucleated cells, removal of nucleated cells with a filter prior to transfusion, and radiation to eliminate mixed-in lymphocytes. In addition, banking of iPS cells with various HLAs enables the construction of a system for stable provision of HLA-matched platelets [55].

Furthermore, Matsubara et al. have discovered that preadipocytes endogenously possess the megakaryocyte inducer p45NF-E2 and can induce differentiation of megakaryocytes into platelets while increasing their own expression of p45NF-E2; they are striving to stably and safely produce platelets for transfusion from preadipocytes for medical applications. However, the number of platelets produced by this method is currently too low for practical use [56].

3 THROMBOPOIETIN

TPO is the most important growth factor in the regulation of megakaryocyte development and platelet production [57].

Several promising novel agents that stimulate TPO receptor and increase platelet levels, such as eltrombopag, romiplostim, and lusutrombopag, are currently in development for the treatment of thrombocytopenia in patients with CLD and cirrhosis [58–61]. The ability to increase platelet levels could significantly reduce the need for platelet transfusions and facilitate the use of interferon-based antiviral therapy and other treatments in patients with liver disease [59,61].

The number of platelets is kept almost constant in our body. TPO is a cytokine that promotes platelet production, but the idea that TPO concentration in the blood is regulated by TPO constantly produced in the liver by TPO receptors on platelets and megakaryocytes (Sponge Theory). However, it was uncertain whether mechanisms regulating TPO production itself existed. Grozovsky et al. found that aged platelets stimulated TPO production in the liver. When platelets age, sialic acid on the surface is lost and it becomes desialylated platelets. The Ashwell-Morell receptor on the surface of hepatocytes binds desialylated platelets and removal of senescent platelets from the blood by endocytosis was shown by the analysis of platelet half-life and platelet hematopoiesis in AMR deficient mice and sialylated enzyme deficient mice. Transfusion experiments of these mice and desialylated platelets also showed that the binding of desialylated platelets to AMR increased expression of TPO mRNA and synthesis of TPO protein in hepatic cells [62].

4 CHRONIC LIVER DISEASE

Chronic liver disease (CLD) refers to a long-term pathological process of continuous destruction of liver parenchyma and its gradual substitution with fibrous tissue, which ultimately results in LC associated with a fatal outcome. CLD has diverse etiologies which include hepatotrophic viruses, chemicals, alcohol and drug abuse, autoimmune disorders, cholestasis, and metabolic diseases [63,64], and is a major cause of morbidity and mortality in many countries [65,66]. Hepatocellular carcinoma (HCC) is a dominant complication of CLD and cirrhosis with the third death rate among malignancies in the world [67]. HCC was shown to significantly correlate with advanced fibrosis as a sustained wound-healing response to hepatitis C virus infection; however, despite the progress in viral hepatitis therapies, they are unable to completely heal hepatocellular injuries and prevent LC [68]. At present, liver transplantation is the only treatment option for end-stage liver failure, but its clinical availability is hindered by serious problems such as donor shortage, surgical complications, and high cost [69]. Therefore other therapeutic approaches are looked into; among them, measures to resolve liver fibrosis have been investigated. Liver fibrosis is caused by a continuous excessive deposition of the extracellular matrix (ECM) in response to chronic liver injury [63,64]. Although advanced liver fibrosis has been considered irreversible, resulting in permanent substitution of hepatocytes with the ECM components, recent reports indicate that certain immunotherapies may promote partial resolution of LC [70]. These studies encourage the development of novel approaches to treat patients with advanced CLD. CLD is a major cause of mortality and morbidity in many countries [65,66].

LC is the end stage of CLD. It carries a poor prognosis and an increased risk of carcinogenesis [67].

Thrombocytopenia, that is, the reduction of platelet count in blood, is a common hematological complication of CLD caused by decreased production of hormone TPO in the damaged liver and/or increased destruction of platelets through phagocytosis in the enlarged spleen, as well as the loss of hematopoietic function in bone marrow due to alcohol abuse or viral infection [71–74].

Therefore thrombocytopenia is thought to have the intimate relation to pathogenesis of CLD and cirrhosis.

5 PLATELETS IN LIVER REGENERATION

Liver regeneration is provided by the proliferation of both parenhymal and nonparenhymal hepatic cells, including hepatocytes, liver sinusoidal endothelial cells (LSECs), biliary epithelial and Kupffer cells, and HSCs, which contribute to the restoration of destroyed hepatic tissue. Cell proliferation is triggered by several growth factors and cytokines such as HGF, TGF-α, tumor necrosis factor-α (TNF-α), EGF, and interleukin-6 (IL-6), which activate their cognate receptors and, consequently, downstream signaling and transcription of the genes associated with cell cycle progression [75–80]. Among the signaling cascades mediating platelet effects on the process of liver regeneration, the most important are TNF-α/nuclear factor-kappa B (NF-κB) [75,76], IL-6/signal transducer and activator of transcription 3 STAT3 [81], and phosphatidylinositol-3-kinase (PI3K)/Akt [80].

Previous studies indicate that platelets can exert positive effects on liver regeneration through cooperation with LSECs and Kupffer cells, and direct interaction with hepatocytes.

LSECs mostly consist of sinusoidal cells which [82], through formation of a continuous thin layer of the sinusoidal endothelium, create a structural barrier between the hepatic parenchyma and blood flowing through the liver [83,84]. LSECs play an important role in the maintenance of hepatic functions by providing exchange of nutrients between circulating blood and hepatocytes because of the presence of open pores beneath the endothelium [85]. In addition, LSECs secrete immunoregulatory cytokines, including HGF, IL-1, IL-6, and interferons affecting liver regeneration. Thus IL-6 secretion increased following hepatectomy [86] triggers STAT3 phosphorylation in hepatocytes, which upregulates the synthesis of acute phase proteins as a part of the mechanism restoring the disturbed physiological homeostasis [87]. The direct contact of platelets with LSECs stimulated LSEC proliferation and accelerated DNA synthesis in hepatocytes by inducing IL-6 secretion, possibly via S1P, a major bioactive lysophospholipid released from platelets [88]. S1P is known as a regulator of diverse cellular activities, including migration, proliferation, and cytoskeletal remodeling, and is known to induce STAT3 activation by stimulating IL-6 secretion [89]. Activated platelets secrete high amounts of S1P which acts on endothelial cells in the processes involving platelet-endothelial interactions, such as thrombosis, angiogenesis, and atherosclerosis [89,90].

Another type of nonparenchymal cells interacting with platelets is Kupffer cells, which constitute over 80% of the tissue macrophages found in the body and act against gastrointestinal bacteria, and microbial debris and endotoxins [91]. Upon activation, Kupffer cells secrete important growth-stimulating cytokines that promote hepatocyte proliferation after hepatectomy and induce processes involved in hepatic tissue restoration [92]. Kupffer cells are the most important source of IL-6 and TNF-α; the latter is increased following hepatectomy, suggesting that this cytokine and its producers Kupffer cells are implicated in the restoration of hepatic function in pathologic conditions. This notion is supported by the observations that anti-TNF-α antibodies suppressed hepatocyte proliferation [93], while TNF-α receptor-deficient mice had delayed liver regeneration after hepatectomy [94,95] because of decreased production of IL-6, which is a key target of TNF-α receptor activation in the regenerating liver [96]. As Kupffer cells are the most active produces of both TNF-α and IL-6 in the liver, it is not surprising that Kupffer cell-depleted mice fail to upregulate TNF-α and IL-6 secretion after hepatectomy [97].

However, the role of Kupffer cells in liver regeneration is controversial. Thus it was shown that the interaction among platelets, Kupffer cells, and leukocytes promotes endothelial cell apoptosis in the liver following ischemia/reperfusion [98]. Depletion of Kupffer cells decreased platelet adherence in sinusoids in rats subjected to ischemia/reperfusion and attenuated damage to liver endothelium [44], which is consistent with the findings that platelets adhering to Kupffer cells during the early period of ischemia/reperfusion promoted hepatocyte apoptosis [99]. Although Nakamura et al. [100] reported that in LPS-injected mice, platelets migrate to the space of Disse, which is mediated by their interaction with Kupffer cells, and then enter hepatocytes, the role of this process in hepatic regeneration is unclear. Further studies are needed to elucidate the mechanism underlying the impact of platelet-Kupffer cell interaction in liver fibrosis. However, it is evident that the contact between platelets and Kupffer cells causes activation of both cells.

Finally, platelets can induce hepatic regeneration by directly interacting with hepatocytes. Thus in thrombocytotic BALB/c mice, platelets accumulate in the liver shortly after liver resection, causing the regeneration of hepatic tissue even following 90% hepatectomy and preventing liver failure by promoting cell cycle progression and metabolic pathways in hepatocytes [101]. Such stimulation of hepatocyte activity is likely a result of platelet accumulation in the sinusoidal space, from where they flow into the space of Disse and directly contact hepatocytes [45,102,103]. These findings suggest that following hepatic injury, platelets quickly migrate to the liver where they, through direct interaction with hepatocytes, activate cell cycle transition-related pathways and induce rapid hepatocyte proliferation. This notion is supported by the study using a co-culture chamber system, which showed that the contact between platelets and hepatocytes triggered the secretion of growth factors, including HGF, IGF-1, and VEGF from platelets, which induced hepatocyte proliferation [1].

We suggest the following mechanistic model explaining the effect of platelets on liver regeneration. Platelets migrating to the injured liver translocate from the liver sinusoids to the space of Disse, where, upon interaction with hepatocytes, they secrete HGF, IGF-1, and VEGF, which induce hepatocyte proliferation resulting in liver regeneration. However, this model may not be fully applicable to humans, because human platelets do not secrete sufficient amounts of HGF [104]; therefore, IGF-1 may be the most important PDGF involved in the restoration of the human liver.

A recent study suggested an additional mechanism underlying platelet stimulation of liver regeneration. Thus Kirschbaum et al. [105] showed that transfer of coding or regulatory RNA could occur between platelets and hepatocytes, promoting hepatocyte proliferation. However, the role of both mechanisms, that is, the release of growth factors and/or RNA transfer from platelets, in liver regeneration needs confirmation in vivo [106,107].

6 PLATELETS IN LIVER CIRRHOSIS

Liver fibrogenesis is triggered by destruction of hepatic cells and represents a wound-healing process leading to excessive deposition of the matrix proteins collagens and elastin, glycoproteins, proteoglycans, and carbohydrates; in the context of chronic liver injury, fibrosis ultimately results in the substitution of liver tissue with ECM, formation of scar tissue, and gradual ceasing of hepatic functions [63,108]. Histologically, liver consists of parenhymal hepatocytes (70%–80%) and nonparenhymal cells such as Kupffer cells, sinusoidal endothelial cells, and stellate cells. Hepatic stellate cells (HSCs) reside in the perisinusoidal space of the liver, also known as the space of Disse, between hepatocytes and sinusoidal endothelial cells and are the major fibrogenic cell type in the liver as they produce a large number of ECM components and secrete TGF-β, a key mediator of liver fibrogenesis [63,108]. In the normal liver, HSCs have a star-like morphology corresponding to a quiescent state, and their primary function is the storage of vitamin A as retinol ester in lipid droplets [109,110]. In response to liver injury, HSCs undergo activation and change into contractile myofibroblastic cells, which proliferate, secrete TGF-β, and increase matrix production. As a result, collagens IV and VI in the space of Disse are progressively replaced by fibrous collagens I and III and fibronectin, characteristic for ECM remodeling and fibrosis [111,112].

In our previous study, we revealed a link between the activation of human HSCs and platelets by showing that platelets and platelet-derived extracts suppressed transdifferentiation of quiescent HSCs into the myofibroblast-like phenotype as well as the production of collagen type I via cAMP signaling [113]. The underlying mechanism is based on the increase of adenosine concentration in HSC milieu due to breakdown of ADP and ATP, which are abundant in platelet dense granules [113]. As a result, adenosine entering HSCs through its cognate receptors prevents their activation and downregulates their ability to secrete TGF-β and deposit the ECM. In addition, interaction between HSCs and platelets promoted the release of platelet-derived HGF, which inhibited the expression of type I collagen in cultured HSCs [114] and attenuated liver fibrosis in mice by decreasing hepatic TGF-β secretion and blocking myofibroblast activation [115]. However, although these findings indicate that platelets can reduce hepatic fibrogenesis through inhibition of HSC activation, it is unclear whether they can be translated to clinical situation, as the production of HGF by human platelets is lower than that in rodent platelets [104].

TPO is the most important factor in the regulation of megakaryocyte proliferation and differentiation into platelets through activation of its cognate receptor c-Mpl, also known as TPO-R [57]. Several agonists of the c-Mpl receptor such as eltrombopag and romiplostim are approved for clinical application as the agents increasing platelet counts in chronic immune thrombocytopenia [57,116]. Moreover, they are currently undergoing clinical trials as treatment options to reduce thrombocytopenia in patients with CLD and LC [58,59,117], as the increase in platelet counts could make these patients eligible for interferon-based antiviral therapy [60,118]. The strategy to treat liver fibrosis in CLD through inhibition of thrombocytopenia was proved feasible in studies showing that TPO improved both platelet counts and liver fibrosis, even in conditions of hepatic cirrhosis [119,120]. Thus in cirrhotic rats with dimethylnitrosamine-induced liver fibrosis and 70% hepatectomy, platelet increase by a single intravenous injection of TPO correlated with the inhibition of HSC activation and decrease of the fibrotic area in the liver, while antiplatelet serum attenuated hepatic regeneration [119]. In another study, mice with liver fibrosis induced by carbon tetrachloride (CCl4) showed improvement after weekly intraperitoneal administration of TPO for 5–8 weeks [120]. Although the mechanistic insights into the correlation of increased platelet counts with the reversal of liver fibrosis are yet to be provided, it can be suggested that platelets may promote hepatocyte proliferation by secreting HGF which is a potent mitogen for hepatocytes through activation of the MET receptor essential for organogenesis and wound healing. Moreover, HGF may contribute to the resolution of fibrosis by modulating levels of TGF-β and matrix metalloproteinases (MMPs) which are the main ECM enzymes degrading collagen. The suggested association between platelets, HGF, and hepatic fibrosis is supported by the findings of Takahashi et al. [121] who showed that transfused human platelets improved CCl4-induced liver fibrosis in severe combined immune deficiency (SCID) mice by increasing HGF levels in the mouse liver, which suppressed HSC activation, induced MMP-9 expression, and inhibited hepatocyte apoptosis.

7 PLATELET TRANSFUSIONS ON CLD AND CIRRHOSIS

Thrombocytopenia is typically treated by platelet transfusion, which is suggested to improve liver function. However, it has not been established whether platelet transfusion can benefit CLD patients with thrombocytopenia since the pathogenesis of thrombocytopenia in CLD is multifactorial; therefore, the published guidelines on platelet transfusion do not cover CLD-related platelet loss [118]. As animal experiments indicate feasibility of using blood transfusion for thrombocytopenia associated with chronic liver injuries, clinical trials have been conducted. A recent study included patients with CLD and cirrhosis (Child-Pugh class A or B) who demonstrated thrombocytopenia with platelet counts between 50,000 and 100,000/μL; the patients were treated with 10 units of platelet concentrate weekly for 12 weeks and followed up for 9 months after the last transfusion [122]. Although the platelet count did not show a significant increase, a marked improvement of liver function was observed, as evidenced by higher serum albumin levels 1 and 3 months posttransfusion and higher serum cholinesterase concentration 9 months posttransfusion; at the same time, serum hyaluronic acid levels indicative of liver fibrosis tended to decrease. However, this clinical trial was a noncontrolled, nonrandomized study based on a small sample size (6 patients); therefore, randomized controlled trials using larger patient cohorts need to be conducted to conclusively determine clinical value of platelet transfusion in CLD.

8 EFFECT OF TPO RECEPTOR AGONIST ELTROMBOPAG IN CLD

Although platelet increment therapies such as splenectomy and platelet transfusion can ameliorate CLD and LC, they may also cause serious side effects [59,123–125]. Thus portal vein thrombosis [59,123], hemorrhage, infection, and injury to the pancreatic tail [124] are among surgical complications encountered during splenectomy, while platelet activation frequently observed in platelet transfusion may result in proinflammatory responses, febrile nonhemolytic reactions, and acute lung injury [125]. Besides, as we indicated earlier, there are reports about detrimental effects of platelets on hepatocytes [44,98,99]. Therefore other approaches to treat thrombocytopenia in CLD should be considered.

Among them, TPO-R agonists eltrombopag and romiplostim, which have been already approved as therapeutics for chronic immune thrombocytopenic purpura, may hold promise also as treatment for CLD and LC. Eltrombopag is a low molecular weight, synthetic nonpeptidyl drug, whereas romiplostim is a peptide containing an IgG Fc fragment in its structure [126]. Eltrombopag has already been tested in a phase II clinical trial involving patients with hepatitis C-associated cirrhosis and concurrent thrombocytopenia to increase platelet counts before starting antiviral therapy with peginterferon and ribavirin, which showed promising results [58]. Although eltrombopag is a TPO-R ligand, it caused little effect on platelet function and did not activate the PI3K/Akt pathway, in contrast to TPO, which caused significant platelet activation [127].

The safety of using TPO-R (c-Mpl) agonists was examined in c-Mpl-expressing leukemia cells. Similar to TPO, eltrombopag did not increase but rather inhibited proliferation of leukemia cells in vitro, suggesting the lack of tumorigenic side effects [128]. Moreover, TPO had no proliferative effect on HCC both in vitro and in vivo [129], while eltrombopag induced cell cycle

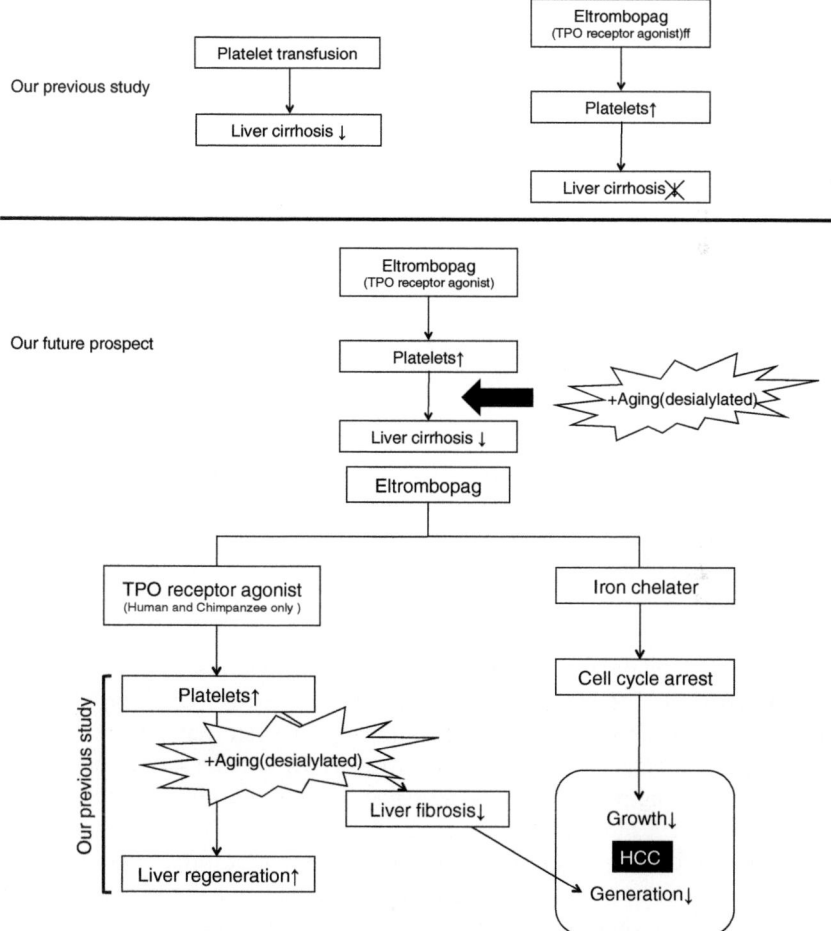

FIGURE 9.1 **Flow charts of our previous study and our future study.** Scheme of the mechanism of eltrombopag regarding its effects on liver regeneration, liver fibrosis and HCC.

arrest in HCC, demonstrating that it may act as a chemotherapeutic agent to inhibit the progression of HCC developed as a complication of CLD and LC [130]. Thus eltrombopag represents a promising novel candidate drug to treat both CLD-related thrombocytopenia and associated malignant neoplasm.

To increase platelets in a continuous manner and avoid PTR, we conducted an exploratory clinical trial and administered eltrombopag for 6 mo. The study included 5 patients with both CLD and a hepatitis C virus infection (Child-Pugh class A) who presented with thrombocytopenia (average platelet count 54×10^9/L). All of the patients administered eltrombopag maintained platelet counts between 10.0 and 15.0×10^{10}/L during the 6-month study, and their serum albumin, cholinesterase, ALT, T-bilirubin, hyaluronic acid, and type IV collagen levels, as well as the PT % and liver volumes, were stable throughout the clinical trial. The liver volumes calculated by computed tomography during the clinical trial were also stable, and no new cancerous lesions were observed.

However, in contrast to our previous clinical study [122], we predicted that the increment values would be small because the liver function was representative of a heterogenous population. Additionally, recent reports stated that the aging of platelets controls TPO production; thus, we hypothesize that the aging of platelets are involved in liver regeneration. Therefore we are planning a novel therapy with TPO receptor agonists by using a desialylated formulation. In other words, increasing the platelet count by using TPO and aging the platelets via administration of a desialylated formulation to promote liver regeneration could be an effective treatment of CLD and LC (Fig. 9.1).

9 CONCLUSION

This chapter suggests the current perspective of novel treatments for liver cirrhosis by using TPO receptor agonists to increase platelet counts in a clinical setting.

There is significant evidence that platelets play a role in improving fibrosis. Upon their release, the ATP and ADP within platelet dense granules are degraded by HSCs into adenosine, which is incorporated into the HSCs.

There are three distinct mechanisms of liver regeneration induced by platelets: a cooperative effect with LSECs; a cooperative effect with Kupffer cells; and a direct effect on hepatocytes. Additionally, platelet transfusion improves liver function in patients with CLD and cirrhosis. Despite its safety and maintenance in increasing platelet counts, administration of the TPO receptor agonist eltrombopag for 6 months did not result in improvements of liver function in patients with CLD. Therefore we are planning a new approach to develop novel strategies with TPO receptor agonists and a desialylated formulation for treating liver diseases for which there are currently no effective treatments except transplantation. Of course, it is necessary to pay sufficient attention to the onset of thrombosis by excessively increasing platelets.

REFERENCES

[1] Matsuo R, Ohkohchi N, Murata S, Ikeda O, Nakano Y, Watanabe M, et al. Platelets strongly induce hepatocyte proliferation with IGF-1 and HGF in vitro. J Surg Res 2008;145(2):279–86.
[2] Nakamura T, Nawa K, Ichihara A, Kaise N, Nishino T. Purification and subunit structure of hepatocyte growth factor from rat platelets. FEBS Lett 1987;224(2):311–6.
[3] Harrison P, Cramer EM. Platelet alpha-granules. Blood Rev 1993;7(1):52–62.
[4] Diegelmann RF, Evans MC. Wound healing: an overview of acute, fibrotic and delayed healing. Front Biosci 2004;9:283–9.
[5] Gruber R, Baron M, Busenlechner D, Kandler B, Fuerst G, Watzek G. Proliferation and osteogenic differentiation of cells from cortical bone cylinders, bone particles from mill, and drilling dust. J Oral Maxillofac Surg 2005;63(2):238–43.
[6] Gerard D, Carlson ER, Gotcher JE, Jacobs M. Effects of platelet-rich plasma at the cellular level on healing of autologous bone-grafted mandibular defects in dogs. J Oral Maxillofac Surg 2007;65(4):721–7.
[7] Man Y, Wang P, Guo Y, Xiang L, Yang Y, Qu Y, et al. Angiogenic and osteogenic potential of platelet-rich plasma and adipose-derived stem cell laden alginate microspheres. Biomaterials 2012;33(34):8802–11.
[8] Suzuki H, Yamazaki H, Tanoue K. Immunocytochemical aspects of platelet membrane glycoproteins and adhesive proteins during activation. Prog Histochem Cytochem 1996;30(1):1–106.
[9] Murata S, Maruyama T, Nowatari T, Takahashi K, Ohkohchi N. Signal transduction of platelet-induced liver regeneration and decrease of liver fibrosis. Int J Mol Sci. Multidisciplinary Digital Publishing Institute 2014;15(4):5412–25.
[10] Broos K, Feys HB, De Meyer SF, Vanhoorelbeke K, Deckmyn H. Platelets at work in primary hemostasis. Blood Rev 2011;25(4):155–67.
[11] Suzuki H, Nakamura S, Itoh Y, Tanaka T, Yamazaki H, Tanoue K. Immunocytochemical evidence for the translocation of alpha-granule membrane glycoprotein IIb/IIIa (integrin alpha IIb beta 3) of human platelets to the surface membrane during the release reaction. Histochemistry 1992;97(5):381–8.
[12] Blair P, Flaumenhaft R. Platelet alpha-granules: basic biology and clinical correlates. Blood Rev 2009;23(4):177–89.
[13] McNicol A, Israels SJ. Platelet dense granules: structure, function and implications for haemostasis. Thromb Res 1999;95(1):1–18.
[14] Polasek J. Platelet secretory granules or secretory lysosomes? Platelets. Informa UK Ltd, UK 2005;16(8):500–1.
[15] Spiegel S, Milstien S. Sphingosine-1-phosphate: an enigmatic signalling lipid. Nat Rev Mol Cell Biol. Nature Publishing Group 2003;4(5):397–407.
[16] McNicol A, Israels SJ. Beyond hemostasis: the role of platelets in inflammation, malignancy and infection. Cardiovasc Hematol Disord Drug Targets 2008;8(2):99–117.
[17] Mehta P. Potential role of platelets in the pathogenesis of tumor metastasis. Blood 1984;63(1):55–63.
[18] Nash GF, Turner LF, Scully MF, Kakkar AK. Platelets and cancer. Lancet Oncol 2002;3(7):425–30.
[19] Miyamoto R, Oda T, Hashimoto S, Kurokawa T, Kohno K, Akashi Y, et al. Platelet × CRP multiplier value as an indicator of poor prognosis in patients with resectable pancreatic cancer. Pancreas 2017;46(1):35–41.

[20] Elzey BD, Sprague DL, Ratliff TL. The emerging role of platelets in adaptive immunity. Cell Immunol 2005;238(1):1–9.
[21] Sowa JM, Crist SA, Ratliff TL, Elzey BD. Platelet influence on T- and B-cell responses. Arch Immunol Ther Exp (Warsz). SP Birkhäuser Verlag Basel 2009;57(4):235–41.
[22] Klinger MHF, Jelkmann W. Role of blood platelets in infection and inflammation. J Interferon Cytokine Res. Mary Ann Liebert, Inc. 2002;22(9):913–22.
[23] Sprague DL, Elzey BD, Crist SA, Waldschmidt TJ, Jensen RJ, Ratliff TL. Platelet-mediated modulation of adaptive immunity: unique delivery of CD154 signal by platelet-derived membrane vesicles. Blood 2008;111(10):5028–36.
[24] Gay LJ, Felding-Habermann B. Contribution of platelets to tumour metastasis. Nat Rev Cancer 2011;11(2):123–34.
[25] Rothwell PM, Wilson M, Price JF, Belch JFF, Meade TW, Mehta Z. Effect of daily aspirin on risk of cancer metastasis: a study of incident cancers during randomised controlled trials. Lancet 2012;379(9826):1591–601.
[26] Rothwell PM, Price JF, Fowkes FGR, Zanchetti A, Roncaglioni MC, Tognoni G, et al. Short-term effects of daily aspirin on cancer incidence, mortality, and non-vascular death: analysis of the time course of risks and benefits in 51 randomised controlled trials. Lancet 2012;379(9826):1602–12.
[27] Holmsen H. Physiological functions of platelets. Ann Med 1989;21(1):23–30.
[28] Mazzucco L, Borzini P, Gope R. Platelet-derived factors involved in tissue repair-from signal to function. Transfus Med Rev 2010;24(3):218–34.
[29] Ranzato E, Balbo V, Boccafoschi F, Mazzucco L, Burlando B. Scratch wound closure of C2C12 mouse myoblasts is enhanced by human platelet lysate. Cell Biol Int. Blackwell Publishing Ltd 2009;33(9):911–7.
[30] Rozman P, Bolta Z. Use of platelet growth factors in treating wounds and soft-tissue injuries. Acta Dermatovenerol Alp Pannonica Adriat 2007;16(4):156–65.
[31] Yamaguchi R, Terashima H, Yoneyama S, Tadano S, Ohkohchi N. Effects of platelet-rich plasma on intestinal anastomotic healing in rats: PRP concentration is a key factor. J Surg Res 2012;173(2):258–66.
[32] Radice F, Yánez R, Gutiérrez V, Rosales J, Pinedo M, Coda S. Comparison of magnetic resonance imaging findings in anterior cruciate ligament grafts with and without autologous platelet-derived growth factors. Arthroscopy 2010;26(1):50–7.
[33] Dugrillon A, Eichler H, Kern S, Klüter H. Autologous concentrated platelet-rich plasma (cPRP) for local application in bone regeneration. Int J Oral Maxillofac Surg 2002;31(6):615–9.
[34] Hartmann EK, Heintel T, Morrison RH, Weckbach A. Influence of platelet-rich plasma on the anterior fusion in spinal injuries: a qualitative and quantitative analysis using computer tomography. Arch Orthop Trauma Surg. Springer-Verlag 2010 Jul;130(7):909–14.
[35] de Vos RJ, Weir A, van Schie HTM, Bierma-Zeinstra SMA, Verhaar JAN, Weinans H, et al. Platelet-rich plasma injection for chronic Achilles tendinopathy: a randomized controlled trial. JAMA. American Medical Association 2010;303(2):144–9.
[36] Rodeo SA, Delos D, Weber A, Ju X, Cunningham ME, Fortier L, et al. What's new in orthopaedic research. J Bone Joint Surg Am 2010;92(14):2491–501.
[37] Nocito A, Georgiev P, Dahm F, Jochum W, Bader M, Graf R, et al. Platelets and platelet-derived serotonin promote tissue repair after normothermic hepatic ischemia in mice. Hepatology. Wiley Subscription Services, Inc., A Wiley Company 2007;45(2):369–76.
[38] Takahashi K, Kurokawa T, Oshiro Y, Fukunaga K, Sakashita S, Ohkohchi N. Postoperative decrease in platelet counts is associated with delayed liver function recovery and complications after partial hepatectomy. Tohoku J Exp Med 2016;239(1):47–55.

[39] Laschke MW, Dold S, Menger MD, Jeppsson B, Thorlacius H. Platelet-dependent accumulation of leukocytes in sinusoids mediates hepatocellular damage in bile duct ligation-induced cholestasis. Br J Pharmacol. Blackwell Publishing Ltd 2008;153(1):148–56.

[40] Zaldivar MM, Pauels K, Hundelshausen von P, Berres M-L, Schmitz P, Bornemann J, et al. CXC chemokine ligand 4 (Cxcl4) is a platelet-derived mediator of experimental liver fibrosis. Hepatology. Wiley Subscription Services, Inc., A Wiley Company 2010;51(4):1345–53.

[41] Khandoga A, Hanschen M, Kessler JS, Krombach F. CD4$^+$ T cells contribute to postischemic liver injury in mice by interacting with sinusoidal endothelium and platelets. Hepatology. Wiley Subscription Services, Inc., A Wiley Company 2006;43(2):306–15.

[42] Khandoga A, Biberthaler P, Messmer K, Krombach F. Platelet-endothelial cell interactions during hepatic ischemia-reperfusion in vivo: a systematic analysis. Microvasc Res 2003;65(2):71–7.

[43] Pak S, Kondo T, Nakano Y, Murata S, Fukunaga K, Oda T, et al. Platelet adhesion in the sinusoid caused hepatic injury by neutrophils after hepatic ischemia reperfusion. Platelets. Informa UK Ltd, UK 2010;21(4):282–8.

[44] Nakano Y, Kondo T, Matsuo R, Hashimoto I, Kawasaki T, Kohno K, et al. Platelet dynamics in the early phase of postischemic liver in vivo. J Surg Res 2008;149(2):192–8.

[45] Murata S, Ohkohchi N, Matsuo R, Ikeda O, Myronovych A, Hoshi R. Platelets promote liver regeneration in early period after hepatectomy in mice. World J Surg 2007;31(4):808–16.

[46] Lang PA, Contaldo C, Georgiev P, El-Badry AM, Recher M, Kurrer M, et al. Aggravation of viral hepatitis by platelet-derived serotonin. Nat Med 2008;14(7):756–61.

[47] Junt T, Schulze H, Chen Z, Massberg S, Goerge T, Krueger A, et al. Dynamic visualization of thrombopoiesis within bone marrow. Science 2007;317(5845):1767–70.

[48] Hitchcock IS, Kaushansky K. Thrombopoietin from beginning to end. Br J Haematol 2014;165(2):259–68.

[49] Chang Y, Auradé F, Larbret F, Zhang Y, Le Couedic J-P, Momeux L, et al. Proplatelet formation is regulated by the Rho/ROCK pathway. Blood 2007;109(10):4229–36.

[50] Thon JN, Montalvo A, Patel-Hett S, Devine MT, Richardson JL, Ehrlicher A, et al. Cytoskeletal mechanics of proplatelet maturation and platelet release. J Cell Biol 2010;191(4):861–74.

[51] Stroncek DF, Rebulla P. Platelet transfusions. Lancet 2007;370(9585):427–38.

[52] Matsunaga T, Tanaka I, Kobune M, Kawano Y, Tanaka M, Kuribayashi K, et al. Ex vivo large-scale generation of human platelets from cord blood CD34$^+$ cells. Stem Cells 2006;24(12):2877–87.

[53] Neildez-Nguyen TMA, Wajcman H, Marden MC, Bensidhoum M, Moncollin V, Giarratana M-C, et al. Human erythroid cells produced ex vivo at large scale differentiate into red blood cells in vivo. Nat Biotechnol 2002;20(5):467–72.

[54] Takahashi K, Tanabe K, Ohnuki M, Narita M, Ichisaka T, Tomoda K, et al. Induction of pluripotent stem cells from adult human fibroblasts by defined factors. Cell 2007;131(5):861–72.

[55] van der Meer PF, Pietersz RNI. Gamma irradiation does not affect 7-day storage of platelet concentrates. Vox Sang 2005;89(2):97–9.

[56] Ono-Uruga Y, Tozawa K, Horiuchi T, Murata M, Okamoto S, Ikeda Y, et al. Human adipose tissue-derived stromal cells can differentiate into megakaryocytes and platelets by secreting endogenous thrombopoietin. J Thromb Haemost 2016;14(6):1285–97.

[57] Wolber E-M, Jelkmann W. Thrombopoiet the novel hepatic hormone. News Physiol Sci 2002;17:6–10.

[58] McHutchison JG, Dusheiko G, Shiffman ML, Rodriguez-Torres M, Sigal S, Bourliere M, et al. Eltrombopag for thrombocytopenia in patients with cirrhosis Associated with hepatitis C. N Engl J Med 2007;357(22):2227–36.

[59] Afdhal NH, Giannini EG, Tayyab G, Mohsin A, Lee J-W, Andriulli A, et al. Eltrombopag before procedures in patients with cirrhosis and thrombocytopenia. N Engl J Med 2012;367(8):716–24.
[60] Kawaguchi T, Komori A, Seike M, Fujiyama S, Watanabe H, Tanaka M, et al. Efficacy and safety of eltrombopag in Japanese patients with chronic liver disease and thrombocytopenia: a randomized, open-label, phase II study. J Gastroenterol 2012;47(12):1342–51.
[61] Katsube T, Ishibashi T, Kano T, Wajima T. Population pharmacokinetic and pharmacodynamic modeling of lusutrombopag, a newly developed oral thrombopoietin receptor agonist, in healthy subjects. Clin Pharmacokinet 2016;55(11):1423–33.
[62] Grozovsky R, Begonja AJ, Liu K, Visner G, Hartwig JH, Falet H, et al. The Ashwell-Morell receptor regulates hepatic thrombopoietin production via JAK2-STAT3 signaling. Nat Med 2015;21(1):47–54.
[63] Bataller R, Brenner DA. Liver fibrosis. J Clin Invest. American Society for Clinical Investigation 2005;115(2):209–18.
[64] Friedman SL. Liver fibrosis – from bench to bedside. J. Hepatol. 2003;38(Suppl. 1):S38–53.
[65] Jaeschke H. Cellular adhesion molecules: regulation and functional significance in the pathogenesis of liver diseases. Am J Physiol. 1997;273(3 Pt 1):G602–11.
[66] Pinzani M, Marra F. Cytokine receptors and signaling in hepatic stellate cells. Semin Liver Dis. Copyright © 2001 by Thieme Medical Publishers, Inc., 333 Seventh Avenue, New York, NY 10001, USA. Tel.: +1(212) 584-4662; 2001;21(3):397–416.
[67] Montalto G, Cervello M, Giannitrapani L, Dantona F, Terranova A, Castagnetta LAM. Epidemiology, risk factors, and natural history of hepatocellular carcinoma. Ann N Y Acad Sci 2006;963(1):13–20.
[68] Tachi Y, Hirai T, Miyata A, Ohara K, Iida T, Ishizu Y, et al. Progressive fibrosis significantly correlates with hepatocellular carcinoma in patients with a sustained virological response. Hepatol Res 2015;45(2):238–46.
[69] Porrett PM, Hsu J, Shaked A. Late surgical complications following liver transplantation. Liver Transpl. Wiley Subscription Services, Inc., A Wiley Company 2009;15 Suppl. 2(S2): S12–8.
[70] Iwamoto T, Terai S, Hisanaga T, Takami T, Yamamoto N, Watanabe S, et al. Bone-marrow-derived cells cultured in serum-free medium reduce liver fibrosis and improve liver function in carbon-tetrachloride-treated cirrhotic mice. Cell Tissue Res. Springer-Verlag 2013;351(3):487–95.
[71] Ishikawa T, Ichida T, Matsuda Y, Sugitani S, Sugiyama M, Kato T, et al. Reduced expression of thrombopoietin is involved in thrombocytopenia in human and rat liver cirrhosis. J Gastroenterol Hepatol 1998;13(9):907–13.
[72] Aster RH. Pooling of platelets in the spleen: role in the pathogenesis of "hypersplenic" thrombocytopenia. J Clin Invest 1966;45(5):645–57.
[73] Peck-Radosavljevic M. Thrombocytopenia in liver disease. Can J Gastroenterol 2000;14 (Suppl. D). 60D–6D.
[74] Kajihara M, Okazaki Y, Kato S, Ishii H, Kawakami Y, Ikeda Y, et al. Evaluation of platelet kinetics in patients with liver cirrhosis: similarity to idiopathic thrombocytopenic purpura. J Gastroenterol Hepatol. Blackwell Publishing Asia 2007;22(1):112–8.
[75] FitzGerald MJ, Webber EM, Donovan JR, Fausto N, Rapid DNA. binding by nuclear factor kappa B in hepatocytes at the start of liver regeneration. Cell Growth Differ 1995;6(4): 417–27.
[76] Cressman DE, Greenbaum LE, Haber BA, Taub R. Rapid activation of post-hepatectomy factor/nuclear factor kappa B in hepatocytes, a primary response in the regenerating liver. J Biol Chem 1994;269(48):30429–35.

[77] Stepniak E, Ricci R, Eferl R, Sumara G, Sumara I, Rath M, et al. c-Jun/AP-1 controls liver regeneration by repressing p53/p21 and p38 MAPK activity. Genes Dev 2006;20(16): 2306–14.

[78] Wang G-L, Salisbury E, Shi X, Timchenko L, Medrano EE, Timchenko NA. HDAC1 promotes liver proliferation in young mice via interactions with C/EBPβ. J Biol Chem 2008;283(38):26179–87.

[79] Factor VM, Seo D, Ishikawa T, Kaposi-Novak P, Marquardt JU, Andersen JB, et al. Loss of c-Met disrupts gene expression program required for G2/M progression during liver regeneration in mice. In: Ng IOL, editor. PLoS One 2010;5(9):e12739.

[80] Jackson LN, Larson SD, Silva SR, Rychahou PG, Chen LA, Qiu S, et al. PI3K/Akt activation is critical for early hepatic regeneration after partial hepatectomy. Am J Physiol Gastrointest Liver Physiol 2008;294(6):G1401–10.

[81] Cressman DE, Diamond RH, Taub R. Rapid activation of the Stat3 transcription complex in liver regeneration. Hepatology 1995;21(5):1443–9.

[82] Knook DL, Sleyster EC. Separation of Kupffer and endothelial cells of the rat liver by centrifugal elutriation. Exp Cell Res 1976;99(2):444–9.

[83] Knock FE, Galt RM, Oester YT, Sylvester R, Haefliger R. Coordinated surgical and drug treatment of cancer. Int Surg 1976;61(5):287–92.

[84] Hisakura K, Murata S, Takahashi K, Matsuo R, Pak S, Ikeda N, et al. Platelets prevent acute hepatitis induced by anti-fas antibody. J Gastroenterol Hepatol. Blackwell Publishing Asia 2011;26(2):348–55.

[85] Braet F, Wisse E. Structural and functional aspects of liver sinusoidal endothelial cell fenestrae: a review. Comp Hepatol. BioMed Central 2002;1(1):1.

[86] Badia JM, Ayton LC, Evans TJ, Carpenter AJ, Nawfal G, Kinderman H, et al. Systemic cytokine response to hepatic resections under total vascular exclusion. Eur J Surg. Taylor & Francis, Ltd 1998;164(3):185–90.

[87] Gauldie J, Richards C, Baumann H. IL6 and the acute phase reaction. Res Immunol 1992;143(7):755–9.

[88] Kawasaki T, Murata S, Takahashi K, Nozaki R, Ohshiro Y, Ikeda N, et al. Activation of human liver sinusoidal endothelial cell by human platelets induces hepatocyte proliferation. J Hepatol 2010;53(4):648–54.

[89] Yatomi Y, Ohmori T, Rile G, Kazama F, Okamoto H, Sano T, et al. Sphingosine 1-phosphate as a major bioactive lysophospholipid that is released from platelets and interacts with endothelial cells. Blood 2000;96(10):3431–8.

[90] Takuwa Y, Okamoto Y, Yoshioka K, Takuwa N. Sphingosine-1-phosphate signaling and biological activities in the cardiovascular system. Biochim Biophys Acta 2008;1781(9): 483–8.

[91] Bilzer M, Roggel F, Gerbes AL. Role of Kupffer cells in host defense and liver disease. Liver Int. Blackwell Publishing Ltd 2006;26(10):1175–86.

[92] Meijer C, Wiezer MJ, Diehl AM, Schouten HJ, Meijer S, van Rooijen N, et al. Kupffer cell depletion by CI2MDP-liposomes alters hepatic cytokine expression and delays liver regeneration after partial hepatectomy. Liver 2000;20(1):66–77.

[93] Akerman P, Cote P, Yang SQ, McClain C, Nelson S, Bagby GJ, et al. Antibodies to tumor necrosis factor-alpha inhibit liver regeneration after partial hepatectomy. Am J Physiol 1992;263(4 Pt 1):G579–85.

[94] Yamada Y, Webber EM, Kirillova I, Peschon JJ, Fausto N. Analysis of liver regeneration in mice lacking type 1 or type 2 tumor necrosis factor receptor: requirement for type 1 but not type 2 receptor. Hepatology 1998;28(4):959–70.

[95] Yamada Y, Kirillova I, Peschon JJ, Fausto N. Initiation of liver growth by tumor necrosis factor: deficient liver regeneration in mice lacking type I tumor necrosis factor receptor. Proc Natl Acad Sci USA. National Academy of Sciences 1997;94(4):1441–2146.
[96] Malik R, Selden C, Hodgson H. The role of non-parenchymal cells in liver growth. Semin Cell Dev Biol 2002;13(6):425–31.
[97] Abshagen K, Eipel C, Kalff JC, Menger MD, Vollmar B. Loss of NF-kappaB activation in Kupffer cell-depleted mice impairs liver regeneration after partial hepatectomy. Am J Physiol Gastrointest Liver Physiol 2007;292(6):G1570–7.
[98] Sindram D, Porte RJ, Hoffman MR, Bentley RC, Clavien PA. Synergism between platelets and leukocytes in inducing endothelial cell apoptosis in the cold ischemic rat liver: a Kupffer cell-mediated injury. FASEB J 2001;15(7):1230–2.
[99] Tamura T, Kondo T, Pak S, Nakano Y, Murata S, Fukunaga K, et al. Interaction between Kupffer cells and platelets in the early period of hepatic ischemia-reperfusion injury—an in vivo study. J Surg Res 2012;178(1):443–51.
[100] Nakamura M, Shibazaki M, Nitta Y, Endo Y. Translocation of platelets into Disse spaces and their entry into hepatocytes in response to lipopolysaccharides, interleukin-1 and tumour necrosis factor: the role of Kupffer cells. J Hepatol 1998;28(6):991–9.
[101] Myronovych A, Murata S, Chiba M, Matsuo R, Ikeda O, Watanabe M, et al. Role of platelets on liver regeneration after 90% hepatectomy in mice. J Hepatol 2008;49(3):363–72.
[102] Murata S, Matsuo R, Ikeda O, Myronovych A, Watanabe M, Hisakura K, et al. Platelets promote liver regeneration under conditions of Kupffer cell depletion after hepatectomy in mice. World J Surg. Springer-Verlag 2008;32(6):1088–96.
[103] Matsuo R, Nakano Y, Ohkohchi N. Platelet administration via the portal vein promotes liver regeneration in rats after 70% hepatectomy. Ann Surg 2011;253(4):759–63.
[104] Nakamura T, Nishizawa T, Hagiya M, Seki T, Shimonishi M, Sugimura A, et al. Molecular cloning and expression of human hepatocyte growth factor. Nature 1989; 342(6248):440–3. Published online: November, 23 1989; DOI: 101038/342440a0. Nature Publishing Group.
[105] Kirschbaum M, Karimian G, Adelmeijer J, Giepmans BNG, Porte RJ, Lisman T. Horizontal RNA transfer mediates platelet-induced hepatocyte proliferation. Blood 2015;126(6):798–806.
[106] Meyer J, Lejmi E, Fontana P, Morel P, Gonelle-Gispert C, Bühler L. A focus on the role of platelets in liver regeneration: do platelet-endothelial cell interactions initiate the regenerative process? J Hepatol 2015;63(5):1263–71.
[107] Lisman T, Kirschbaum M, Porte RJ. The role of platelets in liver regeneration — What don't we know? J Hepatol 2015;63(6):1537–8.
[108] Gressner AM, Weiskirchen R. Modern pathogenetic concepts of liver fibrosis suggest stellate cells and TGF-beta as major players and therapeutic targets. J Cell Mol Med 2006;10(1):76–99.
[109] Friedman SL, Maher JJ, Bissell DM. Mechanisms and therapy of hepatic fibrosis: report of the AASLD Single Topic Basic Research Conference. Hepatology 2000:1403–8.
[110] Flier JS, Underhill LH, Friedman SL. The cellular basis of hepatic fibrosis—mechanisms and treatment strategies. N Engl J Med 1993.
[111] Friedman SL. Mechanisms of hepatic fibrogenesis. Gastroenterology 2008;134(6):1655–69.
[112] Parsons CJ, Takashima M, Rippe RA. Molecular mechanisms of hepatic fibrogenesis. J Gastroenterol Hepatol. Blackwell Publishing Asia 2007;22(s1):S79–84.
[113] Ikeda N, Murata S, Maruyama T, Tamura T, Nozaki R, Kawasaki T, et al. Platelet-derived adenosine 5'-triphosphate suppresses activation of human hepatic stellate cell: in vitro study. Hepatol Res 2011;42(1):91–102.
[114] Kodama T, Takehara T, Hikita H, Shimizu S, Li W, Miyagi T, et al. Thrombocytopenia exacerbates cholestasis-induced liver fibrosis in mice. Gastroenterology 2010;138(7):2487–98.

[115] Xia J-L, Dai C, Michalopoulos GK, Liu Y. Hepatocyte growth factor attenuates liver fibrosis induced by bile duct ligation. Am J Pathol 2006;168(5):1500–12.

[116] Cheng G. Eltrombopag, a thrombopoietin- receptor agonist in the treatment of adult chronic immune thrombocytopenia: a review of the efficacy and safety profile. Ther Adv Hematol. SAGE Publications 2012 Jun;3(3):155–64.

[117] Cooper KL, Fitzgerald P, Dillingham K, Helme K, Akehurst R. Romiplostim and eltrombopag for immune thrombocytopenia: methods for indirect comparison. Int J Technol Assess Health Care. Cambridge University Press 2012;28(3):249–58.

[118] Afdhal N, McHutchison J, Brown R, Jacobson I, Manns M, Poordad F, et al. Thrombocytopenia associated with chronic liver disease. J Hepatol 2008;48(6):1000–7.

[119] Murata S, Hashimoto I, Nakano Y, Myronovych A, Watanabe M, Ohkohchi N. Single administration of thrombopoietin prevents progression of liver fibrosis and promotes liver regeneration after partial hepatectomy in cirrhotic rats. Ann Surg 2008;248(5):821–8.

[120] Watanabe M, Murata S, Hashimoto I, Nakano Y, Ikeda O, Aoyagi Y, et al. Platelets contribute to the reduction of liver fibrosis in mice. J Gastroenterol Hepatol [Internet] 2009;24(1):78–89. Available from: http://onlinelibrary.wiley.com/doi/10.1111/j.1440-1746.2008.05497.x/pdf.

[121] Takahashi K, Murata S, Fukunaga K, Ohkohchi N. Human platelets inhibit liver fibrosis in severe combined immunodeficiency mice. World J Gastroenterol 2013;19(32):5250–60.

[122] Maruyama T, Murata S, Takahashi K, Tamura T, Nozaki R, Ikeda N, et al. Platelet transfusion improves liver function in patients with chronic liver disease and cirrhosis. Tohoku J Exp Med 2013;229(3):213–20.

[123] Stamou KM, Toutouzas KG, Kekis PB, Nakos S, Gafou A, Manouras A, et al. Prospective study of the incidence and risk factors of postsplenectomy thrombosis of the portal, mesenteric, and splenic veins. Arch Surg. American Medical Association 2006;141(7):663–9.

[124] Kojouri K, Vesely SK, Terrell DR, George JN. Splenectomy for adult patients with idiopathic thrombocytopenic purpura: a systematic review to assess long-term platelet count responses, prediction of response, and surgical complications. Blood 2004;104(9):2623–34.

[125] Khan SY, Kelher MR, Heal JM, Blumberg N, Boshkov LK, Phipps R, et al. Soluble CD40 ligand accumulates in stored blood components, primes neutrophils through CD40, and is a potential cofactor in the development of transfusion-related acute lung injury. Blood 2006;108(7):2455–62.

[126] Kuter DJ. New thrombopoietic growth factors. Blood 2007;109(11):4607–16.

[127] Erhardt JA, Erickson-Miller CL, Aivado M, Abboud M, Pillarisetti K, Toomey JR. Comparative analyses of the small molecule thrombopoietin receptor agonist eltrombopag and thrombopoietin on in vitro platelet function. Exp Hematol 2009;37(9):1030–7.

[128] Erickson-Miller CL, Kirchner J, Aivado M, May R, Payne P, Chadderton A. Reduced proliferation of non-megakaryocytic acute myelogenous leukemia and other leukemia and lymphoma cell lines in response to eltrombopag. Leuk Res 2010;34(9):1224–31.

[129] Nozaki R, Murata S, Nowatari T, Maruyama T, Ikeda N, Kawasaki T, et al. Effects of thrombopoietin on growth of hepatocellular carcinoma: Is thrombopoietin therapy for liver disease safe or not? Hepatol Res 2013;43(6):610–20.

[130] Kurokawa T, Murata S, Zheng Y-W, Iwasaki K, Kohno K, Fukunaga K, et al. The Eltrombopag antitumor effect on hepatocellular carcinoma. Int J Oncol 2015;47(5):1696–702.

Chapter 10

Dynamic Tissue Remodeling in Chronic Liver Diseases: Abnormal Proliferation and Differentiation of Hepatocytes and Bile Ducts/Ductules

Yuji Nishikawa
Asahikawa Medical University, Asahikawa, Hokkaido, Japan

1 INTRODUCTION

The liver is well known to have a robust regenerative capacity. Following a 70% partial hepatectomy in rodents, residual liver tissues rapidly proliferate and restore the original liver volume as early as 1 week. In human, the liver can also undergo rapid restoration of the its original mass following a partial hepatectomy, as shown in the donor or recipient livers following living donor liver transplantations. In addition to these compensatory regeneration, following various types of injury, the liver can also rapidly restore its original tissue structures if the supporting framework of the liver are maintained. This is most clearly exemplified by acute CCl_4 injury that causes severe centrilobular necrosis, proliferation and migration of the intact periportal hepatocytes replaces the injured area within 8 days in rodents.

Both hepatocytes and bile ducts/ductules are derived from hepatoblasts that have been induced in the hepatic diverticulum during the liver development [1,2]. Although adult hepatocytes are highly differentiated cells with multiple functions, it is noteworthy that they maintain remarkable proliferative capacity that is almost comparable to that of stem cells. This has been experimentally demonstrated in several repopulation studies, in which transplanted hepatocytes proliferate in chronically injured livers [3,4].

However, in compensatory regeneration, if the liver volume loss is too extensive, such as a 90%–95% partial hepatectomy in rodents, the residual liver can not properly recover the normal liver tissue resulting in death from liver failure [5]. Similar conditions have been experienced as the "small-for-size" syndrome in patients following liver transplantation or extensive liver resection [6]. Furthermore, when the liver is incurred by repetitive or continuous injuries, incited chronic inflammation and fibrosis alter the normal microenvironment of the liver, which then affect morphology and function of regenerated hepatocytes.

In the following sections, various sequelae of failed liver regeneration and mechanisms will be presented. First, I will focus on ductular reaction induced by abnormal differentiation and proliferation of the liver epithelial cell system (hepatocytes and bile duct/ductular cells). Then, possible clonal expansion of selected hepatocytes in liver cirrhosis and its relevance to hepatocarcinogenesis will be discussed.

2 CHRONIC LIVER DISEASES AND CIRRHOSIS

Various noxious stimuli damage the liver and cause chronic inflammatory responses [7]. These include hepatitis virus infection (hepatitis B virus (HBV), hepatitis C virus (HCV)), excessive alcoholic consumption, metabolic syndrome (nonalcoholic steatohepatitis (NASH)), autoimmune injuries (autoimmune hepatitis, primary biliary cholangitis, and primary sclerosing cholangitis), cholestasis, chronic congestion (Budd–Chiari syndrome, veno-occlusive disease), a variety of hereditary metabolic disturbances, and various drugs and toxins [8]. Although chronic liver injury and subsequent inflammatory responses mainly damage liver epithelial cells, hepatocytes and bile duct/ductular cells, the damage to sinusoidal endothelial cells is also detrimental to the liver, since it disturbs local circulation and causes ischemic zonal necrosis [8].

The location of hepatocyte damage is dependent on the etiology of liver injury [7]. In chronic hepatitis due to infection of HBV and HCV, although there are scattered apoptotic death of hepatocytes in the lobules, periportal hepatocytes are most severely affected. The progressive death of periportal hepatocytes results in inflammatory responses with gradual expansion of the portal connective tissues with destruction of the limiting plate, the boundary between the portal connective tissue and hepatocytes. This is the condition called interface hepatitis (piecemeal necrosis). Acute viral hepatitis due to hepatitis A, B, and E viruses can induce extensive necrosis of periportal hepatocytes, but usually followed by complete resolution without scarring. Autoimmune injury affects both periportal and centrilobular hepatocytes with a marked lymphoplasmacytic infiltrate. In alcoholic hepatitis and NASH, centrilobular injury is predominant.

Inflammatory reaction damages vascular structures and the resultant circulatory disturbances induce zonal necrosis of the parenchyma [9]. This parenchymal extinction triggers liver fibrosis. Hepatic stellate cells and portal myofibroblasts are two major cell types that are activated by liver injury and responsible for deposition of collagen. Fibrotic expansion of the portal or centrilobular areas

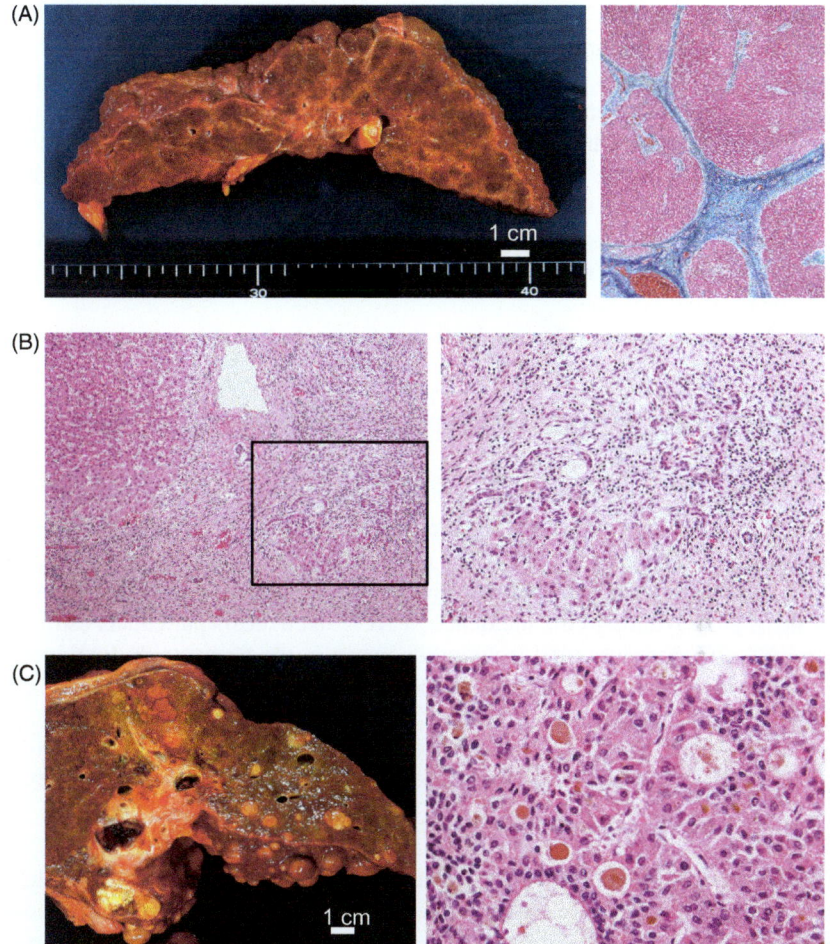

FIGURE 10.1 Ductular reaction and multiple hepatocellular carcinoma in liver cirrhosis. (A) Cirrhosis due to autoimmune hepatitis. Gross appearance of the liver (*left*) and a low-power micrograph of the liver tissue (Masson trichrome staining). (B) Extensive ductular reaction in cirrhosis due to viral hepatitis. *Boxed area* in the *left panels* is magnified in the *right panel*. (C) Multiple hepatocellular carcinomas (HCCs) in cirrhotic liver (burned-out nonalcoholic steatohepatitis). Gross appearance of the liver (*left*) and histology of a tumor (moderately differentiated HCC).

cause portal–portal, portal–central, or central–central bridging by fibrotic septa, resulting in remodeling of the original hepatic lobular structures.

Liver cirrhosis is an end-stage disease status, characterized by a nodular appearance of the liver, composed of regenerative nodules surrounded by fibrous septa (Fig. 10.1A) [10]. The histological features of liver cirrhosis are variable, dependent on the location and severity of hepatocytic injury and subsequent parenchymal extinction, as well as on the causes of liver injury.

According to the sizes of the regenerative nodules, cirrhosis has been classified as macronodular and micronodular. In addition to the decreased volume of functional hepatocytes, disturbed portal circulation and porto-venous shunting aggravate liver functions, eventually leading to hepatic failure.

Although liver cirrhosis has been generally regarded as an irreversible process, there is evidence showing that fibrosis could be regressed if the disease activity is suppressed in experimental animals, as well as in humans [9,11]. Elimination of the causes of cirrhosis, such as abstinence in alcoholic cirrhosis and antiviral therapies in cirrhosis induced by HBV or HCV, retards or reverse cirrhosis [10].

Cirrhosis, as well as chronic liver diseases with fibrosis, are almost invariably associated with an increase of small bile ducts or bile ductules called the ductular reaction (Fig. 10.1B) [12]. The extent of this reaction is parallel to the degree of fibrosis and has been considered to be a useful prognostic factor for chronic liver diseases [13,14].

The most ominous complication of liver cirrhosis is the development of liver tumors, in particular, hepatocellular carcinoma (HCC) (Fig. 10.1C). In HBV-related HCC,the majority (70%–80%) of patients have cirrhosis and the risk of HCC in patients infected with HCV is high in those with advanced fibrosis or cirrhosis [15]. Although the incidence of HCC in cirrhosis patients is estimated to be 2%–7% per year, [10] it would decrease in the future with the development of more effective antiviral treatment, such as direct-acting antivirals for HCV [16].

3 DUCTULAR REACTION AND ITS POSSIBLE CELLULAR ORIGINS

The connective tissue of the normal portal tract contained one or two interlobular bile ducts and is associated with several indistinct bile ductules composed of small cuboidal cells along the clear boundary (limiting plate) between the parenchyma (Fig. 10.2A). The ductular reaction is defined as an increase in the number of bile ductules and has been traditionally classified into three distinct types [17,18]. Type 1 (I) ductular reaction is typically seen in biliary obstruction and characterized by proliferation of bile ductules of uniform shape and size (Fig. 10.2B), which is experimentally induced in rodents by common bile duct ligation (Fig. 10.2C). Type 2 (II) ductular reaction is the most common and also called as atypical ductular reaction due to the irregularity of the ductal shapes which are contrasted with those in type 1 ductular reaction (Fig. 10.2D). Type 3 (III) ductular reaction is characterized by a diffuse and rapid appearance of irregular ductules with stem-like features in fulminant hepatitis, in which the majority of hepatocytes are rapidly degenerated before triggering fibrotic reactions (Fig. 10.2E).

Desmet [18] has proposed to distinguish two subtypes of type 2 ductular reaction according to the location of ductular reaction: type 2A and type 2B,

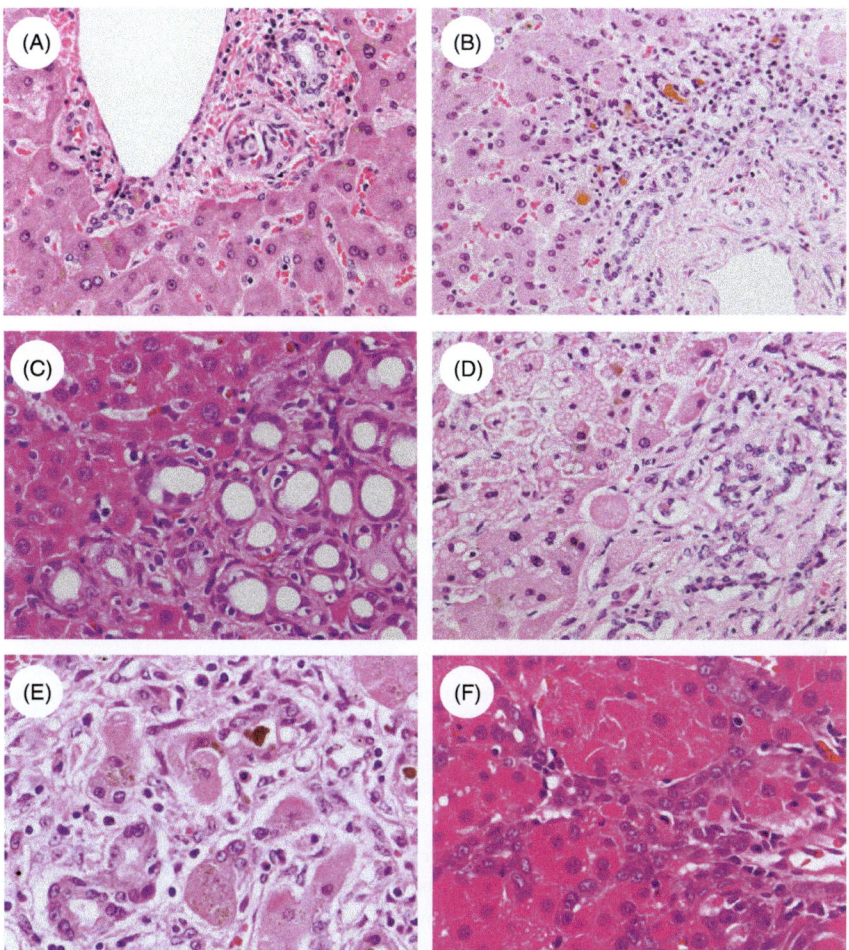

FIGURE 10.2 Various types of ductular reaction. (A) Intact human liver. (B) Type 1 ductular reaction in obstructive jaundice. (C) Type 1 ductular reaction 7 days following bile duct ligation in a rat. (D) Type 2 ductular reaction in viral hepatitis. (E) Type 3 ductular reaction in fulminant hepatitis. (F) Oval cell proliferation induced by acute CCl_4 injury in a 2-aminoacetofluorene (2-AAF)-treated rat (3 days following injury).

periportal and centrilobular, respectively. Type 2A ductular reaction is encountered in various chronic liver diseases, most notably associated with interface hepatitis (Fig. 10.3A and B). They are positive for bile duct specific cytokeratins (CK), such as CK7 (Fig. 10.3C and D) and CK19. Type 2B is seen in the centrilobular area in chronic congestion where no bile duct/ductules are present. Krings et al. [8] have described detailed pathological features of the centrilobular ductular reaction in chronic hepatic venous outflow obstruction, including Budd–Chiari syndrome and veno-occlusive disease. The typical centrilobular

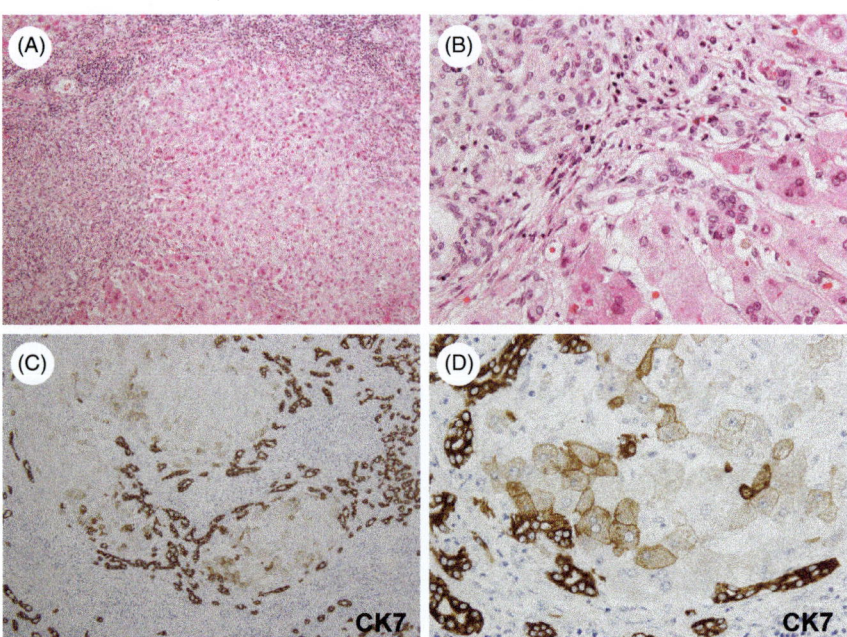

FIGURE 10.3 Periportal ductular reaction (type 2A) in liver cirrhosis due to viral hepatitis. (A and B) Low-power (A) and high-power (B) micrographs. (C and D) Cytokeratin 7 (CK7) immunohistochemistry (C, low-power; D: high-power).

cirrhosis in Budd–Chiari syndrome has been called "reversed lobulation" cirrhosis [19]. As shown in Fig. 10.4, severe centrilobular congestion due to idiopathic pulmonary hypertension induces marked fibrosis around the congested central veins and hepatic veins and there are several ductular structures that are positive for CK7 and, to a lesser extent, CK19. Similar centrilobular ductular reaction is also seen in NASH, in which parenchymal injury and subsequent fibrosis are dominant in the centrilobular area [20].

The cellular origins of ductular reaction in human liver diseases have remained unclear. The proliferation of bile ducts/ductules can be triggered by cholestasis as seen in type 1 ductular reaction [21]. However, in severe and/or prolonged cholestasis, increased ductules also show an irregular morphology reminiscent of type 2 ductular reaction, suggesting that the mechanism may not be simple. Interestingly, type 2 ductular reaction is often accompanied with the expression of bile duct-specific CK (especially, CK7) in the adjacent hepatocytes (Fig. 10.3C and D). These altered hepatocytes have been called "intermediate hepatobiliary cells" and some investigators have considered that they represent activated hepatic stem/progenitor cells for liver regeneration following severe injury [22]. However, this is also possible that these cells are metaplastic (or transdifferentiated) hepatocytes that present bile duct-like phenotype

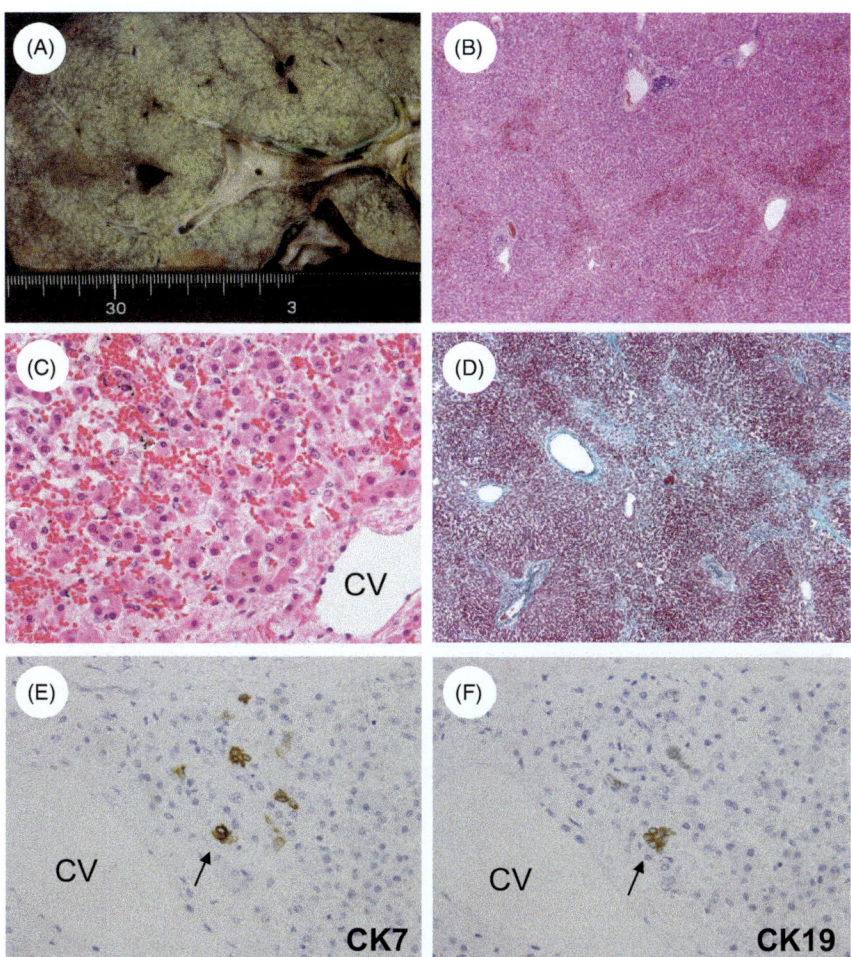

FIGURE 10.4 **Centrilobular ductular reaction (type 2B) in severe chronic congestion of the liver due to idiopathic pulmonary hypertension.** (A) Gross appearance of the liver. (B and C) Low-power (B) and high-power (C) micrographs of the liver tissue. (D) Masson trichrome staining (low-power). (E and F) CK7 (E) and CK19 (F) immunohistochemistry. The *arrow* indicates the same ductular structure. CV, central vein.

in an unfavorable microenvironment. Although it is difficult to determine the direction of differentiation in a tissue specimen, a metaplastic nature of ductular reaction has been suggested by pathological findings of the human liver. Supporting that these ductules have been derived from preexisting hepatocytes, hepatocyte-specific constituents, such as alcoholic hyaline (Mallory–Denk bodies) (Fig. 10.5A), hemosiderin (Fig. 10.5B), and nuclear glycogen (Fig. 10.5C), are found in the increased bile ductules in alcoholic cirrhosis, biliary atresia,

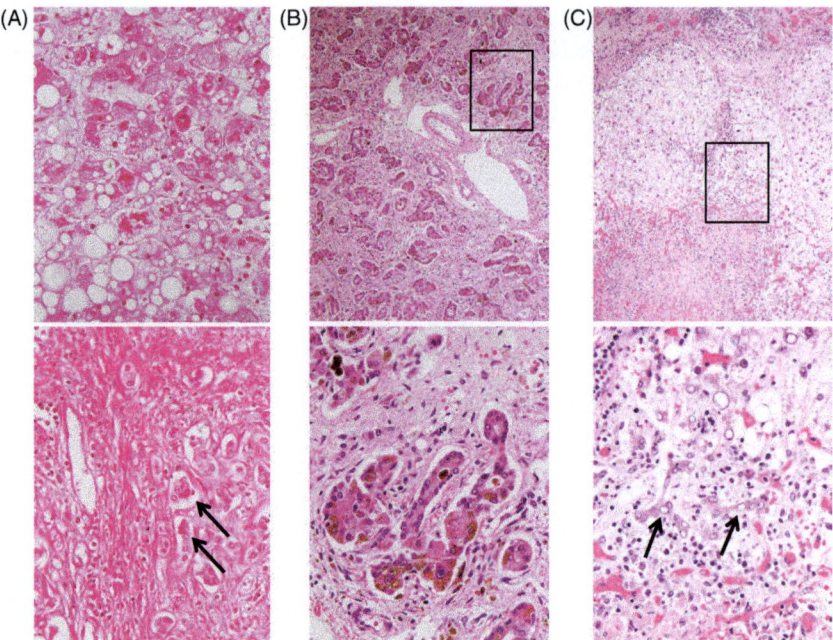

FIGURE 10.5 **Ductular reaction suggesting ductular metaplasia of existing hepatocytes.** (A) Alcoholic cirrhosis. Marked fatty change and Mallory-Denk bodies in the regenerative nodule (*upper panel*) and Mallory-Denk body-containing bile ductules (*arrows*) in the fibrous septa. (B) Biliary atresia. *Boxed area* in the *upper panels* is magnified in the *lower panel*. Hemosiderin pigments in the cytoplasm of the irregular bile ductules with marked cholestasis. (C) Type III glycogen storage disease. *Boxed area* in the *upper panels* is magnified in the *lower panel*. Marked glycogen accumulation in hepatocytes in the regenerative nodules (*upper panel*) and nuclear glycogen in some of the bile ductules (*arrows*).

and glycogen storage disease, respectively. The emergence of ductular structures in type 3 ductular reaction in the almost whole liver in relatively short periods also favors ductular metaplasia of hepatocytes, rather than proliferation of stem/progenitor cells or ductular cells. However, despite their stem-like proliferation capacity, the highly differentiated phenotype of hepatocytes had been regarded to be fixed once matured [23,24].

Many investigators have proposed that ductular reaction, especially type 2A, represents regenerative proliferation of hepatic stem/progenitor cells that may reside at the canal of Hering, where lobular hepatocytes and bile ductules are connected [25]. They are supposed to have a bipotential differentiation capacity toward hepatocytes and bile ducts/ductules, as well as a robust proliferative capacity that would be evident when the liver is severely injured [26]. The prototype hepatic stem/progenitor cells are so-called oval cells, which can be transiently induced in the periportal area by the Solt and Farber protocol for rat hepatocarcinogenesis, which is consisted of a single necrotizing dose of a mutagen (diethylnitrosamine (DEN)), feeding with a hepatocyte growth suppressor (2-acetaminofluorene), and a partial hepatectomy [27]. Oval cells are small epithelial cells with high nuclear/cytoplasmic ratio, express several stem cell markers, such as α-fetoprotein (AFP), bile duct-specific CK, Sox9, and osteopontin, and proliferate in the periportal area (Fig. 10.2F). They share phenotypic characteristics with bile duct/ductular cells rather than hepatoblasts [28]. Similar cells can be induced in mice by feeding them with a diet containing 3,5-diethoxycarbonyl-1,4-dihydrocollidine (DDC), which causes cholestatic cholangiopathy due to ductal accumulation of abnormal porphyrin metabolites.

Sox9-positive putative stem/progenitor cells were reported to significantly contribute to liver homeostasis, as well as regeneration following a variety of liver injuries using a Sox9-CreER/ROSA26R cell lineage tracing system, in which tamoxifen administration was used to activate Cre recombinase activation in a Sox9 promoter-dependent manner [29]. However, subsequent experimental works applying similar cell lineage tracing technologies have failed to reveal such robust regenerative capacity of the putative Sox9-positive hepatic stem/progenitor cells [30–32]. Although the reasons for the discrepancies in cell lineage tracing experiments are unclear, the use of high dose tamoxifen has been shown to affect the results in the systems in which Cre recombinase are fused to a tamoxifen-inducible mutated estrogen receptor [33]. The data obtained in studies using lineage-tracing of osteopontin-positive cells also suggest that the putative hepatic stem/progenitor cells or oval cells may not serve as the facultative hepatic stem/progenitor compartment that can regenerate the liver when parenchymal cell proliferation is hampered by certain pathological conditions [31,34]. They may represent activated bile duct/ductular cells [28], but lineage-tracing of CK19- or hepatocyte nuclear factor (HNF)-1β-positive cells has revealed that the contribution of ductular cells to generation of new hepatocytes is almost negligible [35,36]. Furthermore, some investigators have suggested that hepatic stellate cells might be epithelial progenitors and contribute to ductular reaction [37]. However, subsequent studies have demonstrated that the possibility that they serve as hepatic stem/progenitor cells is highly unlikely [36].

Although ductular reaction may not be due to the activation of putative hepatic stem/progenitor cells, which have been postulated to reside at the periportal area, this does not imply the absence of stem/progenitor-like cells in the liver. Actually, a number of experimental works have isolated of stem/progenitor-like cells that possess robust proliferative activity and bipotential differentiation capacity from the adult intact or injured livers [38–40]. Notably, cells positive for leucine-rich repeat-containing G protein-coupled receptor 5 (Lgr5), a receptor for a Wnt agonist (R-spondin), have been demonstrated to appear in the periportal region upon liver injury, proliferate, and differentiate into mature hepatocytes and bile ducts/ductules [41]. In contrast, centrilobular hepatocytes positive for axin2, which responds to the Wnt-Frizzled signaling, have been shown to contribute to the maintenance of the liver [42]. However, it has also has been reported that a few Lgr5-positive hepatocytes are present in the centrilobular region of intact livers and such centrilobular hepatocytes do not demonstrate a stem cell-like proliferation activity [43]. Furthermore, important roles of periportal hepatocytes, not centrilobular hepatocytes, in liver homeostasis and regeneration have been shown in a recent work [44]. While these reports suggest the significance of the Wnt pathway in liver regeneration and homeostasis, it may be difficult to understand how these findings fit together in a functionally and anatomically coherent way.

Mature hepatocytes might be the source of the putative hepatic stem/progenitor cells. Sox9-positive biphenotypic hepatocytes have been suggested to be derived from mature hepatocytes and supply new hepatocytes [45]. Furthermore, recent studies have proposed that mature hepatocytes could "dedifferentiate" into immature states following liver injury and contribute to liver regeneration, as well as ductular reaction [46,47]. However, it is important to note that, in these experiments, although these "dedifferentiated" hepatocytes express some of the biliary differentiation markers, such as Sox9 and osteopontin, their phenotype might not be typical of that of the hepatoblasts in the developing liver as discussed below. Interestingly, a cocktail of small molecules, Y-27632, A-83-01, and CHIR99021, has shown to convert rat and mouse mature hepatocytes into highly proliferative and bipotent cells [48].

Collectively, while these studies in this field have not yet identified a distinct hepatic stem/progenitor cell compartment in the liver, they strongly suggest that the phenotype of mature hepatocytes may be changeable in response to liver injury and associated alterations of the signaling pathways.

4 PHENOTYPIC PLASTICITY OF MATURE HEPATOCYTES AND ITS INVOLVEMENT IN THE DUCTULAR REACTION

4.1 Ductular Differentiation of Hepatocytes In Vitro

As stated earlier, because the highly differentiated nature of mature hepatocytes, it has long been supposed that the phenotype of hepatocytes is fixed once they have matured. We have attempted to show that hepatocytes are able to differentiate into bile ductular cells in vitro and in vivo. In this section, I will describe the experimental data, including our own, showing the phenotypic plasticity of hepatocytes and discuss its significance.

Hepatocytes can be isolated from the livers of rats or mice by the two-step collagenase perfusion method and maintained in vitro. Although cultured hepatocytes show their highly differentiated phenotype, such as drug metabolizing activities, within a few days, it is rather difficult to maintain the level of the differentiation for longer periods in the traditional monolayer cultures on collagen-coated surfaces. Spheroidal aggregate cultures and cultures on a basement membrane matrix (Matrigel) may prolong the maintenance of the differentiation state. The media containing corticosteroids and oncostatin M (OSM) has been applied to induce hepatocytic differentiation of hepatoblasts cultured on Matrigel [49]. The differentiated phenotype of hepatocytes is affected by both extracellular matrix and soluble factors.

In intact liver, hepatocytes are arranged in cell cords which are supported by net-like delicate reticulin fibers and abut on vascular spaces (sinusoids). Between hepatocytes and sinusoids, there is a space (called the space of Disse), where hepatocytes develop numerous microvilli to facilitate uptake and release of various molecules. Thus, the lack of dense and continuous extracellular matrix including collagen fibers is a requisite for normal hepatocytic functions, contrasting with bile duct cells that are specialized for a bile drainage system, which have distinct basement membranes and are surrounded by connective tissue stroma. In liver fibrosis following hepatocytic injury, hepatic stellate cells reside in the space of Disse are activated and produce abundant collagen, producing an altered microenvironment for residual hepatocytes.

We hypothesized that, if isolated hepatocytes are placed in the microenvironment rich in a collagenous matrix, they might differentiate into bile duct/ductular cells and form ductular structures. Therefore we embedded spheroidal aggregates of rat hepatocytes into a type I collagen gel matrix [50,51]. Hepatocytes underwent branching morphogenesis and eventually formed distinct ductular structures surrounded by basement membranes (Fig. 10.6). In parallel with the rapid loss of expression of hepatocytic markers, such as albumin and HNF-4α1, cultured hepatocytes also expressed markers for bile ducts, such as CK19, glutathione S-transferase placental type. Similar branching morphogenesis and aberrant ductular differentiation have been demonstrated in mice [52]. During the period when the hepatic stem/progenitor hypothesis prevailed, experimental data showing bile ductular differentiation of hepatocytes in vitro, tended to be interpreted as due to contamination of a small number of putative stem/progenitor cells, as well as bile ductal cells. However, we selectively isolated centrilobular hepatocyte fraction, which is devoid of contamination of bile duct cells and putative stem cells, and confirmed that there was no significant differences in branching morphogenesis and bile ductular differentiation that were seen in a entire hepatocyte fraction or a periportal hepatocyte fraction. Furthermore, hepatocytes from a c-Kit mutant rat strain, which is known to be deficient in oval cell activation, demonstrated essentially the same characteristics as that of wild type [51]. Using an organoid cultureof rat hepatocytes in roller bottles, Michalopoulos and colleagues [53,54] also suggested bile duct differentiation of hepatocytes.

The bile duct differentiation of hepatocytes in vitro is dependent on the presence of epidermal growth factor (or hepatocyte growth factor) and insulin [50]. Furthermore, the aberrant differentiation of hepatocytes is enhanced by orthovanadate, a potent protein tyrosine phosphatase inhibitor, that induces prolonged tyrosine phosphorylation of various proteins, suggesting that signaling

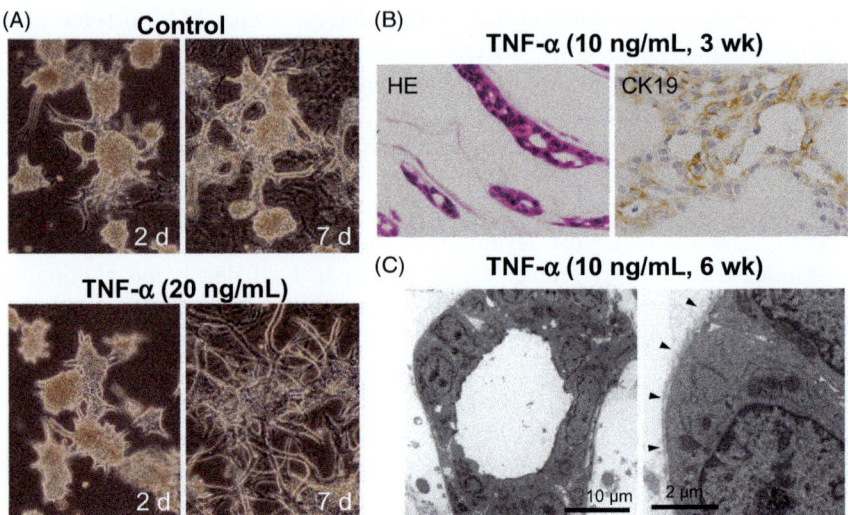

FIGURE 10.6 Ductular differentiation of rat hepatocytes within a collagen gel matrix. (A) Phase-contrast photomicrographs of spheroidal aggregates of hepatocytes 2 and 7 days following collagen gel culture. Tumor necrosis factor (TNF)-α facilitates branching morphogenesis. (B) Ductular structures formed 3 weeks following collagen gel culture (hematoxylin-eosin staining and CK19 immunocytochemistry). (C) Electron micrographs of a ductular structure. *Arrowheads* indicate basement membranes.

pathways regulated by protein tyrosine phosphorylation are involved in the process [55]. Experiments using various inhibitors revealed that the Ras-MEK-ERK and PI3 kinase-AKT pathways are particularly important. Soluble factors in fibroblast conditioned medium facilitate branching morphogenesis of hepatocytes within a collagen gel matrix, although the relevant soluble factors have not been identified [50]. Various inflammatory cytokines also affect the differentiation state of hepatocytes. Most notably, tumor necrosis factor (TNF)-α enhances markedly suppresses hepatocytic differentiation, while enhances branching morphogenesis and ductular differentiation (Fig. 10.6) [56]. Corticosteroids, such as dexamethasone, strongly inhibits ductular differentiation of hepatocytes and maintains their differentiation state [55].

The reversibility of the aberrant differentiation of hepatocytes toward bile duct/ductules is an important issue. The microenvironment, in which hepatocytes were placed, such as collagenous matrices and various growth factors and soluble factors, is considered to be an important determinant. When hepatocytes that have acquired bile ductular phenotype within collagen gels are retrieved by collagenase treatment, transferred to Matrigel, and cultured in a medium containing dexamethasone and OSM (or IL-6), they recovered their original phenotype (Fig. 10.7) [57]. Thus, the aberrant differentiation of hepatocytes is at least partly reversible when the microenvironment renders hepatocytic

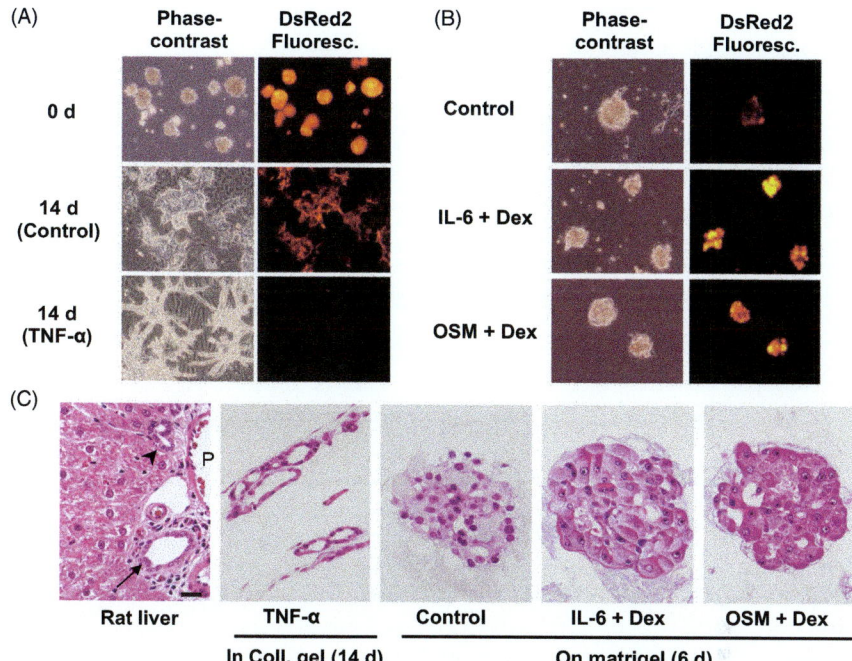

FIGURE 10.7 **Reversibility of ductular differentiation of rat hepatocytes in vitro.** Hepatocytes were isolated from Alb-DsRed2 transgenic rats, in which albumin-expressing cells show DsRed2 fluorescence. (A and B) Phase-contrast micrographs and DsRed2 fluorescence. Complete loss of DsRed2 fluorescence in hepatocytic spheroids within a collagen gel matrix after 14 days in the presence of TNF-α (A) and its recovery after being transferred to Matrigel-coated surfaces in the presence of dexamethasone and interleukin-6 (IL-6) or oncostatin M (OSM) (B). (C) Micrographs showing the processes of ductular differentiation and redifferentiation of hepatocytes in vitro. The *left end panel* show normal rat liver tissue containing a portal vein (P), bile duct (*arrow*), and bile ductule (*arrow head*).

differentiation. We also isolated bile duct/ductular cells from the adult rat liver and cultured on Matrigel with dexamethasone and OSM (or IL-6), and dexamethasone. Although they formed spheroidal aggregates and expressed very low levels of albumin, they did not demonstrate characteristic morphology of hepatocytes. This suggests that the phenotype of bile duct/ductular cells might be fixed after being terminally differentiated.

Recently, we have developed a three-dimensional culture technique of mouse hepatocytes for very long periods by an alternating sequencing of spheroid culture and collagen gel culture (Fig. 10.8). With this technique, hepatocytes can be cultured for more than a year and their phenotypic changes can be monitored. Although the initial spheroidal aggregates of mouse hepatocytes demonstrate branching morphogenesis within a collagen gel matrix, the branches tend to be indistinct after several weeks (Fig. 10.9A). Interestingly,

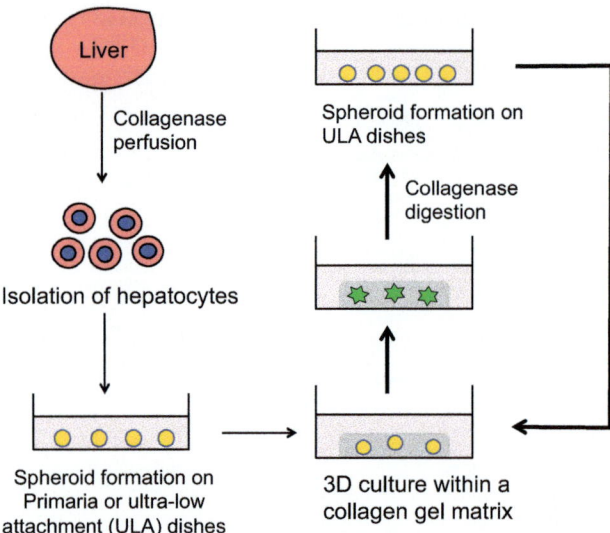

FIGURE 10.8 A method for long-term culture of hepatocytes. An alternating use of spheroid culture and collagen gel culture enables maintenance of mouse hepatocytes for more than a year.

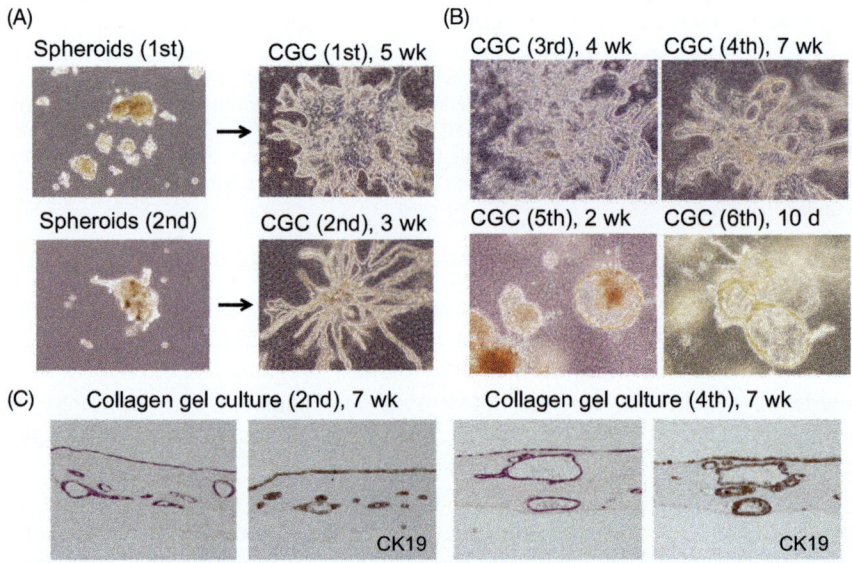

FIGURE 10.9 Ductular and cystic structures formed in long-term cultures of mouse hepatocytes. (A and B) Phase-contrast micrographs of the first and second cycles (A), third to sixth cycles (B) of the alternating spheroid culture-collagen gel culture (CGC). (C) Micrographs of the ductular or cystic structures formed within the collagen gels (hematoxylin-eosin staining and CK19 immunocytochemistry).

the second spheroidal aggregates that are spontaneously formed by cells retrieved from collagen gels by collagenase digestion form more distinct and longer CK19-positive branching structures (Fig. 10.9A). However, with repeating of this cycle of two distinct culture methods, cells within collagen gels gradually lose the capacity of branching morphogenesis and rather form cystic structures, which are also positive for CK19 (Fig. 10.9B and C). The transition of branching morphogenesis to cyst formation might indicate that phenotypic changes progress in bile ductular cells originated from hepatocytes during long-term culture.

In mice, cell lineage tracing systems using ROSA26R mice, in which hepatocyte-specific Cre recombinase expression induces β-galactosidase (β-gal) by Cre/LoxP recombination, can be applied. To confirm the branching ductular structures are derived from hepatocytes, we isolated hepatocytes from *Alb*-Cre × ROSA26R mice and cultured within a collagen gel matrix. *Alb*-Cre × ROSA26R hepatocytes are β-gal positive and the positivity would be maintained even after they differentiate into nonhepatocytic cells, including bile ductular cells. As expected, CK19-positive ductular structures formed within the gels are β-gal positive, which can be highlighted by X-gal histochemistry (Fig. 10.10).

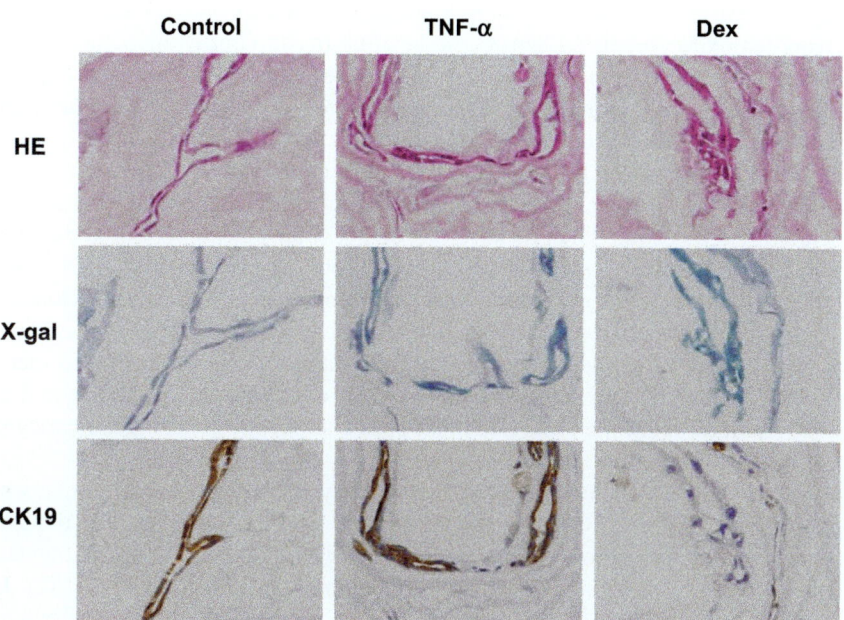

FIGURE 10.10 Hepatocyte origin of ductular structures formed within a collagen gel matrix. Collagen gel culture of β-galactosidase (β-gal)-positive *Alb*-Cre × ROSA26R mouse hepatocytes. X-gal cytochemistry and CK19 immunocytochemistry. Ductular structures are positive for both β-gal and CK19. Dexamethasone (Dex) suppresses CK19 expression.

4.2 Ductular Differentiation of Hepatocytes In Vivo

Whether mature hepatocytes can change their phenotype to bile ductular cells in various diseases has been a matter of debate. To prove bile ductular differentiation of hepatocytes in vivo, the ROSA26R-based hepatocyte lineage-tracing is also useful. We performed lineage-tracing experiments using a liver repopulation model, in which β-gal-positive *Alb*-Cre × ROSA26R hepatocytes were transplanted in the partially hepatectomized livers of retrorsine-treated mice [52]. The pretreatment with an alkaloid retrorsine suppresses the proliferation of host hepatocytes and thus facilitates proliferation and colony formation of transplanted β-gal-positive hepatocytes. Retrorsine treatment and partial hepatectomy induced mild liver injury and ductular reaction, some of which contained β-gal-positive ductular cells. When similarly treated mice were subjected to further liver injury induced by a DDC diet or CCl_4, a prominent ductular reaction occurred and many β-gal-positive ductular cells were found around or inside of the β-gal-positive hepatocyte colonies. These data clearly indicate that mature hepatocytes possess the capacity to differentiate into bile ductular cells in response to microenvironmental changes in vivo.

It is important to determine the relative contribution of the proliferation of bile ducts or ductules and bile ductular differentiation of hepatocytes in the ductular reaction associated with chronic liver diseases. Although several hepatocyte lineage-tracing analyses have been reported, conclusions are unexpectedly various. Using tamoxifen-mediated hepatocyte labeling in *Alb*-CreER mice, Sekiya and Suzuki [58] reported that most cells in ductular reaction in thioacetamide (TAA)-induced liver cirrhosis are derived from hepatocytes. They also used tamoxifen-treated CK19-CreER mice to label bile duct cells and confirmed that contribution of bile duct/ductular cells in ductular reaction is negligible. In contrast, using adenoassociated virus (AAV) 8-mediated hepatocyte labeling system, Malato et al. [59] demonstrated that hepatocytes do not contribute to periportal ductular reaction induced by a DDC diet. Yanger et al. [60] reported that, although DDC-induced ductular reaction is largely derived from existing bile ducts/ductules, it also contains a small number of hepatocyte-derived ductular cells. Furthermore, Tarlow et al. [46] applied a chimeric lineage tracing system and showed that a substantial number of hepatocyte-derived bile duct/ductular cells appeared in DDC-induced liver injury and that those cells could recover original hepatocytic phenotype after cessation of a DDC diet.

In our study, a hepatocyte lineage-tracing system using *Mx1*-Cre mice was applied. In *Mx1*-Cre × ROSA26R mice, the injection of poly(I:C) induces transient Cre expression in the liver, specifically and efficiently in hepatocytes, sparing bile duct cells and other nonparenchymal cells (Fig. 10.11A) [52]. In ductular reaction in the periportal area induced by a DDC diet or methylene dianiline, a bile duct toxin, the reactive bile ductules contained a small number of β-gal-positive cells (Fig. 10.11A). They were estimated to make up 1.9% and 4.7% of the CK19-positive ductular cells in the liver of DDC diet-fed and

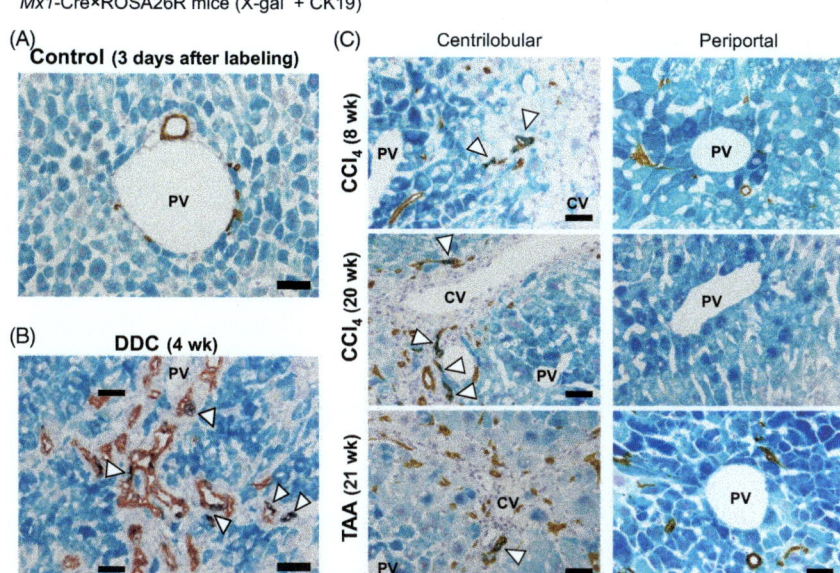

FIGURE 10.11 **Ductular transdifferentiation of hepatocytes and extensive duct/ductular remodeling in periportal and centrilobular ductular reaction.** Hepatocyte lineage-tracing using poly(I:C)-treated *Mx1*-Cre × ROSA26R mice. Combined X-gal histochemistry and CK19 immunohistochemistry. (A) Control liver. (B) Liver with periportal ductular reaction induced by 3,5-diethoxycarbonyl-1,4-dihydrocollidine (DDC) (periportal area). (C) Cirrhotic liver in CCl_4- or thioacetamide (TAA)-treated mice (centrilobular and periportal areas). *Arrow heads* indicate β-gal-positive ductules. Scale bar = 40 μm.

DAPM-treated mice, respectively. We also found that a small number of β-gal-positive cells are present in type 1 ductular reaction following common bile duct ligation. Furthermore, after a long time had elapsed since labeling, a small but significant increase in β-gal-positive cells occurred, reaching 4.2% of CK19-positive ductular cells. These findings clearly demonstrate that involvement of periportal hepatocytes in bile duct homeostasis and periportal ductular reaction, but their contribution is rather limited.

We also evaluated the cellular origins of ductular reactions in the centrilobular area, where no bile ductular cells are present [52]. Chronic administration of CCl_4 or TAA selectively damage centrilobular hepatocytes, causing inflammation and fibrosis, and ductular reaction in the centrilobular area. The CK19-positive ductular cells in the injured area are composed of β-gal-positive and -negative cells, which were often connected each other. In the early period of CCl_4 injury (8 weeks), 20.6% of ductular cells were β-gal-positive, but this proportion decreased thereafter as the ductular reaction progressed (Fig. 10.11B). The proportion of β-gal-positive cells was small (4.3%) in TAA-induced ductular reaction (Fig. 10.11B). Our results indicate that hepatocytes both in the periportal and centrilobular areas can contribute to ductular reaction following liver injury, as well as to homeostasis of bile ducts/ductules.

4.3 Transdifferentiation Versus Dedifferentiation of Mature Hepatocytes

It has been clear that hepatocytes show phenotypic plasticity, at least a capacity to differentiate into bile duct/ductular cells, even after maturation. As described earlier, some investigators have interpreted this phenomenon as "dedifferentiation," in which a terminally differentiated cell reverting back to a less-differentiated stage from within its own lineage [61]. Hepatoblasts express many fetal/neonatal genes, such as *AFP*, Delta-like 1 (*DLK1*), insulin-like growth factor 2 (*IGF2*), and glypican 3 (*GPC3*). Thus, to show that the process is the result of "dediffrentiation," it is necessary to demonstrate at least some of the features of hepatoblasts in hepatocyte-derived bile ductular cells. However, putative hepatocyte-derived stem/progenitor cells, including the recently proposed "hybrid hepatocytes," [47] have not been shown to express these genes, as well as their products.

We also confirmed that no mRNA expression of fetal/neonatal genes is detected in CCl_4-induced mouse liver cirrhosis tissues that contain a large amount of ductular reaction (see Fig. 10.14). Immunohistochemistry for AFP and DLK1 revealed that ductular cells in ductular reaction in CCl_4- or TAA-induced cirrhosis in mice did not express these proteins. Furthermore, the ductular differentiation of rat and mouse hepatocytes within collagen gels is not associated with the mRNA expression of *AFP* and *DLK1* [52,55]. The expression of bile duct markers and ductular differentiation of hepatocytes following liver injury in vivo and bile ductular differentiation of hepatocytes in vitro might not fulfill the criteria of dedifferentiation.

We consider that the aberrant differentiation of hepatocytes is, in fact, "transdifferentiation," which is defined as lineage switching of one cell type to another cell type [61], in this case, partial transition of hepatocytic phenotype to that of bile duct/ductular cells. Thus, ductular differentiation of hepatocytes in naturally occurring processes, which only express bile duct/ductular markers (Sox9, osteopontin, and CK19), might be regarded as transdifferentiation. However, it is not yet clear that the transdifferentiation is executed by a direct transversion or mediated by a transient dedifferentiation step.

5 REMODELING OF THE INTRAHEPATIC BILE DUCT SYSTEM IN CHRONIC LIVER DISEASES

Here I focus on another important aspect of the findings of the hepatocyte lineage-tracing experiments: remodeling of the intrahepatic bile duct system. Although hepatocytes indeed contribute to ductular reaction, both periportal and centrilobular, the majority of the emerging ductular cells are derived from existing bile ducts/ductules. This raises an particularly intriguing question regarding to the extensive centrilobular ductular reaction (type 2B) induced by chronic treatment with CCl_4 or TAA in mice. Since the centrilobular area normally contain no bile duct/ductular cells, where did they come from? Surprisingly, as the centrilobular ductular reaction progressed, the bile ducts/ductules in the portal tract appeared to migrate toward the injured centrilobular area, leaving the portal tract without duct/ductular structures (Fig. 10.11B). Three-dimensional analyses of the structures of the centrilobular ductular reaction revealed that the increased bile ducts/ductules were present as extensive networks within the lobules, most of them reaching the centrilobular area. Retrograde injection of India ink confirmed that there were connections between the aberrant centrilobular ductules and preexisting biliary system. Furthermore, the hepatic arteries, which are usually confined within the portal connective tissue, were often located in the midst of liver lobules. By establishing a technique to visualize the details of the bile duct trees by infusion of black ink into the

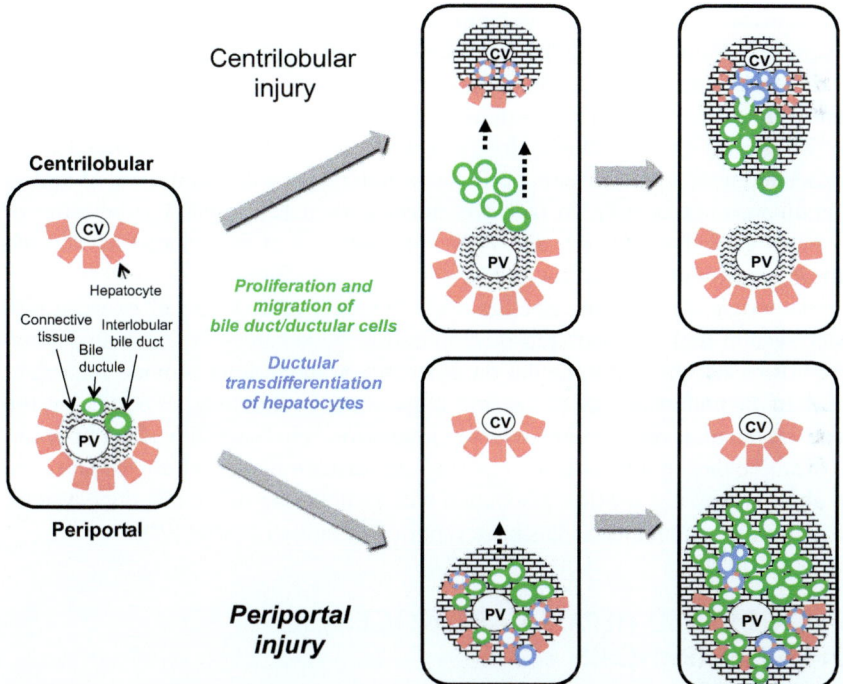

FIGURE 10.12 Schematic representation of two distinct origins for duct/ductular cells in the centrilobular and periportal ductular reaction. Hepatocytes transdifferentiate into bile duct/ductular cells in response to microenvironmental changes induced by chronic liver injury. Concomitantly, existing bile ductular cells actively proliferate and migrate. The migration of the duct/ductules is especially prominent in the centrilobular ductular reaction, in which bile duct/ductules of distinct origins establish anatomical connections.

common bile duct, Kaneko et al. [62] have also demonstrated that extensive bile duct remodeling occurs following chronic liver injury.

In CCl_4-injured liver of poly(I:C)-treated *Mx1*Cre × ROSA26R mice, the anatomical connection between β-gal-positive (hepatocyte origin) and -negative ductular cells (portal bile duct/ductule origin) has been proven by examination of serial sections, indicating that ductular cells of different origins actually communicate each other. Immunohistochemistry for a proliferation marker, phospho-histone H3, demonstrated that, while β-gal-positive ductular cells are virtually nonproliferative, β-gal-negative ductular cells show weak proliferative activity. Thus, centrilobular liver injury triggers proliferation and migration of portal bile duct/ductular cells to the inflamed and fibrosed area and the migrated bile duct/ductular cells establish the connection with bile ductules derived hepatocytes which undergo transdifferentiation due to microenvironmental changes associated with the injury (Fig. 10.12). In periportal ductular reaction, proliferation of portal bile duct/ductules and ductular transdifferentiation of hepatocytes ensue from inflammation and fibrosis induced by the injuries to periportal hepatocytes and/or bile duct/ductules (Fig. 10.12). These results provide clearer mechanistic insights into how centrilobular and periportal ductular reaction progress without assuming the presence of hepatic stem/progenitor cells.

The changes in extracellular matrices might be related to the extensive remodeling of the bile duct system in chronic liver injuries. The remodeling of the bile duct system during liver regeneration has been demonstrated to be associated with an increase of S100-A4, which induces the expression of matrix metalloproteinases [63]. In fact, we detected increased mRNA expression of *S100a4*, as well as *Mmp2*, and *Timp1* in cirrhotic livers induced by chronic administration of CCl_4 or TAA [52].

Although it is unclear whether such dynamic remodeling of existing bile duct system and ductular transdifferentiation of hepatocytes occur in human liver diseases, the experimental data described above may provide important clues to the understanding of the pathology of the human liver, in particular, the type 2B ductular reaction in "reversed lobulation" cirrhosis. It is also important to examine the significance of circulatory abnormalities, especially the presence of aberrant hepatic arteries, associated with remodeling of the bile duct system, since similar phenomenon has been reported in human NASH [20].

6 ENHANCED HEPATOCARCINOGENESIS IN LIVER CIRRHOSIS

6.1 Clonality of Regenerative Nodules in Liver Cirrhosis

It has long been considered that regenerative nodules in liver cirrhosis are formed by regenerative, therefore most probably polyclonal, proliferation of residual hepatocytes following the remodeling of hepatic lobules by portal-portal, portal-central, or central-central bridging fibrosis due to parenchymal extinction [10]. Hepatocytes are known to possess a robust regenerative activity and almost all the hepatocytes have been shown to enter the cell cycle after a partial hepatectomy in rodent models [64]. However, they also have a powerful system for cell growth restriction to prevent overgrowth of the liver following injury, although the exact cellular mechanisms have yet been elucidated [65]. It is unclear whether all hepatocytes are able to respond to repetitive or continuous stimuli for regeneration, such as exposure to growth factors. In fact, hepatocytes could show some characteristics for replicative senescence in chronic liver injury in which hepatocytes are exposed to excessive growth stimulation [66].

Clonal analyses of regenerative nodules of human liver cirrhosis have been reported by several groups. Using microdissection techniques and analysis of X-chromosome inactivation by examining the methylation status of the human androgen receptor (HUMARA), Paradis et al. [67] concluded that 51% of regenerative nodules are monoclonal. However, a similar study performed by Piao et al. [68] demonstrated all the large regenerative nodules (0.6–1.2 cm in diameter) examined are polyclonal. Aihara et al. [69] analyzed X-chromosome-linked phosphoglycerokinase gene and found that 43% of the regenerative nodules are monoclonal. Gong et al. [70] demonstrated that 48.3% of regenerative nodules with small cell change are monoclonal. More recent analysis performed by Lin et al. [71] using mitochondrial DNA mutations (cytochrome c oxidase) has found that 18% of regenerative nodules are homogeneously deficient, whereas only 3% of nodules have a mixed phenotype.

Despite the technical limitations inherent to the methods that can be applied for clonal analyses of human tissues, above results imply that a substantial proportion of the regenerative nodules in human liver cirrhosis are formed by selective proliferation of particular hepatocytes, rather than fibrotic dissection and subsequent proliferation of preexisting hepatocytes, each of which has comparative proliferating activity. The recognition that monoclonal hepatocyte proliferation may involve in the formation of regenerative nodules is important in considering the subsequent hepatocarcinogenesis.

6.2 Experimental Evidence for the Selective Proliferation of Subsets of Hepatocytes in Chronic Liver Injury

In mouse models of liver cirrhosis induced by chronic CCl_4 or TAA treatment, multiple liver tumors with hepatocytic differentiation (hepatocellular adenoma or HCC) appear almost invariably after 25 weeks of treatment, recapitulating the cirrhosis–carcinoma sequence in human (Fig. 10.13A) [73]. Thus, it is important to examine whether the regenerative nodules in mouse cirrhosis models also have monoclonal features. Although clonal analyses of regenerative nodules have not yet been reported in mouse models, we have recently obtained evidence for a monoclonal feature of regenerative nodules in studies employing an AAV-mediated hepatocyte lineage-tracing studies.

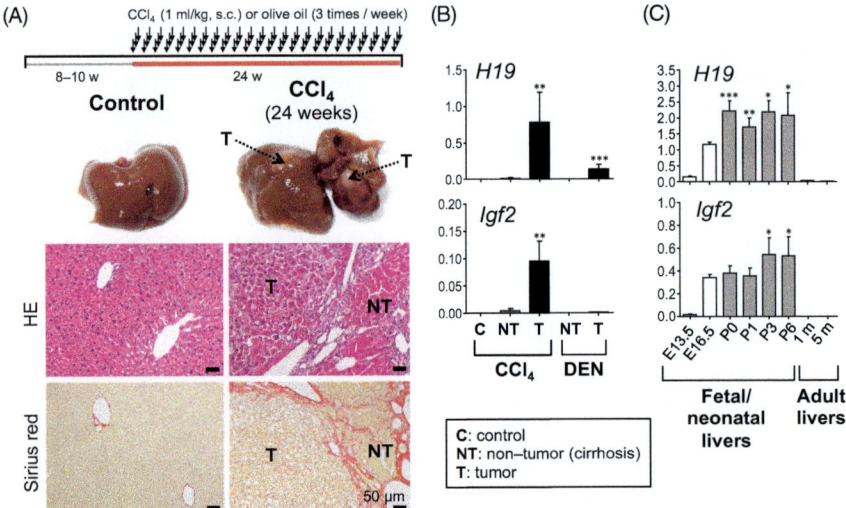

FIGURE 10.13 Fetal/neonatal gene expression in mouse liver tumors generated in CCl$_4$-induced cirrhosis. (A) Experimental protocol, gross appearances, and histology (hematoxylin-eosin, sirius red) of the tumors (T) and nontumor tissue (NT). (B) mRNA expression of *H19* and *IGF2* genes in CCl$_4$-induced tumors and diethylnitrosamine (DEN)-induced tumors (quantitative reverse transcription-polymerase chain reaction (RT-qPCR)). (C) mRNA expression of *H19* and *IGF2* genes in fetal (E13.5 and E16.5), neonatal (P0, P1, P3, and P6), and adult (1 and 5-month-old) mouse livers (RT-qPCR).

AAV serotype 8 (AAV8),[1] is known to efficiently and specifically, but transiently, infect mouse hepatocytes in vivo and has been frequently used in hepatocyte lineage-tracing experiments [59]. In ROSA26R mice, virtually all hepatocytes become β-gal-positive by infection of AAV8-*Tbg*-Cre, in which Cre is expressed under the control of the hepatocyte-specific thyroxin-binding globulin (TBG) promoter, while bile ducts and ductules, as well as other nonparenchymal cells, are completely negative (Fig. 10.14A). Following the hepatocyte labeling by AAV8-*Tbg*-Cre infection, the animals were subjected to chronic liver injury induced by CCl$_4$. After 20 weeks, in the cirrhotic livers, β-gal-positive, hepatocyte-derived bile ductules were found, confirming the presence of ductular transdifferentiation of hepatocytes that we found in experiments using the poly(I:C)-*Mx1*-Cre/ROSA26R model (Fig. 10.14B). However, surprisingly there were many regenerative nodules that were partly or entirely β-gal-negative (Fig. 10.14B), in addition to nodules composed of β-gal-positive hepatocytes. Given the high efficiency of hepatocyte labeling, this is an unexpected finding. Using a similar AAV8-mediated hepatocyte lineage-labeling system, an emergence of small numbers of β-gal-negative hepatocytes (approximately 1% of all hepatocytes) has been described following liver injuries, including partial hepatectomy, chronic CCl$_4$ injury (6 weeks),

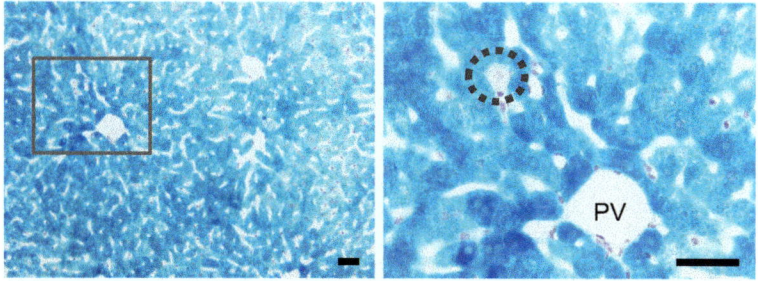

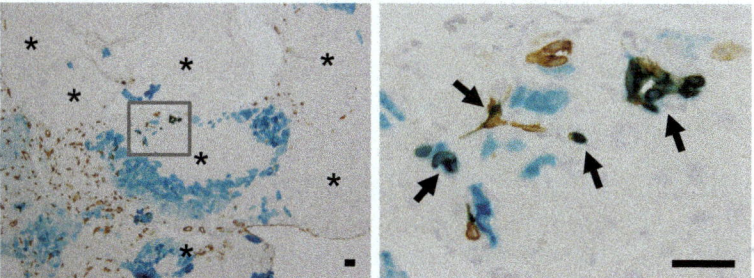

FIGURE 10.14 Clonal proliferation of subsets of hepatocytes in liver cirrhosis. Hepatocyte lineage-tracing using ROSA26R mice infected with AAV8-*Tbg*-Cre. (A) Control liver (X-gal histochemistry). *Boxed area* in the *left panels* is magnified in the *right panel*. A β-gal-negative hepatocyte is *circled* in the *right panel*. (B) Cirrhotic liver in CCl$_4$-treated mice (combined X-gal histochemistry and CK19 immunohistochemistry). *Boxed area* in the *left panels* is magnified in the *right panel*. Asterisks indicate β-gal-negative regenerative nodules. *Arrows* indicate β-gal-positive ductules. Scale bar = 40 μm.

bile duct ligation, and DDC diet [59]. Since the authors did not find any β-gal-negative hepatocytes in the livers after labeling in the absence of liver injury, they concluded that β-gal-negative hepatocytes are probably of hepatic stem/progenitor cell origin. However, in our hepatocyte lineage-labeling system, β-gal-negative hepatocytes can be detected sporadically as single cells within the entire liver lobules, albeit at very low frequency (Fig. 10.14A). This indicates that β-gal-negative hepatocytes may have a growth advantage over β-gal-positive hepatocytes and form regenerative nodules, although the reason for the advantage is not clear. Our preliminary results support clonal proliferation of subsets of hepatocytes in regenerative nodules in liver cirrhosis, although future clonal analyses are required to conclude.

The lack of ductular structures within regenerative nodules, either β-gal-positive or -negative, also favors the possible involvement of monoclonal proliferation of subsets of hepatocytes (Fig. 10.14B). The increased ductules appear to be concentrated between the regenerative nodules, as if they are crowded out from the original locations (portal or periportal) to the centrilobular area by the expanding regenerative nodules. Thus, monoclonal expansion of the regenerative nodules may contribute to the remodeling of the bile duct system.

6.3 Dedifferentiation of Hepatocytes and Hepatocarcinogenesis

Recent liver cell lineage-tracing experiments in mice have demonstrated that hepatocytic tumors induced by DEN treatment or hepatocyte-specific genetic manipulations are originated from hepatocytes, not from hepatic stem/progenitor cells or other cellular compartments of the liver, such as hepatic stellate cells [34]. Generally, malignant tumors, as well as most benign tumors, show phenotypes with various degrees of dedifferentiation, which is represented as cellular and structural atypia. HCC also frequently displays less differentiated phenotypes, by which pathological diagnosis is made. Thus, it is plausible that mature hepatocytes possess the capacity for dedifferentiation, in addition to that for transdifferentiation.

In an attempt to find genes specifically involved in mouse hepatocarcinogenesis with or without liver cirrhosis, we compared the mRNA expression profiles of mouse liver tumors induced by repeated injection of CCl_4 or a single DEN injection using a cDNA microarray and quantitative RT-PCR [72]. We identified 15 genes whose mRNA expression was increased predominantly in either CCl_4-induced or DEN-induced liver tumors with hepatocytic differentiation (hepatocellular adenoma or HCC) as well as genes that were increased comparably in both (Fig. 10.13B). The expression of these genes was not activated in cirrhotic tissues in CCl_4-induced hepatocarcinogenesis (Fig. 10.13B). Surprisingly, all but one of the genes, including *IGF2*, *H19*, *AFP*, and *GPC3*, were highly expressed in the fetal/neonatal period, suggesting strongly that hepatocytes undergo dedifferentiation in hepatocarcinogenesis (Fig. 10.13C).

Using a combination of the Sleeping Beauty transposon-mediated oncogene integration and hydrodynamic tail vein injection, a rapid mouse hepatocarcinogenesis model has been established [73]. Although it is demonstrated that the genes are delivered and integrated only in hepatocytes in this method, typical cholangiocarcinoma has been generated when the AKT signaling pathway is coactivated with the YAP or the Notch signaling pathway [74–76]. The Notch signaling pathway, a downstream effector of YAP, is known to play important roles in the development of intrahepatic bile ducts [77,78]. The presence of hepatocyte-derived cholangiocarcinoma further confirms the transdifferentiation capacity of mature hepatocytes. Furthermore, we have found that the introduction of Myc and mutant HRAS into hepatocytes generate hepatoblastoma-like tumors with a high nuclear-cytoplasmic ratio [79]. The tumor cells express various fetal/neonatal genes, including *IGF2*, *H19*, *AFP*, as well as *DLK1*, compatible with their hepatoblast-like features. These findings suggest that mature hepatocytes are able to dedifferentiate and become tumorigenic

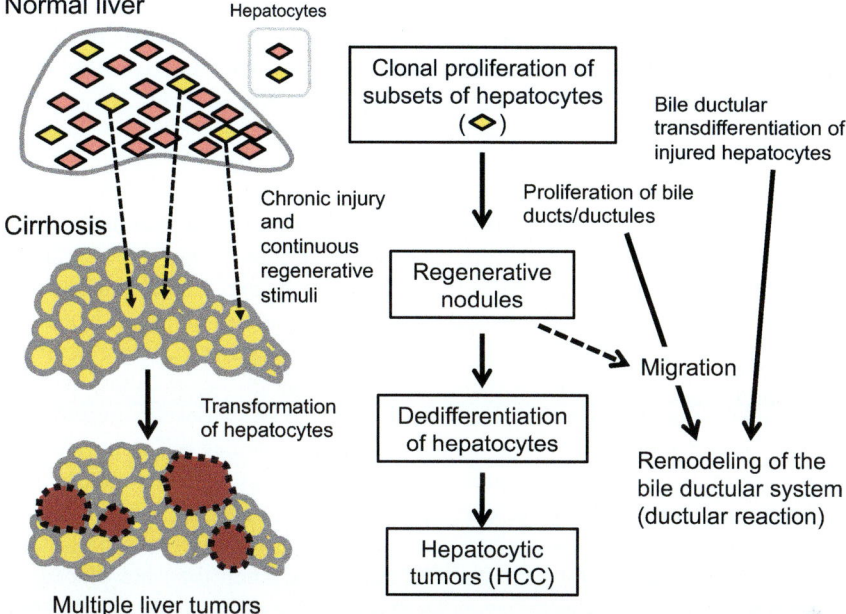

FIGURE 10.15 Hypothetical cellular mechanisms involved in the cirrhosis-carcinoma sequence and ductular reaction. Hepatocytes may not be homogeneous in terms of proliferative activity in response to repetitive or continuous growth stimuli in chronic liver injury. Regenerative nodules in liver cirrhosis may be formed by monoclonal proliferation of subsets of hepatocytes, rendering the nodules preneoplastic. Ductular reaction can be regarded as a combination of ductular transdifferentiation of hepatocytes and the remodeling of existing bile duct/ductular system and may not necessarily be an activation of hepatic stem/progenitor cells.

under the influences of the particular combinations of activated oncogenic signaling pathways.

Hepatocytes in the adult liver may be heterogeneous in the response to repetitive or continuous growth stimuli and regenerative nodules may be formed by selective proliferation of subsets of hepatocytes (Fig. 10.15). In fact, the presence of subpopulations of hepatocytes that are more proliferative has been suggested in vitro and in vivo [80,81]. It is possible that the long sought hepatic stem/progenitor cells might be such proliferative hepatocytes with the capacity to enter the cell cycle repeatedly or indefinitely. It is of note that telomerase activation is one of the most frequent and early genetic alterations in human HCC [82]. These hepatocytes that comprise regenerative nodules may be highly susceptible to genetic or epigenetic alterations of their genome in the setting of chronic inflammation causing oxidative stress, leading to multiple tumors with some features of dedifferentiation, which is induced by the activation of various oncogenes and/or inactivation of tumor suppressor genes (Fig. 10.15). The associated ductular reaction representing remodeling of the bile duct system and transdifferentiation of hepatocytes could be considered to be adaptive responses to bile ducts/ductules and hepatocytes to the adverse microenvironment, rather than the activation of hepatic stem/progenitor compartment, and therefore may not be directly involved in the hepatocarcinogenesis (Fig. 10.15).

7 CONCLUSIONS

In the early phase of the liver development, the liver bud formed at the hepatic diverticulum is consisted of primitive hepatoblasts that do not express albumin. Following hepatocyte specification induced by the expression of transcription factors, such as HNF-4α1, hepatoblasts start to express albumin. Not only mature hepatocytes, but also intrahepatic bile duct/ductular cells are derived from the albumin-expressing hepatoblasts (or immature hepatocytes), which have already committed to the hepatocytic lineage [83], despite expressing various fetal/neonatal genes (Fig. 10.16). Bile duct/ductular differentiation of hepatoblasts takes place at the interface between expanding hepatoblasts and the developing portal connective tissue [1].

In this chapter, I have described that mature hepatocytes retain the capacity to transdifferentiate into bile duct/ductular cells when they are placed in a microenvironment rich in collagenous matrices and relevant soluble factors (growth factors, cytokines) and that, at least in vitro, this process is partially

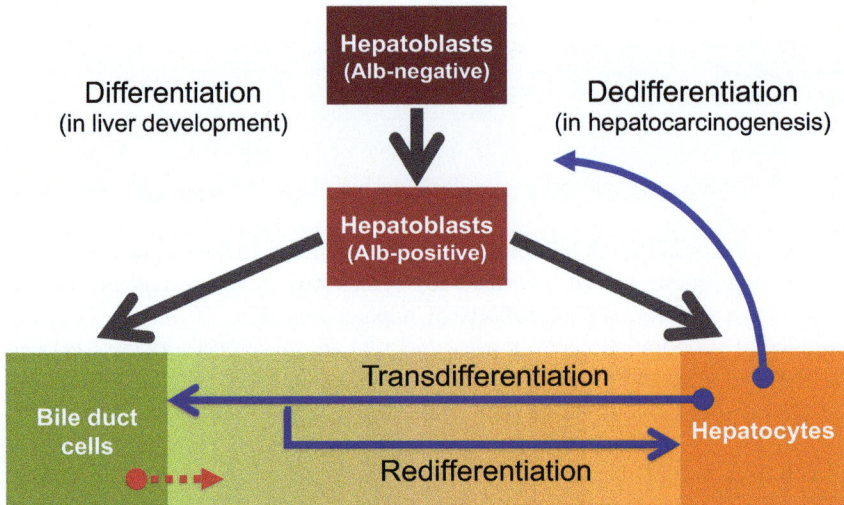

FIGURE 10.16 **The development of the liver epithelial system: its relevance to the phenotypic plasticity of mature hepatocytes.** Both hepatocytes and intrahepatic bile ducts/ductules are derived from albumin-expressing hepatoblasts during development. Mature hepatocytes can transdifferentiate to bile duct/ductular cells and this process may be partly reversible. Upon activation of particular oncogenic pathways, hepatocytes can dedifferentiate to generate tumors of their own lineage, as well as of bile duct/ductular lineage.

reversible (Fig. 10.16). Furthermore, mature hepatocytes are dedifferentiated to become neoplastic with expression of various fetal/neonatal genes, and the resultant tumor phenotypes can be not only hepatocytic, but also bile duct/ductular or mixed (hepatobiliary), reflecting their transdifferentiation capacity (Fig. 10.16). To elucidate the cellular and molecular mechanisms involved in the cirrhosis-HCC sequence, it is important to examine the possible heterogeneity in the proliferative activity of hepatocytes in response to chronic liver injury, rendering the regenerative nodules monoclonal and thus preneoplastic. Although the phenotype of fully matured bile duct/ductular cells appears to be more fixed (Fig. 10.16), they exhibit robust proliferative and remodeling activity in both periportal and centrilobular liver injury. Recent studies have also shown that these cells can be transformed to generate biliary tumors (e.g., cholangiocarcinoma) by genetic alterations [84,85].

ACKNOWLEDGMENTS

I would like to thank all the members of my laboratory past and present, as well as my former colleagues of the Department of Pathology at Akita University Graduate School of Medicine, who helped me to obtain the data reviewed in this chapter. This work was in part supported by grants from the Ministry of Education, Culture, Sports, Science, and Technology of Japan (#18590362, #21590426, and #24390092) and the Akiyama Life Science Foundation (2015).

REFERENCES

[1] Shiojiri N, Nagai Y. Preferential differentiation of the bile ducts along the portal vein in the development of mouse liver. Anat Embryol (Berl) 1992;185:17–24.

[2] Zhao R, Duncan SA. Embryonic development of the liver. Hepatology 2005;41:956–67.

[3] Overturf K, al-Dhalimy M, Ou CN, Finegold M, Grompe M. Serial transplantation reveals the stem-cell-like regenerative potential of adult mouse hepatocytes. Am J Pathol 1997;151:1273–80.

[4] Overturf K, Al-Dhalimy M, Finegold M, Grompe M. The repopulation potential of hepatocyte populations differing in size and prior mitotic expansion. Am J Pathol 1999;155:2135–43.

[5] Makino H, Togo S, Kubota T, Morioka D, Morita T, Kobayashi T, et al. A good model of hepatic failure after excessive hepatectomy in mice. J Surg Res 2005;127:171–6.

[6] Tanaka K, Ogura Y. "Small-for-size graft" and "small-for-size syndrome" in living donor liver transplantation. Yonsei Med J 2004;45:1089–94.

[7] Burt AD, Portmann BC, Ferrell LD, editors. MacSween's Pathology of the Liver. Churchill Livingstone Elsevier; 2007.

[8] Krings G, Can B, Ferrell L. Aberrant centrizonal features in chronic hepatic venous outflow obstruction: centrilobular mimicry of portal-based disease. Am J Surg Pathol 2014;38:205–14.

[9] Wanless IR, Nakashima E, Sherman M. Regression of human cirrhosis. Morphologic features and the genesis of incomplete septal cirrhosis. Arch Pathol Lab Med 2000;124:1599–607.

[10] Schuppan D, Afdhal NH. Liver cirrhosis. Lancet 2008;371:838–51.

[11] Perez-Tamayo R. Cirrhosis of the liver: a reversible disease? Pathol Annu 1979;14(Pt 2):183–213.

[12] Desmet V, Roskams T, Van Eyken P. Ductular reaction in the liver. Pathol Res Pract 1995;191:513–24.

[13] Williams MJ, Clouston AD, Forbes SJ. Links between hepatic rosis, ductular reaction, and progenitor cell expansion. Gastroenterology 2014;146:349–56.

[14] Richardson MM, Jonsson JR, Powell EE, Brunt EM, Neuschwander-Tetri BA, Bhathal PS, et al. Progressive fibrosis in nonalcoholic steatohepatitis: association with altered regeneration and a ductular reaction. Gastroenterology 2007;133:80–90.
[15] El-Serag HB. Hepatocellular carcinoma. N Engl J Med 2011;365:1118–27.
[16] Kanwal F, Kramer J, Asch SM, Chayanupatkul M, Cao Y, El-Serag HB. Risk of hepatocellular cancer in HCV patients treated with direct-acting antiviral agents. Gastroenterology 2017;153: 996–1005. e1001.
[17] Alvaro D, Mancino MG, Glaser S, Gaudio E, Marzioni M, Francis H, et al. Proliferating cholangiocytes: a neuroendocrine compartment in the diseased liver. Gastroenterology 2007;132:415–31.
[18] Desmet VJ. Ductal plates in hepatic ductular reactions. Hypothesis and implications. I. Types of ductular reaction reconsidered. Virchows Arch 2011;458:251–9.
[19] Tanaka M, Wanless IR. Pathology of the liver in Budd–Chiari syndrome: portal vein thrombosis and the histogenesis of veno-centric cirrhosis, veno-portal cirrhosis, and large regenerative nodules. Hepatology 1998;27:488–96.
[20] Gill RM, Belt P, Wilson L, Bass NM, Ferrell LD. Centrizonal arteries and microvessels in nonalcoholic steatohepatitis. Am J Surg Pathol 2011;35:1400–4.
[21] Slott PA, Liu MH, Tavoloni N. Origin, pattern, and mechanism of bile duct proliferation following biliary obstruction in the rat. Gastroenterology 1990;99:466–77.
[22] Ziol M, Nault JC, Aout M, Barget N, Tepper M, Martin A, et al. Intermediate hepatobiliary cells predict an increased risk of hepatocarcinogenesis in patients with hepatitis C virus-related cirrhosis. Gastroenterology 2010;139:335–43.
[23] Fausto N, Campbell JS. The role of hepatocytes and oval cells in liver regeneration and repopulation. Mech Dev 2003;120:117–30.
[24] Shafritz DA, Oertel M, Menthena A, Nierhoff D, Dabeva MD. Liver stem cells and prospects for liver reconstitution by transplanted cells. Hepatology 2006;43:S89–98.
[25] Miyajima A, Tanaka M, Itoh T. Stem/progenitor cells in liver development, homeostasis, regeneration, and reprogramming. Cell Stem Cell 2014;14:561–74.
[26] Fausto N. Liver regeneration and repair: hepatocytes, progenitor cells, and stem cells. Hepatology 2004;39:1477–87.
[27] Solt DB, Medline A, Farber E. Rapid emergence of carcinogen-induced hyperplastic lesions in a new model for the sequential analysis of liver carcinogenesis. Am J Pathol 1977;88:595–618.
[28] Shiojiri N, Lemire JM, Fausto N. Cell lineages and oval cell progenitors in rat liver development. Cancer Res 1991;51:2611–20.
[29] Furuyama K, Kawaguchi Y, Akiyama H, Horiguchi M, Kodama S, Kuhara T, et al. Continuous cell supply from a Sox9-expressing progenitor zone in adult liver, exocrine pancreas and intestine. Nat Genet 2011;43:34–41.
[30] Carpentier R, Suner RE, van Hul N, Kopp JL, Beaudry JB, Cordi S, et al. Embryonic ductal plate cells give rise to cholangiocytes, periportal hepatocytes, and adult liver progenitor cells. Gastroenterology 2011;141:1432–8.
[31] Espanol-Suner R, Carpentier R, Van Hul N, Legry V, Achouri Y, Cordi S, et al. Liver progenitor cells yield functional hepatocytes in response to chronic liver injury in mice. Gastroenterology 2012;143:1564–75. e1567.
[32] Tarlow BD, Finegold MJ, Grompe M. Clonal tracing of Sox9+ liver progenitors in mouse oval cell injury. Hepatology 2014;60:278–89.
[33] Lemaigre FP. Determining the fate of hepatic cells by lineage tracing: facts and pitfalls. Hepatology 2015;61:2100–3.

[34] Mu X, Espanol-Suner R, Mederacke I, Affo S, Manco R, Sempoux C, et al. Hepatocellular carcinoma originates from hepatocytes and not from the progenitor/biliary compartment. J Clin Invest 2015;125:3891–903.
[35] Yanger K, Knigin D, Zong Y, Maggs L, Gu G, Akiyama H, et al. Adult hepatocytes are generated by self-duplication rather than stem cell differentiation. Cell Stem Cell 2014;15:340–9.
[36] Schaub JR, Malato Y, Gormond C, Willenbring H. Evidence against a stem cell origin of new hepatocytes in a common mouse model of chronic liver injury. Cell Rep 2014;8:933–9.
[37] Yang L, Jung Y, Omenetti A, Witek RP, Choi S, Vandongen HM, et al. Fate-mapping evidence that hepatic stellate cells are epithelial progenitors in adult mouse livers. Stem Cells 2008;26:2104–13.
[38] Suzuki A, Sekiya S, Onishi M, Oshima N, Kiyonari H, Nakauchi H, et al. Flow cytometric isolation and clonal identification of self-renewing bipotent hepatic progenitor cells in adult mouse liver. Hepatology 2008;48:1964–78.
[39] Lu WY, Bird TG, Boulter L, Tsuchiya A, Cole AM, Hay T, et al. Hepatic progenitor cells of biliary origin with liver repopulation capacity. Nat Cell Biol 2015;17:971–83.
[40] Huch M, Gehart H, van Boxtel R, Hamer K, Blokzijl F, Verstegen MM, et al. Long-term culture of genome-stable bipotent stem cells from adult human liver. Cell 2015;160:299–312.
[41] Huch M, Dorrell C, Boj SF, van Es JH, Li VS, van de Wetering M, et al. In vitro expansion of single Lgr5(+) liver stem cells induced by Wnt-driven regeneration. Nature 2013; 494:247-50.
[42] Wang B, Zhao L, Fish M, Logan CY, Nusse R. Self-renewing diploid Axin2(+) cells fuel homeostatic renewal of the liver. Nature 2015;524:180–5.
[43] Planas-Paz L, Orsini V, Boulter L, Calabrese D, Pikiolek M, Nigsch F, et al. The RSPO-LGR4/5-ZNRF3/RNF43 module controls liver zonation and size. Nat Cell Biol 2016;18:467–79.
[44] Pu W, Zhang H, Huang X, Tian X, He L, Wang Y, et al. Mfsd2a+ hepatocytes repopulate the liver during injury and regeneration. Nat Commun 2016;7:13369.
[45] Tanimizu N, Nishikawa Y, Ichinohe N, Akiyama H, Mitaka T. Sry HMG box protein 9-positive (Sox9 +) epithelial cell adhesion molecule-negative (EpCAM-) biphenotypic cells derived from hepatocytes are involved in mouse liver regeneration. J Biol Chem 2014; 289:7589-98.
[46] Tarlow BD, Pelz C, Naugler WE, Wakefield L, Wilson EM, Finegold MJ, et al. Bipotential adult liver progenitors are derived from chronically injured mature hepatocytes. Cell Stem Cell 2014;15:605–18.
[47] Font-Burgada J, Shalapour S, Ramaswamy S, Hsueh B, Rossell D, Umemura A, et al. Hybrid periportal hepatocytes regenerate the injured liver without giving rise to cancer. Cell 2015;162:766–79.
[48] Katsuda T, Kawamata M, Hagiwara K, Takahashi RU, Yamamoto Y, Camargo FD, et al. Conversion of terminally committed hepatocytes to culturable bipotent progenitor cells with regenerative capacity. Cell Stem Cell 2017;20:41–55.
[49] Kamiya A, Kojima N, Kinoshita T, Sakai Y, Miyaijma A. Maturation of fetal hepatocytes in vitro by extracellular matrices and oncostatin M: induction of tryptophan oxygenase. Hepatology 2002;35:1351–9.
[50] Nishikawa Y, Tokusashi Y, Kadohama T, Nishimori H, Ogawa K. Hepatocytic cells form bile duct-like structures within a three-dimensional collagen gel matrix. Exp Cell Res 1996;223:357–71.
[51] Nishikawa Y. Transdifferentiation of mature hepatocytes into bile duct/ductule cells within a collagen gel matrix. Methods Mol Biol 2012;826:153–60.
[52] Nagahama Y, Sone M, Chen X, Okada Y, Yamamoto M, Xin B, et al. Contributions of hepatocytes and bile ductular cells in ductular reactions and remodeling of the biliary system after chronic liver injury. Am J Pathol 2014;184:3001–12.

[53] Block GD, Locker J, Bowen WC, Petersen BE, Katyal S, Strom SC, et al. Population expansion, clonal growth, and specific differentiation patterns in primary cultures of hepatocytes induced by HGF/SF, EGF and TGF alpha in a chemically defined (HGM) medium. J Cell Biol 1996;132:1133–49.

[54] Limaye PB, Bowen WC, Orr AV, Luo J, Tseng GC, Michalopoulos GK. Mechanisms of hepatocyte growth factor-mediated and epidermal growth factor-mediated signaling in transdifferentiation of rat hepatocytes to biliary epithelium. Hepatology 2008;47:1702–13.

[55] Nishikawa Y, Doi Y, Watanabe H, Tokairin T, Omori Y, Su M, et al. Transdifferentiation of mature rat hepatocytes into bile duct-like cells in vitro. Am J Pathol 2005;166:1077–88.

[56] Nishikawa Y, Sone M, Nagahama Y, Kumagai E, Doi Y, Omori Y, et al. Tumor necrosis factor-alpha promotes bile ductular transdifferentiation of mature rat hepatocytes in vitro. J Cell Biochem 2013;114:831–43.

[57] Sone M, Nishikawa Y, Nagahama Y, Kumagai E, Doi Y, Omori Y, et al. Recovery of mature hepatocytic phenotype following bile ductular transdifferentiation of rat hepatocytes in vitro. Am J Pathol 2012;181:2094–104.

[58] Sekiya S, Suzuki A. Hepatocytes, rather than cholangiocytes, can be the major source of primitive ductules in the chronically injured mouse liver. Am J Pathol 2014;184:1468–78.

[59] Malato Y, Naqvi S, Schurmann N, Ng R, Wang B, Zape J, et al. Fate tracing of mature hepatocytes in mouse liver homeostasis and regeneration. J Clin Invest 2011;121:4850–60.

[60] Yanger K, Zong Y, Maggs LR, Shapira SN, Maddipati R, Aiello NM, et al. Robust cellular reprogramming occurs spontaneously during liver regeneration. Genes Dev 2013;27:719–24.

[61] Jopling C, Boue S, Izpisua Belmonte JC. Dedifferentiation, transdifferentiation and reprogramming: three routes to regeneration. Nat Rev Mol Cell Biol 2011;12:79–89.

[62] Kaneko K, Kamimoto K, Miyajima A, Itoh T. Adaptive remodeling of the biliary architecture underlies liver homeostasis. Hepatology 2015;61:2056–66.

[63] Meng F, Francis H, Glaser S, Han Y, DeMorrow S, Stokes A, et al. Role of stem cell factor and granulocyte colony-stimulating factor in remodeling during liver regeneration. Hepatology 2012;55:209–21.

[64] Michalopoulos GK, DeFrances MC. Liver regeneration. Science 1997;276:60–6.

[65] Michalopoulos GK. Hepatostat: liver regeneration and normal liver tissue maintenance. Hepatology 2017;65:1384–92.

[66] Hoare M, Das T, Alexander G. Ageing, telomeres, senescence, and liver injury. J Hepatol 2010;53:950–61.

[67] Paradis V, Dargere D, Bonvoust F, Rubbia-Brandt L, Ba N, Bioulac-Sage P, et al. Clonal analysis of micronodules in virus C-induced liver cirrhosis using laser capture microdissection (LCM) and HUMARA assay. Laboratory Invest 2000;80:1553–9.

[68] Piao Z, Park YN, Kim H, Park C. Clonality of large regenerative nodules in liver cirrhosis. Liver 1997;17:251–6.

[69] Aihara T, Noguchi S, Sasaki Y, Nakano H, Imaoka S. Clonal analysis of regenerative nodules in hepatitis C virus-induced liver cirrhosis. Gastroenterology 1994;107:1805–11.

[70] Gong L, Li YH, Su Q, Chu X, Zhang W. Clonality of nodular lesions in liver cirrhosis and chromosomal abnormalities in monoclonal nodules of altered hepatocytes. Histopathology 2010;56:589–99.

[71] Lin WR, Lim SN, McDonald SA, Graham T, Wright VL, Peplow CL, et al. The histogenesis of regenerative nodules in human liver cirrhosis. Hepatology 2010;51:1017–26.

[72] Chen X, Yamamoto M, Fujii K, Nagahama Y, Ooshio T, Xin B, et al. Differential reactivation of fetal/neonatal genes in mouse liver tumors induced in cirrhotic and non-cirrhotic conditions. Cancer Sci 2015;106:972–81.

[73] Chen X, Calvisi DF. Hydrodynamic transfection for generation of novel mouse models for liver cancer research. Am J Pathol 2014;184:912–23.

[74] Li X, Tao J, Cigliano A, Sini M, Calderaro J, Azoulay D, et al. Co-activation of PIK3CA and Yap promotes development of hepatocellular and cholangiocellular tumors in mouse and human liver. Oncotarget 2015;6:10102–15.

[75] Fan B, Malato Y, Calvisi DF, Naqvi S, Razumilava N, Ribback S, et al. Cholangiocarcinomas can originate from hepatocytes in mice. J Clin Invest 2012;122:2911–5.

[76] Yamamoto M, Xin B, Watanabe K, Ooshio T, Fujii K, Chen X, et al. Oncogenic determination of a broad spectrum of phenotypes of hepatocyte-derived mouse liver tumors. Am J Pathol 2017;187:2711–25.

[77] Crosby HA, Nijjar SS, de Goyet Jde V, Kelly DA, Strain AJ. Progenitor cells of the biliary epithelial cell lineage. Semin Cell Dev Biol 2002;13:397–403.

[78] Yimlamai D, Christodoulou C, Galli GG, Yanger K, Pepe-Mooney B, Gurung B, et al. Hippo pathway activity influences liver cell fate. Cell 2014;157:1324–38.

[79] Xin BY M, Fujii K, Ooshio T, Chen X, Okada Y, Watanabe K, et al. Critical role of Myc activation in mouse hepatocarcinogenesis induced by the activation of AKT and RAS pathways. Oncogene 2017;36:5087–97.

[80] Mitaka T, Mikami M, Sattler GL, Pitot HC, Mochizuki Y. Small cell colonies appear in the primary culture of adult rat hepatocytes in the presence of nicotinamide and epidermal growth factor. Hepatology 1992;16:440–7.

[81] Gordon GJ, Coleman WB, Hixson DC, Grisham JW. Liver regeneration in rats with retrorsine-induced hepatocellular injury proceeds through a novel cellular response. Am J Pathol 2000;156:607–19.

[82] Zucman-Rossi J, Jeannot E, Nhieu JT, Scoazec JY, Guettier C, Rebouissou S, et al. Genotype-phenotype correlation in hepatocellular adenoma: new classification and relationship with HCC. Hepatology 2006;43:515–24.

[83] Shiojiri N, Katayama H. Secondary joining of the bile ducts during the hepatogenesis of the mouse embryo. Anat Embryol (Berl) 1987;177:153–63.

[84] Yamada D, Rizvi S, Razumilava N, Bronk SF, Davila JI, Champion MD, et al. IL-33 facilitates oncogene-induced cholangiocarcinoma in mice by an interleukin-6-sensitive mechanism. Hepatology 2015;61:1627–42.

[85] Ikenoue T, Terakado Y, Nakagawa H, Hikiba Y, Fujii T, Matsubara D, et al. A novel mouse model of intrahepatic cholangiocarcinoma induced by liver-specific Kras activation and Pten deletion. Sci Rep 2016;6:23899.

Chapter 11

The Role of Stem Cells in the Hepatobiliary System and in Cancer Development: a Surgeon's Perspective

Naoto Koike
Seirei Sakura Citizen Hospital, Sakura, Yokohama City University School of Medicine, Yokohama, Japan

1 INTRODUCTION

The hepatobiliary system is essential for digestion and usually includes the liver and biliary tract. The liver is the largest organ in the abdomen and plays important roles in homeostasis, including metabolism, glycogen storage, drug detoxication, production of various serum proteins, and bile secretion [1]. Hepatic cells, especially hepatocytes, can rapidly proliferate after acute liver injury (e.g., acute hepatitis, hepatic toxins, and hepatectomy) to repair damage [2]. However, severe and/or chronic liver damages impair the proliferative capacity of hepatocytes. Liver stem/progenitor cells (LPCs) are suspected to appear and undergo a massive expansion under this severe condition. However, origin of and mechanism of activation of LPCs in humans remain to be elucidated.

High-level skill is often necessary for surgeries concerning the hepatobiliary system. Radical liver resection of advanced liver cancer is a principal potentially curative treatment. Insufficient future liver remnant volume (FLRV) is the main cause of primary unresectable liver cancer. Recently, as clinical trials, portal vein embolization (PVE) with application of autologous stem cells has been attempted to increase insufficient FLRV. Liver cirrhosis (LC) is one of the leading causes of death in the world, and currently the only therapeutic option for end-stage liver disease (e.g., acute liver failure, cirrhosis, chronic hepatitis, cholestatic diseases, metabolic diseases, and malignant neoplasms) is orthotropic liver transplantation (OLT) [3]. A critical shortage of donor organs for treating end-stage organ failure highlights the urgent need for regenerative medicine using LPCs or induced pluripotent stem cells (iPSCs).

CSCs have been detected in several tumor types and may be viable therapeutic targets. A small subset of cancer cells with CSC characteristics are thought to be responsible for tumor initiation, drug and radiation resistance, invasive growth, metastasis, and tumor relapse, which are the main causes of cancer-related deaths [4]. There is accumulating evidence that CSCs are involved in the aggressiveness of primary hepatobiliary cancers [5]. CSCs are thought to derive from transformation of normal stem/progenitor cells and/or from the reprogramming of adult cells that convert them to stem/progenitor cells. Therefore the markers are similar in both normal stem/progenitor cells and CSCs in the hepatobiliary system [6]. Hepatobiliary CSCs markers are currently used as prognostic factors after surgical therapy and studied as candidates of therapeutic targets.

In this chapter, from a surgeon's perspective, I describe the recent progress in liver regeneration and CSC research relevant to the hepatobiliary system and viable therapeutic approaches based on stem cell biology.

2 ANATOMY AND DEVELOPMENT OF THE HEPATOBILIARY SYSTEM (FIG. 11.1)

The liver is the largest and pivotal organ in the abdomen and is the center of metabolism, detoxification, and digestion. The biliary tract comprises the intrahepatic bile duct, which begins at the canal of Hering in the liver and continues with bile ductules, interlobular ducts, septal, area, and segmental ducts, as well as the extrahepatic bile ducts [7]. Area and segmental ducts are considered to be large intrahepatic bile ducts, whereas septal ducts represent an intermediate link between the large and interlobular biliary systems [8]. The ramifying intrahepatic bile duct is merged into the right and left hepatic ducts, which are the first portion of the extrahepatic bile duct. The right and left hepatic ducts are combined at the hepatic hilum into the common hepatic duct. The common hepatic duct leads to the common bile duct and hepatopancreatic ampulla, successively. The hepatopancreatic ampulla drains into the duodenum via the papilla of Vater. The gallbladder with the cystic duct is located between the hepatic and bile ducts. The liver consists of two types of epithelial cells, hepatocytes and biliary epithelial cells called cholangiocytes. Most of the metabolic and synthetic functions of the liver are carried out by hepatocytes. The biliary tract is lined by cholangiocytes, and the large intrahepatic bile duct and extrahepatic bile duct contain peribiliary glands (PBG) deep within the duct walls. This biliary tract is the drainage system of the liver parenchyma and of the PBGs. The bile secreted by hepatocytes and secretion from PBGs drain into the biliary tract lumen and eventually into the duodenal lumen [9].

Organogenesis of the liver and biliary tract occur from the ventral foregut endoderm near the cardiac mesenchyme as hepatic diverticulum [10]. The commitment of endoderm cells to the liver is dictated by the following two crucial cytokines: fibroblast growth factor (FGF) from the developing cardiac

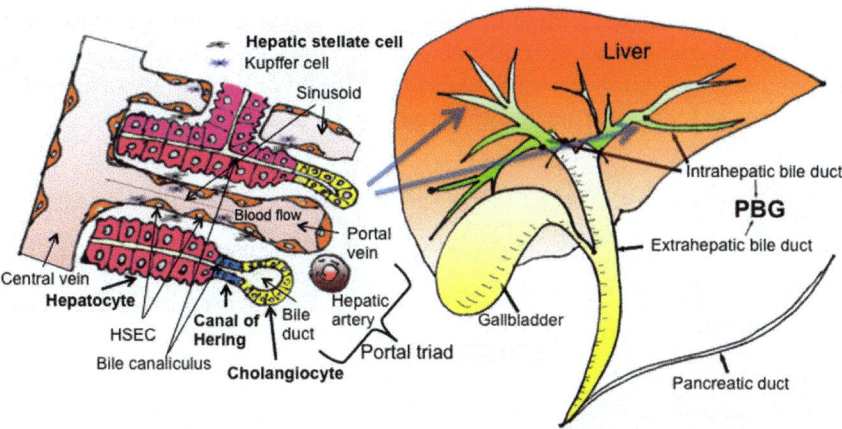

FIGURE 11.1 Schematic overview of the hepatobiliary system (*right*) and liver lobule (*left*). The hepatobiliary system composed of the liver and biliary tract. The biliary tract comprises the intrahepatic and extrahepatic bile ducts. Peribiliary glands (PBGs) contain stem cell niches within the biliary tract. The portal triad consists of portal vein, hepatic artery, and bile duct. Blood from portal vein and artery flows toward central vein through the sinusoid surrounded by hepatic sinusoidal endothelial cells (HSEC). Bile is produced by hepatocyte into the bile canaliculus and flows toward bile duct. Kupffer cells, the resident macrophage of the liver, are present in the sinusoid. The hepatic stellate cell is located near the HSEC in the space of Disse. The origin of liver stem/progenitor cells (LPCs) is still not clear. Candidates of the LPCs are indicated by bold characters. Under steady condition, hepatocyte and cholangiocyte slowly proliferate and maintain homeostasis. In traditional theory, the origin of LPCs is presumed to be located in the canal of Hering.

mesenchyme [11] and bone morphogenetic protein (BMP) from the septum transversum mesenchyme (STM) [12]. The foregut endoderm cells specified for hepatic commitment begin to express transcription factors, hematopoietically expressed homeobox (HEX), and hepatocyte nuclear factor 4α (HNF4α), as well as the liver-specific genes α-fetoprotein (AFP) and albumin (ALB), and delaminate from the endoderm and invade as cords into the surrounding STM. These cells are common progenitor cells, which differentiate into two epithelial cell types, hepatocytes and cholangiocytes, and are called "hepatoblasts" during liver development [1,13]. The hepatoblast surrounding the intrahepatic branches of the portal vein near the liver hilum form the so-called ductal plate. The intrahepatic bile ducts develop from the ductal plate. Along with remodeling of the plate, excess plate cells undergo regression, but some plate cells migrate into definitive portal tracts and form primitive intrahepatic bile ducts. The entire process of bile duct development progresses from near the hepatic hilum and from the larger to the smaller portal tracts, as well as in the liver. Intriguingly, the PBGs around the intrahepatic large bile ducts also derive from the ductal plate [9]. The extrahepatic bile duct also arises from the ventral endoderm of the foregut, but its origin is different from the liver and intrahepatic bile ducts [7]. Both hepatic ducts and part of the cystic duct develop from the cephalic end of the diverticulum, while the caudal segment develops into the gallbladder, which is part of the cystic duct and the common hepatic duct. Although the extrahepatic and intrahepatic bile ducts have different origins, they connect to each other to form a continuous biliary network.

3   STEM CELLS IN THE DEVELOPMENT AND REGENERATION OF HEPATOBILIARY SYSTEM

3.1   Stem Cells in Liver Development

Stem cells are found in all multicellular organisms and are likely to be present as discrete populations in most tissues. Stem cells are defined as cells that have the ability to perpetuate themselves through self-renewal and to generate mature cells of a particular tissue through differentiation [14]. During the process of embryogenesis, such stem cells differentiate via a variety of signaling and tissue environments and form each organ. Most stem cells lose their pluripotency and, therefore are rare in most mature organs. However, these small numbers of somatic stem cells play pivotal roles in organ development, tissue homeostasis, and tissue damage repair.

Hepatoblasts are presumed to be LPCs as aforementioned. Among stem cells in many tissues, hematopoietic stem cells (HSCs) have been the most extensively studied by combining fluorescently labeled monoclonal antibodies (mAbs) for cell markers and fluorescence-activated cell sorting (FACS) methodologies [15]. Taniguchi et al. have identified HSCs in the adult mouse liver by combining mAbs and FACS in 1996 [16]. Because the fetal liver is a major hematopoietic organ, a combination of negative selection by blood cell markers, cluster of differentiation (CD) 45 (common leukocyte antigen) and TER119 (erythroid cell antigen), and positive selection by cell surface markers has been utilized for FACS to successfully isolate hepatoblasts. Suzuki et al. developed a single-cell-based assay called the hepatic colony-forming unit in culture (H-CFU-C) and showed that the $CD45^-$, $TER119^-$, $c\text{-}KIT^-$, $CD49f^+$, and $CD29^+$ fractions of E13.5 mouse livers contained colony-forming cells which have the potential to differentiate into distinct hepatocytes and cholangiocytes [17]. Furthermore, they showed that the minor fractions of $CD45^-$, $TER119^-$, $c\text{-}KIT^-$, $c\text{-}Met^+$, $CD49f^{+/low}$, and $CD29^+$, which is only 0.3 % in the fetal liver, achieved the most enrichment in H-CFU-C. These cell fractions remained pluripotent for over 6 months in vitro and reconstituted hepatic cords and bile duct-like structure after implantation in the mouse [18]. On the other hand, Minguet et al. reported that the $c\text{-}KIT^{+/low}$ (CD45 TER119)$^-$ population of E11 mouse livers has the hepatic progenitor cell characteristics [19]. Tanimizu et al. isolated delta-like 1 homolog (DLK1) positive cells by FACS or an automatic magnetic cell sorter from E14.5 mouse livers and reported that colony-forming $DLK1^+$ cells had both hepatocyte and biliary epithelial cell lineages [20]. Tanaka et al. also reported that epithelial cell adhesion molecule

(EpCAM)$^+$ DLK1$^+$ cells in E11.5 mouse livers contained early hepatoblasts with high proliferation potential [21]. These suggested that mouse LPCs were not composed of a single population. Kubota et al. isolated classical major histocompatibility complex (MHC) class I, RT1A^{l-}, OX18$^{+/low}$, and intercellular adhesion molecule 1 (ICAM-1)$^-$ cells from E13 rat livers and revealed that these cells contained LPCs [22]. In humans, isolated EpCAM-positive cells by magnetic immunoselection using mAb HEA-125 from fetal and postnatal donor liver cell suspensions contained LPCs [23].

3.2 Stem Cells in the Adult Liver

The matured, adult liver had relatively low cell turnover rates under normal physiological conditions [24]. However, once the liver has suffered from injury, it regenerates rapidly. After a partial hepatectomy, regeneration of the liver is mainly achieved by hypertrophy and hyperplasia of residual, healthy hepatocytes [1,25,26]. Therefore it was unclear if stem cells were required for liver maintenance under normal conditions, such as that observed for intestinal epithelium and hematopoietic cells. A number of studies have investigated LPCs. Oval cells are referred as prototypes of LPCs in rodent liver regeneration. Oval cells first were reported in 1956 as the small round cells in the portal area induced by partial hepatectomy or administration of carbon tetrachloride after blocking hepatocyte proliferation with 2-acetylaminofluorene (2-AAF) [27]. Oval cells are induced in mice via a different method (a 3,5-diethoxycarbonyl-1,4-dihydrocollidine (DDC)-containing diet) [28]. Oval cells express both ALB and cytokeratin 19 (K19), which are hepatocytic and cholangiocytic markers, respectively, and are believed to differentiate into hepatocytic and biliary lineages, similar to hepatoblasts in the embryonic liver. Oval cells emerge in the portal area when the liver is injured, thus causing defective hepatocyte proliferation [29]. Oval cells accompany the proliferation of epithelium at the interface of the biliary tract and hepatic cords (canal of Hering), which is known as the ductular reaction [30–32]. These ductular reactions are even observed in patients with severe chronic hepatitis or LC. These oval cells and proliferating cells with ductular reactions generated around the canal of Hering might contribute to liver regeneration following severe damage. Kaneko et al. used novel methods for delineating the 3D structure of the biliary tracts in the mouse liver, which included a simple visualization method with ink injection from the extrahepatic bile duct into the intrahepatic bile duct, section analysis in combination with ink injection to test the tubular connection, and immunostaining of K19. In the results, they showed proliferating epithelial cells with ductular reactions tightly linked with biliary tracts and proposed that LPCs after liver injury were reflective of extensive remodeling of the biliary structure [33].

Several studies have identified possible LPC-specific markers that can distinguish LPCs from cholangiocytes. DLK [34,35] in the rat and TROP2 in the mouse [36] have been determined as specific markers of LPCs. Several oval cell markers such as A6, CD133, CD44, and EpCAM are also expressed in cholangiocytes [37]. However, CD133- [38, 39] or EpCAM-positive cells [36] in normal livers contain stem cell like populations, which have the following characteristics (1) clonogenic with high growth potential, (2) able to differentiate into both hepatocyte and cholangiocyte lineages in vitro, and (3) capable of repopulating the liver upon transplantation. There has been successful isolation of LPC-like populations from mouse, rat, and human mature livers based on oval cell and cholangiocyte markers. These cells cannot be identified as liver stem cells as they can only be functionally defined in vitro; therefore it remains to be elucidated whether and where they exist in vivo and their behavior in the liver.

Genetic lineage tracing and long-term label-retaining assay methods have been used to identify and characterize stem cells in vivo. These assays have allowed extensive characterization of several tissue-specific stem cells, intestinal stem cells, dermal stem cells, and hair follicle stem cells [40–42]. However, the slow turnover rate of hepatocytes makes it practically difficult to apply long-term label-retaining assay to identify stem cells in the normal adult liver. Kuwabara et al. performed label-retaining assay with acetaminophen liver injury providing a loading dose for bromodeoxyuridine (BrdU) incorporation of LPCs and a "chase" for washout of BrdU in transit amplifying progenitor cells. In this study, four possible hepatic stem cell niches were identified: the canal of Hering, intralobular bile ducts, periductal "null" mononuclear cells, and peribiliary hepatocytes [43]. Based on radiolabeled nucleotide-incorporation assays in the rat, a liver model for hepatocyte turnover, the "streaming liver hypothesis" has been established [44]. Many recent studies have employed Cre/loxP-mediated genetic marking and lineage tracing systems to characterize the mode of tissue maintenance in the adult mouse liver. Furuyama et al. supported this theory via a genetic lineage tracing study based on the cholangiocyte marker sex-determining region Y-box 9 (SOX9) [45]. The SOX9-labeled cells gradually spread out to hepatocytes from the periportal toward pericentral regions. However, subsequent studies by other groups using different types of genetic lineage tracing systems have provided conflicting results with the streaming theory [46]. Several studies revealed that most of all new hepatocytes come from preexisting hepatocytes [47–50]. Contrary, biliary cells assumed to be the major contributors to hepatocyte regeneration after extreme hepatocyte loss were observed in zebrafish liver regeneration models [51]. Studies by Kordes et al. have suggested that hepatic stellate cells may function as multipotent progenitor cells that generate functional hepatocytes and cholangiocytes as gleaned from rat livers [52] (Fig. 11.1). Font-Burgada et al. reported the presence of suspected hybrid hepatocytes with potential to differentiate into hepatocytes and cholangiocytes in the area of portal triad, and these cells extensively proliferated and replenished liver volume after chronic hepatocyte-depleting injuries [53]. Wang et al. identified hepatocytes expressing Wnt-responsive gene Axin2 adjacent to central vein endothelial cells and reported that these cells subserved homeostatic hepatocyte renewal in normal livers [54]. Given the varied results, as reported in the investigation regarding LPCs in the adult liver, the origin and existence of LPCs still remains to be determined.

3.3 Stem Cells in the Extrahepatic Biliary Tract

Recently, the existence of stem and/or progenitor cells in the extrahepatic bile duct has been reported and warranted much attention [9,55]. Koike et al. reported cell proliferation kinetics of the extrahepatic bile ducts using flash and cumulative BrdU labeling methods and immunohistochemical techniques. After a single injection of BrdU (flash labeling), labeled cells appeared in the lower portion of the gland-like downgrowth of the epithelium in the extrahepatic bile ducts. A gradual accumulation of the labeled cells at the surface epithelium was observed during cumulative labeling. They suggested that epithelial cells of extrahepatic bile duct are renewed at the lower portion of gland-like downgrowth of the epithelium (Fig. 11.2). The renewal time of the epithelial cells was shorter at the bile duct of the duodenal side than the hepatic side of the extrahepatic bile duct [56]. Because the gland-like downgrowth of rat extrahepatic bile ducts contained no acinar cells, the structure was not a true gland. In contrast, PBGs of mouse and human are tubuloalveolar glands with mucinous and serous glandular acini located in the deeper tissue of the bile duct walls [57]. Nakanuma et al. hypothesized that the PBGs of the extrahepatic bile ducts might be the compartment that harbors regenerative cells [8]. Experiments in mice have suggested that the extrahepatic bile ducts contained local progenitor cells that were c-Kit positive [58]. Cardinale et al. reported that multipotent stem/progenitors are present in human PBGs of extrahepatic biliary tracts expressing endodermal transcription factors and stem/progenitor surface markers (Fig. 11.1). These cells could differentiate in vitro into mature hepatocytes,

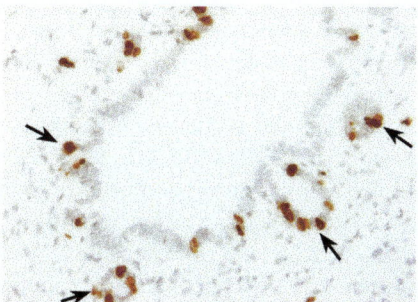

FIGURE 11.2 **Photomicrograph of immunohistochemical staining of the extrahepatic bile duct with monoclonal antibody against BrdU.** The specimen was obtained from the rat which received 6 times injections of BrdU given at 5 hours intervals. Many labeled cells at the lower portion of the downgrowth of the epithelium and some labeled cells at the bottom of shallow downgrowth (*arrows*) are observed (Objective X40).

cholangiocytes, or functional islets (which include β cells). Transplantation of these cells into quiescent livers of immunocompromised mice resulted in functional human hepatocytes and cholangiocytes, whereas transplantation into the fat pads of mice with streptozocin-induced diabetes resulted in functional islets that released human insulin, which significantly reduced hyperglycemia. They concluded extrahepatic bile ducts were a source of multipotent stem or progenitor cells for hepatic and pancreatic lineages [59–61]. Sutton et al. performed double immunostaining of K19 and c-Kit for human specimens to identify and localize cholangiocytes coexpressing putative progenitor cell markers. They suspected that PBGs of the extrahepatic bile ducts could be considered as a local niche of biliary progenitor cells [62]. These stem and progenitor cells are presumed to play key roles in pathophysiology and oncogenesis, as well as regeneration of the liver and biliary tracts.

3.4 The Role of Stem Cells in Hepatobiliary Surgery

End-stage liver disease and liver failure are major health problems worldwide, leading to high mortality and morbidity and high healthcare costs. Alcohol, hepatitis viruses, and metabolic diseases induce LC, which is a key cause of end-stage liver disease and liver failure [63,64]. Currently, OLT is the only appropriate therapy for end-stage liver disease patients. However, critical shortage of donor organs for treating of liver failure, high costs, and risk of organ rejection are the major obstacles to liver transplantation. Therefore alternative methods, with the potential to substitute for liver transplantation or reduce the amount of time for patients awaiting transplantation, are urgently required. Koike et al. engineered 3D microvessels using a coculture of human umbilical vein endothelial cells (HUVECs) and mouse mesenchymal stem cells (MSCs) in vitro. They transplanted micro-vessels constructs into the mouse cranial windows and observed the process in which this artificial vessel linked and acquired blood flow from the host vascular system under the intravital microscopy [65]. This engineered microvessel is fundamental technology for tissue engineering. Takebe et al. successfully established 3D liver buds in cocultures of human iPSCs with HUVECs, as well as human MSCs in vitro (Fig. 11.3). They generated vascularized and functional human liver in mice via transplantation of liver buds created in vitro [66]. Their study highlighted the enormous therapeutic potential using in-vitro-grown organ-bud transplantation to treat organ failure. More basic and clinical studies are warranted to assess the safety of these technologies.

However, cell therapy is already confirmed as an alternative therapy to OLT, and various types of cells, including stem cells, have been studied and been applied to end-stage liver disease and liver failure [67–69]. To date, autologous stem cells derived from bone marrow are the only stem cell type to have been clinically investigated. Terai et al. reported intravenous administration of mononuclear cells separated from autologous bone marrow improved liver function

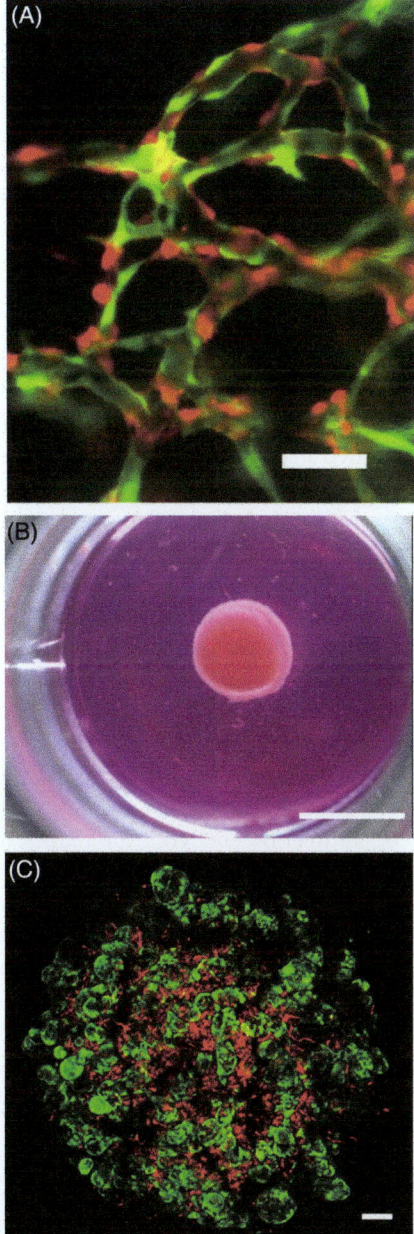

FIGURE 11.3 (A) Human umbilical vein endothelial cells (HUVECs) and human mesenchymal stem cells (human MSCs) were seeded in three-dimensional constructs and then implanted in mice. Three-dimensional, intravital microscopy image of engineered vessels 7 weeks after implantation into mice (*green*, HUVECs expressing enhanced green fluorescent protein (EGFP); *red*, hMSC expressing Kusabira Orange). Perfused vessels constructed with HUVECs are covered and stabilized by hMSC. Scale bar, 50 µm. (B) Gross observation of a three-dimensional human iPSC-liver bud (LB) in cocultures of human iPSC-hepatic endoderm cells (HEs) with HUVECs and human MSCs. Scale bar, 5 mm. (C) The fluorescence imaging of a human iPSC-LB. Presence of human iPSC-HEs and nascent endothelial networks inside human iPSC-LB. *Green*, human iPSC-HE; *red*, HUVEC. Scale bars, 100 µm.

of LC patients [70]. Pai et al. expanded autologous CD34$^+$ cells obtained by leukapheresis followed by granulocyte colony-stimulating factor (G-CSF) mobilization and injected these cells into the hepatic artery of severely alcoholic LC patients. Clinical and biochemical improvement was observed in this patient cohort [71]. Several experimental studies with animal models showed a transdifferentiation of transplanted stem cells into hepatocytes [72,73]. However, transdifferentiation of transplanted autologous stem cells into hepatocytes is thought to be a rare event in humans, but rather this process facilitates hepatic regeneration by supporting resident hepatocyte functions that promote vascular and macrophage-led matrix remodeling and immune modulation [67,74].

Hepatectomy is a principal radical treatment for hepatobiliary malignant diseases. However, extended hepatectomy, defined as resection of at least five hepatic segments, is sometimes necessary to achieve tumor-negative resection margins [75,76]. The liver has a large capacity for regeneration after resection. However, below a critical level of FLRV, extended hepatectomy is accompanied by a significant increase of postoperative liver failure [77,78]. There are currently several methods that can increase insufficient FLRV before extended hepatectomy [79]. These are two-stage hepatectomy [80], PVE [81–84], portal vein ligation (PVL) [85] and subsequent liver resection, PVE with a combination of transarterial chemotherapy, and associating liver partition and portal vein ligation for staged hepatectomy (ALPPS) [86,87]. These procedures have been widely used as a part of multimodal therapy with the aim to increase secondary extended liver resections. Recently, use of autologous stem cells to increase the regenerative capacity of the liver parenchyma and reduce the interval necessary to increase FLRV has been attempted. The cells are administered into the side of the liver without malignant tumor after PVE. A combination of PVE with autologous stem cells is safe for patients, without any severe complications. Five groups have already reported the results of this procedure (Table 11.1). In comparison with the PVE alone, this combination method has better results in terms of the time required for increase of FLRV [88]. There are two possible techniques for collecting autologous stem cells. First, bone marrow from the crista iliaca posterior superior can be used. The bone marrow aspirate is centrifuged and the CD133$^+$ stem cells separated [89,90]. Other groups utilize stimulation of autologous stem cells using G-CSF applied subcutaneously for 4 days. Then, large-scale leukapheresis is performed [91–94]. After administration of autologous stem cells, the growth of the contralateral liver lobe is monitored using computed tomography (CT) volumetry, and follow-up laboratory test, including liver functions, are performed. Franchi et al. reported a case of two-stage hepatectomy after PVE and autologous stem cell administration for unresectable liver metastases [95]. However, for the combination of PVE and application of autologous stem cells, there remains the potential danger of tumor growth progression, not only in the liver but also in other organs. There are previous reports noting the progression of tumor growth after PVE, but this is controversial [96,97]. Tumor tissue contains a great number of stimulants

TABLE 11.1 A Combination of PVE With Autologous Stem Cells Application

Author (year)	Disease	Cell source	Delivery of stem cell/timing of stem cells administration	Number of treatment group	Number of control group	Timing of FLRV evaluation outcome	Outcome	References
Ludvik et al. (2017)	CLM	GCSF mobilized peripheral blood	Under laparotomy, by a transileocolic approach—one or two day after PVE	20	PVE alone = 20	23 day	Significant FLRV increase	[94]
Treska et al. (2014)	CLM	GCSF mobilized peripheral blood	Under laparotomy, by a transileocolic approach—one day after PVE	11	PVE alone = 14	3 weeks	Significant FLRV increase	[93]
Han et al. (2014)	HCC	GCSF mobilized peripheral blood	Percutaneous ipsilateral approach—same day as PVE	3	Mononuclear cell = 4; PVE alone = 3	2 weeks	Significant FLRV increase	[92]
Canepa et al. (2013)	CLM, HCC, and CC	GCSF mobilized peripheral blood	Percutaneous ipsilateral approach—same day as PVE	6	PVE alone = 10	30 days	Significant FLRV increase	[91]
am Esch et al. (2012)	CLM, HCC, CC, GBC etc.,	Bone marrow from the crista iliaca posterior superior	Under laparotomy, by a transileocolic approach—same day as PVE	11	PVE alone = 11; Control = 18	2 weeks	Significant better prognosis	[90]

CC, cholangiocarcinoma; *CLM*, colorectal liver metastases; *FLRV*, future liver remnant volume; *GBC*, gallbladder carcinoma; *G-CSF*, granulocyte colony-stimulating factor; *HCC*, hepatocellular carcinoma; *PVE*, portal vein embolization; *Ref*, references.

(growth factors, matrix metalloproteinases, and cytokines) that can lead to the migration of stem cells into tumor tissue, which might then stimulate the growth and spread of the tumor. Future research is necessary, with a focus on the mechanisms of tumor proliferation in connection with this method [88].

The biliary epithelium is continuously exposed to highly cytotoxic bile acids and pathogens and thus is at persistent risk for injury. The PBGs of large bile ducts have been identified as a niche of progenitor cells that contribute to regeneration of biliary epithelium after injury as aforementioned. op den Dries et al. reported that injury of PBGs before transplantation of the liver was strongly associated with the occurrence of biliary strictures after transplantation [98]. The stem/progenitor cells of PBGs can differentiate into hepatocytes, cholangiocytes, or pancreatic islets, depending on the microenvironment, and this has implications for regenerative medicine, although its practical use has yet to be investigated thoroughly. In recent years, complications involving the biliary system are increasing. Biliary stenosis is the most frequently experienced complication. Various invasive and noninvasive techniques are now available to treat biliary stenosis, but the stenosis is sometimes refractory with a high risk of recurrence. If an artificial bile duct functionally identical to the natural organ is a feasible option, it can be used as a substitute for the pathological bile duct. Some studies investigating the regeneration of extrahepatic bile ducts have been reported [99–102]. Bioabsorbable materials were used for scaffolds of artificial bile ducts in animal studies. Although the stem/progenitor cells were not applied as the source of regenerative bile ducts, bile ducts were regenerated with bile ducts epithelium and accessory glands, which seemed to be PBGs within several weeks after implantation. However, the origins of epithelial cells and accessory glands observed in regenerative bile ducts have not yet been clarified.

4 CSCs IN HEPATOBILIARY CANCERS

4.1 Tumorigenesis and Aggressiveness of Hepatobiliary Cancers

Liver cancer represents 6% and 9% of the global cancer incidence and mortality burden worldwide, respectively. With an estimated 746,000 deaths in 2012, liver cancer is the second most common cause of death from cancer worldwide [103]. Cancers derived from the liver and extrahepatic bile duct are often aggressive and lead to poor prognoses. Surgical resection is a principal curative therapy; however, these cancers have usually developed beyond the scope of operation when discovered [104,105]. Chronic hepatitis due to viral hepatitis type B or C, alcoholic hepatitis, and nonalcoholic steatohepatitis (NASH) are major etiological causes of liver and biliary tract cancers. Therefore the prevention and development of therapies for chronic hepatitis are very important to reduce hepatobiliary cancer deaths [103].

Of primary hepatic malignancies, hepatocellular carcinoma (HCC), composed of malignant hepatocytes, represents about 80% of tumors.

Cholangiocarcinoma (CC), a gland-forming adenocarcinoma deriving from the biliary tracts, is the next most common cancer type. Combined hepatocellular–cholangiocarcinoma (CHC) is rare, and these tumors contain mixed elements of both HCC and CC [103]. However, CHC was recently subclassified due to the advanced studies of primary liver cancer with stem/progenitor cell features [106,107]. Most types of extrahepatic bile duct carcinoma are CC.

Genetic lineage tracing and long-term label-retaining assay methods have allowed extensive characterization of several tissue-specific stem cells, intestinal stem cells, dermal stem cells, and hair follicle stem cells as previously mentioned. In view of the key roles of stem/progenitor cells in the rapid turnover of tissues, it is acceptable that these are also the cells of origin in cancer of the intestine, skin, and hematopoietic systems [108–110]. In contrast, the matured adult liver is usually a slow turnover organ under normal physiological conditions, and the presence of LPCs still remains unknown. The CHC subtype with stem cell feature is suspected to be histologically generated from malignant transformation of LPCs, but this type of liver cancer is very rare, and transformation is not still clear. Mu et al. used complementary fate-tracing approaches to label the progenitor/biliary compartment and hepatocytes in murine hepatocarcinogenesis. In this study, HCC was exclusively originated from hepatocytes in mouse models. HCCs sometimes expressed CSC markers as if their origin was LPCs. However, the authors speculated that a progenitor signature of HCCs did not reflect liver progenitor origin, but rather dedifferentiation of hepatocyte-derived tumor cells [111]. Studies of transdifferentiation have also greatly progressed, as well as LPC research [112,113]. A recent mouse model of hepatocyte fate tracing demonstrated that CC could be derived not only from cholangiocytes [114] but also from hepatocytes [115,116] (Fig. 11.4). Such evidence suggested that hepatobiliary cancers have multiple cellular origins, including differentiated hepatocytes, intrahepatic biliary epithelial cells (cholangiocytes), LPCs, and biliary tract stem/progenitor cells, and PBGs. Multisteps for hepatobiliary carcinogenesis arise from multiple cellular origins, with the accumulation of genetic and epigenetic alterations and several molecules playing important roles in this process [52,117].

Cancers of liver and biliary tract origins are heterogeneous in their phenotypic traits, and as a result, clinical outcomes are difficult to manage. Cancers have small subpopulations of cells termed CSC. The hierarchical CSC model is that CSCs have the ability for unlimited cell division and comprise symmetric (self-renewal) and asymmetric (lineage restriction and differentiation) capabilities, which results in heterogeneous cell progenies [5,6,118]. The molecules related to CSCs contributed prominently to its aggressiveness, metastasis, and radiochemoresistance. There is accumulating evidence that CSCs have also been found in hepatobiliary cancers. Therefore unveiling molecular characteristics of hepatobiliary CSCs is important in the development of novel target therapies to substitute surgical therapies (Table 11.2). CSCs, especially liver CSCs, have been successfully identified using functional and cell surface markers via FACS [119–121]. Most of these markers are expressed in normal LPCs and are known as oncofetal markers [122].

224 Stem Cells and Cancer in Hepatology

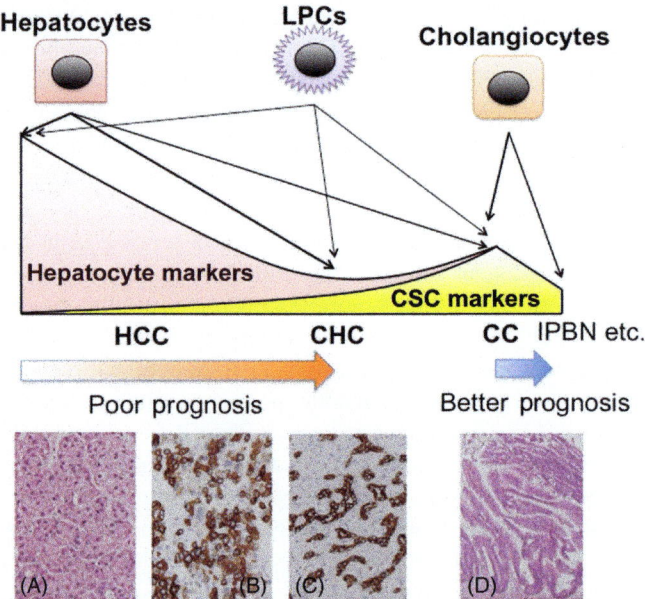

FIGURE 11.4 **Origins and prognosis of hepatobiliary cancers.** Most of the primary hepatic malignancies are hepatocellular carcinoma (HCC), composed of malignant hepatocytes, and the next common is cholangiocarcinoma (CC), a gland-forming adenocarcinoma deriving from the biliary tracts. Combined hepatocellular–cholangiocarcinoma (CHC) is a rare primary cancer. Intraductal-growth type of CC (e.g., intraductal papillary neoplasms (IPBN)) is rare and has better prognosis. The hepatobiliary cancers should have multiple cellular origins, including matured hepatocytes, cholangiocytes, and LPCs. In general, the hepatobiliary cancers especially HCC with stem/progenitor cell features, characterized by CSC markers, were shown to be highly aggressive and resulted in poor prognosis. (A) microphotograph of well differentiated HCC, (B) K19 immunostaining of poorly differentiated HCC, (C) K7 immunostaining of CHC, and (D) microphotograph of IPBN.

5 CSC MARKERS AS THERAPEUTIC TARGETS

5.1 Side Population

Side population (SP) cell sorting was initially applied for the identification of HSCs and has been used to enrich stem cell compartments in diverse tissues and organs. The SP subpopulation was determined via the ability to efflux the dye Hoechst 33342 through an adenosine triphosphate (ATP)-binding cassette (ABC) membrane transporter [123]. SP phenotype has widely been applied to the detection of CSCs in various cancers as well. In HCC, the existence of CSCs was first reported as this SP subpopulation [124]. They also identified polycomb complex protein, B lymphoma Moloney murine leukemia virus insertion region 1 homolog (BMI1), as a critical regulator of self-renewal and

TABLE 11.2 CSC Markers in Hepatobiliary Cancers

Type of cancer	Markers	Molecular background and mechanism	Target inhibitor	References
HCC	SP	ABC transporter	BMI1 siRNA, and ABC transporter inhibitor	[124–127]
	ALDH1[high]	ALDH inhibitor		[133]
	α2δ1 subunit	Calcium influx and ERK	Antibody 1B50-1 for α1δ2	[182]
	CD133	Akt, BCL-2, MAPK/PI3K, and neurotensin-interleukin-8	Akt inhibitor, Anti-CD133 antibody, and antisense oligonucleotides	[137,138, 141–143]
	OV6	Wnt/β-catenin	MicroRNA targeting β-catenin	[146]
	CD90	c-KIT, AMPK/mTOR, and β3 integrin	CD44 blockade, imatinib, shRNA or antibody targeting CD90 and β3 integrin, OSU-CG5	[148–150]
	EpCAM (CD326)	Wnt/β-catenin, SALL4, and CDH4	RNAi for EpCAM, HDAC inhibitor, PARP inhibitor, immunotoxin (VB4-845), and targeting EpCAM	[153,159–161]
	CD44 and CD133			[151]
	CD44s	TGF-β mediated EMT		[152]
	CD13	ROS-induced DNA damage reduction	Aminopeptidase N inhibitor (Ubenimex)	[162]
	CD24	STAT3-mediated NANOG regulation	STAT inhibitor, miR-21	[163,165]
	CD47	CTSS/PAR2 loop	CD47 antibody (B6H12)	[167]
	K19 (CK19)	TGF-β/SMAD	TGF-βR inhibitors	[173]
	TLR4			[179]
	SOX9	TGF-β/SMAD		[177]

(Continued)

TABLE 11.2 CSC Markers in Hepatobiliary Cancers (cont.)

Type of cancer	Markers	Molecular background and mechanism	Target inhibitor	References
CC, GBC	CD44 and CD133	ABCG2, Hedgehog (Gli1)	Hedgehog or ABC transporter inhibitor	[121]
	CD24, CD44, and EpCAM			[120]
	CD133	Oct4, NANOG, Nestin, and Akt	Arsenic trioxide, Akt inhibitor	[121,144,145]
	SP	TGF-β induced EMT, ABCG2, and ROS	ROS generator	[131,132]
	CD274 (PD-L1)^{low}	ALDH, ROS		[181]

ABC, Adenosine triphosphate (ATP)-binding cassette; *ABCG2*, ABC subfamily G, member 2; *ALDH*, aldehyde dehydrogenase; *AMPK*, adenosine monophosphate-activated protein kinase; *BCL*, B cell lymphoma; *BMI1*, B lymphoma Moloney murine leukemia virus insertion region 1 homolog; *CC*, cholangiocarcinoma; *CD*, cluster of differentiation; *CHD4*, chromodomain-helicase-Deoxyribonucleic acid-binding protein 4; *CTSS*, cathepsin S; *EMT*, epithelial-mesenchymal transition; *EpCAM*, epithelial cell adhesion molecule; *ERK*, p-extracellular signal-regulated kinases; *GBC*, gallbladder carcinoma; *Gli*, glioma-associated oncogene; *HCC*, hepatocellular carcinoma; *HDAC*, histone deacetylase; *MAPK*, mitogen-activated protein kinases; *mTOR*, mammalia target of rapamycin; *Oct*, octamer transcription factor; *PAR*, protease-activated receptor; *PARP*, polymeric adenosine diphosphate ribose (poly (ADP-ribose) polymerase; *PD-L*, programmed death ligand); *PI3K*, Phosphatidylinositol-4,5-bisphosphate 3-kinase; *PI3K*, phosphatidylinositol-4,5-bisphosphate 3-kinase; *RNAi*, RNA interference; *ROS*, reactive oxygen species; *SALL*, sal-like protein; *shRNA*, short hairpin ribonucleic acid; *siRNA*, small interfering RNA; *SOX*, sex determining region Y-box; *SP*, side population; *STAT*, signal transducer and activator of transcription; *TGF*, transforming growth factor; *TLR*, tall-like receptor.

tumorigenicity in the SP subpopulation [125,126]. It has been reported that BMI1 gene silencing inhibits the proliferation and invasiveness of HCC cells and increases their sensitivity to 5-fluorouracil (5-FU) [127]. Although SP subpopulation cells did not reveal a specific molecular target for therapeutic development, increased ABC transporter expression, including ABC subfamily G, member 2 (ABCG2), supported the therapeutic potential of targeting the SP population [128,129]. Because ABCG2, as an important multidrug resistance transporter, had the efflux capability of various chemotherapy drugs, it may also contribute to drug resistance of cancer cells [130]. ABCG2 has been indicated as a potential CSCs marker for HCC [129]. SP technology was used to isolate a small subpopulation of SP cells in a human gallbladder cancer (GBC) cell line as well [131]. Li et al. reported that 1,3,8-trihydroxy-6-methylanthraquinone (emodin), a reactive oxygen species (ROS) generator, was an effective agent targeting CSCs such as SP cells of GBC, either alone or acting as a chemotherapy enhancer [132].

Enhanced aldehyde dehydrogenase (ALDH) activity is a common hallmark between normal stem cells and CSC. The ALDH^{high} characteristics have been used to detect CSCs in HCC [133]. However, controversial data were recently reported stating that ALDH1A1-overexpressing cells may be differentiated cells rather than cancer stem cells in HCC [134].

5.2 CD133

The CD133 antigen is a five-transmembrane domain glycoprotein, also known as a prominin-1, initially identified as a HSC marker and present in various human embryonic epithelia, including neural tube, gut, kidney, and liver [135]. CD133 is also a common CSC marker [136]. CD133 was isolated from human HCC cells in 2006 [137,138] and GBC cells in 2010 [121] as well. Ma et al. isolated CD133$^+$ HCC cells and reported chemoresistance of CD133$^+$ CSCs through the preferential activation of certain survival pathways, including the Akt signaling and BCL-2 pathways [139]. Piao et al. reported that CD133$^+$ HCC CSCs modulate radioresistance through mitogen-activated protein kinase/phosphatidylinositol-4, 5-bisphosphate 3-kinase (MAPK/PI3K) pathway in human HCC [140]. Tang et al. reported that CD133$^+$ liver CSCs exhibit a greater ability to self-renew, induce tumor angiogenesis, and initiate tumors through neurotensin-interleukin (IL)-8 signaling [141]. CD133 may be a potential therapeutic target for the treatment of HCC. Direct suppression of CD133 by antisense oligonucleotides altered the cell cycle and affected HCC cell growth [142]. CD133 targeting was achieved by the administration of an anti-CD133 antibody conjugated to a potent cytotoxic drug, monomethyl auristatin F (MMAF) [143]. In GBC cells, CD44$^+$ and CD133$^+$ cells demonstrated a high degree of chemoresistance, possibly due to upregulation of ABCG2 and the transcription factor glioma-associated oncogene (Gli) 1 [121]. Ai et al. reported that arsenic trioxide (As$_2$O$_3$) effectively induces

CD133$^+$ GBC apoptosis. Furthermore, the ectopic expression of CD133 attenuated the apoptotic effect of As$_2$O$_3$ on cells through activation of Akt signaling pathways. Collectively, As$_2$O$_3$ effectively targets CD133 in GBC [144]. Li et al. reported that Akt inhibitor Wortmannin markedly inhibited the migration ability of CD133-overexpressing GBC cells [145].

5.3 OV6

Yang et al. demonstrated that hepatic progenitor marker OV6$^+$ cells possessed a greater ability to form tumors in vivo and showed a substantial resistance to standard chemotherapy compared with OV6$^-$ tumor cells. The chemoresistance of OV6$^+$ HCC progenitor-like cells could be reversed by lentivirus-delivered stable expression of microRNA targeting β-catenin. These results highlighted that therapies targeted to the Wnt/β-catenin signaling may provide a specific method to disrupt this resistance mechanism to improve overall tumor control with chemotherapy [146].

5.4 CD90

CD90 (Thy-1) is a 25–37 kDa glycosylphosphatidylinositol (GPI)-anchored glycoprotein expressed mainly in leukocytes and is involved in cell–cell and cell–matrix interactions [147]. Yang et al. reported that CD90 was a potential marker for liver CSCs, because the CD45$^-$ CD90$^+$ cells isolated from tumor tissues and blood samples of liver cancer patients had the capacity to generate tumor nodules in immunodeficient mice, whereas the CD90$^-$ or CD45$^-$ CD90$^-$ cells did not. The CD90$^+$ CD44$^+$ cells, further isolated from CD90$^+$ cells, contributed prominently to its aggressiveness and metastasis. The prevention of local and systemic tumor formation by CD44 blockade highlighted the potential of CD44 as a therapeutic target for CD90$^+$ CSCs [148]. Yamashita et al. reported that CD90$^+$ HCC cells possessed features of vascular endothelial cells and had highly metastatic potential. CD90$^+$ HCC cells showed abundant expression of c-KIT and chemosensitivity to imatinib mesylate in vitro [149]. Chen et al. showed that the expression of CD133 is higher in CD90-positive cells than that in CD90-negative tumor cells, and signaling analyses revealed that adenosine monophosphate (AMP)-activated protein kinase (AMPK)/mammalia target of rapamycin (mTOR) and β3 integrin are required for the induction of CD133 by CD90. Their experiments also revealed that the tumorigenicity of cancer cells could be blocked by short hairpin ribonucleic acid (shRNA) or antibodies targeting CD90, β3 integrin, and CD133. Energy restriction mimetic agent OSU-CG5 effectively decreased CD90-positive cells and tumor growth. These results demonstrated that the CD90-integrin-AMPK-CD133 signal axis was essential for the growth of liver cancer, and this signal pathway might be a viable therapeutics target of liver cancer [150].

5.5 CD44

CD133 is a reported potential CSC marker of liver cancer. Zhu et al. demonstrated that both CD133- and CD44-positive HCC cells possessed properties similar to CSCs, including extensive proliferation, self-renewal, and the ability to give rise to differentiated progeny, as well as initiated tumor growth in immunodeficient mice at very low cell numbers, when their $CD133^+$ $CD44^-$ counterparts did not. They concluded that CSCs in HCC were characterized by coexpression of CD133 and CD44 [151]. Mima et al. reported that overexpression of the standard isoform of CD44 (CD44s) promoted tumor invasiveness, increased the expression of a mesenchymal marker vimentin, and regulated the transforming growth factor (TGF)-β–mediated mesenchymal phenotype in HCC cells. Clinically, overexpression of CD44s was associated with low expression of E-cadherin, high expression of vimentin, and poor prognosis in HCC patients. The findings suggested that CD44s plays a critical role in the TGF-β–mediated mesenchymal phenotype and therefore represents a potential therapeutic target for HCC [152]. In GBCs, $CD44^+$ $CD133^+$ cells demonstrated a high degree of chemoresistance, possibly due to upregulation of ABCG2 and the transcription factor Gli1-like CD133 [121].

5.6 EpCAM

EpCAM is a type I transmembrane glycoprotein, known as CD326, characterized as one of the specific markers of LPC as mentioned previously. EpCAM is expressed in many human cancers with an epithelial origin. Yamashita et al. reported EpCAM is a downstream effector of the Wnt/β-catenin signaling pathway, whereas RNA interference-based blockage (RNAi) of EpCAM, a Wnt/β-catenin signaling target, attenuated the activities of these cells [153]. They also used EpCAM as a prognostic marker and proposed an HCC classification system based on the stem/maturation status of the tumor by EpCAM and AFP expression status. The system showed that the $EpCAM^+/AFP^+$ subtype of HCC significantly correlated with a poor prognosis for HCC patients [154]. In addition, Sal-like protein 4 (SALL4), a member of the zinc finger transcription factor family, plays a critical role in maintaining embryonic stem cell pluripotency and self-renewal [155]. Oikawa et al. reported that SALL4 is expressed in normal murine hepatoblasts and controls the lineage commitment of hepatoblasts, not only via inhibiting their hepatocyte-lineage differentiation but also driving their differentiation toward cholangiocytes [156]. Expression of SALL4 in HCC cells was reported to be an indicator of stem cells and a prognostic marker, and overexpression of SALL4 increased tumor growth and chemoresistance [157]. SALL4 was recently found to directly interact with the epigenetic modulator nucleosome remodeling and histone deacetylase (NuRD) complex [158]. Zeng et al. showed that SALL4 regulated stemness of EpCAM-positive HCC cells, and the histone deacetylase (HDAC) inhibitor successfully suppressed proliferation of SALL4-positive HCC cells [159]. Their study further showed that chromodomain-helicase-Deoxyribonucleic acid (DNA)-binding protein 4 (CHD4), a member of the NuRD complex, together with HDAC regulates chromatin remodeling and HDAC activities, and is activated in EpCAM-positive CSCs. To inhibit the functions of CHD4 that are mediated through HDAC and polymeric adenosine diphosphate ribose (poly (ADP-ribose)) polymerase (PARP), they evaluated the effect of the HDAC inhibitor suberohydroxamic acid (SBHA) and the PARP inhibitor AG-014699. The combination of SBHA and AG-014699 was most effective for treatment of EpCAM-positive HCC [160]. Ogawa et al. focused on the expression of EpCAM in human HCC cell lines and analyzed the effects of VB4-845 (Oportuzumab monatox), an immunotoxin targeting EpCAM, on human HCC cell lines. They demonstrated that the combination of VB4-845 plus 5-FU showed significant regression of tumors compared with the control and suggested that EpCAM-targeted therapy might offer a promising and novel approach for the treatment of HCC in patients with poorer prognosis [161].

5.7 CD13

Haraguchi et al. have identified CD13 (aminopeptidase N) as a novel liver CSC marker, which was enriched in SP populations isolated from HCC cell lines. Further analysis showed that the CD13$^+$ HCC cells predominated in the G0 phase of the cell cycle as a semiquiescent CSC marker. CD13$^+$ HCC cells produced CD90$^+$ HCC cells and demonstrated enhanced tumorigenicity and self-renewal. Subsequent studies suggested that CD13 reduced ROS-induced DNA damage after chemotherapy and radiotherapy and protected cells from apoptosis. Importantly, combination of a CD13 inhibitor (a CD31 neutralizing antibody or CD13 inhibitor (Ubenimex)) with 5-FU treatment drastically reduced tumor volume compared with either agent alone in mouse xenograft models. The investigators proposed that combining a CD13 inhibitor with a ROS-inducing chemo/radiation therapy might improve the treatment of liver cancer [162].

5.8 CD24

Lee et al. have identified CD24, a mucin-like cell surface glycoprotein, as a functional liver CSC marker in a chemoresistant HCC xenograft tumors. They assessed the quantitative polymerase chain reaction (qPCR) data for effects of CD24 on stemness-associated genes by both cell sorting and a lentiviral-knockdown approach to determine the major downstream mediator of CD24 in tumor initiation and self-renewal. NANOG significantly correlated with CD24 in HCC cell lines and clinical samples and was identified as a downstream effector of CD24 signaling, which acted through signal transducer and activator of transcription (STAT) 3 activation [163]. Wang et al. showed that CD24$^+$ CD44$^+$

EpCAM^{high} cells isolated from three extrahepatic CC tumor grafts exhibited higher tumorigenicity potential compared to CD24$^-$ CD44$^-$ EpCAM$^{low/-}$ cells. These studies demonstrated that these tumorigenic extrahepatic CC cells exhibited stem cell properties of self-renewal and had the ability to produce heterogeneous progeny. They reported the identification of a CSC population in extrahepatic CC characterized by CD24, CD44, and EpCAM phenotypes [120]. MicroRNA-21 (miR-21) was reported to be significantly overexpressed in HCC and intrahepatic CC [164]. Zhang et al. showed that anti-miR-21 treatment reduced liver tumor growth and prevented tumor development accompanied with a decrease in liver fibrosis and a concomitant reduction of CD24$^+$ liver progenitor cells in mice. Therefore they suggested that anti-miR-21 might be effective at targeting CSCs (CD24$^+$ cells) [165].

5.9 CD47

CD47, an integrin-associated protein (IAP), is a universally expressed member of the immunoglobulin superfamily that plays a critical role in self-recognition. Recently, CD47 was recognized an immune checkpoint molecule [166]. Lee et al. found upregulation of CD47 in chemoresistant hepatospheres when compared with differentiated progenies. Their studies demonstrated that CD47 contributed to tumor initiation, self-renewal, metastasis, and significantly affected patients' clinical outcome as liver CSCs. They also showed that knockdown of CD47 by a lentiviral-based shRNA approach suppressed CSC characteristics and identified cathepsin S (CTSS) as an important downstream effector of CD47-mediated HCC tumorigenicity, metastasis, and self-renewal. Using the cell line and patient-derived xenograft models, suppression of CD47 by a morpholino approach inhibited growth of HCC and sensitized cells to the effect of chemotherapy through blockade of CTSS/protease-activated receptor 2 (PAR2) signaling [167]. They reported in another study that the anti-CD47 antibody in combination with doxorubicin exerted maximal effects on tumor suppression, as compared with doxorubicin or anti-CD47 antibody alone, using a patient-derived HCC xenograft mouse model. In conclusion, they suggested that anti-CD47 antibody treatment could complement chemotherapy, which may be a promising therapeutic strategy for the treatment of HCC patients [168].

5.10 K19

During embryogenesis, K19 (CK19) is expressed on ALB-positive LPCs as a marker. Normal cholangiocytes express K19, but matured hepatocytes usually do not as previously mentioned. K19 is known to be a marker of poor prognosis in HCC as per several recent studies [107,169,170]. Kim et al. reported that K19 plays an important role of invasiveness of HCC cells through the upregulation of the epithelial-mesenchymal transition (EMT) associated gene [171]. Govaere et al. showed that K19-positive cells have chemoresistance characteristics [172].

Moreover, Kawai et al. studied whether K19-positive cells have CSC properties in HCC and thus could be a new therapeutic target [173]. They transfected a transgene vector that expressed green fluorescence protein (EGFP) under the control of the human K19 promoter into four HCC cell lines to characterize K19$^+$ HCC cells as liver CSCs and isolated K19$^+$ cells by FACS. K19$^+$ cells showed self-renewal and differentiation into K19$^-$ cells. K19$^+$ cells displayed high proliferation capacity and 5-FU resistance in vitro and generated large tumors at a high frequency in vivo. Their experiments demonstrated that K19$^+$ cells were involved in EMT and the activation of TGF-β/SMAD signaling, and these properties were suppressed by K19 knockdown using small interfering RNA (siRNA) or treatment with a TGF-β receptor 1 (TGF-βR1) inhibitor (LY2157299). Treatment with LY2157299 also showed high therapeutic effects against K19$^+$ tumor in a mouse xenograft model. LY2157299 is now in clinical trials of not only liver cancer but also several human solid cancers [174,175].

5.11 SOX9

SOX9 is involved in organogenesis of the liver and pancreas. During hepatogenesis, SOX9 expression is confined to the bile duct, while hepatocytes do not express SOX9, and this expression pattern persists in adulthood. Therefore SOX9 is available as a cholangiocyte marker of the normal adult hepatobiliary system. Liu et al. reported that SOX9 is highly expressed in NANOG-positive HCC cells and SOX9 regulates self-renewal/tumorigenicity in HCC [176]. Kawai et al. studied whether SOX9 cells have CSC properties in HCC and could be a new therapeutic target using the same procedures that they used for identification of K19 as liver CSCs. SOX9$^+$ cells showed self-renewal properties, high proliferation capacity, EMT characters, and the activation of TGF-β/SMAD signaling as K19. The authors evaluated SOX9 expression in surgical specimens of HCC and measured osteopontin (OPN) levels in the patients' serum. SOX9$^+$ patients had significantly worse recurrence-free survival and higher serum OPN levels than SOX9$^-$ patients. Thus SOX9 was proposed as a novel CSC marker, and OPN was as a useful surrogate marker of SOX9 in HCC patients [177].

5.12 TLR4

Toll-like receptor 4 (TLR4), acting as a receptor for lipopolysaccharide (LPS), has a pivotal role in the regulation of immune responses to infection. TLR4 is also expressed on several types of tumor cells and plays an important role in carcinogenesis, metastasis, and cancer progression [178]. Liu et al. reported that stem-like features were found in TLR4 positive HCC cells. TLR4$^+$ cells frequently expressed other CSC markers (e.g., CD133 and EpCAM); presented a higher capacity of migration and invasion than TLR4$^-$ cells; elevated EMT characteristics; and displayed enhanced colony ability, chemoresistance, and antiapoptosis. Clinically, TLR4 expression on surgical specimens was significantly correlated with HCC metastasis, early recurrence, and poor survival. They proposed that TLR4 was a potential CSC marker and biomarker of HCC [179].

5.13 CD274

CD274, also known as programmed cell death ligand 1(PD-L1), is known as an immune checkpoint molecule, which induces immunosuppressive tumor microenvironments. PD-L1 and its ligand PD-1 are major targets of immune checkpoint therapy, which enhance antitumor immune responses of T cells [180]. In contrast to this immunosuppressive effect, Tamai et al. reported a novel function of CD274 as negative regulator of CSC-related phenotypes in human CC. They indicated that $CD274^{low}$ cells possessed several CSC-related characteristics, such as ALDH activity, ROS species production, and a dormant state in the cell cycle as per in-vitro experiments. They also showed that $CD274^{low}$ cells were highly tumorigenic compared with $CD274^{high}$ cells in mouse xenograft models. They also demonstrated in a clinical study using surgical specimens of human CC that the CD274 low-expression group showed worse prognosis when compared with the CD274 high-expression group [181].

5.14 α2δ1

α2δ1 was found to play an essential role in modulating calcium oscillation amplitude, which may be important in maintaining the properties of CSCs. Zhao et al. generated a mAb, 1B50-1, raised against recurrent HCC cells. 1B50-1 bound to α2δ1 isoform5 and $1B50-1^+$ cells were a restricted subset of highly tumorigenic cells among $CD13^+$, $CD133^+$, and $EpCAM^+$ liver CSCs. They identified p-extracellular signal-regulated kinases (ERK) 1/2 as a key downstream target of α2δ1 function in liver CSCs. Treatment of 1B50-1 reduced CSC properties and showed therapeutic effects in mouse xenograft models. Therefore α2δ1 was proposed as a potential CSC marker of HCC and thus a therapeutic target [182].

6 CSCs MARKERS AS ROLES OF PROGNOSTIC FACTORS

CSCs target therapies are expected to improve the prognosis of aggressive hepatobiliary malignancies, and studies in this field provide future perspective. In primary liver cancer, prognosis of CC without intraductal-growth type was presumed to be worse than HCC prognosis and similar to CHC prognosis [183–185]. According to the latest World Health Organization (WHO) classification, CHC is divided into classical type and subtypes with stem cell features. Moreover, the latter is subdivided into typical subtype, intermediate cell subtype, and cholangiocellular subtype [106]. After this latest revision of the classification, several important works of CHC with stem cell features were published, and several clinicopathological differences were reported between each subtype but prognostic significances were not determined [184,186,187]. However, in general, the hepatobiliary cancers with stem/progenitor cell features, characterized by stem/progenitor cell-related gene expression or immunophenotypes, were shown to be highly aggressive and resulted in poor prognosis as aforementioned. Therefore expression status of CSC markers, via histological analyses, is widely accepted as a useful predictor of clinicopathological outcome and prognosis in hepatobiliary cancer patients after surgery. The studies in which CSC markers are associated with prognosis of hepatobiliary cancers are listed in Table 11.3. Most of the studies were retrospective analyses of postsurgical therapy of HCC. Compared to conventional HCCs, HCCs with CSC markers more frequently demonstrated infiltrative growth patterns, vascular invasion, and more intratumoral fibrous stroma, high serum AFP levels, and early recurrences after surgery [171,188]. K19 reports using immunohistochemical or gene level analyses were most frequently published. Most studies reported that K19 was associated with tumor aggressiveness and poor prognosis [113,169–172,189–196]. Choi et al. examined CSC marker mRNA levels for K19, EpCAM, and CD44 in peripheral blood before and after operation using real-time reverse transcription PCR (RT-PCR) and the ratio of postoperative to preoperative mRNA levels for each marker was calculated. They showed that a high ratio for both K19 and CD44 mRNA was an independent poor prognostic factor for relapse-free survival (RFS) [191]. Yang et al. reported that K10 (CK10) alone, or in combination with K19, could be a novel predictor for poor prognosis of HCC patients after curative resection by immunohistochemistry and protein and gene level analyses [195]. Chan et al. reported that EpCAM was an independent factor for disease-free survival at all stages, and CD133 also became an independent factor at stage I. However, they reported that K19 and CD56 were not independent prognostic factors as gleaned from immunohistochemical analysis after surgical therapy [197]. There were several reports revealed CSC markers CD133 and EpCAM as roles of prognostic markers. Song et al. and Sasaki et al. reported that expression of CD133 was associated with poor prognosis of HCC using immunohistochemical studies [198,199]. Yang et al. measured CSC marker genes K7 (CK7), K19, EpCAM, OV6, ABCG2, CD133, Nestin, and CD44 of surgical specimens after curative resection of HCC by RT-PCR and tissue microarray and confirmed CD133, CD44, and Nestin as independent predictors for RFS and overall survival (OS) based on Cox regression model [200]. As mentioned previously, Yamashita et al. proposed an HCC classification system based on the stem/maturation status of the tumor as determined via EpCAM and AFP expression status [154]. Sun et al. prospectively identified EpCAM-positive circulating tumor cells (CTCs) via PCR and showed EpCAM-positive CTCs as novel predictors for tumor recurrence in HCC patients after surgery [201]. Guo et al. immunohistochemically explored associations of three CSC markers CD133, CD90, and EpCAM in resected

TABLE 11.3 CSC Markers and Prognoses

Type of cancer	CSC markers	Prognoses	Methods	References
HCC	B/S group, W/B group	B/S type poor prognosis than negative	IHC	[169]
	EpCAM	Poor prognosis	IHC	[197,202,203]
		Poor prognosis	GA	[201]
		EpCAM$^-$AFP$^+$ correlated with poor prognosis, EpCAM$^+$AFP$^-$ correlated with good prognosis	GA, IHC	[154]
	SALL4	Poor prognosis	IHC	[169]
		Poor prognosis	GA, IHC	[159,205]
		Poor prognosis	GA	[157,204]
	K19	Poor prognosis	IHC	[113,169,190, 192–194,196]
		Poor prognosis	GA, IHC	[171]
		Poor prognosis	GA	[191]
	K19 and K10	Poor prognosis	GA, IHC	[195]
	HNF1β	Poor prognosis	IHC	[209]
	USP22	Poor prognosis	GA, IHC	[208]
	Survivin	Poor prognosis	GA, IHC	[208]
	CD90	Poor prognosis	IHC	[202]
	CD133	Poor prognosis	IHC	[198, 199]
		Poor prognosis	GA	[200]
	CD44	Poor prognosis	GA	[200]
	CD44s	Poor prognosis	IHC	[152]
	Lin28B	Poor prognosis	GA	[207]
	ALDH1	Poor prognosis	IHC	[206]
	Nestin	Poor prognosis	GA	[200]
ICC	EpCAM	Poor prognosis	IHC	[212]
	SOX2	Poor prognosis	GA, IHC	[213]
	CD133	Better prognosis	IHC	[215]
		Poor prognosis	IHC	[214]
ECC	Notch1-3	Poor prognosis	IHC	[216]
	HES1	Poor prognosis	IHC	[216]

B/S group, The biliary/stem cell markers positive group; *ECC*, extrahepatic cholangiocarcinoma; *GA*, gene analyses; *HNF1β*, hepatocyte nuclear factor 1β; *ICC*, intrahepatic cholangiocarcinoma; *IHC*, immunohistochemistry; *USP22*, ubiquitin-specific protease 22; *W/S group*, Wnt/β-catenin-related markers positive group.

HCC specimens with clinicopathological characteristics, early recurrence, and survival time. In this study, their multivariate analysis identified only CD90 expression as significantly associated with early recurrence, and log-rank analysis identified expression of both CD90 and EpCAM as significantly associated with survival time of HCC patients. Moreover, Cox regression identified EpCAM expression as an independent predictor of survival time, but CD133 expressions were not significantly correlated with prognostic factors [202]. To note, liver transplantation is the most effective therapy for cirrhosis-associated HCC, but its utility is limited due to posttransplant tumor recurrence. Zeng et al. performed immunohistochemical staining of EpCAM, CD44, CD90, and CD133 for surgical specimens of 39 HCC liver explants (23 with no treatment and 16 after transcatheter arterial chemoembolization (TACE)). They showed the presence of high EpCAM staining was associated with tumor recurrence in the TACE group [203]. SALL4 is a recently proposed marker for a progenitor subclass of HCCs with aggressive behavior [204,205]. Tsujikawa et al. conducted immunohistochemical expression analysis of biliary/stem cell markers (K19, SALL4, EpCAM, and CD133), and Wnt/β-catenin signaling–related molecules (β-catenin and glutamine synthetase) in HCCs surgically resected from 142 patients and analyzed the results with respect to clinicopathological features. They reported that the biliary/stem cell marker-positive group exhibited poor tumor differentiation, increased frequency of portal vein invasion and/or intrahepatic metastasis, and had shorter RFS than negative groups [169]. Other gene or immunohistochemical expression of CSC markers, ALDH1, Lin28B, Survivin, Ubiquitin-specific protease 22 (USP22), and HNF-1β was reported to be associated with prognosis of HCC patients [206–209]. There are only a few studies regarding CC [210–212]. HNF-1β, which was associated with prognosis after surgery of HCC patients, was not correlated with survival after surgery of intrahepatic CC patients [209]. Gu et al. reported that SOX2 expression is associated with aggressive behavior and poor OS in intrahepatic CC patients [213]. Regarding CD133, Shimada et al. reported that CD133 expression tended to be related to higher incidences of intrahepatic metastasis and was independently related to worse prognosis similarly to HCC [214]. However, Fan et al. reported opposite results. Particularly, the positive expression of CD133 on tumor cells was significantly correlated with well or moderately differentiated CC and predicted a better prognosis for the patients [215]. Notch and hairy and enhancer of split 1 (HES1) was reported to be associated with poor survival in extrahepatic CC patients [216].

7 OTHER NOVEL THERAPIES FOR HEPATOBILIARY CSCs

The majority of hepatobiliary cancers are resistant to conventional chemotherapy and radiation therapy. Therefore development of novel therapies is crucial. Several target therapies for hepatobiliary CSCs have been noted in previous sections of this review. There is also research aiming to develop therapies for

CSC-related signaling, including the Hedgehog pathway, Wnt/β-catenin signaling, Ras/MAPK pathway, Akt/mTOR pathway, and Notch signaling.

7.1 Immunotherapy for CSCs of Hepatobiliary Cancers

Recently, effective immunotherapies, such as immune checkpoint inhibitors and gene-modified T-cell therapy, were developed and have led to important clinical advances [180]. Wang et al. reported that a phase I trial of active, specific immunotherapy with autologous dendritic cells pulsed with autologous irradiated HCC stem cells after surgery. This is only a feasibility and safety trial, not a therapeutic efficacy trial, for hepatitis B virus (HBV)-positive patients with HCC. They showed a success rate of 100%, with no significant toxicity, including no worsening of hepatic inflammation, function, or enzymes in HBV-positive HCC patients [217]. Morita et al. reported novel immunotherapy targeted for colon CSCs. They identified a novel cancer testis antigen, olfactory receptor family 7 subfamily C member 1 (OR7C1) from SP cells of colorectal cancer. OR7C1 has essential roles in maintenance of colon CSCs, and high expression of OR7C1 correlated with poor prognosis in colorectal patients. Targeting CSCs via OR7C1-specific cytotoxic T lymphocyte (CTL) clones showed potent antitumor effects in murine models [218]. This is a unique and safe therapy for aggressive and chemoresistant CSCs. If immune targets are isolated in hepatobiliary CSCs as well, this CSCs targeting immunotherapy may become a new weapon for preventing recurrences of refractory hepatobiliary cancers postsurgery.

7.2 Drug Repositioning

Recent reports revealed that repositioning drugs sometimes exhibit antitumor effects. Several drugs show anti-CSCs action. These drugs are usually widely used, safe, cheap, and they may be available to prevent recurrences after surgery.

Metformin, a first-line drug for diabetes, has been shown to decrease cancer incidence and mortality, including HCCs. Saito et al. showed that metformin treatment decreased the number of EpCAM-positive HCC cells and impaired sphere-forming ability and self-renewal capability [219]. However, Xin et al. reported that antitumor effects of metformin was not associated with inhibition of HCC CSC growth using liver label retaining cancer cells method [220]. Therefore the mechanism of the antitumor effects of metformin remains unclear [4,221].

Disulfiram is an inhibitor of ALDH. This is widely used in patients with alcohol dependence. Chiba et al. reported that disulfiram treatment significantly decreased the number of EpCAM-, CD133-, and CD13-positive HCC CSCs and inhibited tumor growth in a dose-dependent manner using a xenograft transplant murine model. However, this anti-CSC and antitumor effect was not depended on ALDH. They suspected that disulfiram impaired the tumorigenicity of HCC

CSCs through activation of the ROS-p38 pathway and in part through the downregulation of glypican3 (GPC3) [222].

Sulfasalazine has been used to treat rheumatoid arthritis and inflammatory bowel disease. Sulfasalazine is an inhibitor for the cystine/glutamate (xCT) transporter. Thanee et al. examined xCT-targeted CD44v-CSC chemotherapy using sulfasalazine. Sulfasalazine treatment influenced CC cell proliferation and survival. They suggested that xCT-targeting drug might improve CC therapy by sensitization to the available drug (e.g., gemcitabine) via blocking the mechanism of the ROS defensive system [223].

Celecoxib, a cyclooxygenase-2 (COX-2) inhibitor and nonsteroidal antiinflammatory drug, can prevent several types of cancer, including HCC. Chu et al.'s histological analysis revealed that celecoxib therapy reduced the abundance of $CD44^+/CD133^+$ human CSCs in HCC tissues. They suggested that celecoxib suppressed cancer stemness and progression of HCC via activation of peroxisome proliferator-activated receptor (PPAR) γ/phosphatase and tensin homolog (PTEN) signaling [224].

8 CONCLUSIONS

The mechanism of regeneration of adult liver remains to be elucidated, and the origin and existence of LPCs is still controversial. The number of organs for OLT is limited. Cell therapy has been established to improve liver function. However, novel therapies, including small liver constructs for end-stage liver failure, should be developed using hepatobiliary stem/progenitor cells. Prognosis of hepatobiliary cancer after surgery is still dismal. Further efforts are required to improve prognosis, and clarification of the genetic and epigenetic changes of CSCs is critically important.

ACKNOWLEDGMENTS

I thank Drs. Hideki Taniguchi, Takanori Takebe, and Makio Kawakami for their insightful advice and comments and Mr. Hirosuke Nakajima for technical assistance. This work was supported by Grants-in-Aid for scientific research (B) from the Ministry of Education, Culture, Sports and Science and Technology of Japan (26293291).

LIST OF ABBREVIATIONS

CSC	Cancer stem cell and/or initiating cell
LPC	Liver stem/progenitor cell
FLRV	Future liver remnant volume
PVE	Portal vein embolization
HSC	Hematopoietic stem cell
LC	Liver cirrhosis
OLT	Orthotropic liver transplantation
iPSC	Induced pluripotent stem cells
PBG	Peribiliary gland

FGF	Fibroblast growth factor
BMP	Bone morphogenetic protein
STM	Septum transversum mesenchyme
HEX	Hematopoietically expressed homeobox
HNF	Hepatocyte nuclear factor
AFP	α-fetoprotein
ALB	Albumin
mAb	Monoclonal antibody
FACS	Fluorescence-activated cell sorting
CD	Cluster of differentiation
H-CFU-C	Hepatic colony-forming unit in culture
DLK1	Delta-like 1 homolog
EpCAM	Epithelial cell adhesion molecule
MHC	Major histocompatibility complex
ICAM1	Intercellular adhesion molecule 1
AAF	Acetylaminofluorene
DDC	3,5-diethoxycarbonyl-1,4-dihydrocollidine
CK	Cytokeratin
BrdU	Bromodeoxyuridine
SOX	Sex determining region Y-box
HUVEC	Human umbilical vein endothelial cell
MSC	Mesenchymal stem cell
G-CSF	Granulocyte colony-stimulating factor
PVL	Portal vein ligation
ALPPS	Associating liver partition and portal vein ligation for staged hepatectomy
CT	Computed tomography
NASH	Steatohepatitis
HCC	Hepatocellular carcinoma
CC	Cholangiocarcinoma
CHC	Combined hepatocellular–cholangiocarcinoma
SP	Side population
ATP	Adenosine triphosphate
ABC	Adenosine triphosphate-binding cassette
BMI1	B lymphoma Moloney murine leukemia virus insertion region 1 homolog
5-FU	5-fluorouracil
ABCG	Adenosine triphosphate-binding cassette subfamily G
GBC	Gallbladder cancer
Emodin	1,3,8-Trihydroxy-6-methylanthraquinone
ROS	Reactive oxygen species
ALDH	Aldehyde dehydrogenase
BCL	B cell lymphoma
MAPK	Mitogen-activated protein kinases
PI3K	Phosphatidylinositol-4, 5-bisphosphate 3-kinase
IL	Interleukin
MMAF	Monomethyl auristatin F
Gli	Glioma-associated oncogene

GPI	Glycosylphosphatidylinositol
AMP	Adenosine monophosphate
AMPK	Adenosine monophosphate-activated protein kinase
mTOR	Mammalia target of rapamycin
shRNA	Short hairpin ribonucleic acid
TGF	Transforming growth factor
RNAi	Ribonucleic acid interference-based blockage
SALL4	Sal-like protein 4
NuRD	Nucleosome remodeling and histone deacetylase
HDAC	Histone deacetylase
DNA	Deoxyribonucleic acid
CHD4	Chromodomain-helicase-Deoxyribonucleic acid-binding protein 4
poly (ADP-ribose)	Polymeric adenosine diphosphate ribose
PARP	Polymeric adenosine diphosphate ribose polymerase
SBHA	Suberohydroxamic acid
VB4-845	Oportuzumab monatox
CD13	Aminopeptidase N
qPCR	Quantitative polymerase chain reaction
STAT	Signal transducer and activator of transcription
miR	MicroRNA
IAP	Integrin-associated protein
CTSS	Cathepsin S
PAR2	Protease-activated receptor 2
EMT	Epithelial-mesenchymal transition
EGFP	Green fluorescence protein
siRNA	Small interfering RNA
TGF-βR1	TGF-β receptor 1
OPN	Osteopontin
TLR4	Toll like receptor 4
LPS	Lipopolysaccharide
PD-L1	Programmed cell death ligand 1
ERK	p-Extracellular signal-regulated kinases
WHO	World Health Organization
RT-PCR	Reverse transcription polymerase chain reaction
RFS	Relapse-free survival
OS	Overall survival
CSC	Circulating tumor cell
TACE	Transcatheter arterial chemoembolization
USP	Ubiquitin-specific protease
HES1	Hairy and enhancer of split 1
HBV	Hepatitis B virus
OR7C1	Olfactory receptor family 7 subfamily C member 1
CTL	Cytotoxic T lymphocytes
GPC3	Glypican3
xCT	Cystine/glutamate
COX-2	Cyclooxygenase-2
PPAR	Peroxisome proliferator-activated receptors
PTEN	Phosphatase and tensin homolog

REFERENCES

[1] Miyajima A, Tanaka M, Itoh T. Stem/progenitor cells in liver development, homeostasis, regeneration, and reprogramming. Cell Stem Cell 2014;14(5):561–74.
[2] Michalopoulos GK, DeFrances MC. Liver regeneration. Science 1997;276(5309):60–6.
[3] Hansel MC, Davila JC, Vosough M, Gramignoli R, Skvorak KJ, Dorko K, et al. The use of induced pluripotent stem cells for the study and treatment of liver diseases. Curr Protoc Toxicol 2016;67. 14.3.1-.3.27.
[4] Taniguchi H, Moriya C, Igarashi H, Saitoh A, Yamamoto H, Adachi Y, et al. Cancer stem cells in human gastrointestinal cancer. Cancer Sci 2016;107(11):1556–62.
[5] Tanaka S. Cancer stem cells as therapeutic targets of hepato-biliary-pancreatic cancers. J Hepatobiliary Pancreat Sci 2015;22(7):531–7.
[6] Oikawa T. Cancer stem cells and their cellular origins in primary liver and biliary tract cancers. Hepatology 2016;64(2):645–51.
[7] Roskams TA, Theise ND, Balabaud C, Bhagat G, Bhathal PS, Bioulac-Sage P, et al. Nomenclature of the finer branches of the biliary tree: canals, ductules, and ductular reactions in human livers. Hepatology 2004;39(6):1739–45.
[8] Nakanuma Y, Hoso M, Sanzen T, Sasaki M. Microstructure and development of the normal and pathologic biliary tract in humans, including blood supply. Microsc Res Tech 1997;38(6):552–70.
[9] Nakanuma Y. A novel approach to biliary tract pathology based on similarities to pancreatic counterparts: is the biliary tract an incomplete pancreas? Pathol Int 2010;60(6):419–29.
[10] Shiojiri N. Development and differentiation of bile ducts in the mammalian liver. Microsc Res Tech 1997;39(4):328–35.
[11] Jung J, Zheng M, Goldfarb M, Zaret KS. Initiation of mammalian liver development from endoderm by fibroblast growth factors. Science 1999;284(5422):1998–2003.
[12] Rossi JM, Dunn NR, Hogan BL, Zaret KS. Distinct mesodermal signals, including BMPs from the septum transversum mesenchyme, are required in combination for hepatogenesis from the endoderm. Genes Dev 2001;15(15):1998–2009.
[13] Zong Y, Stanger BZ. Molecular mechanisms of liver and bile duct development. Wiley Interdiscip Rev Dev Biol 2012;1(5):643–55.
[14] Reya T, Morrison SJ, Clarke MF, Weissman IL. Stem cells, cancer, and cancer stem cells. Nature 2001;414(6859):105–11.
[15] Osawa M, Hanada K, Hamada H, Nakauchi H. Long-term lymphohematopoietic reconstitution by a single CD34-low/negative hematopoietic stem cell. Science 1996;273(5272):242–5.
[16] Taniguchi H, Toyoshima T, Fukao K, Nakauchi H. Presence of hematopoietic stem cells in the adult liver. Nature Med 1996;2(2):198–203.
[17] Suzuki A, Zheng Y, Kondo R, Kusakabe M, Takada Y, Fukao K, et al. Flow-cytometric separation and enrichment of hepatic progenitor cells in the developing mouse liver. Hepatology 2000;32(6):1230–9.
[18] Suzuki A, Zheng YW, Kaneko S, Onodera M, Fukao K, Nakauchi H, et al. Clonal identification and characterization of self-renewing pluripotent stem cells in the developing liver. J Cell Biol 2002;156(1):173–84.
[19] Minguet S, Cortegano I, Gonzalo P, Martinez-Marin JA, de Andres B, Salas C, et al. A population of c-Kit(low)(CD45/TER119)- hepatic cell progenitors of 11-day postcoitus mouse embryo liver reconstitutes cell-depleted liver organoids. J Clin Invest 2003;112(8):1152–63.
[20] Tanimizu N, Nishikawa M, Saito H, Tsujimura T, Miyajima A. Isolation of hepatoblasts based on the expression of Dlk/Pref-1. J Cell Sci 2003;116(Pt 9):1775–86.

[21] Tanaka M, Okabe M, Suzuki K, Kamiya Y, Tsukahara Y, Saito S, et al. Mouse hepatoblasts at distinct developmental stages are characterized by expression of EpCAM and DLK1: drastic change of EpCAM expression during liver development. Mech Dev 2009;126(8–9):665–76.
[22] Kubota H, Reid LM. Clonogenic hepatoblasts, common precursors for hepatocytic and biliary lineages, are lacking classical major histocompatibility complex class I antigen. Proc Natl Acad Sci USA 2000;97(22):12132–7.
[23] Schmelzer E, Zhang L, Bruce A, Wauthier E, Ludlow J, Yao HL, et al. Human hepatic stem cells from fetal and postnatal donors. J Exp Med 2007;204(8):1973–87.
[24] Magami Y, Azuma T, Inokuchi H, Kokuno S, Moriyasu F, Kawai K, et al. Cell proliferation and renewal of normal hepatocytes and bile duct cells in adult mouse liver. Liver 2002;22(5):419–25.
[25] Espanol-Suner R, Carpentier R, Van Hul N, Legry V, Achouri Y, Cordi S, et al. Liver progenitor cells yield functional hepatocytes in response to chronic liver injury in mice. Gastroenterology 2012;143(6):1564–75.
[26] Miyaoka Y, Ebato K, Kato H, Arakawa S, Shimizu S, Miyajima A. Hypertrophy and unconventional cell division of hepatocytes underlie liver regeneration. Curr Biol 2012;22(13):1166–75.
[27] Farber E. Similarities in the sequence of early histological changes induced in the liver of the rat by ethionine, 2-acetylamino-fluorene, and 3′-methyl-4-dimethylaminoazobenzene. Cancer Res 1956;16(2):142–8.
[28] Akhurst B, Croager EJ, Farley-Roche CA, Ong JK, Dumble ML, Knight B, et al. A modified choline-deficient, ethionine-supplemented diet protocol effectively induces oval cells in mouse liver. Hepatology 2001;34(3):519–22.
[29] Fausto N. Liver regeneration and repair: hepatocytes, progenitor cells, and stem cells. Hepatology 2004;39(6):1477–87.
[30] Alison MR, Golding M, Sarraf CE, Edwards RJ, Lalani EN. Liver damage in the rat induces hepatocyte stem cells from biliary epithelial cells. Gastroenterology 1996;110(4):1182–90.
[31] Theise ND, Saxena R, Portmann BC, Thung SN, Yee H, Chiriboga L, et al. The canals of Hering and hepatic stem cells in humans. Hepatology 1999;30(6):1425–33.
[32] Roskams TA, Libbrecht L, Desmet VJ. Progenitor cells in diseased human liver. Semin Liver Dis 2003;23(4):385–96.
[33] Kaneko K, Kamimoto K, Miyajima A, Itoh T. Adaptive remodeling of the biliary architecture underlies liver homeostasis. Hepatology 2015;61(6):2056–66.
[34] Jensen CH, Jauho EI, Santoni-Rugiu E, Holmskov U, Teisner B, Tygstrup N, et al. Transit-amplifying ductular (oval) cells and their hepatocytic progeny are characterized by a novel and distinctive expression of delta-like protein/preadipocyte factor 1/fetal antigen 1. Am J Pathol 2004;164(4):1347–59.
[35] Tanimizu N, Tsujimura T, Takahide K, Kodama T, Nakamura K, Miyajima A. Expression of Dlk/Pref-1 defines a subpopulation in the oval cell compartment of rat liver. Gene Expr Patterns 2004;5(2):209–18.
[36] Okabe M, Tsukahara Y, Tanaka M, Suzuki K, Saito S, Kamiya Y, et al. Potential hepatic stem cells reside in EpCAM+ cells of normal and injured mouse liver. Development 2009;136(11):1951–60.
[37] Koike H, Taniguchi H. Characteristics of hepatic stem/progenitor cells in the fetal and adult liver. J Hepatobiliary Pancreat Sci 2012;19(6):587–93.
[38] Suzuki A, Sekiya S, Onishi M, Oshima N, Kiyonari H, Nakauchi H, et al. Flow cytometric isolation and clonal identification of self-renewing bipotent hepatic progenitor cells in adult mouse liver. Hepatology 2008;48(6):1964–78.

[39] Kamiya A, Kakinuma S, Yamazaki Y, Nakauchi H. Enrichment and clonal culture of progenitor cells during mouse postnatal liver development in mice. Gastroenterology 2009;137(3):1114–26.
[40] Barker N, van Oudenaarden A, Clevers H. Identifying the stem cell of the intestinal crypt: strategies and pitfalls. Cell Stem Cell 2012;11(4):452–60.
[41] Blanpain C, Fuchs E. Epidermal homeostasis: a balancing act of stem cells in the skin. Nat Rev Mol Cell Biol 2009;10(3):207–17.
[42] Fuchs E. The tortoise and the hair: slow-cycling cells in the stem cell race. Cell 2009;137(5):811–9.
[43] Kuwahara R, Kofman AV, Landis CS, Swenson ES, Barendswaard E, Theise ND. The hepatic stem cell niche: identification by label-retaining cell assay. Hepatology 2008;47(6): 1994–2002.
[44] Zajicek G, Oren R, Weinreb M Jr. The streaming liver. Liver 1985;5(6):293–300.
[45] Furuyama K, Kawaguchi Y, Akiyama H, Horiguchi M, Kodama S, Kuhara T, et al. Continuous cell supply from a Sox9-expressing progenitor zone in adult liver, exocrine pancreas and intestine. Nat Genet 2011;43(1):34–41.
[46] Carpentier R, Suner RE, van Hul N, Kopp JL, Beaudry JB, Cordi S, et al. Embryonic ductal plate cells give rise to cholangiocytes, periportal hepatocytes, and adult liver progenitor cells. Gastroenterology 2011;141(4):1432–8.
[47] Malato Y, Naqvi S, Schurmann N, Ng R, Wang B, Zape J, et al. Fate tracing of mature hepatocytes in mouse liver homeostasis and regeneration. J Clin Invest 2011;121(12):4850–60.
[48] Sekiya S, Suzuki A. Hepatocytes, rather than cholangiocytes, can be the major source of primitive ductules in the chronically injured mouse liver. Am J Pathol 2014;184(5):1468–78.
[49] Yanger K, Zong Y, Maggs LR, Shapira SN, Maddipati R, Aiello NM, et al. Robust cellular reprogramming occurs spontaneously during liver regeneration. Genes Dev 2013;27(7): 719–24.
[50] Tarlow BD, Pelz C, Naugler WE, Wakefield L, Wilson EM, Finegold MJ, et al. Bipotential adult liver progenitors are derived from chronically injured mature hepatocytes. Cell Stem Cell 2014;15(5):605–18.
[51] He J, Lu H, Zou Q, Luo L. Regeneration of liver after extreme hepatocyte loss occurs mainly via biliary transdifferentiation in zebrafish. Gastroenterology 2014;146(3):789–800.
[52] Kordes C, Sawitza I, Gotze S, Herebian D, Haussinger D. Hepatic stellate cells contribute to progenitor cells and liver regeneration. J Clin Invest 2014;124(12):5503–15.
[53] Font-Burgada J, Shalapour S, Ramaswamy S, Hsueh B, Rossell D, Umemura A, et al. Hybrid periportal hepatocytes regenerate the injured liver without giving rise to cancer. Cell 2015;162(4):766–79.
[54] Wang B, Zhao L, Fish M, Logan CY, Nusse R. Self-renewing diploid Axin2(+) cells fuel homeostatic renewal of the liver. Nature 2015;524(7564):180–5.
[55] Nakanuma Y, Harada K, Sasaki M, Sato Y. Proposal of a new disease concept "biliary diseases with pancreatic counterparts". Anatomical and pathological bases. Histol Histopathol 2014;29(1):1–10.
[56] Koike N, Saitoh K, Todoroki T, Kawamoto T, Iwasaki Y, Nakamura K. Cell proliferation kinetics of the bile duct epithelium of the rat. Cell Prolif 1993;26(2):183–93.
[57] Nakanuma Y, Katayanagi K, Terada T, Saito K. Intrahepatic peribiliary glands of humans. I. Anatomy, development and presumed functions. J Gastroenterol Hepatol 1994;9(1):75–9.
[58] Irie T, Asahina K, Shimizu-Saito K, Teramoto K, Arii S, Teraoka H. Hepatic progenitor cells in the mouse extrahepatic bile duct after a bile duct ligation. Stem Cells Dev 2007;16(6): 979–87.

[59] Cardinale V, Wang Y, Carpino G, Cui CB, Gatto M, Rossi M, et al. Multipotent stem/progenitor cells in human biliary tree give rise to hepatocytes, cholangiocytes, and pancreatic islets. Hepatology 2011;54(6):2159–72.

[60] Cardinale V, Wang Y, Carpino G, Mendel G, Alpini G, Gaudio E, et al. The biliary tree—a reservoir of multipotent stem cells. Nat Rev Gastroenterol Hepatol 2012;9(4):231–40.

[61] Carpino G, Cardinale V, Onori P, Franchitto A, Berloco PB, Rossi M, et al. Biliary tree stem/progenitor cells in glands of extrahepatic and intraheptic bile ducts: an anatomical in situ study yielding evidence of maturational lineages. J Anat 2012;220(2):186–99.

[62] Sutton ME, op den Dries S, Koster MH, Lisman T, Gouw AS, Porte RJ. Regeneration of human extrahepatic biliary epithelium: the peribiliary glands as progenitor cell compartment. Liver Int 2012;32(4):554–9.

[63] Kumar A, Pati NT, Sarin SK. Use of stem cells for liver diseases-current scenario. J Clin Exp Hepatol 2011;1(1):17–26.

[64] Schuppan D, Afdhal NH. Liver cirrhosis. Lancet 2008;371(9615):838–51.

[65] Koike N, Fukumura D, Gralla O, Au P, Schechner JS, Jain RK. Tissue engineering: creation of long-lasting blood vessels. Nature 2004;428(6979):138–9.

[66] Takebe T, Sekine K, Enomura M, Koike H, Kimura M, Ogaeri T, et al. Vascularized and functional human liver from an iPSC-derived organ bud transplant. Nature 2013;499(7459):481–4.

[67] Stutchfield BM, Forbes SJ, Wigmore SJ. Prospects for stem cell transplantation in the treatment of hepatic disease. Liver Transpl 2010;16(7):827–36.

[68] Esrefoglu M. Role of stem cells in repair of liver injury: experimental and clinical benefit of transferred stem cells on liver failure. World J Gastroenterol 2013;19(40):6757–73.

[69] Saito T, Tomita K, Haga H, Okumoto K, Ueno Y. Bone marrow cell-based regenerative therapy for liver cirrhosis. World J Methodol 2013;3(4):65–9.

[70] Terai S, Ishikawa T, Omori K, Aoyama K, Marumoto Y, Urata Y, et al. Improved liver function in patients with liver cirrhosis after autologous bone marrow cell infusion therapy. Stem Cells 2006;24(10):2292–8.

[71] Pai M, Zacharoulis D, Milicevic MN, Helmy S, Jiao LR, Levicar N, et al. Autologous infusion of expanded mobilized adult bone marrow-derived CD34+ cells into patients with alcoholic liver cirrhosis. Am J Gastroenterol 2008;103(8):1952–8.

[72] Jang YY, Collector MI, Baylin SB, Diehl AM, Sharkis SJ. Hematopoietic stem cells convert into liver cells within days without fusion. Nat Cell Biol 2004;6(6):532–9.

[73] Ghodsizad A, Fahy BN, Waclawczyk S, Liedtke S, Gonzalez Berjon JM, Barrios R, et al. Portal application of human unrestricted somatic stem cells to support hepatic regeneration after portal embolization and tumor surgery. ASAIO J 2012;58(3):255–61.

[74] Togel F, Westenfelder C. Adult bone marrow-derived stem cells for organ regeneration and repair. Dev Dyn 2007;236(12):3321–31.

[75] Iwatsuki S, Starzl TE. Personal experience with 411 hepatic resections. Ann Surg 1988;208(4):421–34.

[76] Lai EC, Ng IO, You KT, Choi TK, Fan ST, Mok FP, et al. Hepatectomy for large hepatocellular carcinoma: the optimal resection margin. World J Surg 1991;15(1):141–5.

[77] Belghiti J, Hiramatsu K, Benoist S, Massault P, Sauvanet A, Farges O. Seven hundred forty-seven hepatectomies in the 1990s: an update to evaluate the actual risk of liver resection. J Am Coll Surg 2000;191(1):38–46.

[78] Jarnagin WR, Gonen M, Fong Y, DeMatteo RP, Ben-Porat L, Little S, et al. Improvement in perioperative outcome after hepatic resection: analysis of 1,803 consecutive cases over the past decade. Ann Surg 2002;236(4):397–406.

[79] Clavien PA, Petrowsky H, DeOliveira ML, Graf R. Strategies for safer liver surgery and partial liver transplantation. N Engl J Med 2007;356(15):1545–59.
[80] Adam R, Miller R, Pitombo M, Wicherts DA, de Haas RJ, Bitsakou G, et al. Two-stage hepatectomy approach for initially unresectable colorectal hepatic metastases. Surg Oncol Clin N Am 2007;16(3):525–36.
[81] Makuuchi M, Takayasu K, Takuma T. Preoperative transcatheter embolization of the portal venous branch for patients receiving extended lobectomy due to the bile duct carcinoma. J Jpn Surg Assoc 1984;45(12):1558–64.
[82] Kinoshita H, Sakai K, Hirohashi K, Igawa S, Yamasaki O, Kubo S. Preoperative portal vein embolization for hepatocellular carcinoma. World J Surg 1986;10(5):803–8.
[83] Makuuchi M, Thai BL, Takayasu K, Takayama T, Kosuge T, Gunven P, et al. Preoperative portal embolization to increase safety of major hepatectomy for hilar bile duct carcinoma: a preliminary report. Surgery 1990;107(5):521–7.
[84] Hemming AW, Reed AI, Howard RJ, Fujita S, Hochwald SN, Caridi JG, et al. Preoperative portal vein embolization for extended hepatectomy. Ann Surg 2003;237(5):686–91.
[85] Capussotti L, Muratore A, Baracchi F, Lelong B, Ferrero A, Regge D, et al. Portal vein ligation as an efficient method of increasing the future liver remnant volume in the surgical treatment of colorectal metastases. Arch Surg 2008;143(10):978–82.
[86] Schnitzbauer AA, Lang SA, Goessmann H, Nadalin S, Baumgart J, Farkas SA, et al. Right portal vein ligation combined with in situ splitting induces rapid left lateral liver lobe hypertrophy enabling 2-staged extended right hepatic resection in small-for-size settings. Ann Surg 2012;255(3):405–14.
[87] Nadalin S, Capobianco I, Li J, Girotti P, Konigsrainer I, Konigsrainer A. Indications and limits for associating liver partition and portal vein ligation for staged hepatectomy (ALPPS). Lessons Learned from 15 cases at a single centre. Z Gastroenterol 2014;52(1):35–42.
[88] Treska V. Methods to increase future liver remnant volume in patients with primarily unresectable colorectal liver metastases: current state and future perspectives. Anticancer Res 2016;36(5):2065–71.
[89] Furst G, Schulte am Esch J, Poll LW, Hosch SB, Fritz LB, Klein M, et al. Portal vein embolization and autologous CD133+ bone marrow stem cells for liver regeneration: initial experience. Radiology 2007;243(1):171–9.
[90] am Esch JS, Schmelzle M, Furst G, Robson SC, Krieg A, Duhme C, et al. Infusion of CD133+ bone marrow-derived stem cells after selective portal vein embolization enhances functional hepatic reserves after extended right hepatectomy: a retrospective single-center study. Ann Surg 2012;255(1):79–85.
[91] Canepa MC, Quaretti P, Perotti C, Vercelli A, Rademacher J, Peloso A, et al. Autologous CD133+ cells augment the effect of portal embolization. Minerva Chir 2013;68(2):163–8.
[92] Han HS, Ahn KS, Cho JY, Yoon YS, Yoon CJ, Park KU, et al. Autologous stem cell transplantation for expansion of remnant liver volume with extensive hepatectomy. Hepato-gastroenterology 2014;61(129):156–61.
[93] Treska V, Liska V, Fichtl J, Lysak D, Mirka H, Bruha J, et al. Portal vein embolisation with application of haematopoietic stem cells in patients with primarily or non-resectable colorectal liver metastases. Anticancer Res 2014;34(12):7279–85.
[94] Ludvik J, Duras P, Treska V, Matouskova T, Bruha J, Fichtl J, et al. Portal vein embolization with contralateral application of stem cells facilitates increase of future liver remnant volume in patients with liver metastases. Cardiovasc Intervent Radiol 2017;40(5):690–6.

[95] Franchi E, Canepa MC, Peloso A, Barbieri L, Briani L, Panyor G, et al. Two-stage hepatectomy after autologous CD133+ stem cells administration: a case report. World J Surg Oncol 2013;11(1):192.
[96] Williamson JM, Thairu N, Katsoulas N, Stamp G, Ahmad R, du Potet E, et al. Impact of portal vein embolization on expression of cancer stem cell markers in regenerated liver and colorectal liver metastases. Scand J Gastroenterol 2010;45(12):1472–9.
[97] de Graaf W, van den Esschert JW, van Lienden KP, van Gulik TM. Induction of tumor growth after preoperative portal vein embolization: is it a real problem? Ann Surg Oncol 2009;16(2):423–30.
[98] op den Dries S, Westerkamp AC, Karimian N, Gouw AS, Bruinsma BG, Markmann JF, et al. Injury to peribiliary glands and vascular plexus before liver transplantation predicts formation of non-anastomotic biliary strictures. J Hepatol 2014;60(6):1172–9.
[99] Aikawa M, Miyazawa M, Okada K, Toshimitsu Y, Torii T, Otani Y, et al. Regeneration of extrahepatic bile duct--possibility to clinical application by recognition of the regenerative process. J Smooth Muscle Res 2007;43(6):211–8.
[100] Miyazawa M, Aikawa M, Okada K, Toshimitsu Y, Okamoto K, Koyama I, et al. Regeneration of extrahepatic bile ducts by tissue engineering with a bioabsorbable polymer. J Artif Organs 2012;15(1):26–31.
[101] Perez Alonso AJ, Del Olmo Rivas C, Romero IM, Canizares Garcia FJ, Poyatos PT. Tissue-engineering repair of extrahepatic bile ducts. J Surg Res 2013;179(1):18–21.
[102] Tao L, Li Q, Ren H, Chen B, Hou X, Mou L, et al. Repair of extrahepatic bile duct defect using a collagen patch in a Swine model. Artif Organs 2015;39(4):352–60.
[103] Theise ND. Liver cancer. In: Stewart BW, Wild CP, editors. World Cancer Report 2014. Lyon: The International Agency for Research on Cancer; 2014. p. 403–12.
[104] Todoroki T, Takahashi H, Koike N, Kawamoto T, Kondo T, Yoshida S, et al. Outcomes of aggressive treatment of stage IV gallbladder cancer and predictors of survival. Hepatogastroenterology 1999;46(28):2114–21.
[105] Todoroki T, Kawamoto T, Koike N, Takahashi H, Yoshida S, Kashiwagi H, et al. Radical resection of hilar bile duct carcinoma and predictors of survival. Br J Surg 2000;87(3): 306–13.
[106] Theise ND, Nakashima O, Park YN, Nakanuma Y. Combined hepatocellular-cholangiocarcinoma. In: Bosman FT, Carneiro F, Hruban RH, Theise ND, editors. WHO Classfication of Tumours of the Digestive System. 4th ed. Lyon: Internationl Agency for Research on Cancer; 2010. p. 225–7.
[107] Kumagai A, Kondo F, Sano K, Inoue M, Fujii T, Hashimoto M, et al. Immunohistochemical study of hepatocyte, cholangiocyte and stem cell markers of hepatocellular carcinoma: the second report: relationship with tumor size and cell differentiation. J Hepatobiliary Pancreat Sci 2016;23(7):414–21.
[108] Barker N, Ridgway RA, van Es JH, van de Wetering M, Begthel H, van den Born M, et al. Crypt stem cells as the cells-of-origin of intestinal cancer. Nature 2009;457(7229):608–11.
[109] Lapouge G, Youssef KK, Vokaer B, Achouri Y, Michaux C, Sotiropoulou PA, et al. Identifying the cellular origin of squamous skin tumors. Proc Natl Acad Sci USA 2011;108(18):7431–6.
[110] Shlush LI, Zandi S, Mitchell A, Chen WC, Brandwein JM, Gupta V, et al. Identification of pre-leukaemic haematopoietic stem cells in acute leukaemia. Nature 2014;506(7488): 328–33.
[111] Mu X, Espanol-Suner R, Mederacke I, Affo S, Manco R, Sempoux C, et al. Hepatocellular carcinoma originates from hepatocytes and not from the progenitor/biliary compartment. J Clin Invest 2015;125(10):3891–903.

[112] Nishikawa Y, Doi Y, Watanabe H, Tokairin T, Omori Y, Su M, et al. Transdifferentiation of mature rat hepatocytes into bile duct-like cells in vitro. Am J Pathol 2005;166(4):1077–88.
[113] Yoneda N, Sato Y, Kitao A, Ikeda H, Sawada-Kitamura S, Miyakoshi M, et al. Epidermal growth factor induces cytokeratin 19 expression accompanied by increased growth abilities in human hepatocellular carcinoma. Lab Invest 2011;91(2):262–72.
[114] Guest RV, Boulter L, Kendall TJ, Minnis-Lyons SE, Walker R, Wigmore SJ, et al. Cell lineage tracing reveals a biliary origin of intrahepatic cholangiocarcinoma. Cancer Res 2014;74(4):1005–10.
[115] Sekiya S, Suzuki A. Intrahepatic cholangiocarcinoma can arise from Notch-mediated conversion of hepatocytes. J Clin Invest 2012;122(11):3914–8.
[116] Fan B, Malato Y, Calvisi DF, Naqvi S, Razumilava N, Ribback S, et al. Cholangiocarcinomas can originate from hepatocytes in mice. J Clin Invest 2012;122(8):2911–5.
[117] Wei M, Lu L, Lin P, Chen Z, Quan Z, Tang Z. Multiple cellular origins and molecular evolution of intrahepatic cholangiocarcinoma. Cancer Lett 2016;379(2):253–61.
[118] Kreso A, Dick JE. Evolution of the cancer stem cell model. Cell Stem Cell 2014;14(3): 275–91.
[119] Chiba T, Iwama A, Yokosuka O. Cancer stem cells in hepatocellular carcinoma: therapeutic implications based on stem cell biology. Hepatol Res 2016;46(1):50–7.
[120] Wang M, Xiao J, Shen M, Yahong Y, Tian R, Zhu F, et al. Isolation and characterization of tumorigenic extrahepatic cholangiocarcinoma cells with stem cell-like properties. Int J Cancer 2011;128(1):72–81.
[121] Shi C, Tian R, Wang M, Wang X, Jiang J, Zhang Z, et al. CD44+ CD133+ population exhibits cancer stem cell-like characteristics in human gallbladder carcinoma. Cancer Biol Ther 2010;10(11):1182–90.
[122] Yamashita T, Kaneko S. Orchestration of hepatocellular carcinoma development by diverse liver cancer stem cells. J Gastroenterol 2014;49(7):1105–10.
[123] Hirschmann-Jax C, Foster AE, Wulf GG, Nuchtern JG, Jax TW, Gobel U, et al. A distinct "side population" of cells with high drug efflux capacity in human tumor cells. Proc Natl Acad Sci USA 2004;101(39):14228–33.
[124] Chiba T, Kita K, Zheng YW, Yokosuka O, Saisho H, Iwama A, et al. Side population purified from hepatocellular carcinoma cells harbors cancer stem cell-like properties. Hepatology 2006;44(1):240–51.
[125] Chiba T, Zheng YW, Kita K, Yokosuka O, Saisho H, Onodera M, et al. Enhanced self-renewal capability in hepatic stem/progenitor cells drives cancer initiation. Gastroenterology 2007;133(3):937–50.
[126] Chiba T, Miyagi S, Saraya A, Aoki R, Seki A, Morita Y, et al. The polycomb gene product BMI1 contributes to the maintenance of tumor-initiating side population cells in hepatocellular carcinoma. Cancer Res 2008;68(19):7742–9.
[127] Zhang R, Xu LB, Yue XJ, Yu XH, Wang J, Liu C. BMI1 gene silencing inhibits the proliferation and invasiveness of human hepatocellular carcinoma cells and increases their sensitivity to 5-fluorouracil. Oncol Rep 2013;29(3):967–74.
[128] Lee TK, Cheung VC, Ng IO. Liver tumor-initiating cells as a therapeutic target for hepatocellular carcinoma. Cancer Lett 2013;338(1):101–9.
[129] Zhang G, Wang Z, Luo W, Jiao H, Wu J, Jiang C. Expression of potential cancer stem cell Marker ABCG2 is associated with malignant behaviors of hepatocellular carcinoma. Gastroenterol Res Pract 2013;2013:782581.
[130] Zhou S, Morris JJ, Barnes Y, Lan L, Schuetz JD, Sorrentino BP. Bcrp1 gene expression is required for normal numbers of side population stem cells in mice, and confers

relative protection to mitoxantrone in hematopoietic cells in vivo. Proc Natl Acad Sci USA 2002;99(19):12339–44.
[131] Zhang Z, Zhu F, Xiao L, Wang M, Tian R, Shi C, et al. Side population cells in human gallbladder cancer cell line GBC-SD regulated by TGF-beta-induced epithelial-mesenchymal transition. J Huazhong Univ Sci Technolog Med Sci 2011;31(6):749–55.
[132] Li XX, Dong Y, Wang W, Wang HL, Chen YY, Shi GY, et al. Emodin as an effective agent in targeting cancer stem-like side population cells of gallbladder carcinoma. Stem Cells Dev 2013;22(4):554–66.
[133] Ma S, Chan KW, Lee TK, Tang KH, Wo JY, Zheng BJ, et al. Aldehyde dehydrogenase discriminates the CD133 liver cancer stem cell populations. Mol Cancer Res 2008;6(7):1146–53.
[134] Tanaka K, Tomita H, Hisamatsu K, Nakashima T, Hatano Y, Sasaki Y, et al. ALDH1A1-overexpressing cells are differentiated cells but not cancer stem or progenitor cells in human hepatocellular carcinoma. Oncotarget 2015;6(28):24722–32.
[135] Yin AH, Miraglia S, Zanjani ED, Almeida-Porada G, Ogawa M, Leary AG, et al. AC133, a novel marker for human hematopoietic stem and progenitor cells. Blood 1997;90(12):5002–12.
[136] Singh SK, Hawkins C, Clarke ID, Squire JA, Bayani J, Hide T, et al. Identification of human brain tumour initiating cells. Nature 2004;432(7015):396–401.
[137] Suetsugu A, Nagaki M, Aoki H, Motohashi T, Kunisada T, Moriwaki H. Characterization of CD133+ hepatocellular carcinoma cells as cancer stem/progenitor cells. Biochem Biophys Res Commun 2006;351(4):820–4.
[138] Ma S, Chan KW, Hu L, Lee TK, Wo JY, Ng IO, et al. Identification and characterization of tumorigenic liver cancer stem/progenitor cells. Gastroenterology 2007;132(7):2542–56.
[139] Ma S, Lee TK, Zheng BJ, Chan KW, Guan XY. CD133+ HCC cancer stem cells confer chemoresistance by preferential expression of the Akt/PKB survival pathway. Oncogene 2008;27(12):1749–58.
[140] Piao LS, Hur W, Kim TK, Hong SW, Kim SW, Choi JE, et al. CD133+ liver cancer stem cells modulate radioresistance in human hepatocellular carcinoma. Cancer Lett 2012;315(2):129–37.
[141] Tang KH, Ma S, Lee TK, Chan YP, Kwan PS, Tong CM, et al. CD133(+) liver tumor-initiating cells promote tumor angiogenesis, growth, and self-renewal through neurotensin/interleukin-8/CXCL1 signaling. Hepatology 2012;55(3):807–20.
[142] Yao J, Zhang T, Ren J, Yu M, Wu G. Effect of CD133/prominin-1 antisense oligodeoxynucleotide on in vitro growth characteristics of Huh-7 human hepatocarcinoma cells and U251 human glioma cells. Oncol Rep 2009;22(4):781–7.
[143] Smith LM, Nesterova A, Ryan MC, Duniho S, Jonas M, Anderson M, et al. CD133/prominin-1 is a potential therapeutic target for antibody-drug conjugates in hepatocellular and gastric cancers. Br J Cancer 2008;99(1):100–9.
[144] Ai Z, Pan H, Suo T, Lv C, Wang Y, Tong S, et al. Arsenic oxide targets stem cell marker CD133/prominin-1 in gallbladder carcinoma. Cancer Lett 2011;310(2):181–7.
[145] Li C, Wang C, Xing Y, Zhen J, Ai Z. CD133 promotes gallbladder carcinoma cell migration through activating Akt phosphorylation. Oncotarget 2016;7(14):17751–9.
[146] Yang W, Yan HX, Chen L, Liu Q, He YQ, Yu LX, et al. Wnt/beta-catenin signaling contributes to activation of normal and tumorigenic liver progenitor cells. Cancer Res 2008;68(11):4287–95.
[147] Rege TA, Hagood JS. Thy-1 as a regulator of cell-cell and cell-matrix interactions in axon regeneration, apoptosis, adhesion, migration, cancer, and fibrosis. FASEB J 2006;20(8):1045–54.

[148] Yang ZF, Ho DW, Ng MN, Lau CK, Yu WC, Ngai P, et al. Significance of CD90+ cancer stem cells in human liver cancer. Cancer Cell 2008;13(2):153–66.
[149] Yamashita T, Honda M, Nakamoto Y, Baba M, Nio K, Hara Y, et al. Discrete nature of EpCAM+ and CD90+ cancer stem cells in human hepatocellular carcinoma. Hepatology 2013;57(4):1484–97.
[150] Chen WC, Chang YS, Hsu HP, Yen MC, Huang HL, Cho CY, et al. Therapeutics targeting CD90-integrin-AMPK-CD133 signal axis in liver cancer. Oncotarget 2015;6(40):42923–37.
[151] Zhu Z, Hao X, Yan M, Yao M, Ge C, Gu J, et al. Cancer stem/progenitor cells are highly enriched in CD133+ CD44+ population in hepatocellular carcinoma. Int J Cancer 2010;126(9):2067–78.
[152] Mima K, Okabe H, Ishimoto T, Hayashi H, Nakagawa S, Kuroki H, et al. CD44s regulates the TGF-beta-mediated mesenchymal phenotype and is associated with poor prognosis in patients with hepatocellular carcinoma. Cancer Res 2012;72(13):3414–23.
[153] Yamashita T, Ji J, Budhu A, Forgues M, Yang W, Wang HY, et al. EpCAM-positive hepatocellular carcinoma cells are tumor-initiating cells with stem/progenitor cell features. Gastroenterology 2009;136(3):1012–24.
[154] Yamashita T, Forgues M, Wang W, Kim JW, Ye Q, Jia H, et al. EpCAM and alpha-fetoprotein expression defines novel prognostic subtypes of hepatocellular carcinoma. Cancer Res 2008;68(5):1451–61.
[155] Elling U, Klasen C, Eisenberger T, Anlag K, Treier M. Murine inner cell mass-derived lineages depend on Sall4 function. Proc Natl Acad Sci USA 2006;103(44):16319–24.
[156] Oikawa T, Kamiya A, Kakinuma S, Zeniya M, Nishinakamura R, Tajiri H, et al. Sall4 regulates cell fate decision in fetal hepatic stem/progenitor cells. Gastroenterology 2009;136(3):1000–11.
[157] Oikawa T, Kamiya A, Zeniya M, Chikada H, Hyuck AD, Yamazaki Y, et al. Sal-like protein 4 (SALL4), a stem cell biomarker in liver cancers. Hepatology 2013;57(4):1469–83.
[158] Lu J, Jeong HW, Kong N, Yang Y, Carroll J, Luo HR, et al. Stem cell factor SALL4 represses the transcriptions of PTEN and SALL1 through an epigenetic repressor complex. PLoS One 2009;4(5):e5577.
[159] Zeng SS, Yamashita T, Kondo M, Nio K, Hayashi T, Hara Y, et al. The transcription factor SALL4 regulates stemness of EpCAM-positive hepatocellular carcinoma. J Hepatol 2014;60(1):127–34.
[160] Nio K, Yamashita T, Okada H, Kondo M, Hayashi T, Hara Y, et al. Defeating EpCAM(+) liver cancer stem cells by targeting chromatin remodeling enzyme CHD4 in human hepatocellular carcinoma. J Hepatol 2015;63(5):1164–72.
[161] Ogawa K, Tanaka S, Matsumura S, Murakata A, Ban D, Ochiai T, et al. EpCAM-targeted therapy for human hepatocellular carcinoma. Ann Surg Oncol 2014;21(4):1314–22.
[162] Haraguchi N, Ishii H, Mimori K, Tanaka F, Ohkuma M, Kim HM, et al. CD13 is a therapeutic target in human liver cancer stem cells. J Clin Invest 2010;120(9):3326–39.
[163] Lee TK, Castilho A, Cheung VC, Tang KH, Ma S, Ng IO. CD24(+) liver tumor-initiating cells drive self-renewal and tumor initiation through STAT3-mediated NANOG regulation. Cell Stem Cell 2011;9(1):50–63.
[164] Karakatsanis A, Papaconstantinou I, Gazouli M, Lyberopoulou A, Polymeneas G, Voros D. Expression of microRNAs, miR-21, miR-31, miR-122, miR-145, miR-146a, miR-200c, miR-221, miR-222, and miR-223 in patients with hepatocellular carcinoma or intrahepatic cholangiocarcinoma and its prognostic significance. Mol Carcinog 2013;52(4):297–303.

[165] Zhang J, Jiao J, Cermelli S, Muir K, Jung KH, Zou R, et al. miR-21 inhibition reduces liver fibrosis and prevents tumor development by inducing apoptosis of CD24+ progenitor cells. Cancer Res 2015;75(9):1859–67.
[166] Liu X, Kwon H, Li Z, Fu YX. Is CD47 an innate immune checkpoint for tumor evasion? J Hematol Oncol 2017;10(1):12.
[167] Lee TK, Cheung VC, Lu P, Lau EY, Ma S, Tang KH, et al. Blockade of CD47-mediated cathepsin S/protease-activated receptor 2 signaling provides a therapeutic target for hepatocellular carcinoma. Hepatology 2014;60(1):179–91.
[168] Lo J, Lau EY, So FT, Lu P, Chan VS, Cheung VC, et al. Anti-CD47 antibody suppresses tumour growth and augments the effect of chemotherapy treatment in hepatocellular carcinoma. Liver Int 2016;36(5):737–45.
[169] Tsujikawa H, Masugi Y, Yamazaki K, Itano O, Kitagawa Y, Sakamoto M. Immunohistochemical molecular analysis indicates hepatocellular carcinoma subgroups that reflect tumor aggressiveness. Hum Pathol 2016;50:24–33.
[170] Miltiadous O, Sia D, Hoshida Y, Fiel MI, Harrington AN, Thung SN, et al. Progenitor cell markers predict outcome of patients with hepatocellular carcinoma beyond Milan criteria undergoing liver transplantation. J Hepatol 2015;63(6):1368–77.
[171] Kim H, Choi GH, Na DC, Ahn EY, Kim GI, Lee JE, et al. Human hepatocellular carcinomas with "Stemness"-related marker expression: keratin 19 expression and a poor prognosis. Hepatology 2011;54(5):1707–17.
[172] Govaere O, Komuta M, Berkers J, Spee B, Janssen C, de Luca F, et al. Keratin 19: a key role player in the invasion of human hepatocellular carcinomas. Gut 2014;63(4):674–85.
[173] Kawai T, Yasuchika K, Ishii T, Katayama H, Yoshitoshi EY, Ogiso S, et al. Keratin 19, a cancer stem cell marker in human hepatocellular carcinoma. Clin Cancer Res 2015;21(13):3081–91.
[174] Brandes AA, Carpentier AF, Kesari S, Sepulveda-Sanchez JM, Wheeler HR, Chinot O, et al. A Phase II randomized study of galunisertib monotherapy or galunisertib plus lomustine compared with lomustine monotherapy in patients with recurrent glioblastoma. Neuro-Oncol 2016;18(8):1146–56.
[175] Fujiwara Y, Nokihara H, Yamada Y, Yamamoto N, Sunami K, Utsumi H, et al. Phase 1 study of galunisertib, a TGF-beta receptor I kinase inhibitor, in Japanese patients with advanced solid tumors. Cancer Chemother Pharmacol 2015;76(6):1143–52.
[176] Liu C, Liu L, Chen X, Cheng J, Zhang H, Shen J, et al. Sox9 regulates self-renewal and tumorigenicity by promoting symmetrical cell division of cancer stem cells in hepatocellular carcinoma. Hepatology 2016;64(1):117–29.
[177] Kawai T, Yasuchika K, Ishii T, Miyauchi Y, Kojima H, Yamaoka R, et al. SOX9 is a novel cancer stem cell marker surrogated by osteopontin in human hepatocellular carcinoma. Sci Rep 2016;6:30489.
[178] Shi M, Chen X, Ye K, Yao Y, Li Y. Application potential of toll-like receptors in cancer immunotherapy: systematic review. Medicine 2016;95(25). e3951.
[179] Liu WT, Jing YY, Yu GF, Han ZP, Yu DD, Fan QM, et al. Toll like receptor 4 facilitates invasion and migration as a cancer stem cell marker in hepatocellular carcinoma. Cancer Lett 2015;358(2):136–43.
[180] Sharma P, Allison JP. The future of immune checkpoint therapy. Science 2015;348(6230):56–61.
[181] Tamai K, Nakamura M, Mizuma M, Mochizuki M, Yokoyama M, Endo H, et al. Suppressive expression of CD274 increases tumorigenesis and cancer stem cell phenotypes in cholangiocarcinoma. Cancer Sci 2014;105(6):667–74.

[182] Zhao W, Wang L, Han H, Jin K, Lin N, Guo T, et al. 1B50-1, a mAb raised against recurrent tumor cells, targets liver tumor-initiating cells by binding to the calcium channel alpha2delta1 subunit. Cancer Cell 2013;23(4):541–56.
[183] Zhou XD, Tang ZY, Fan J, Zhou J, Wu ZQ, Qin LX, et al. Intrahepatic cholangiocarcinoma: report of 272 patients compared with 5,829 patients with hepatocellular carcinoma. J Cancer Res Clin Oncol 2009;135(8):1073–80.
[184] Shibahara J, Hayashi A, Misumi K, Sakamoto Y, Arita J, Hasegawa K, et al. Clinicopathologic characteristics of hepatocellular carcinoma with reactive ductule-like components, a subset of liver cancer currently classified as combined hepatocellular-cholangiocarcinoma with stem-cell features, typical subtype. Am J Surg Pathol 2016;40(5):608–16.
[185] Yoon YI, Hwang S, Lee YJ, Kim KH, Ahn CS, Moon DB, et al. Postresection outcomes of combined hepatocellular carcinoma-cholangiocarcinoma, hepatocellular carcinoma and intrahepatic cholangiocarcinoma. J Gastrointest Surg 2016;20(2):411–20.
[186] Akiba J, Nakashima O, Hattori S, Tanikawa K, Takenaka M, Nakayama M, et al. Clinicopathologic analysis of combined hepatocellular-cholangiocarcinoma according to the latest WHO classification. Am J Surg Pathol 2013;37(4):496–505.
[187] Sasaki M, Sato H, Kakuda Y, Sato Y, Choi JH, Nakanuma Y. Clinicopathological significance of 'subtypes with stem-cell feature' in combined hepatocellular-cholangiocarcinoma. Liver Int 2015;35(3):1024–35.
[188] Kim H, Park YN. Hepatocellular carcinomas expressing 'stemness'-related markers: clinicopathological characteristics. Dig Dis 2014;32(6):778–85.
[189] Yuan RH, Jeng YM, Hu RH, Lai PL, Lee PH, Cheng CC, et al. Role of p53 and beta-catenin mutations in conjunction with CK19 expression on early tumor recurrence and prognosis of hepatocellular carcinoma. J Gastrointest Surg 2011;15(2):321–9.
[190] Uenishi T, Kubo S, Yamamoto T, Shuto T, Ogawa M, Tanaka H, et al. Cytokeratin 19 expression in hepatocellular carcinoma predicts early postoperative recurrence. Cancer Sci 2003;94(10):851–7.
[191] Choi GH, Kim GI, Yoo JE, Na DC, Han DH, Roh YH, et al. Increased expression of circulating cancer stem cell markers during the perioperative period predicts early recurrence after curative resection of hepatocellular carcinoma. Ann Surg Oncol 2015;22(Suppl. 3):S1444–52.
[192] Yamamoto T, Uenishi T, Ogawa M, Ichikawa T, Hai S, Sakabe K, et al. Immunohistologic attempt to find carcinogenesis from hepatic progenitor cell in hepatocellular carcinoma. Dig Surg 2005;22(5):364–70.
[193] Durnez A, Verslype C, Nevens F, Fevery J, Aerts R, Pirenne J, et al. The clinicopathological and prognostic relevance of cytokeratin 7 and 19 expression in hepatocellular carcinoma. A possible progenitor cell origin. Histopathology 2006;49(2):138–51.
[194] Aishima S, Nishihara Y, Kuroda Y, Taguchi K, Iguchi T, Taketomi A, et al. Histologic characteristics and prognostic significance in small hepatocellular carcinoma with biliary differentiation: subdivision and comparison with ordinary hepatocellular carcinoma. Am J Surg Pathol 2007;31(5):783–91.
[195] Yang XR, Xu Y, Shi GM, Fan J, Zhou J, Ji Y, et al. Cytokeratin 10 and cytokeratin 19: predictive markers for poor prognosis in hepatocellular carcinoma patients after curative resection. Clin Cancer Res 2008;14(12):3850–9.
[196] Tsuchiya K, Komuta M, Yasui Y, Tamaki N, Hosokawa T, Ueda K, et al. Expression of keratin 19 is related to high recurrence of hepatocellular carcinoma after radiofrequency ablation. Oncology 2011;80(3–4):278–88.

[197] Chan AW, Tong JH, Chan SL, Lai PB, To KF. Expression of stemness markers (CD133 and EpCAM) in prognostication of hepatocellular carcinoma. Histopathology 2014;64(7): 935–50.
[198] Song W, Li H, Tao K, Li R, Song Z, Zhao Q, et al. Expression and clinical significance of the stem cell marker CD133 in hepatocellular carcinoma. Int J Clin Pract 2008;62(8):1212–8.
[199] Sasaki A, Kamiyama T, Yokoo H, Nakanishi K, Kubota K, Haga H, et al. Cytoplasmic expression of CD133 is an important risk factor for overall survival in hepatocellular carcinoma. Oncol Rep 2010;24(2):537–46.
[200] Yang XR, Xu Y, Yu B, Zhou J, Qiu SJ, Shi GM, et al. High expression levels of putative hepatic stem/progenitor cell biomarkers related to tumour angiogenesis and poor prognosis of hepatocellular carcinoma. Gut 2010;59(7):953–62.
[201] Sun YF, Xu Y, Yang XR, Guo W, Zhang X, Qiu SJ, et al. Circulating stem cell-like epithelial cell adhesion molecule-positive tumor cells indicate poor prognosis of hepatocellular carcinoma after curative resection. Hepatology 2013;57(4):1458–68.
[202] Guo Z, Li LQ, Jiang JH, Ou C, Zeng LX, Xiang BD. Cancer stem cell markers correlate with early recurrence and survival in hepatocellular carcinoma. World J Gastroenterol 2014;20(8):2098–106.
[203] Zeng Z, Ren J, O'Neil M, Zhao J, Bridges B, Cox J, et al. Impact of stem cell marker expression on recurrence of TACE-treated hepatocellular carcinoma post liver transplantation. BMC Cancer 2012;12:584.
[204] Yong KJ, Chai L, Tenen DG. Oncofetal gene SALL4 in aggressive hepatocellular carcinoma. N Engl J Med 2013;369(12):1171–2.
[205] Liu TC, Vachharajani N, Chapman WC, Brunt EM. SALL4 immunoreactivity predicts prognosis in Western hepatocellular carcinoma patients but is a rare event: a study of 236 cases. Am J Surg Pathol 2014;38(7):966–72.
[206] Suzuki E, Chiba T, Zen Y, Miyagi S, Tada M, Kanai F, et al. Aldehyde dehydrogenase 1 is associated with recurrence-free survival but not stem cell-like properties in hepatocellular carcinoma. Hepatol Res 2012;42(11):1100–11.
[207] Cheng SW, Tsai HW, Lin YJ, Cheng PN, Chang YC, Yen CJ, et al. Lin28B is an oncofetal circulating cancer stem cell-like marker associated with recurrence of hepatocellular carcinoma. PLoS One 2013;8(11):e80053.
[208] Tang B, Liang X, Tang F, Zhang J, Zeng S, Jin S, et al. Expression of USP22 and Survivin is an indicator of malignant behavior in hepatocellular carcinoma. Int J Oncol 2015;47(6):2208–16.
[209] Yu DD, Jing YY, Guo SW, Ye F, Lu W, Li Q, et al. Overexpression of hepatocyte nuclear factor-1beta predicting poor prognosis is associated with biliary phenotype in patients with hepatocellular carcinoma. Sci Rep 2015;5:13319.
[210] Leelawat K, Thongtawee T, Narong S, Subwongcharoen S, Treepongkaruna SA. Strong expression of CD133 is associated with increased cholangiocarcinoma progression. World J Gastroenterol 2011;17(9):1192–8.
[211] Nanashima A, Hatachi G, Tsuchiya T, Matsumoto H, Arai J, Abo T, et al. Clinical significances of cancer stem cells markers in patients with intrahepatic cholangiocarcinoma who underwent hepatectomy. Anticancer Res 2013;33(5):2107–14.
[212] Sulpice L, Rayar M, Turlin B, Boucher E, Bellaud P, Desille M, et al. Epithelial cell adhesion molecule is a prognosis marker for intrahepatic cholangiocarcinoma. J Surg Res 2014;192(1):117–23.
[213] Gu MJ, Jang BI. Clinicopathologic significance of Sox2. CD44 and CD44v6 expression in intrahepatic cholangiocarcinoma. Pathol Oncol Res 2014;20(3):655–60.

[214] Shimada M, Sugimoto K, Iwahashi S, Utsunomiya T, Morine Y, Imura S, et al. CD133 expression is a potential prognostic indicator in intrahepatic cholangiocarcinoma. J Gastroenterol 2010;45(8):896–902.
[215] Fan L, He F, Liu H, Zhu J, Liu Y, Yin Z, et al. CD133: a potential indicator for differentiation and prognosis of human cholangiocarcinoma. BMC Cancer 2011;11:320.
[216] Aoki S, Mizuma M, Takahashi Y, Haji Y, Okada R, Abe T, et al. Aberrant activation of Notch signaling in extrahepatic cholangiocarcinoma: clinicopathological features and therapeutic potential for cancer stem cell-like properties. BMC Cancer 2016;16(1):854.
[217] Wang X, Bayer ME, Chen X, Fredrickson C, Cornforth AN, Liang G, et al. Phase I trial of active specific immunotherapy with autologous dendritic cells pulsed with autologous irradiated tumor stem cells in hepatitis B-positive patients with hepatocellular carcinoma. J Surg Oncol 2015;111(7):862–7.
[218] Morita R, Hirohashi Y, Torigoe T, Ito-Inoda S, Takahashi A, Mariya T, et al. Olfactory receptor family 7 subfamily C member 1 is a novel marker of colon cancer-initiating cells and is a potent target of immunotherapy. Clin Cancer Res 2016;22(13):3298–309.
[219] Saito T, Chiba T, Yuki K, Zen Y, Oshima M, Koide S, et al. Metformin, a diabetes drug, eliminates tumor-initiating hepatocellular carcinoma cells. PLoS One 2013;8(7):e70010.
[220] Xin HW, Ambe CM, Miller TC, Chen JQ, Wiegand GW, Anderson AJ, et al. Liver label retaining cancer cells are relatively resistant to the reported anti-cancer stem cell drug metformin. J Cancer 2016;7(9):1142–51.
[221] DePeralta DK, Wei L, Ghoshal S, Schmidt B, Lauwers GY, Lanuti M, et al. Metformin prevents hepatocellular carcinoma development by suppressing hepatic progenitor cell activation in a rat model of cirrhosis. Cancer 2016;122(8):1216–27.
[222] Chiba T, Suzuki E, Yuki K, Zen Y, Oshima M, Miyagi S, et al. Disulfiram eradicates tumor-initiating hepatocellular carcinoma cells in ROS-p38 MAPK pathway-dependent and -independent manners. PloS one 2014;9(1):e84807.
[223] Thanee M, Loilome W, Techasen A, Sugihara E, Okazaki S, Abe S, et al. CD44 variant-dependent redox status regulation in liver fluke-associated cholangiocarcinoma: A target for cholangiocarcinoma treatment. Cancer Sci 2016;107(7):991–1000.
[224] Chu TH, Chan HH, Kuo HM, Liu LF, Hu TH, Sun CK, et al. Celecoxib suppresses hepatoma stemness and progression by up-regulating PTEN. Oncotarget 2014;5(6):1475–90.

Chapter 12

Stem/Progenitor Cells in Chronically Injured Liver and the Surrounding Microenvironment

Michitaka Matsuda, Minoru Tanaka
National Center for Global Health and Medicine, Tokyo, Japan

1 INTRODUCTION

It has long been known that the liver has remarkable regenerative capacity, which made it possible to transplant the liver from living donor. The mechanism of this type of regeneration has been intensively investigated by using well-established experimental model such as partial hepatectomy (PHx) [1,2]. After surgical resection, residual liver rapidly grows to return to the original size. This phenomenon depends on hypertrophy and mitosis of the residual hepatocytes [3,4], followed by the proliferation of nonparenchymal cells [5]. Therefore it is sometimes called compensatory hypertrophy [6]. The histological reconstitution of the liver mass is very rapid, which is achieved in 7–10 days in rodents [1] and about 6–8 weeks in human [7]. Decades of excellent studies using PHx model have provided us with the understanding about the mechanism of regeneration and elucidated various key factors contributing to this process, for example, HGF, EGF, and IL-6 [1,8].

On the other hand, chronic liver injury is driven by a variety of causative agents such as alcohol, obesity, metabolic disorder, and viral infection. Chronic liver injury and subsequent cirrhosis is a major health problem worldwide. The disease is associated with high morbidity due to loss of liver function, and the only curative treatment for end-stage disease is liver transplantation [9,10]. Therefore promotion or restoration of regenerative capacity in the chronically injured liver is an attractive target for the therapy [11]. However, compared to PHx model, the mechanism of regeneration in various liver diseases, including chronic liver disease, seems to be more complicated for some reasons. First, PHx model is basically "clean" model because the residual liver hardly accompanies injury or inflammation. In contrast, cellular death and secondary immune cell infiltration commonly occurs under pathological condition of most liver diseases. These immune cells and mediators that are supposed to play an important role in regeneration vary depending on the cause of liver diseases [12]. Second, regeneration observed in PHx model is primary regeneration, because regenerative capacity of residual hepatocytes seems to be intact. In contrast, following prolonged or severe injury, hepatocyte-mediated regeneration is impaired [13], and the regenerative ability of liver becomes overwhelmed with aberrant architecture and severe fibrosis [14,15]. The mechanisms underlying such abnormal regeneration in chronically injured liver are less well described. The circumstances such as cellular death, immune response, and restriction of hepatocyte replication highlight the need for another mechanism, that is, stem/progenitor cell-mediated regeneration. In the severely injured liver, unique cell population emerges from periportal area. This phenomenon, called "ductular reaction," has been described in many forms of chronic human liver disease [16–18] and in severe acute liver disease [19]. Ductular reaction has been believed to take part in liver stem/progenitor cell, which can contribute to regeneration in diseased liver [20].

In general, stem cells are defined by their ability of self-renewing and differentiating into multiple lineages [21–23]. The concept of liver stem/progenitor cell originates from the area of developmental biology, and the liver stem/progenitor cell has been defined as the cell that gives rise to both hepatocytes and biliary epithelial cells (BECs or cholangiocytes), the two types of liver epithelial cells [20]. While hepatoblasts in the developing liver have been extensively characterized as such, it has been an issue of intensive debates whether adult liver really contains such stem/progenitor cell to contribute to the recovery from disease [20]. In addition, although there are many reports describing liver stem cells, in most cases, the experiments used to define liver stem cells have not necessarily fulfilled the strict criteria of stem cell such as multipotent differentiation capacity based on clonality and long-term repopulating capacity in vivo. Therefore "liver progenitor cell (LPC)" is more appropriate to represent stem-like characteristics in the field of hepatology, and hence we use the term "LPC" in descriptions of potential stem/progenitor cells in the liver.

Extensive research efforts have been made to elucidate the role of LPC for liver regeneration for a long time. Many excellent studies using fluorescence-activated cell sorting (FACS)-based isolation of LPC have supported the existence of LPC in injured liver. By contrast, more recent studies using genetic labeling system argue the contribution of LPC to regeneration. In addition to the nature of LPCs, the mechanisms of their regulation have attracted many researchers' interests, and several components such as secretory factors, signaling molecules, and extracellular matrix (ECM) have been investigated extensively. In this review, we first describe a historical view of the studies on liver regeneration to clarify the definition of hepatic stem/progenitor cell, and secondly argue the character and nature of LPC. Finally, the contribution of LPC to regeneration in various surrounding environment will be discussed in the last chapter.

2 HISTORICAL VIEW OF LIVER STEM/PROGENITOR CELL STUDIES

Historically, the mechanisms of LPC-mediated regeneration have been intensively investigated by using various experimental models of liver injury in rodents. A proto-type of LPC is "oval cell," a small cell with an ovoid nucleus and a high nuclear-to-cytoplasmic ratio, which was initially observed in injured liver of rat under the restriction of hepatic replication by chemical administration [24,25]. It was believed that oval cells have bipotential capacity of differentiation, because these cells have been noted to coexpress markers of hepatocyte (e.g., albumin) and BEC (e.g., keratin-19) [26–28]. By contrast, the procedure used for oval cell induction in rat was not applicable to the case in mouse model because of the difference of metabolic reactivity to chemical agents between species. Thereafter, some distinct mouse models have been developed to induce oval cell accompanied by ductular reaction [29–31]. These models have been extensively utilized to investigate the nature of oval cell because they certainly induced ductular reaction, including the cells morphologically similar to rat oval cells. However, it was unclear whether the oval cells derived from each experimental setting exhibit similar properties, because these models were related to ductular reaction, but distinct in the etiology of liver diseases. Furthermore, in many cases, BEC marker-positive cells forming atypical ducts were regarded as oval cells in rodents and considered stem/progenitor cells irrespective of their bipotential differentiation or functional relevance to regeneration, which is a cause for confusion in the field of LPC research. Thus while several indirect observations suggested the existence of LPC in chronically injured liver, the definitive evidence that the cell contributes to liver regeneration via differentiation into functional hepatic cells had remained unclear for a long time.

Nonetheless, much efforts have been made to identify putative LPC by prospective isolation using specific cell surface markers based on bidirectional differentiation capacity and clonogenicity in vitro [32–35]. In addition, recent technical advances on genetic manipulation such as fate mapping or lineage tracing experiments help accumulate knowledge on the entity of LPCs in regenerating liver and their interaction with the surrounding environment. The potentiality of LPC-mediated liver regeneration will be further discussed in the following chapter.

3 CHARACTERIZATION OF LPC

Herein, we describe the character of LPC from various viewpoints, for example, origin, mouse models and associated diseases in human, and molecular markers.

3.1 The Deduced Origin of LPC

Adult normal liver consists of minimal functional unit called hepatic lobule (Fig. 12.1). Hepatic lobule is pentagonal or hexagonal in shape and consists

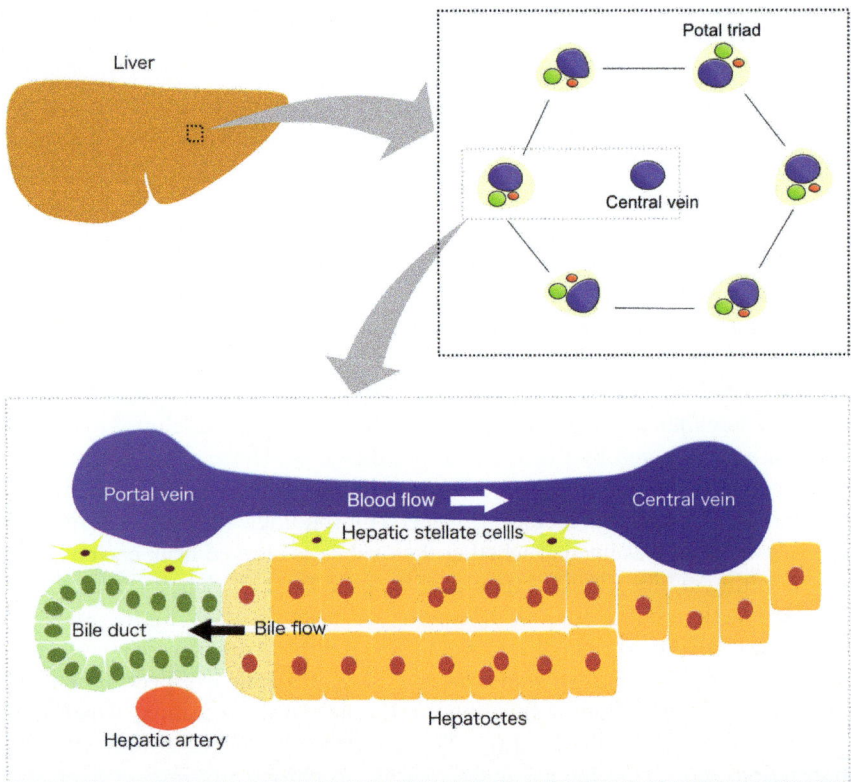

FIGURE 12.1 Schematic view of liver lobule. The portal triad consists of the portal vein, hepatic artery, and bile ducts. Blood from portal vein and hepatic artery flow to the central vein via sinusoids. Bile produced by hepatocytes is collected into bile duct. Hepatic stellate cells reside in the space of Disse, a location between hepatocytes and sinusoids, and periportal fibroblast exists around portal area.

of afferent central vein located in the center and surrounded by 5–6 "portal triad" containing hepatic artery, portal vein, and bile duct. While oxidized arterial blood comes into the liver through hepatic artery, materials absorbed from intestine flow into the liver through portal vein. Both blood flows are mixed and go through sinusoids to the central vein, connecting to systemic circulation. The location of LPC origin has been of great interest to many researchers. Historically, based on the histological analyses using rat models, the canal of Hering, the junctional structure connecting the hepatic plate with bile ducts, has been suggested to harbor a precursor for oval cell (Fig. 12.2). The location interposed between hepatocytes and cholangiocytes would be ideal if oval cells are required for bilineage differentiation. This is consistent with the fact that oval cells invariably emerge from periportal area, expressing BEC markers in rodent models. In human, in situ antigenic profiling

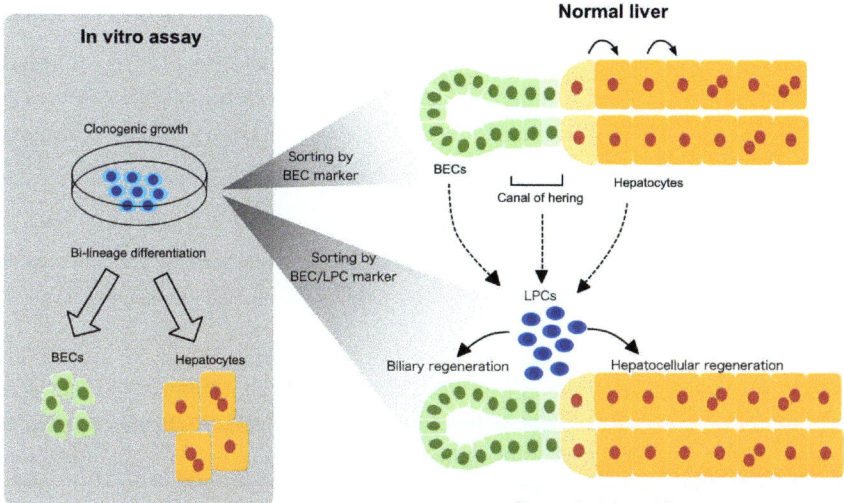

FIGURE 12.2 **The homeostasis of hepatocyte in steady state is achieved predominantly by hepatocellular proliferation.** Under severe liver injury, the capacity of hepatocyte replication is impaired and unique duct-like cell expands around portal area, which is called "ductular reaction." Ductular reaction is considered to contain LPCs that can differentiate into hepatocytes, as well as biliary epithelial cells (BECs). To date, several candidates are assumed to be the origin of LPC: BECs, hepatocytes, and canal of Hering, a junction between these two epithelial cells. Cell sorting-based purification and subsequent culture experiments demonstrated that BEC marker-positive population contains "potential" liver stem cell defined by clonogenicity and bidirectional differentiation.

showed that EpCAM+ progenitor cells are located at the canal of Hering [36]. Furthermore, mitochondrial DNA mutation analysis showed that clonal proliferative units existed in human liver, most likely originating from a periportal niche [37]. Thus the canal of Hering was considered to be in accordance with these previous studies in various respects, and came to be widely accepted as a plausible origin for LPC. Considering the fact that the isolated BEC-marker positive cells include stem-like cells exhibiting clonogenicity and bipotential differentiation capacity in vitro as described in the forthcoming paragraphs, it was reasonable to suppose that potential liver stem cells would be embedded in or associated with bile ducts. However, there was no direct proof so far because of the difficulty in anatomical and histological discrimination of the canal of Hering in rodents and the lack of specific markers to isolate the component cells in the normal liver.

3.2 Experimental Rodent Models and Human Liver Diseases Related to LPC

To elucidate the nature of LPC, many experimental models of chronic liver injury have been established.

3.2.1 Rat Model

In classical papers by Thorgeirsson [24] and Farber [25], it was reported that hepatocyte proliferation was inhibited by the chemical 2-acetylaminofluorene (AAF) in rat models of carcinogenesis. When 2-AAF treatment was combined with surgical resection or hepatotoxin carbon tetrachloride (CCl_4) injury, liver showed prompt regeneration. In the paper, [3H]thymidine was administered to the 2-AAF treated rat at 6 days following PHx. The [3H]thymidine labeled the only epithelial cell that was proliferating in the liver at the time. This small round-shaped primitive cell was named "oval cell," which is derived from its characteristic morphology. When the rat was subsequently sacrificed from 9 to 13 days, [3H]thymidine was identified in hepatocytes in the rat, suggesting the differentiation of oval cell into hepatocytes. This result indicated that oval cell might play possible roles as the source of stem/progenitor cell for regeneration, as well as precancerous cell, especially when hepatocyte proliferation is inhibited under severe liver injuries. In addition, such progenitor-like cells were discovered in the other rat models of liver injury using D-galactosamine or retrorsine [38,39]. Although rat oval cells have been characterized intensively, histological analysis has been preceded in many experiments due to the limitation of genetic manipulation and cell isolation technique in rat models.

3.2.2 Mouse Model

The 2-AAF/PHx model is the most reliable procedure to induce oval cells. However, this protocol was not applicable in mouse, because mouse hepatocyte is lacking in the enzyme that metabolizes 2-AAF to hepatotoxic metabolite. Therefore several other injury models have been developed to induce oval cell in mouse. The 3,5-diethoxycarbonyl-1, 4-dihydro-collidine (DDC) is a porphyrinogenic hepatotoxin, which causes the accumulation of porphyrin crystals in hepatocyte and small bile duct [40]. The DDC-fed mice partially show the pathology of sclerosing cholangitis, leading to severe cholestasis, pericholangitis, periportal fibrosis, and ductular reactions. Therefore the DDC model has been used in many studies to examine the character and the role of LPC. When the ductular reaction was inhibited in this model, liver damage was exacerbated following hepatocyte necrosis [41], indicating the protective role during chronic injury. Another typical and well-established protocol is choline-deficient ethionine-supplemented (CDE) diet model, also known as a kind of nonalcoholic steatohepatitis (NASH) model to date, which causes steatosis, hepatocellular injury, and subsequent ductular reaction [29]. Additionally, several other models of chronic liver injury caused by hepatotoxins (e.g., thioacetamide, carbon tetrachloride, and dipin [42]) or bile duct ligation have been proposed to induce ductular reaction.

3.2.3 Human Liver Diseases

While bile duct is observed only in portal triad in the normal liver, it has been long known that dramatic increase of biliary trees and architectural change occur. This observations based on histological analysis of the liver from patients with chronic liver injury, including chronic hepatitis, alcoholic hepatitis, and NASH [43]. The term "ductular reaction" was originally coined by Popper et al. in 1957 and has been thereafter used in the hepatology literature. Importantly, the level of increased duct-like cell is related to disease severity [44]. These cells often exhibit typically immature and intermediate phenotypes between hepatocyte and cholangiocytes as determined by morphology and molecular marker expression. Therefore they have been considered bipotential progenitor cell population [26,45,46]. There are some indirect evidences suggesting the close relationship between ductular reaction and liver regeneration. Morphological study claimed that newly regenerated hepatocyte buds in damaged parenchymal area could be derived from progenitor cells [47]. Mitochondrial DNA mutation analysis showed that regenerative nodules and adjacent duct could be derived from clonal population [48]. The analysis of liver sample biopsied from patients with chronic hepatitis revealed that telomere lengths in EpCAM(+) hepatocytes in cirrhosis were higher than EpCAM(−) hepatocytes, and relatively shorter than those in the EpCAM(+) hepatobiliary cells corresponding to ductular reaction [49]. These observations in human disease are highly suggestive, yet, not conclusive to define bipotential stem cell in ductular reaction.

The correlation between ductular reaction and the severity of hepatic damage in chronic hepatitis was seen to be common to rodent models and human liver diseases. Although rodent models are useful to promote the elucidation of LPC in ductular reaction, it is still controversial whether the ductular reactions contain bipotential LPC contributing to liver regeneration in several rodent studies. In this regard, it should be noted that the etiology, cellular status, and intrahepatic environment apparently differ among rodent models. In fact, while hepatocyte proliferation is inhibited by 2-AAF in rat model, hepatocytes can proliferate after chronic liver injury in most of the mouse models. Considering that liver fibrosis or cellular senescence in patients with cirrhosis or chronic hepatitis could impair hepatocyte replication, the results derived from different rodent models should be carefully interpreted to evaluate the potential of LPC in liver regeneration.

3.3 Characterization of LPCs by Molecular Marker

Based on the notion that ductular reaction represents the emergence of stem/progenitor cells, oval cell had attracted much interest of the scientists for a long time. In line with the interest, the focus of oval cell study gradually shifted from histopathological analysis to isolation and characterization of LPC by molecular markers. A number of markers (e.g., cytokeratin 19, EpCAM, and CD133) have been proposed in mouse, rat, and human LPCs so far [33–35,50,51]. Sox9, Osteopontin, and the MIC1-1C3 antigen are also regarded as equivalent LPC markers, at least in mice [32,52–54]. By using antibodies against cell surface markers (e.g., EpCAM, CD133, or MIC-1C3), the isolated cells from injured liver certainly exhibited clonogenicity and multilineage (i.e., hepatocytic and cholagiocytic) differentiation capacity in vitro, which are hallmarks of stem cell. Although these evidences suggested the presence of stem cell compartment in the liver, it was still unclear whether these in vitro assays substantially reflect stem cells contributing to regeneration in vivo. In addition, most of these markers were shared with BEC and not specific to LPC. Therefore it was also unclear whether BECs or a specialized population of BEC marker-positive cell are endowed with such stem-like characteristics. A few possible LPC markers have been reported to distinguish LPC from cholangiocyte. Trop2 (Tacstd-2), a paralog of EpCAM (Tacstd-1), has been found to be expressed on most oval cells in injured liver of mouse upon DDC administration, but not on cholangiocytes in the normal liver. Therefore Trop2 expression would be useful to classify ductular cells into LPC or BEC, although the fate of Trpo2-positive cells by genetic labeling experiment has not been reported as of yet. By contrast, transcriptional factor Foxl1 has been reported as a marker for bipotential hepatic progenitor. Intriguingly, transgenic (Tg) mouse line expressing Cre recombinase under the control of Foxl1 promoter has been generated to demonstrate that genetically labeled Foxl1-expressing cells give rise to both hepatocytes and cholangiocytes [55]. However, the activity of Cre recombinase used in this Tg mouse is regulated constitutively but not temporarily, which cannot exclude the possibility that Foxl1 is ectopically induced in hepatocyte or cholangiocyte by the stimulus of liver damage. It is also unclear whether the newly labeled hepatocytes and cholangiocytes are derived from a clonal population or distinct Folx1-expressing progenitors. Even thinking about it, however, Foxl1-Cre Tg mouse makes sense to explore LPC compartment during regeneration in vivo. Lgr5, a well-established stem cell marker in certain types of tissue, including intestine, has also been identified in the injured liver. The knockin mice with insertion of *LacZ* or *IRES-CreERT2* cassette in the *Lgr5* locus have been used to demonstrate that Lgr5-LacZ+ cells appear near bile duct upon liver injury and that damage-induced Lgr5+ cells can give rise to hepatocytes and BECs in vivo. It has also been shown that single Lgr5+ cells from damaged liver can be clonally expanded in vitro as organoids, which is transplantable *in vivo* [56]. However, the relationship between Lgr5+ cells and the cells expressing other LPC markers has remained unclear. More importantly, none of the cells expressing such authentic LPC markers as Trop2, Foxl1, and Lgr5 were found in the normal liver, as judged by at least histological analyses using immunostaining or reporter activity. The evidences suggested that these markers are not expressed almost or at all prior to damage, but induced in "facultative" stem/progenitor cell population after hepatic injury.

By contrast, several works have suggested that LPCs can be derived from hepatocyte under chronic liver injury [57,58]. In these lineage-tracing experiments, hepatocyte-specific labeling by using the tropism of adeno-associated virus serotype or *Albumin* promoter was employed to elucidate the source of oval cells. It was also reported that biphenotypic hepatocytes, which express both hepatocytic marker (HNF4α) and cholangiocytic marker (Sox9), but not EpCAM, appeared near the ductular structures in DDC-injured liver [59]. The sorted Sox9+EpCAM− hepatocytes proliferated and could differentiate into both functional hepatocytes and ductal cells in vitro under appropriate culture conditions. Notably, in a study using chimeric liver generated from the transplantation of genetically labeled Fah+/+ hepatocytes to Fah−/− mouse recipients, Tarlow et al. demonstrated that oval cells induced by DDC injury were derived from both donor hepatocytes and recipient BECs [60]. More interestingly, they showed that these two ductal populations exhibited transcriptionally and functionally different properties and that hepatocyte-derived LPCs could revert back to hepatocytes after injury subsides. All of these reports strongly suggested the conversion of hepatocytes to the ductal cells or LPCs under the pathological condition. Therefore these cells could be a candidate for the origin of LPCs, though it was unclear which hepatocytes possess such plasticity and where they are located in uninjured liver.

4 MICROENVIRONMENT SURROUNDING LPC IN INJURED LIVER

While the exact role and origin of LPCs remained unclear, numerous presumptive evidences have suggested their contribution to liver regeneration. Herein, we describe the possible roles of environmental factors for LPC during liver regeneration.

4.1 Signaling Pathways Related to LPC Regulation

In addition to the characteristic and origin of LPCs, there have been many reports on the regulatory mechanisms underlying their expansion and differentiation in the liver. Because LPCs are activated in response to acute severe or chronic liver injury, inflammatory cytokine would be a good candidate. In fact, several cytokines, such as TNF-alpha, IL-6, IFN-γ, have been reported to be involved in LPC proliferation [29,61,62]. Notably, TNF-Related WEAK Inducer of Apotosis (TWEAK) is one of the most crucial factors for LPC expansion. TWEAK stimulates oval cell proliferation through its receptor Fn14. Mice lacking Fn14 or treated with neutralizing anti-TWEAK antibody showed suppression of the ductular reaction in DDC model, while overexpression of TWEAK in the liver by using Tg mouse or adenovirus-mediated gene transfer induced LPC proliferation [63]. Similarly, FGF7 has been reported to be implicated in LPC expansion in DDC model. The ductular reaction was induced by overexpression of FGF7 in the uninjured liver using Tg mouse, while it was suppressed in the liver of Fgf7 KO mouse upon DDC injury [64]. While both TWEAK and FGF7 exhibit a mitotic effect on LPCs, it was reported that the ductular reactions induced by TWEAK and FGF7 overexpression showed the morphological differences in the biliary architectures, suggesting their distinct effects on LPCs [65]. It is also unclear whether TWEAK and FGF7 are involved in the differentiation of LPC into hepatocyte during liver regeneration in vivo, although newly formed ducts in the ductular reaction are functionally connected with portal bile duct for biliary excretion [65].

On the other hand, HGF and EGF have also been reported to regulate the proliferation and/or differentiation of LPCs. These growth factors are well-known mitogens during the process of compensatory hypertrophy and fetal development. The study using HGF receptor c-Met conditional knockout mice indicated that HGF/c-MET signal is necessary for the differentiation of LPCs into hepatocytes [66]. In contrast, the liver-specific genetic ablation of Egfr facilitated LPC-mediated liver regeneration by differentiating LPC toward hepatocyte lineage [67]. EGFR induced NOTCH1 to promote cholangiocyte differentiation and branching morphogenesis, while suppressing hepatocyte commitment. This is contrast to c-MET signaling, a strong inducer of hepatocyte differentiation by activating AKT and STAT3. These evidences suggest HGF and EGF cooperatively regulate the fate decision of LPC.

The Notch signal that is known to promote biliary differentiation during liver development has also been shown to play a relevant role in the ductular reaction [57,68]. Liver-specific loss of Notch pathway-related genes resulted in defective ductular reaction in DDC model by preventing biliary reprogramming or tubule formation. By contrast, the Wnt/β-catenin pathway is widely considered critical for stem cell regulation in many organs/tissues. The signal has been reported to be involved in LPC regulation in the injured liver [69–73]. Remarkably, Boulter et al. have proposed the potential relationship between Notch and Wnt pathways in LPC regulation. Activated myofibroblasts surrounding LPC express Jagged-1, a ligand for Notch family receptors to induce the ductular reaction via Notch signal, whereas activated macrophages engulfing debris of dead hepatocyts produce Wnt3a to promote the specification of LPC to hepatocyte by inducing Numb, a negative regulator of Notch pathway [73]. (Fig. 12.3). More recently, it has been reported that Hippo/YAP signaling plays a role in controlling the fate and phenotypic plasticity of hepatocyte via *Notch2* gene expression [74].

Additionally, it has been suggested that ECM such as collagens and laminins play a possible role in LPC regulation. Actually, in a several human disease and experimental settings, ductular reaction is strongly correlated with fibrosis progression, although there is still debate about cause and effect [16,44,75]. Thus many studies have identified reliable factors to influence LPC behavior. However, in many cases, it is still unclear to what extent liver regeneration is mediated by LPC in each experimental setting.

4.2 Contribution of LPC to Liver Regeneration

Genetic marking of cells has been applied extensively for characterizing stem/progenitor cells in various organs/tissues and examining their roles for

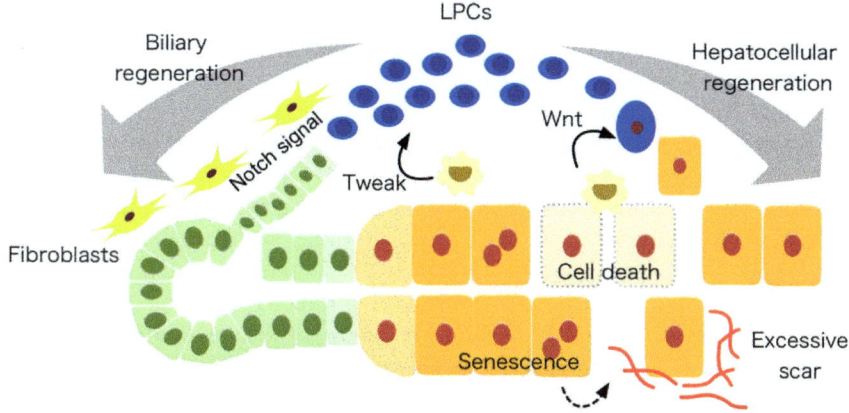

FIGURE 12.3 In the chronically injured liver, normal hepatocyte proliferation is impaired with senescence, and LPCs expand from portal area by multiple signals, such as macrophage-derived Tweak. LPC-mediated liver regeneration can be characterized into hepatocellular and biliary regeneration. Macrophages phagocyte debris from damaged hepatocyte and express Wnt, which cues LPCs in pro-hepatocyte differentiation. In biliary regeneration, fibroblasts influence their maturation through Notch signaling pathway.

homeostasis and regeneration. The combination of cell type-specific Cre driver strain and loxP-regulated reporter strain made it possible to trace the fate of the specifically labeled cells in vivo. Based on a prevailing idea that LPCs are derived from biliary compartment as mentioned earlier, many BEC marker genes have been tested as Cre driver to label potential LPCs in lineage-tracing experiments. Using a knockin mouse wherein CreERT2 is inserted into the Sox9 locus, Furuyama et al. reported tamoxifen-inducible Cre-mediated cell tracking of Sox9-expressing cell in the liver [76]. In this mouse, BECs but not hepatocytes were genetically labeled in the uninjured adult liver by tamoxifen administration. Surprisingly, the labeled cells at the periportal area gradually spread towards the central regions, differentiating into hepatocytes even under uninjured conditions. The evidence suggested that Sox9-expressing precursor pool in the biliary compartment could supply hepatocytes extensively and continuously for homeostasis, as well as regeneration after injury. Unfortunately, however, most of the other experiments using BEC-marker promoter failed to show convincing regeneration from the ductal source [77,78]. For example, in the lineage-tracing experiments using inducible Cre strains under control of osteopontin (Opn) or HNF1β regulatory region, it was shown that 2.45% or 1.86% of hepatocytes were derived from biliary ducts/LPCs expressing Opn or HNF1β following CDE injury, respectively [79,80]. The lineage-tracing experiment using inducible cytokeratin 19 (*Krt19*) promoter CreER knockin mouse demonstrated that BECs or oval cells did not virtually exhibit progenitor cell activity [77]. Moreover, in another complementary study using specific labeling of hepatocyte by adeno-associated virus-mediated gene transfer of Cre recombinase, it was reported that 1.3% of hepatocytes were derived from nonlabeled cells, that is, nonhepatocyte source following chronic CCl_4 injury [4]. More recently, Font-Burgada et al. showed that a subpopulation of hepatocytes expressing several BEC-enriched genes, including low amounts of Sox9, exist at the periportal region of uninjured liver [81]. The cell, which was named "hybrid hepatocyte (HybHP)," contributed to liver regeneration by producing fully differentiated and functional hepatocytes under sustained hepatic damage. In addition, under cholestatic liver injury in DDC model, a small part of HybHP could generate ductal cells, which constitute a minor proportion of oval cells. Therefore HybHP seems to be a potent origin for a subset of oval cells that can contribute to liver regeneration by reverting back to hepatocytes after ductular reaction. Considering that, together with the fact that LPC emerge from the periportal area, there should be a specified circumstance to define the plastic state of hepatocyte around portal vein. Most recently, it has been shown that a combination of three chemical inhibitors against signaling molecules without genetic modifications can convert rat and mouse mature hepatocytes into proliferative bipotent cells in vitro [82]. These findings may provide a clue to explore the regulatory mechanisms underlying the plasticity of hepatocyte in vivo. Taken together, these evidences strongly suggest a major contribution of preexisting hepatocyte and/or LPC of hepatocytic origin to liver regeneration and homeostasis in terms of the replenishment of hepatocytes.

Nonetheless, the potentiality of LPC originating from biliary component should not be ruled out. Oval cell was originally discovered from rat liver injury model under the restriction of hepatocyte replication. Similarly, in advanced human liver disease, there is often widespread hepatocyte senescence, causing irreversible block to hepatocyte replication. By contrast, in mouse models used for oval cell induction, hepatocyte proliferation seems not to be inhibited [60]. Therefore mouse models may be too mild to exploit the potential of LPCs of biliary origin compared with the original models in rats or chronic liver diseases in humans. A recent study challenged the limitation by developing a mouse model wherein the E3 ubiquitin ligase Mdm2 is inducibly deleted in more than 98% of hepatocytes, causing apoptosis, necrosis, and senescence with nearly all hepatocytes. Using this mouse as a recipient, it has been shown that the transplanted LPCs of biliary origin exhibited repopulation capacity under such situation [83]. More recently, the same group reported that BECs act as facultative liver stem cell under impaired hepatocyte regeneration in two independent mouse models: beta1-integrin knockdown or p21 overexpression in hepatocyte. Both models showed impaired hepatocyte proliferation, wherein clusters of hepatocytes derived from BECs were observed after ductular reaction in $CK19^{CreERT}$ $tdTomato^{LSL}$ lineage-tracing model [84]. The study clearly showed that hepatocyte could be regenerated from nonhepatocyte origin, especially ductal cell, suggesting that full potentials of LPC would be exerted under the strained condition with impaired hepatocyte proliferation. Similarly, in the zebrafish studies, the near-complete ablation of hepaotcytes brought out a substantial contribution of BECs to hepatocytes through the dedifferentiation of BECs into LPC-like cells [85,86]. Further analysis is required to clarify whether a specialized subpopulation or all of BECs have capacity to supply new hepatocytes. These works do not rule out the contribution of other candidates of LPC such as hybrid hepatocyte to hepatocyte regeneration, raising the possibility that the priority of LPC origin for regeneration may vary depending on the severity of hepatic impairment. The origin of LPC should be further evaluated for therapeutic application by using the models mimicking human pathology of liver diseases.

5 CONCLUSIONS AND FUTURE DIRECTIONS

Ductular reaction has been believed to be a sign of LPC-mediated liver regeneration for a long time. Because oval cells in the ductular reaction exhibited many characteristics of BEC, including morphology, molecular markers, and functional structures, the potential LPC of biliary origin has been intensively investigated in numerous experiments using rodents. However, recent studies using mouse lineage-tracing models have revealed that LPCs of biliary origin seem to present a minor contribution to liver regeneration. Moreover, considering the fact that hepatocytes can convert to ductal cells in certain injury models, including classic DDC-model, and that periportal hepatocytes with some BEC characteristics have high regenerative capacity, it is convincing that LPCs of hepatocytic origin primarily contribute to liver regeneration in chronic injury. However, it should be also noted that conventional mouse models may not be enough to extract the full capacities of LPCs of ductal origin. Concomitantly, these evidences indicate that not only LPCs but also hepatocytes include a heterogeneous population ranging from BEC-like to hepatocyte-like cells. New insights into the molecular mechanisms regulating dedifferentiation or reprograming of hepatocyte and/or BEC, as well as redifferentiation of LPC into parenchymal cells, would provide an opportunity for LPC-mediated liver regeneration therapy.

REFERENCES

[1] Michalopoulos GK, DeFrances MC. Liver regeneration. Science 1997;276:60–6.
[2] Fausto N, Campbell JS, Riehle KJ. Liver regeneration. Hepatology 2006;43:S45–53.
[3] Miyaoka Y, Ebato K, Kato H, Arakawa S, Shimizu S, Miyajima A. Hypertrophy and unconventional cell division of hepatocytes underlie liver regeneration. Curr Biol 2012;22:1166–75.
[4] Malato Y, Naqvi S, Schurmann N, Ng R, Wang B, Zape J, et al. Fate tracing of mature hepatocytes in mouse liver homeostasis and regeneration. J Clin Invest 2011;121:4850–60.
[5] Grisham JW. A morphologic study of deoxyribonucleic acid synthesis and cell proliferation in regenerating rat liver; autoradiography with thymidine-H3. Cancer Res 1962;22:842–9.
[6] Itoh T, Miyajima A. Liver regeneration by stem/progenitor cells. Hepatology 2014;59: 1617–26.

[7] Yamanaka N, Okamoto E, Kawamura E, Kato T, Oriyama T, Fujimoto J, et al. Dynamics of normal and injured human liver regeneration after hepatectomy as assessed on the basis of computed tomography and liver function. Hepatology 1993;18:79–85.
[8] Kang LI, Mars WM, Michalopoulos GK. Signals and cells involved in regulating liver regeneration. Cells 2012;1:1261–92.
[9] Tsochatzis EA, Bosch J, Burroughs AK. Liver cirrhosis. Lancet 2014;383:1749–61.
[10] Schuppan D, Afdhal NH. Liver cirrhosis. Lancet 2008;371:838–51.
[11] Forbes SJ, Rosenthal N. Preparing the ground for tissue regeneration: from mechanism to therapy. Nat Med 2014;20:857–69.
[12] Heymann F, Tacke F. Immunology in the liver–from homeostasis to disease. Nat Rev Gastroenterol Hepatol 2016;13:88–110.
[13] Marshall A, Rushbrook S, Davies SE, Morris LS, Scott IS, Vowler SL, et al. Relation between hepatocyte G1 arrest, impaired hepatic regeneration, and fibrosis in chronic hepatitis C virus infection. Gastroenterology 2005;128:33–42.
[14] Issa R, Zhou X, Trim N, Millward-Sadler H, Krane S, Benyon C, et al. Mutation in collagen-1 that confers resistance to the action of collagenase results in failure of recovery from CCl4-induced liver fibrosis, persistence of activated hepatic stellate cells, and diminished hepatocyte regeneration. FASEB J 2003;17:47–9.
[15] Bhushan B, Walesky C, Manley M, Gallagher T, Borude P, Edwards G, et al. Pro-regenerative signaling after acetaminophen-induced acute liver injury in mice identified using a novel incremental dose model. Am J Pathol 2014;184:3013–25.
[16] Clouston AD, Powell EE, Walsh MJ, Richardson MM, Demetris AJ, Jonsson JR. Fibrosis correlates with a ductular reaction in hepatitis C: roles of impaired replication, progenitor cells and steatosis. Hepatology 2005;41:809–18.
[17] Gouw AS, Clouston AD, Theise ND. Ductular reactions in human liver: diversity at the interface. Hepatology 2011;54:1853–63.
[18] Richardson MM, Jonsson JR, Powell EE, Brunt EM, Neuschwander-Tetri BA, Bhathal PS, et al. Progressive fibrosis in nonalcoholic steatohepatitis: association with altered regeneration and a ductular reaction. Gastroenterology 2007;133:80–90.
[19] Katoonizadeh A, Nevens F, Verslype C, Pirenne J, Roskams T. Liver regeneration in acute severe liver impairment: a clinicopathological correlation study. Liver Int 2006;26:1225–33.
[20] Miyajima A, Tanaka M, Itoh T. Stem/progenitor cells in liver development, homeostasis, regeneration, and reprogramming. Cell Stem Cell 2014;14:561–74.
[21] Barker N, van Oudenaarden A, Clevers H. Identifying the stem cell of the intestinal crypt: strategies and pitfalls. Cell Stem Cell 2012;11:452–60.
[22] Blanpain C, Fuchs E. Epidermal homeostasis: a balancing act of stem cells in the skin. Nat Rev Mol Cell Biol 2009;10:207–17.
[23] Fuchs E. The tortoise and the hair: slow-cycling cells in the stem cell race. Cell 2009;137:811–9.
[24] Fujio K, Evarts RP, Hu Z, Marsden ER, Thorgeirsson SS. Expression of stem cell factor and its receptor, c-kit, during liver regeneration from putative stem cells in adult rat. Lab Invest 1994;70:511–6.
[25] Ghoshal AK, Mullen B, Medline A, Farber E. Sequential analysis of hepatic carcinogenesis. Regeneration of liver after carbon tetrachloride-induced liver necrosis when hepatocyte proliferation is inhibited by 2-acetylaminofluorene. Lab Invest 1983;48:224–30.
[26] Fausto N. Liver regeneration and repair: hepatocytes, progenitor cells, and stem cells. Hepatology 2004;39:1477–87.

[27] Tian YW, Smith PG, Yeoh GC. The oval-shaped cell as a candidate for a liver stem cell in embryonic, neonatal and precancerous liver: identification based on morphology and immunohistochemical staining for albumin and pyruvate kinase isoenzyme expression. Histochem Cell Biol 1997;107:243–50.
[28] Hixson DC, Brown J, McBride AC, Affigne S. Differentiation status of rat ductal cells and ethionine-induced hepatic carcinomas defined with surface-reactive monoclonal antibodies. Exp Mol Pathol 2000;68:152–69.
[29] Akhurst B, Croager EJ, Farley-Roche CA, Ong JK, Dumble ML, Knight B, et al. A modified choline-deficient, ethionine-supplemented diet protocol effectively induces oval cells in mouse liver. Hepatology 2001;34:519–22.
[30] Preisegger KH, Factor VM, Fuchsbichler A, Stumptner C, Denk H, Thorgeirsson SS. Atypical ductular proliferation and its inhibition by transforming growth factor beta1 in the 3,5-diethoxycarbonyl-1,4-dihydrocollidine mouse model for chronic alcoholic liver disease. Lab Invest 1999;79:103–9.
[31] Engelhardt NV, Factor VM, Yasova AK, Poltoranina VS, Baranov VN, Lasareva MN. Common antigens of mouse oval and biliary epithelial cells. Expression on newly formed hepatocytes. Differentiation 1990;45:29–37.
[32] Dorrell C, Erker L, Schug J, Kopp JL, Canaday PS, Fox AJ, et al. Prospective isolation of a bipotential clonogenic liver progenitor cell in adult mice. Genes Dev 2011;25:1193–203.
[33] Okabe M, Tsukahara Y, Tanaka M, Suzuki K, Saito S, Kamiya Y, et al. Potential hepatic stem cells reside in EpCAM+ cells of normal and injured mouse liver. Development 2009;136:1951–60.
[34] Suzuki A, Sekiya S, Onishi M, Oshima N, Kiyonari H, Nakauchi H, et al. Flow cytometric isolation and clonal identification of self-renewing bipotent hepatic progenitor cells in adult mouse liver. Hepatology 2008;48:1964–78.
[35] Rountree CB, Barsky L, Ge S, Zhu J, Senadheera S, Crooks GM. A CD133-expressing murine liver oval cell population with bilineage potential. Stem Cells 2007;25:2419–29.
[36] Zhang L, Theise N, Chua M, Reid LM. The stem cell niche of human livers: symmetry between development and regeneration. Hepatology 2008;48:1598–607.
[37] Fellous TG, Islam S, Tadrous PJ, Elia G, Kocher HM, Bhattacharya S, et al. Locating the stem cell niche and tracing hepatocyte lineages in human liver. Hepatology 2009;49:1655–63.
[38] Yovchev MI, Grozdanov PN, Zhou H, Racherla H, Guha C, Dabeva MD. Identification of adult hepatic progenitor cells capable of repopulating injured rat liver. Hepatology 2008;47:636–47.
[39] Ichinohe N, Ishii M, Tanimizu N, Kon J, Yoshioka Y, Ochiya T, et al. Transplantation of Thy1+ cells accelerates liver regeneration by enhancing the growth of small hepatocyte-like progenitor cells via IL17RB signaling. Stem Cells 2017;35:920–31.
[40] Delire B, Starkel P, Leclercq I. Animal models for fibrotic liver diseases: what we have, what we need, and what is under development. J Clin Transl Hepatol 2015;3:53–66.
[41] Kim KH, Chen CC, Alpini G, Lau LF. CCN1 induces hepatic ductular reaction through integrin alphavbeta(5)-mediated activation of NF-kappaB. J Clin Invest 2015;125:1886–900.
[42] Factor VM, Radaeva SA, Thorgeirsson SS. Origin and fate of oval cells in dipin-induced hepatocarcinogenesis in the mouse. Am J Pathol 1994;145:409–22.
[43] Roskams TA, Theise ND, Balabaud C, Bhagat G, Bhathal PS, Bioulac-Sage P, et al. Nomenclature of the finer branches of the biliary tree: canals, ductules, and ductular reactions in human livers. Hepatology 2004;39:1739–45.
[44] Lowes KN, Brennan BA, Yeoh GC, Olynyk JK. Oval cell numbers in human chronic liver diseases are directly related to disease severity. Am J Pathol 1999;154:537–41.

[45] Roskams TA, Libbrecht L, Desmet VJ. Progenitor cells in diseased human liver. Semin Liver Dis 2003;23:385–96.
[46] Turanyi E, Dezso K, Csomor J, Schaff Z, Paku S, Nagy P. Immunohistochemical classification of ductular reactions in human liver. Histopathology 2010;57:607–14.
[47] Stueck AE, Wanless IR. Hepatocyte buds derived from progenitor cells repopulate regions of parenchymal extinction in human cirrhosis. Hepatology 2015;61:1696–707.
[48] Lin WR, Lim SN, McDonald SA, Graham T, Wright VL, Peplow CL, et al. The histogenesis of regenerative nodules in human liver cirrhosis. Hepatology 2010;51:1017–26.
[49] Yoon SM, Gerasimidou D, Kuwahara R, Hytiroglou P, Yoo JE, Park YN, et al. Epithelial cell adhesion molecule (EpCAM) marks hepatocytes newly derived from stem/progenitor cells in humans. Hepatology 2011;53:964–73.
[50] Yovchev MI, Grozdanov PN, Joseph B, Gupta S, Dabeva MD. Novel hepatic progenitor cell surface markers in the adult rat liver. Hepatology 2007;45:139–49.
[51] Schmelzer E, Zhang L, Bruce A, Wauthier E, Ludlow J, Yao HL, et al. Human hepatic stem cells from fetal and postnatal donors. J Exp Med 2007;204:1973–87.
[52] Carpentier R, Suner RE, van Hul N, Kopp JL, Beaudry JB, Cordi S, et al. Embryonic ductal plate cells give rise to cholangiocytes, periportal hepatocytes, and adult liver progenitor cells. Gastroenterology 2011;141:1432–8. 1438 e1431–1434.
[53] Dorrell C, Erker L, Lanxon-Cookson KM, Abraham SL, Victoroff T, Ro S, et al. Surface markers for the murine oval cell response. Hepatology 2008;48:1282–91.
[54] Matsuo A, Yoshida T, Yasukawa T, Miki R, Kume K, Kume S. Epiplakin1 is expressed in the cholangiocyte lineage cells in normal liver and adult progenitor cells in injured liver. Gene Express Patterns 2011;11:255–62.
[55] Sackett SD, Li Z, Hurtt R, Gao Y, Wells RG, Brondell K, et al. Foxl1 is a marker of bipotential hepatic progenitor cells in mice. Hepatology 2009;49:920–9.
[56] Huch M, Dorrell C, Boj SF, van Es JH, Li VS, van de Wetering M, et al. In vitro expansion of single Lgr5+ liver stem cells induced by Wnt-driven regeneration. Nature 2013;494:247–50.
[57] Yanger K, Zong Y, Maggs LR, Shapira SN, Maddipati R, Aiello NM, et al. Robust cellular reprogramming occurs spontaneously during liver regeneration. Genes Dev 2013;27:719–24.
[58] Sekiya S, Suzuki A. Hepatocytes, rather than cholangiocytes, can be the major source of primitive ductules in the chronically injured mouse liver. Am J Pathol 2014;184:1468–78.
[59] Tanimizu N, Nishikawa Y, Ichinohe N, Akiyama H, Mitaka T, Sry HMG. box protein 9-positive (Sox9+) epithelial cell adhesion molecule-negative (EpCAM−) biphenotypic cells derived from hepatocytes are involved in mouse liver regeneration. J Biol Chem 2014;289:7589–98.
[60] Tarlow BD, Pelz C, Naugler WE, Wakefield L, Wilson EM, Finegold MJ, et al. Bipotential adult liver progenitors are derived from chronically injured mature hepatocytes. Cell Stem Cell 2014;15:605–18.
[61] Knight B, Yeoh GC, Husk KL, Ly T, Abraham LJ, Yu C, et al. Impaired preneoplastic changes and liver tumor formation in tumor necrosis factor receptor type 1 knockout mice. J Exp Med 2000;192:1809–18.
[62] Yeoh GC, Ernst M, Rose-John S, Akhurst B, Payne C, Long S, et al. Opposing roles of gp130-mediated STAT-3 and ERK-1/2 signaling in liver progenitor cell migration and proliferation. Hepatology 2007;45:486–94.
[63] Jakubowski A, Ambrose C, Parr M, Lincecum JM, Wang MZ, Zheng TS, et al. TWEAK induces liver progenitor cell proliferation. J Clin Invest 2005;115:2330–40.
[64] Takase HM, Itoh T, Ino S, Wang T, Koji T, Akira S, et al. FGF7 is a functional niche signal required for stimulation of adult liver progenitor cells that support liver regeneration. Genes Dev 2013;27:169–81.

[65] Kaneko K, Kamimoto K, Miyajima A, Itoh T. Adaptive remodeling of the biliary architecture underlies liver homeostasis. Hepatology 2015;61:2056–66.
[66] Ishikawa T, Factor VM, Marquardt JU, Raggi C, Seo D, Kitade M, et al. Hepatocyte growth factor/c-met signaling is required for stem-cell-mediated liver regeneration in mice. Hepatology 2012;55:1215–26.
[67] Kitade M, Factor VM, Andersen JB, Tomokuni A, Kaji K, Akita H, et al. Specific fate decisions in adult hepatic progenitor cells driven by MET and EGFR signaling. Genes Dev 2013;27:1706–17.
[68] Fiorotto R, Raizner A, Morell CM, Torsello B, Scirpo R, Fabris L, et al. Notch signaling regulates tubular morphogenesis during repair from biliary damage in mice. J Hepatol 2013;59:124–30.
[69] Apte U, Thompson MD, Cui S, Liu B, Cieply B, Monga SP. Wnt/beta-catenin signaling mediates oval cell response in rodents. Hepatology 2008;47:288–95.
[70] Hu M, Kurobe M, Jeong YJ, Fuerer C, Ghole S, Nusse R, et al. Wnt/beta-catenin signaling in murine hepatic transit amplifying progenitor cells. Gastroenterology 2007;133:1579–91.
[71] Itoh T, Kamiya Y, Okabe M, Tanaka M, Miyajima A. Inducible expression of Wnt genes during adult hepatic stem/progenitor cell response. FEBS Lett 2009;583:777–81.
[72] Yang W, Yan HX, Chen L, Liu Q, He YQ, Yu LX, et al. Wnt/beta-catenin signaling contributes to activation of normal and tumorigenic liver progenitor cells. Cancer Res 2008;68:4287–95.
[73] Boulter L, Govaere O, Bird TG, Radulescu S, Ramachandran P, Pellicoro A, et al. Macrophage-derived Wnt opposes Notch signaling to specify hepatic progenitor cell fate in chronic liver disease. Nat Med 2012;18:572–9.
[74] Yimlamai D, Christodoulou C, Galli GG, Yanger K, Pepe-Mooney B, Gurung B, et al. Hippo pathway activity influences liver cell fate. Cell 2014;157:1324–38.
[75] Grzelak CA, Martelotto LG, Sigglekow ND, Patkunanathan B, Ajami K, Calabro SR, et al. The intrahepatic signalling niche of hedgehog is defined by primary cilia positive cells during chronic liver injury. J Hepatol 2014;60:143–51.
[76] Furuyama K, Kawaguchi Y, Akiyama H, Horiguchi M, Kodama S, Kuhara T, et al. Continuous cell supply from a Sox9-expressing progenitor zone in adult liver, exocrine pancreas and intestine. Nat Genet 2011;43:34–41.
[77] Yanger K, Knigin D, Zong Y, Maggs L, Gu G, Akiyama H, et al. Adult hepatocytes are generated by self-duplication rather than stem cell differentiation. Cell Stem Cell 2014;15: 340–9.
[78] Schaub JR, Malato Y, Gormond C, Willenbring H. Evidence against a stem cell origin of new hepatocytes in a common mouse model of chronic liver injury. Cell Rep 2014;8:933–9.
[79] Espanol-Suner R, Carpentier R, Van Hul N, Legry V, Achouri Y, Cordi S, et al. Liver progenitor cells yield functional hepatocytes in response to chronic liver injury in mice. Gastroenterology 2012;143:1564–75. e1567.
[80] Rodrigo-Torres D, Affo S, Coll M, Morales-Ibanez O, Millan C, Blaya D, et al. The biliary epithelium gives rise to liver progenitor cells. Hepatology 2014;60:1367–77.
[81] Font-Burgada J, Shalapour S, Ramaswamy S, Hsueh B, Rossell D, Umemura A, et al. Hybrid periportal hepatocytes regenerate the injured liver without giving rise to cancer. Cell 2015;162:766–79.
[82] Katsuda T, Kawamata M, Hagiwara K, Takahashi RU, Yamamoto Y, Camargo FD, et al. Conversion of terminally committed hepatocytes to culturable bipotent progenitor cells with regenerative capacity. Cell Stem Cell 2017;20:41–55.
[83] Lu WY, Bird TG, Boulter L, Tsuchiya A, Cole AM, Hay T, et al. Hepatic progenitor cells of biliary origin with liver repopulation capacity. Nat Cell Biol 2015;17:971–83.

[84] Raven A, Lu WY, Man TY, Ferreira-Gonzalez S, O'Duibhir E, Dwyer BJ, et al. Cholangiocytes act as facultative liver stem cells during impaired hepatocyte regeneration. Nature 2017;547:350–4.

[85] He J, Lu H, Zou Q, Luo L. Regeneration of liver after extreme hepatocyte loss occurs mainly via biliary transdifferentiation in zebrafish. Gastroenterology 2014;146:789–800. e788.

[86] Choi TY, Ninov N, Stainier DY, Shin D. Extensive conversion of hepatic biliary epithelial cells to hepatocytes after near total loss of hepatocytes in zebrafish. Gastroenterology 2014;146:776–88.

Chapter 13

The Stem Cells in Liver Cancers and the Controversies

Hiroyuki Tomita*, Tomohiro Kanayama*, Ayumi Niwa*, Kei Noguchi*, Takuji Tanaka**, Akira Hara*

*Gifu University Graduate School of Medicine, Gifu, Japan; **Research Center of Diagnostic Pathology (RC-DiP), Gifu Municipal Hospital, Gifu, Japan

1 INTRODUCTION

The cancer stem cell (CSC) concept is attractive to the clinicians and researchers and assumes hierarchical cellular structure of a tumor, analogous to normal tissue. Self-renewal and the potential to yield various cell types are the main attributes of stem cells and shared by CSCs. Normal tissue is characterized by a fixed number of cells. Dying mature cells are replaced by new-born mature cells derived from progenitors. This process is strictly controlled by mutual interactions between every cell forming the tissue although this process is disrupted in carcinogenesis. CSCs are the only cells being able to self-renew and produce a heterogeneous tumor cell population. CSCs are thought to derive from normal stem cells; however, it is not a unique hypothesis in carcinogenesis. On the other hand, primary liver cancer is of two major types: hepatocellular carcinoma (HCC) and cholangiocarcinoma (CCA). Although there are some differences between HCC and CCA, both are closely associated with chronic liver damage. By persistent inflammatory processes, cell damage and high cellular turnover are elicited, and the hepatocytes and cholangiocytes, also including liver stem/progenitor cells, are thought to be transformed to malignant cells. In this chapter, we discuss the characteristics and roles of stem cells and in the major categories of primary liver cancer. Further, we also highlight the importance of the cellular origin of liver cancer.

2 STEM CELL CONCEPT OF LIVER (HEPATOCYTE AND CHOLANGIOCYTE)

2.1 Liver Function and Architecture

The liver is the largest solid organ in human body. It plays wide roles for maintaining homeostasis, such as metabolism, glycogen pool, drug detoxification, protein production, and bile acid secretion. There are two cell types in the liver: parenchymal cells and non-parenchymal cells. In parenchymal cells, hepatocytes are in charge of most of the metabolic and synthetic functions of the liver. Cholangiocytes, which form the bile duct, are major liver epithelial cells. Non-parenchymal cells that constitute the liver are hepatic sinusoidal endothelial cells, Kupffer cells located at the luminal side of sinusoid, and stellate cells at the space of Disse. These cells modulate the function of hepatocytes through cell–cell interaction.

2.2 The Concept of Liver Stem/Progenitor Cell

Stem cells are functionally defined by the capacity of self-renewal and generation of differentiated cells. Concerning liver stem/progenitor cells, besides the self-renewability, the bipotential to differentiate into hepatocytes and cholangiocytes is the characteristic property. Liver stem/progenitor cells carries out important roles in development, homeostasis, and regeneration. Thus, the liver conserves two stem/progenitor cell systems: fetal liver stem/progenitor cells relating to development, and adult liver stem/progenitor cells with respect to homeostasis and regeneration.

2.3 Fetal Liver Stem/Progenitor Cell

The liver development in mice begins at embryonic day (E) 8.5 from the foregut endoderm [1]. Gadue et al. [2] showed ENDM1 and ENDM2 expression specific to mouse foregut ventral endoderm. Endoderm cells positive for ENDM1 and ENDM2 in immunohistochemistry demonstrate the potential to generate cells of the hepatic lineage and downregulation along specification to the hepatic differentiation. The foregut endoderm cells destined for hepatic fate start to express transcription factors HEX and HNF4α as well as the liver-specific genes α-fetoprotein (*Afp*) and albumin (*Alb*), and transmigrate as cords into the surrounding septum transversum mesenchyme. By E9.0, the ventral area of the foregut close to the cardiac mesoderm and septum transversum thickens to organize the liver diverticulum. Afterward, the diverticulum thickens and conversion from a monolayer of cuboidal endoderm cells into a multilayer of pseudostratified cells begins. These cells are common progenitor cells, which give rise to both hepatocytes and cholangiocytes, and are called "hepatoblasts" during liver development. Recently, the combination of specific cell-sorting markers has been employed to identify fetal liver stem/progenitor cells as previous description in Chapter 11. E-cadherin and LIV2 are also useful cell-surface markers to isolate epithelial cells expressed in E12.5 [3–6]. CD24a and Neighbor of Punc E11 (NOPE) are also keyed out as sorting markers [7]. HNF4α$^+$ liver stem/progenitor cells express epithelial cell adhesion molecule (EpCAM) in mice as early as E9.5. The EpCAM$^+$ cells isolated from the human fetal liver were shown to include multipotent precursors of liver stem/progenitor cells [8].

2.4 Adult Liver Stem/Progenitor Cell in Homeostasis

The epithelial turnover of human intestine occurs every 3–5 days, and new cells are constantly supplied stem/progenitor cell system. On the other hand, the normal turnover time of mature hepatocytes is over a period of more than several months [9]. Therefore, it is questionable whether any stem cell is required for liver maintenance, at least under steady state. The slow turnover rate makes it difficult to employ the long-term label-retaining assay to trace stem/progenitor cells in vivo. The terminal segment of the biliary duct system, the canals of Hering, has been thought to constitute stem/progenitor cell niche. But there is conflicting evidence of support and opposition. Malato et al. [10] transfected a Cre-expressing adeno-associated viral vector into an R26R-EYFP mouse to label hepatocytes. The results suggested that newly formed hepatocytes derive from preexisting hepatocytes, and contribution of non-hepatocytic lineage cells was not observed. Recently, a new stem/progenitor cell population Axin2-positive diploid cells around the central vein is identified as mentioned in Chapter 11.

2.5 Liver Stem/Progenitor Cell and Regeneration after Injury

The liver has a remarkable capacity to regenerate. Traditionally, liver regeneration has been thought to depend primarily on the proliferation of adult hepatocytes [11,12]. This paradigm arose from the conventional experimental approach, the partial hepatectomy model. Such surgical removal of particular lobe of the liver cause less damage on the residual liver, and in the course of liver regeneration, hypertrophy of hepatocytes is also observed [13]. However, recent genetic lineage-tracing studies have demonstrated that a small portion of newborn hepatocytes originates from candidate stem/progenitor cells [10,14]. In addition to surgical treatment, injury models with intoxication have been investigated. Intoxication of carbon tetrachloride (CCl_4) induces acute liver injury. Transgenic mouse with Cre-expressing adeno-associated viral vector was administered CCl_4, and this lineage-tracing model suggested the contribution of stem/progenitor cells to restoration [10]. Another conditional transgenic mouse also supported the involvement of stem/progenitor cells in the regeneration process after acute injury [14].

In contrast to the restoration caused by acute liver damage, severe and chronic liver damage brings on a defect of the proliferation of mature hepatocytes. Thus, in the process of regeneration from such severe and chronic liver injury, adult liver stem/progenitor cells are supposed to play a key role. The proliferation of small epithelial cells in the periportal, named as oval cells, was observed during serious liver injury induced by toxic chemicals in rodents. The properties of the oval cell have already been described in Chapter 11.

There are several specific markers for detecting cells including of postnatal stem/progenitor cells. Some of them are the same as fetal stem/progenitor cell sorting markers such as EpCAM and CD133. Other reported markers are LGR5 [15], $CD13^+$ $CD49f^+$ $CD133^+$ [16], $CD133^+$ $MIC1-1C3^+$ [17], and CD24 [18].

2.6 Transdifferentiation between Hepatocyte and Cholangiocyte

Hepatocytes and cholangiocytes are regarded to be differentiated from single stem/progenitor cells, and they show potential to transdifferentiate into other liver epithelial cell types. Tarlow et al. [11] labeled SOX9-positive cells by lineage-tracing in mice, analyzed the formation of organoids in culture, monitored the responses of cells in mice on a choline-deficient ethionine diet or diets containing 3,5-diethoxycarbonyl-1,4-dihydrocollidine, and traced cells transferred into fumarylacetoacetate hydrolase (Fah) deficient mice. Hepatocytes from normal, immune-compatible donors could be transplanted and successfully recolonized the liver of these mice; <1% of hepatocytes were derived from SOX9-positive precursors [19]. The cholangiocytes induced from hepatocytes kept the expressions of some hepatocyte-specific genes such as *Hnf4* and showed low EpCAM expression [20]. Lu et al. [21] reported the conversion of cholangiocytes to hepatocytes in transgenic mice in which hepatocyte *Mdm2* (an E3 ubiquitin ligase gene) loss was induced. Huch et al. [19] isolated EpCAM$^+$ bile duct cells from the human liver. The cells were developed into organoids, induced to transdifferentiate in culture, and expressed hepatocyte-specific genes. Cholangiocytes isolated from liver biopsies from patients with α1-antitrypsin deficiency and Alagille syndrome also differentiated into hepatocytes in the organoid cultures, but still carried markers of the patients' diseases. However, it is important to note that in these previous studies, the transdifferentiation of cholangiocytes to hepatocytes was observed in vivo, and the hepatocyte phenotype seen before transplantation into the mice continued to be observed after the cells were transplanted. It seems, therefore, reasonable to conclude that under most conditions of normal liver regeneration or toxic injury, hepatocytes and cholangiocytes proliferate and retain their phenotype. This restoration process is strongly supported by previous studies in rodents.

3 HEPATOCELLULAR CARCINOMA

3.1 The Characteristics of HCC

HCC represents the major histological subtype of liver cancers, accounting for 85%–90% of all primary liver cancers. Its occurrence reaches a peak at approximately 70 years of age [22]. The incidence of HCC is highest in men, and this difference is even higher in populations with elevating incidence of HCC [23]. The differences in the geographical distribution of HCC are due to the differences in exposure to the hepatitis viruses and different environmental pathogens; therefore the incidence is highest in East Asia, Sub-Saharan Africa, and Melanesia, with 85% of the total number of cases [24], although in most industrialized countries the incidence is low, except in the South of Europe [25]. HCC is thought to arise from hepatocytes comprising the liver parenchyma. The ratio of HCC preceded by liver cirrhosis is 80% of all cases. Liver cirrhosis is mainly caused by chronic liver injury, leading to the subsequent regeneration of liver cells and formation of abnormal structural nodules with surrounding fibrosis. The common causes of cirrhosis include chronic viral hepatitis infection such as hepatitis B and C viruses; metabolic liver diseases such as non-alcoholic fatty liver disease, hemochromatosis, α1-antitrypsin deficiency, Wilson's disease, and alcoholic liver disease; and autoimmune diseases [26].

3.2 CSC Markers in HCC

Treatments for HCC have been developed and increased, that is, surgery, radiofrequency ablation therapy, and chemotherapy. Although there is a case that treatment may be seen initially successful, eventually it results in failure [27]. CSC theory is suitable for the above phenomenon and this theory has been developed by many researchers. CSC markers contain functional and cell surface markers. CD133 (prominin-1), CD90 (THY-1), CD44, CD326 (EpCAM), CD24, CD13, CD47, and OV6 are common cell surface markers to detect the CSCs of HCC [28]. For detailed descriptions about these markers please see Chapter 11. In addition, some functional markers are used to detect cells having CSC potency, such as aldehyde dehydrogenase 1 (ALDH1) and side population (SP) [29]. Further, HCC cells with low 26S proteasome activity and low reactive oxygen species (ROS) levels are shown to act as CSCs [30]. "Side population" (SP) cells are the first subpopulation to be detected as having CSC potency in HCC. They were identified using the DNA-binding dye Hoechst 33342 in three hepatoma cell lines (HepG2, Hep3B, and Huh-7 cell) [31]. Huh-7 and Hep3B cell lines were found to contain SP cells; however, HepG2 cells did not show a SP [31].

3.3 Heterogeneity in CSC of HCC

Currently, we can detect CSCs in HCC by using some CSC markers. However, it is important to awake that HCCs have heterogeneity among patients similar to most malignancies. The intratumoral heterogeneity, particularly the variability of the CSC markers, determines specific behavior and prognosis of the tumor.

Cancer is usually considered to be unicellular in origin, however, with the object of tumor progression, the more aggressive subtype which arises with genetic change is efficient [32]. According to a CSC evolution model proposed [33], CSC with genetic mutations generates many daughter cells that serve heterogeneity. EpCAM-positive cells and CD90-positive cells are mutually exclusive in primary HCC [34]. EpCAM-positive cells have features of epithelial cells, and CD90-positive cells have features of vascular endothelial cells. Some of the CSC markers could also be used as the indicator of prognosis of HCC, which have been mentioned in Chapter 11. This fact indicates that CSC may represent a heterogeneous cell population [28].

3.4 Therapy for HCC

While chemotherapy and radiation can eliminate tumor cells in proliferating cell cycles, CSCs are intrinsically resistant to these treatments. Thus, strategy for the targeted therapy to affect the self-renewal, survival, and niche properties of CSCs.

The inhibition of CSC-specific pathways is a promising therapeutic approach. For example, the self-renewal function of CSCs of HCCs is dependent on the polycomb group protein BMI1, which has been reported to be involved in the regulation of CSCs [35]. Pharmacological disruption of EZH2, a major component of polycomb repressive complex 2 (PRC2), by the S-adenosylhomocysteine hydrolase inhibitor (3-deazaneplanocin A) has been impaired both the self-renewal and tumor-initiating capabilities of HCCs [36].

Dysregulation of epigenetic mechanisms, such as DNA methylation and histone modification, is associated with cancer development and progression. Growing evidence indicates the efficacy of epigenetic agents in eradicating CSCs in HCC [37]. Zebularine is a DNA methyltransferase (DNMT) inhibitor and was shown to decline CSC ability of self-renewal and tumorigenicity in HCC cells cultured [38]. SALL4 plays a pivotal role in transcription repression by recruiting histone deacetylase (HDAC)-containing nucleosome remodeling and HDAC complex. HDAC inhibitors, such as trichostatin A and vorinostat, have been reported to prevent the cell growth of SALL4-overexpressing HCC cell lines compared with that of SALL-negative HCC cell lines [39]. These studies indicate that epigenetic treatment with DNMT inhibitors and HDAC inhibitors may be a potential approach for the eradication of CSCs in HCC.

Monoclonal antibodies targeting CSC-specific antigens have been considered to be another approach for the eradication of CSCs, and efficacy is reported against CD13, EpCAM, and CD133 for the eradication of CSCs in HCC cells [40–42]. Because these markers are also observed in normal liver tissues and normal tissue SCs, the safety of the monoclonal antibodies should be assessed.

Overexpression of hepatocyte nuclear factor α (HNF4α) is an important regulator of hepatocyte differentiation, and resulted in a decrease in the number of tumorigenic CD90-positive and CD133-positive cells [43]. Oncostatin M (OSM) has been observed similarly to induce the differentiation of EpCAM-positive CSCs of HCCs through the OSM receptor-signaling pathway [44].

Niche microenvironments, where both CSCs and normal tissue SCs reside, is also a promising target to treat HCC. In glioblastoma of brain, CSCs exist in vascular niches where they are maintained in an undifferentiated state by endothelial cells [45]. A multi-kinase inhibitor, such as sorafenib, is the most pivotal molecular target drug clinically approved to treat advanced HCC. This drug blocks tumor cell proliferation by affecting the RAF/mitogen-activated protein kinase kinase/extracellular signal-regulated kinase signaling cascade and exerts an anti-angiogenic effect by targeting tyrosine kinase receptors, including vascular endothelial growth factor receptor and platelet-derived growth factor receptor [46]. While its role of the CSC niche in HCC has not been clarified, sorafenib may contribute to the elimination of CSCs in HCC.

4 CHOLANGIOCARCINOMA (CCA)

4.1 The Characteristics of CCA

CCA is an epithelial cell malignancy and shows markers of cholangiocytes, which arises from varying locations within the biliary tree. CCA is anatomically classified as intrahepatic, perihilar, and distal CCAs. Intrahepatic CCA is defined as the bile duct from proximal to the second-degree area in the liver. Perihilar CCA is set as the bile duct from the second-degree to the junction of the cystic duct and common bile duct. Distal CCA is defined as the bile duct from the junction to the ampulla of Vater. Intrahepatic CCA is less than 10%, perihilar is 50%, and distal is 40% of cases [47]. Combined hepatocellular CCA accounts for 1% of CCA cases. Intrahepatic CCA is increasing in western countries [48,49]. The increasing of both the recognition and incidence have led to rising interest in CCAs.

Cirrhosis, viral hepatitis B, and C have been recognized as risk factors for CCA, particularly, intrahepatic CCA. Viral hepatitis C is considered to be a risk factor for CCA with the strongest association for intrahepatic CCA in the studies of the United States and Europe [50]. On the other hand, it has been reported that viral hepatitis B is a more consistent risk factor for intrahepatic CCA in the studies of South Korea and China [50,51]. A meta-analysis of several case-control studies about risk factors for intrahepatic CCA demonstrates that the combined odds ratios (ORs) [95% confidence interval (CI)] of cirrhosis, hepatitis B, and hepatitis C were 22.92 (18.24–28.79), 5.10 (2.91–8.95), and 4.84 (2.41–9.71), respectively [52]. Some genetic polymorphism has been identified that increase risk of development of CCA and associated with DNA repair, cellular protection against toxins, or immunological surveillance [53]. Hepatolithiasis and biliary enteric drainage, susceptible patients to enteric bacteria bile duct colonization, and infections are reported as additional risk factors for CCA [54]. The role of alcohol and smoking exposure has been controversial [53]. Obesity is associated with the increased risk of CCA [55], although long-term cohort studies are needed for confirming.

4.2 CSC Markers in CCA

The cell-surface antigen, such as CD133, CD24, CD44, and EpCAM are involved as CSC markers in CCAs. CD133, known as prominin-1, expresses in normal epithelial SCs and CCA [56]. CD133-positive cells showed higher invasiveness than CD133-negative cells. Shimada et al. [57] have reported that CD133 expression in 29 patients with intrahepatic CCA and the 5-year survival rate in the CD133-positive group was worse compared with the CD133-negative group. However, Fan et al. [58] have reported contrasting results; CD133 expression was associated with a higher tumor differentiation status in 54 CCA specimens. Moreover, positive CD133 expression is significantly associated with a better prognosis. The details for the roles of these markers in CCA carcinogenesis have been described in Chapter 11.

4.3 Therapy for CCA

Surgical resection is the main curative therapy for improving patients in CCA [59]. The 5-year survival rate after surgical resection is about 35% for intrahepatic CCA and approximately 40% for perihilar CCA [59–61]. The first line of the chemotherapy is gemcitabine with cisplatin, which is used in advanced and metastatic CCAs [62,63]. The role of radiation or chemoradiation treatment in CCA remains to be defined. The patterns of recurrence following resection of hilar or distal CCA play a significant role in defining the appropriate strategy for adjuvant therapy [64].

The CSC target therapy has been challenged in vivo. CD133 inhibits cell growth of Hep3B human hepatoma cell line and abrogated tumor growth in vivo experiments. Small-interfering RNA (siRNA) inhibits EpCAM, which decreases tumorigenicity in hepatic progenitor cells [65]. Moreover, CCA cell lines were inhibited by CD44 siRNA on invasiveness and CSCs migration [66]. CD24 inhibition decreased the invasive ability of CCA cells [67]. These data suggest that the therapy linked to the surface markers is a new strategy for a CSC target therapy for CCA.

5 THE CONTROVERSIES IN STEM CELLS OF HCC AND CCA

Historically, it is considered that hepatocytes give rise to HCCs and cholangiocytes give rise to CCAs. These convincing assumptions are that tumor arises from stem cell populations. Recently, cancer also derives from adult and mature cells which give rise to cells with stem/progenitor cell traits [68]. It has never been found animal models that phenocopy the human liver carcinogenesis with the sequential steps of chronic inflammation and cirrhosis. Consequently, the cellular origin of primary liver cancers is still controversial (Figure 13.1) [69,70].

Currently, liver cancer origins from the malignant transformation of normal stem/progenitor cells are categorized. HCCs originate from hepatocytes and IH-CCAs from cholangiocytes. The combined HCC-CCAs, cholangiolocellular carcinomas, and hepatoblastomas are considered to originate from normal hepatic/progenitors that are located in intrahepatic lesions within the canals of Hering by the portal triads and have the ability to differentiate into both hepatocytes and cholangiocytes [3,71]. Recently, biliary tree stem cells (BTSCs) have established to be precursors of both liver and pancreas and are found in peribiliary glands, which are stem cell niches that are located throughout the biliary

The cellular origin of HCC and CCA

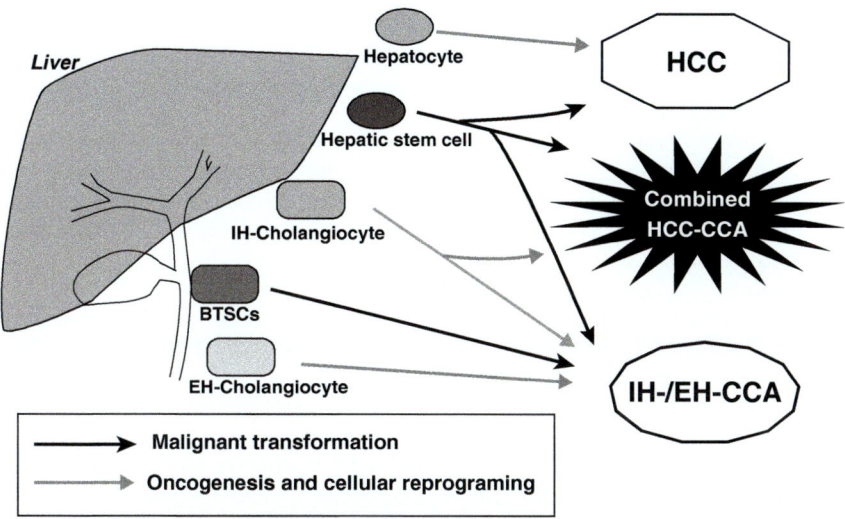

FIGURE 13.1　The cellular origin of HCC and CCA.

tree [72,73]. The typical onset sites of extrahepatic CCAs are at the locations of peribiliary glands in the biliary tree, which include the hilum, the branching points of the biliary tree, and periampullary region.

Several studies suggest that the reprogramming of the differentiated parenchymal lineages contribute to the generation of the phenotypes observed in adult primary liver cancer [74]. Stem cells and/or progenitors may not be the only source of CSCs, and that CSC can be derived from reprogrammed differentiated precursors (e.g., adult hepatocytes), capable of supporting both inter- and intratumoral phenotypic and genetic heterogeneity observed in most human cancers [75].

CONCLUSION

Growing evidences suggest that primary liver cancers derive from the transformation of stem/progenitor cells or from mature cells subjected to reprogramming that results in stem/progenitor cell features. The analysis of cellular origin and molecular mechanisms underlying the complexity of CSCs could be available for new therapies against liver cancers.

ACKNOWLEDGMENTS

The authors wish to thank all the members related to this chapter for their cooperation. *Grant support*: This study was supported by grants from the Ministry of Education, Culture, Sports, Science, and Technology of Japan, nos. 15K11289 (A.H.) and 26430111 (H.T.). *Conflicts of interest*: The authors do not have any conflicts of interest.

REFERENCES

[1] Tremblay KD, Zaret KS. Distinct populations of endoderm cells converge to generate the embryonic liver bud and ventral foregut tissues. Dev Biol 2005;280:87–99.
[2] Gadue P, Gouon-Evans V, Cheng X, Wandzioch E, Zaret KS, Grompe M, et al. Generation of monoclonal antibodies specific for cell surface molecules expressed on early mouse endoderm. Stem Cells 2009;27:2103–13.
[3] Miyajima A, Tanaka M, Itoh T. Stem/progenitor cells in liver development, homeostasis, regeneration, and reprogramming. Cell Stem Cell 2014;14:561–74.
[4] Nierhoff D, Ogawa A, Oertel M, Chen YQ, Shafritz DA. Purification and characterization of mouse fetal liver epithelial cells with high in vivo repopulation capacity. Hepatology 2005;42:130–9.
[5] Watanabe T, Nakagawa K, Ohata S, Kitagawa D, Nishitai G, Seo J, et al. SEK1/MKK4-mediated SAPK/JNK signaling participates in embryonic hepatoblast proliferation via a pathway different from NF-kappaB-induced anti-apoptosis. Dev Biol 2002;250:332–47.
[6] Nitou M, Sugiyama Y, Ishikawa K, Shiojiri N. Purification of fetal mouse hepatoblasts by magnetic beads coated with monoclonal anti-e-cadherin antibodies and their in vitro culture. Exp Cell Res 2002;279:330–43.
[7] Nierhoff D, Levoci L, Schulte S, Goeser T, Rogler LE, Shafritz DA. New cell surface markers for murine fetal hepatic stem cells identified through high density complementary DNA microarrays. Hepatology 2007;46:535–47.
[8] Schmelzer E, Zhang L, Bruce A, Wauthier E, Ludlow J, Yao HL, et al. Human hepatic stem cells from fetal and postnatal donors. J Exp Med 2007;204:1973–87.
[9] Magami Y, Azuma T, Inokuchi H, Kokuno S, Moriyasu F, Kawai K, et al. Cell proliferation and renewal of normal hepatocytes and bile duct cells in adult mouse liver. Liver 2002;22:419–25.
[10] Malato Y, Naqvi S, Schurmann N, Ng R, Wang B, Zape J, et al. Fate tracing of mature hepatocytes in mouse liver homeostasis and regeneration. J Clin Invest 2011;121:4850–60.
[11] Tarlow BD, Finegold MJ, Grompe M. Clonal tracing of Sox9+ liver progenitors in mouse oval cell injury. Hepatology 2014;60:278–89.
[12] Fausto N, Campbell JS. The role of hepatocytes and oval cells in liver regeneration and repopulation. Mech Dev 2003;120:117–30.
[13] Miyaoka Y, Ebato K, Kato H, Arakawa S, Shimizu S, Miyajima A. Hypertrophy and unconventional cell division of hepatocytes underlie liver regeneration. Curr Biol 2012;22:1166–75.
[14] Furuyama K, Kawaguchi Y, Akiyama H, Horiguchi M, Kodama S, Kuhara T, et al. Continuous cell supply from a Sox9-expressing progenitor zone in adult liver, exocrine pancreas and intestine. Nat Genet 2011;43:34–41.
[15] Huch M, Dorrell C, Boj SF, van Es JH, Li VS, van de Wetering M, et al. In vitro expansion of single Lgr5+ liver stem cells induced by Wnt-driven regeneration. Nature 2013;494:247–50.
[16] Kamiya A, Kakinuma S, Yamazaki Y, Nakauchi H. Enrichment and clonal culture of progenitor cells during mouse postnatal liver development in mice. Gastroenterology 2009;137:1114–26. 1126.e1–14.
[17] Dorrell C, Erker L, Schug J, Kopp JL, Canaday PS, Fox AJ, et al. Prospective isolation of a bipotential clonogenic liver progenitor cell in adult mice. Genes Dev 2011;25:1193–203.
[18] Qiu Q, Hernandez JC, Dean AM, Rao PH, Darlington GJ. CD24-positive cells from normal adult mouse liver are hepatocyte progenitor cells. Stem Cells Dev 2011;20:2177–88.
[19] Huch M, Gehart H, van Boxtel R, Hamer K, Blokzijl F, Verstegen MM, et al. Long-term culture of genome-stable bipotent stem cells from adult human liver. Cell 2015;160:299–312.

[20] Tarlow BD, Pelz C, Naugler WE, Wakefield L, Wilson EM, Finegold MJ, et al. Bipotential adult liver progenitors are derived from chronically injured mature hepatocytes. Cell Stem Cell 2014;15:605–18.
[21] Lu WY, Bird TG, Boulter L, Tsuchiya A, Cole AM, Hay T, et al. Hepatic progenitor cells of biliary origin with liver repopulation capacity. Nat Cell Biol 2015;17:971–83.
[22] Janevska D, Chaloska-Ivanova V, Janevski V. Hepatocellular carcinoma: risk factors, diagnosis and treatment. Open Access Maced J Med Sci 2015;3:732–6.
[23] Pascual S, Herrera I, Irurzun J. New advances in hepatocellular carcinoma. World J Hepatol 2016;8:421–38.
[24] Parkin DM, Bray F, Ferlay J, Pisani P. Global cancer statistics, 2002. CA Cancer J Clin 2005;55:74–108.
[25] Bosetti C, Levi F, Boffetta P, Lucchini F, Negri E, La Vecchia C. Trends in mortality from hepatocellular carcinoma in Europe, 1980–2004. Hepatology 2008;48:137–45.
[26] Grandhi MS, Kim AK, Ronnekleiv-Kelly SM, Kamel IR, Ghasebeh MA, Pawlik TM. Hepatocellular carcinoma: from diagnosis to treatment. Surg Oncol 2016;25:74–85.
[27] Romano M, De Francesco F, Pirozzi G, Gringeri E, Boetto R, Di Domenico M, et al. Expression of cancer stem cell biomarkers as a tool for a correct therapeutic approach to hepatocellular carcinoma. Oncoscience 2015;2:443–56.
[28] Chiba T, Iwama A, Yokosuka O. Cancer stem cells in hepatocellular carcinoma: therapeutic implications based on stem cell biology. Hepatol Res 2016;46:50–7.
[29] Tomita H, Tanaka K, Tanaka T, Hara A. Aldehyde dehydrogenase 1A1 in stem cells and cancer. Oncotarget 2016;7:11018–32.
[30] Muramatsu S, Tanaka S, Mogushi K, Adikrisna R, Aihara A, Ban D, et al. Visualization of stem cell features in human hepatocellular carcinoma reveals in vivo significance of tumor–host interaction and clinical course. Hepatology 2013;58:218–28.
[31] Haraguchi N, Utsunomiya T, Inoue H, Tanaka F, Mimori K, Barnard GF, et al. Characterization of a side population of cancer cells from human gastrointestinal system. Stem Cells 2006;24:506–13.
[32] Nowell PC. The clonal evolution of tumor cell populations. Science 1976;194:23–8.
[33] Kreso A, Dick JE. Evolution of the cancer stem cell model. Cell Stem Cell 2014;14:275–91.
[34] Yamashita T, Honda M, Nakamoto Y, Baba M, Nio K, Hara Y, et al. Discrete nature of EpCAM+ and CD90+ cancer stem cells in human hepatocellular carcinoma. Hepatology 2013;57:1484–97.
[35] Wang MC, Li CL, Cui J, Jiao M, Wu T, Jing LI, et al. BMI-1, a promising therapeutic target for human cancer. Oncol Lett 2015;10:583–8.
[36] Suva ML, Riggi N, Janiszewska M, Radovanovic I, Provero P, Stehle JC, et al. EZH2 is essential for glioblastoma cancer stem cell maintenance. Cancer Res 2009;69:9211–8.
[37] Marquardt JU, Thorgeirsson SS. SnapShot: Hepatocellular carcinoma. Cancer Cell 2014;25:550. e551.
[38] Raggi C, Factor VM, Seo D, Holczbauer A, Gillen MC, Marquardt JU, et al. Epigenetic reprogramming modulates malignant properties of human liver cancer. Hepatology 2014;59:2251–62.
[39] Zeng SS, Yamashita T, Kondo M, Nio K, Hayashi T, Hara Y, et al. The transcription factor SALL4 regulates stemness of EpCAM-positive hepatocellular carcinoma. J Hepatol 2014;60:127–34.
[40] Haraguchi N, Ishii H, Mimori K, Tanaka F, Ohkuma M, Kim HM, et al. CD13 is a therapeutic target in human liver cancer stem cells. J Clin Invest 2010;120:3326–39.

[41] Ogawa K, Tanaka S, Matsumura S, Murakata A, Ban D, Ochiai T, et al. EpCAM-targeted therapy for human hepatocellular carcinoma. Ann Surg Oncol 2014;21:1314–22.
[42] Smith LM, Nesterova A, Ryan MC, Duniho S, Jonas M, Anderson M, et al. CD133/prominin-1 is a potential therapeutic target for antibody-drug conjugates in hepatocellular and gastric cancers. Br J Cancer 2008;99:100–9.
[43] Yin C, Lin Y, Zhang X, Chen YX, Zeng X, Yue HY, et al. Differentiation therapy of hepatocellular carcinoma in mice with recombinant adenovirus carrying hepatocyte nuclear factor-4alpha gene. Hepatology 2008;48:1528–39.
[44] Yamashita T, Honda M, Nio K, Nakamoto Y, Yamashita T, Takamura H, et al. Oncostatin M renders epithelial cell adhesion molecule-positive liver cancer stem cells sensitive to 5-fluorouracil by inducing hepatocytic differentiation. Cancer Res 2010;70:4687–97.
[45] Gilbertson RJ, Rich JN. Making a tumour's bed: glioblastoma stem cells and the vascular niche. Nat Rev Cancer 2007;7:733–6.
[46] Wilhelm SM, Carter C, Tang L, Wilkie D, McNabola A, Rong H, et al. BAY 43-9006 exhibits broad spectrum oral antitumor activity and targets the RAF/MEK/ERK pathway and receptor tyrosine kinases involved in tumor progression and angiogenesis. Cancer Res 2004;64:7099–109.
[47] DeOliveira ML, Cunningham SC, Cameron JL, Kamangar F, Winter JM, Lillemoe KD, et al. Cholangiocarcinoma: thirty-one-year experience with 564 patients at a single institution. Ann Surg 2007;245:755–62.
[48] Khan SA, Emadossadaty S, Ladep NG, Thomas HC, Elliott P, Taylor-Robinson SD, et al. Rising trends in cholangiocarcinoma: is the ICD classification system misleading us? J Hepatol 2012;56:848–54.
[49] McLean L, Patel T. Racial and ethnic variations in the epidemiology of intrahepatic cholangiocarcinoma in the United States. Liver Int 2006;26:1047–53.
[50] Welzel TM, Mellemkjaer L, Gloria G, Sakoda LC, Hsing AW, El Ghormli L, et al. Risk factors for intrahepatic cholangiocarcinoma in a low-risk population: a nationwide case-control study. Int J Cancer 2007;120:638–41.
[51] Lee TY, Lee SS, Jung SW, Jeon SH, Yun SC, Oh HC, et al. Hepatitis B virus infection and intrahepatic cholangiocarcinoma in Korea: a case-control study. Am J Gastroenterol 2008;103:1716–20.
[52] Palmer WC, Patel T. Are common factors involved in the pathogenesis of primary liver cancers? A meta-analysis of risk factors for intrahepatic cholangiocarcinoma. J Hepatol 2012;57:69–76.
[53] Tyson GL, El-Serag HB. Risk factors for cholangiocarcinoma. Hepatology 2011;54:173–84.
[54] Tocchi A, Mazzoni G, Liotta G, Lepre L, Cassini D, Miccini M. Late development of bile duct cancer in patients who had biliary-enteric drainage for benign disease: a follow-up study of more than 1,000 patients. Ann Surg 2001;234:210–4.
[55] Li JS, Han TJ, Jing N, Li L, Zhang XH, Ma FZ, et al. Obesity and the risk of cholangiocarcinoma: a meta-analysis. Tumour Biol 2014;35:6831–8.
[56] Wang M, Xiao J, Shen M, Yahong Y, Tian R, Zhu F, et al. Isolation and characterization of tumorigenic extrahepatic cholangiocarcinoma cells with stem cell-like properties. Int J Cancer 2011;128:72–81.
[57] Shimada M, Sugimoto K, Iwahashi S, Utsunomiya T, Morine Y, Imura S, et al. CD133 expression is a potential prognostic indicator in intrahepatic cholangiocarcinoma. J Gastroenterol 2010;45:896–902.

[58] Fan L, He F, Liu H, Zhu J, Liu Y, Yin Z, et al. CD133: a potential indicator for differentiation and prognosis of human cholangiocarcinoma. BMC Cancer 2011;11:320.
[59] de Jong MC, Nathan H, Sotiropoulos GC, Paul A, Alexandrescu S, Marques H, et al. Intrahepatic cholangiocarcinoma: an international multi-institutional analysis of prognostic factors and lymph node assessment. J Clin Oncol 2011;29:3140–5.
[60] Zaydfudim VM, Rosen CB, Nagorney DM. Hilar cholangiocarcinoma. Surg Oncol Clin N Am 2014;23:247–63.
[61] Fendrich V, Langer P, Celik I, Bartsch DK, Zielke A, Ramaswamy A, et al. An aggressive surgical approach leads to long-term survival in patients with pancreatic endocrine tumors. Ann Surg 2006;244:845–51. discussion 852–3.
[62] Valle J, Wasan H, Palmer DH, Cunningham D, Anthoney A, Maraveyas A, et al. Cisplatin plus gemcitabine versus gemcitabine for biliary tract cancer. N Engl J Med 2010;362:1273–81.
[63] Okusaka T, Nakachi K, Fukutomi A, Mizuno N, Ohkawa S, Funakoshi A, et al. Gemcitabine alone or in combination with cisplatin in patients with biliary tract cancer: a comparative multicentre study in Japan. Br J Cancer 2010;103:469–74.
[64] Jarnagin WR, Ruo L, Little SA, Klimstra D, D'Angelica M, DeMatteo RP, et al. Patterns of initial disease recurrence after resection of gallbladder carcinoma and hilar cholangiocarcinoma: implications for adjuvant therapeutic strategies. Cancer 2003;98:1689–700.
[65] Yamashita T, Ji J, Budhu A, Forgues M, Yang W, Wang HY, et al. EpCAM-positive hepatocellular carcinoma cells are tumor-initiating cells with stem/progenitor cell features. Gastroenterology 2009;136:1012–24.
[66] Nathan H, Pawlik TM, Wolfgang CL, Choti MA, Cameron JL, Schulick RD. Trends in survival after surgery for cholangiocarcinoma: a 30-year population-based SEER database analysis. J Gastrointest Surg 2007;11:1488–96. discussion 1496–7.
[67] Keeratichamroen S, Leelawat K, Thongtawee T, Narong S, Aegem U, Tujinda S, et al. Expression of CD24 in cholangiocarcinoma cells is associated with disease progression and reduced patient survival. Int J Oncol 2011;39:873–81.
[68] Oikawa T. Cancer stem cells and their cellular origins in primary liver and biliary tract cancers. Hepatology 2016;64:645–51.
[69] Marquardt JU, Andersen JB, Thorgeirsson SS. Functional and genetic deconstruction of the cellular origin in liver cancer. Nat Rev Cancer 2015;15:653–67.
[70] Sell S. Cellular origin of hepatocellular carcinomas. Semin Cell Dev Biol 2002;13:419–24.
[71] Kuwahara R, Kofman AV, Landis CS, Swenson ES, Barendswaard E, Theise ND. The hepatic stem cell niche: identification by label-retaining cell assay. Hepatology 2008;47:1994–2002.
[72] Lanzoni G, Oikawa T, Wang Y, Cui CB, Carpino G, Cardinale V, et al. Concise review: clinical programs of stem cell therapies for liver and pancreas. Stem Cells 2013;31:2047–60.
[73] Cardinale V, Wang Y, Carpino G, Cui CB, Gatto M, Rossi M, et al. Multipotent stem/progenitor cells in human biliary tree give rise to hepatocytes, cholangiocytes, and pancreatic islets. Hepatology 2011;54:2159–72.
[74] Matter MS, Marquardt JU, Andersen JB, Quintavalle C, Korokhov N, Stauffer JK, et al. Oncogenic driver genes and the inflammatory microenvironment dictate liver tumor phenotype. Hepatology 2016;63:1888–99.
[75] Plaks V, Kong N, Werb Z. The cancer stem cell niche: how essential is the niche in regulating stemness of tumor cells? Cell Stem Cell 2015;16:225–38.

FURTHER READING

[76] Suzuki A, Zheng Y, Kondo R, Kusakabe M, Takada Y, Fukao K, et al. Flow-cytometric separation and enrichment of hepatic progenitor cells in the developing mouse liver. Hepatology 2000;32:1230–9.
[77] Minguet S, Cortegano I, Gonzalo P, Martinez-Marin JA, de Andres B, Salas C, et al. A population of c-Kit(low)(CD45/TER119)- hepatic cell progenitors of 11-day postcoitus mouse embryo liver reconstitutes cell-depleted liver organoids. J Clin Invest 2003;112:1152–63.
[78] Suzuki A, Iwama A, Miyashita H, Nakauchi H, Taniguchi H. Role for growth factors and extracellular matrix in controlling differentiation of prospectively isolated hepatic stem cells. Development 2003;130:2513–24.
[79] Kakinuma S, Ohta H, Kamiya A, Yamazaki Y, Oikawa T, Okada K, et al. Analyses of cell surface molecules on hepatic stem/progenitor cells in mouse fetal liver. J Hepatol 2009;51:127–38.
[80] Tanimizu N, Nishikawa M, Saito H, Tsujimura T, Miyajima A. Isolation of hepatoblasts based on the expression of Dlk/Pref-1. J Cell Sci 2003;116:1775–86.
[81] Tanaka M, Okabe M, Suzuki K, Kamiya Y, Tsukahara Y, Saito S, et al. Mouse hepatoblasts at distinct developmental stages are characterized by expression of EpCAM and DLK1: drastic change of EpCAM expression during liver development. Mech Dev 2009;126:665–76.
[82] Wang B, Zhao L, Fish M, Logan CY, Nusse R. Self-renewing diploid Axin2(+) cells fuel homeostatic renewal of the liver. Nature 2015;524:180–5.
[83] Okabe M, Tsukahara Y, Tanaka M, Suzuki K, Saito S, Kamiya Y, et al. Potential hepatic stem cells reside in EpCAM+ cells of normal and injured mouse liver. Development 2009;136:1951–60.
[84] Yovchev MI, Grozdanov PN, Zhou H, Racherla H, Guha C, Dabeva MD. Identification of adult hepatic progenitor cells capable of repopulating injured rat liver. Hepatology 2008;47:636–47.
[85] Suzuki A, Sekiya S, Onishi M, Oshima N, Kiyonari H, Nakauchi H, et al. Flow cytometric isolation and clonal identification of self-renewing bipotent hepatic progenitor cells in adult mouse liver. Hepatology 2008;48:1964–78.
[86] Rountree CB, Barsky L, Ge S, Zhu J, Senadheera S, Crooks GM. A CD133-expressing murine liver oval cell population with bilineage potential. Stem Cells 2007;25:2419–29.
[87] Font-Burgada J, Shalapour S, Ramaswamy S, Hsueh B, Rossell D, Umemura A, et al. Hybrid periportal hepatocytes regenerate the injured liver without giving rise to cancer. Cell 2015;162:766–79.
[88] Dolle L, Best J, Empsen C, Mei J, Van Rossen E, Roelandt P, et al. Successful isolation of liver progenitor cells by aldehyde dehydrogenase activity in naive mice. Hepatology 2012;55:540–52.
[89] O'Brien CA, Kreso A, Jamieson CH. Cancer stem cells and self-renewal. Clin Cancer Res 2010;16:3113–20.
[90] Yoon SK. The biology of cancer stem cells and its clinical implication in hepatocellular carcinoma. Gut Liver 2012;6:29–40.
[91] Yang ZF, Ho DW, Ng MN, Lau CK, Yu WC, Ngai P, et al. Significance of CD90+ cancer stem cells in human liver cancer. Cancer Cell 2008;13:153–66.
[92] Xiang Y, Yang T, Pang BY, Zhu Y, Liu YN. The progress and prospects of putative biomarkers for liver cancer stem cells in hepatocellular carcinoma. Stem Cells Int 2016;2016:7614971.
[93] Zhu Z, Hao X, Yan M, Yao M, Ge C, Gu J, et al. Cancer stem/progenitor cells are highly enriched in CD133+CD44+ population in hepatocellular carcinoma. Int J Cancer 2010;126:2067–78.

[94] Schmelzer E, Reid LM. EpCAM expression in normal, non-pathological tissues. Front Biosci 2008;13:3096–100.
[95] Kim JW, Ye Q, Forgues M, Chen Y, Budhu A, Sime J, et al. Cancer-associated molecular signature in the tissue samples of patients with cirrhosis. Hepatology 2004;39:518–27.
[96] Yamashita T, Budhu A, Forgues M, Wang XW. Activation of hepatic stem cell marker EpCAM by Wnt-beta-catenin signaling in hepatocellular carcinoma. Cancer Res 2007;67:10831–9.
[97] Liu LL, Fu D, Ma Y, Shen XZ. The power and the promise of liver cancer stem cell markers. Stem Cells Dev 2011;20:2023–30.
[98] Petrovic N, Schacke W, Gahagan JR, O'Conor CA, Winnicka B, Conway RE, et al. CD13/APN regulates endothelial invasion and filopodia formation. Blood 2007;110:142–50.
[99] Hashida H, Takabayashi A, Kanai M, Adachi M, Kondo K, Kohno N, et al. Aminopeptidase N is involved in cell motility and angiogenesis: its clinical significance in human colon cancer. Gastroenterology 2002;122:376–86.
[100] Dunsford HA, Sell S. Production of monoclonal antibodies to preneoplastic liver cell populations induced by chemical carcinogens in rats and to transplantable Morris hepatomas. Cancer Res 1989;49:4887–93.
[101] Marrero JA. Hepatocellular carcinoma. Curr Opin Gastroenterol 2005;21:308–12.
[102] Yang W, Yan HX, Chen L, Liu Q, He YQ, Yu LX, et al. Wnt/beta-catenin signaling contributes to activation of normal and tumorigenic liver progenitor cells. Cancer Res 2008;68:4287–95.
[103] Lu JW, Chang JG, Yeh KT, Chen RM, Tsai JJ, Hu RM. Overexpression of Thy1/CD90 in human hepatocellular carcinoma is associated with HBV infection and poor prognosis. Acta Histochem 2011;113:833–8.
[104] Song W, Li H, Tao K, Li R, Song Z, Zhao Q, et al. Expression and clinical significance of the stem cell marker CD133 in hepatocellular carcinoma. Int J Clin Pract 2008;62:1212–8.
[105] Sasaki A, Kamiyama T, Yokoo H, Nakanishi K, Kubota K, Haga H, et al. Cytoplasmic expression of CD133 is an important risk factor for overall survival in hepatocellular carcinoma. Oncol Rep 2010;24:537–46.
[106] Chan AW, Tong JH, Chan SL, Lai PB, To KF. Expression of stemness markers (CD133 and EpCAM) in prognostication of hepatocellular carcinoma. Histopathology 2014;64:935–50.
[107] van Zijl F, Zulehner G, Petz M, Schneller D, Kornauth C, Hau M, et al. Epithelial-mesenchymal transition in hepatocellular carcinoma. Future Oncol 2009;5:1169–79.
[108] Hirohashi K, Tanaka H, Kanazawa A, Kubo S, Ohno K, Tsukamoto T, et al. Living-related liver transplantation in a patient with end-stage hepatolithiasis and a biliary-bronchial fistula. Hepatogastroenterology 2004;51:822–4.
[109] Yang XR, Xu Y, Yu B, Zhou J, Li JC, Qiu SJ, et al. CD24 is a novel predictor for poor prognosis of hepatocellular carcinoma after surgery. Clin Cancer Res 2009;15:5518–27.
[110] Riener MO, Vogetseder A, Pestalozzi BC, Clavien PA, Probst-Hensch N, Kristiansen G, et al. Cell adhesion molecules P-cadherin and CD24 are markers for carcinoma and dysplasia in the biliary tract. Hum Pathol 2010;41:1558–65.
[111] Agrawal S, Kuvshinoff BW, Khoury T, Yu J, Javle MM, LeVea C, et al. CD24 expression is an independent prognostic marker in cholangiocarcinoma. J Gastrointest Surg 2007;11: 445–51.
[112] de Boer CJ, van Krieken JH, Janssen-van Rhijn CM, Litvinov SV. Expression of Ep-CAM in normal, regenerating, metaplastic, and neoplastic liver. J Pathol 1999;188:201–6.

Chapter 14

Liver Cancer Stem Cells

Jin Ding*,**, Wei-Fen Xie*
*Changzheng Hospital, Second Military Medical University, Shanghai, China; **The Eastern Hepatobiliary Surgery Hospital, Second Military Medical University, Shanghai, China

1 CHARACTERISTICS OF LIVER CANCER STEM CELLS

1.1 Self-renewal

Liver cancer is a common malignant tumor with extremely poor prognosis [1–4], which is considered to be associated with liver cancer stem cells (LCSCs) [5–7]. Self-renewal is a key feature of stem cells or cancer stem cells and is responsible for the maintenance of stemness and propagation of stem cells. Through asymmetric cell division, stem cells or cancer stem cells maintain their numbers, while their numbers are expanded through symmetric cell division. Compared with stem cells, CSCs have a longer proliferative period and undergo more symmetrical cell divisions, which lead to their enhanced proliferation and aberrant differentiation [1]. During mitosis of LCSCs, the CSC marker CD133 is co-segregated with the template DNA of the daughter CSC, while the other daughter hepatocellular carcinoma (HCC) cell expresses markers of differentiated cells.

1.2 Multiple Differentiation Capacity

Another distinct characteristic of stem cells or CSCs is their multiple differentiation capacity. It is accepted that the transformation of liver progenitor cells (LPCs) to LCSCs can give rise to combined HCC–cholangiocellular carcinoma (HCC-CCA). HCC-CCA contains features of both hepatocellular carcinoma and cholangiocellular carcinoma, indicating the multiple differentiation capacity of LCSCs [2]. Accumulating studies have demonstrated that many cases of HCCs are histologically heterogeneous and express stem cell markers, suggesting a liver progenitor/stem cell origin [3, 4]. Yamashita et al. proposed the utility of EpCAM for the stratification of liver cancer and found that EpCAM$^+$ HCC cells exhibit CSC-like features, including self-renewal and differentiation potential [5].

1.3 Constant Latency and Chemoresistance

It has been accepted that current conventional chemotherapies predominantly target the rapidly proliferating liver cancer cells but exerts little effect on the relatively quiescent LCSC population. The existence of LCSCs may be largely responsible for liver cancer relapse after conventional chemotherapy. In fact, LCSCs expressing CD133, EpCAM, CD24, or side population (SP) have been reported to be responsible for chemoresistance of liver cancer through various signaling pathways. We and others have demonstrated that numerous stemness-associated transcription factors are involved in the propagation and chemoresistance of LCSCs [6]. Chemotherapy surviving drug-resistant LCSCs subsequently initiate the recurrence of liver cancer [7].

1.4 Heterogeneity of Liver Cancer Stem Cells

One other interesting and important property of LCSCs is their evident heterogeneity. Distinct LCSC populations exhibit different tumorigenic/metastatic potentials and chemoresistance when isolated using different LCSC markers. Yamashita et al. determined the expression of LCSC markers including CD133, EpCAM, and CD90 in six HCC cell lines [5]. They detected a subpopulation of EpCAM$^+$ and CD133$^+$ LCSCs in AFP$^+$ HCC cell lines, but no CD90$^+$ cells were detected. Interestingly, in AFP$^-$ HCC cell lines, they detected a subpopulation of CD90$^+$ cells but not EpCAM$^+$ or CD133$^+$ cells [5]. In a certain HCC cell line, the proportion of different LCSC populations could be largely different. For example, we found that EpCAM$^+$ cells in LM3 cells accounted for approximately 20%–40% of the cells, but the proportion of CD133$^+$ LCSCs was merely approximately 2%. These observations suggest that CSC markers are not equally expressed in HCC cells. Previous reports showed that the expression patterns of LCSC markers could be attributed to the heterogeneity of activated stemness-associated signaling pathways and stem cell-associated transcription factors [8].

1.5 Evasion of Immune Clearance

Recently, Calderaro et al. reported the expression of PD-L1 in a limited subset of tumors and observed a correlation between PD-L1 expression and HCC progenitor subtype. Expression of CK19 was detected in >5% of this subset of tumors, and these liver cancers were considered to be initiated by LCSCs derived from the transformed LPCs. Their data indicated that the expression of PD-L1 might have contributed to the aggressive behavior by inhibiting antitumor immunity and was correlated with poor prognosis of patients [9]. In another recent study, Zhao et al. showed that LCSCs could recruit M2 macrophages by enhancing the transcription of CCL2/CSF1 through the activation of YAP, which facilitated the liver cancer cells to evade immune clearance [10].

1.6 Metastatic Potential

It has been accepted that migrating cancer cells exhibit certain properties of CSCs such as quiescence and anchorage-independent cell proliferation. The acquisition of migratory capacity by CSCs during the epithelial-mesenchymal transition (EMT) process may facilitate their migration and invasion from the primary tumor [11, 12]. Recently, Lee et al. reported that hepatic transmembrane 4 L six family member 5 (TM4SF5) could enhance the self-renewing ability and circulating tumor cell (CTC) properties during metastasis of HCC [13]. Consistently, metastasis-initiating cells were detected in both primary tumors and metastatic nodules as well as in metastatic cancer cell lines.

2 IDENTIFICATION AND ISOLATION OF LIVER CANCER STEM CELLS

2.1 Surface Biomarkers

LCSCs can be identified and isolated by their distinct cell surface markers (Table 14.1). To date, there have been only a few cell surface proteins that have been established as LCSC markers, including CD133, CD90, CD44, OV6, EpCAM, CD13, CD24, DLK1, $\alpha2\delta1$, ICAM-1, CD47, and Lgr5 (Figure 14.1). Ma et al. first reported CD133 as an LCSC marker and proposed that CD133 could be involved in the self-renewal of LCSCs [14, 15]. Yang et al. were the first to report CD90 as an LCSC marker and elucidated the CSC-like properties of $CD90^+$ LCSCs using liver cancer cell lines [16]. CD44 itself is not an LCSC marker, but it has been regarded as an LCSC marker in combination with CD133 [17] or CD90 [18]. Yang et al. reported that OV6 could be another LCSC marker because $OV6^+$ HCC cells possess greater tumorigenicity as well as chemoresistance than $OV6^-$ HCC cells [19]. Yamashita et al. reported EpCAM as an LCSC marker, and their study showed that $EpCAM^+$ HCC cells exhibit a potent ability for self-renewal and tumor initiation [5]. We and others have observed $EpCAM^+$ cells in the cirrhotic livers of rats and patients, and these cells might be the origin of LCSCs and could be utilized for the early diagnosis of HCC. Haraguchi et al. proposed CD13 as an LCSC marker and observed that $CD13^+$ cells were enriched in SP of HCC cells. Moreover, their study demonstrated that most of the $CD13^+$ HCC cells were maintained in G1/G0 phase of the cell cycle, consistent with the quiescent feature of CSCs [20]. Han and co-workers found that DLK^+ HCC cells also behaved as LCSCs exhibiting potent chemoresistance and self-renewal [21]. CD24 and ICAM-1 were both accepted as functional LCSC markers that could promote HCC development by upregulating the expression of Nanog, which is a key stemness-associated transcription factor [22, 23]. Zhao et al. sorted liver CSCs using a new antibody (1B50-1) recognizing isoform 5 of the cell surface calcium channel $\alpha2\delta1$ subunit. They also verified that $\alpha2\delta1$ potentiated liver CSCs by activating pro-survival pathways via a calcium-dependent

TABLE 14.1 Information List of Identified Liver Cancer Stem Cells

Markers	Author	Year	Cell line/sample	Function	Mechanism
CD133	S. Ma	2007	Cell line/patient HCC	Self-renewal, tumorigenicity, chemoresistance, invasiveness	Neurotensin/IL-8/CXCL1 signaling
EpCAM	T. Yamashita	2009	Cell line/patient HCC	Self-renewal, invasiveness, tumor formation	Activation of Wnt signaling
CD24	T.K. Lee	2011	Cell line/patient HCC	Tumor formation, self-renewal, chemoresistance, metastasis	STAT3-mediated Nanog upregulation
OV6	W. Yang	2008	Cell line/patient HCC	Tumorigenicity, chemoresistance	Activation of Wnt signaling
CD90	Z. Yang	2008	Cell line/patient HCC	Tumor formation, metastasis	Not clear
CD13	N. Haraguchi	2010	Cell line/patient HCC	Self-renewal, cell proliferation, tumor formation	Reduction of ROS-induced DNA damage
CD47	Irene Oi Lin Ng	2014	Cell line/patient HCC	Tumor initiation, self-renewal, chemoresistance, invasiveness	Regulating liver TICs through the NF-kB/CTSS/PAR2 loop
ICAM-1	S.R. Liu	2013	Cell line/patient HCC	Self-renewal, tumor formation	Transcriptionally regulating Nanog expression
DLK1	X. Xu	2012	Cell line	Self-renewal, tumorigenicity, chemoresistance,	Not clear
CD44	Z. Yang	2008	Cell line/patient HCC	Tumorigenicity, chemoresistance, metastasis	Regulation of redox status through xCT
α2δ1	Z.Q. Zhang	2013	Cell line/patient HCC	Self-renewal, tumor formation, cell proliferation	Regulated by calcium influx
Lgr5	B. Wang	2015	Cell line/patient HCC	Self-renewal, chemoresistance	Through Prickle1 and APC-enhanced β-catenin activity

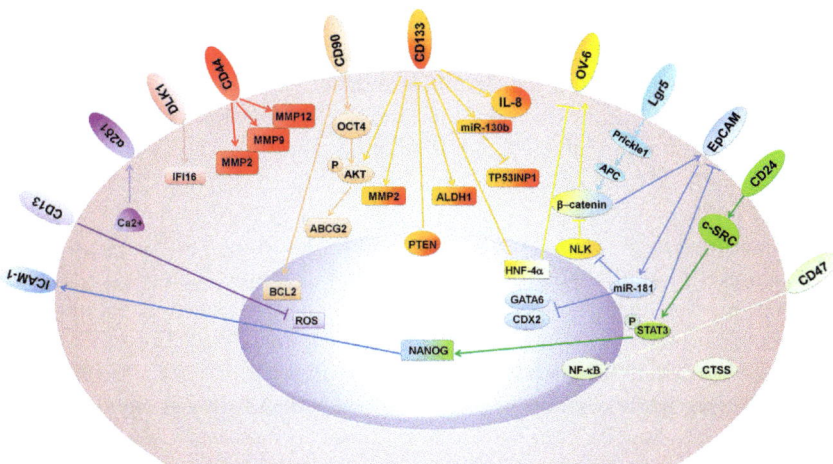

FIGURE 14.1 Schematic presentation of liver cancer stem cell markers with mechanistic correlation.

mechanism. Lee et al. reported CD47 as another LCSC marker preferentially expressed in LCSCs compared with liver cancer cells. Their data showed that CD47$^+$ HCC cells were important for initiation and metastasis of HCC in vivo and correlated with patient survival [25]. Lei et al. reported that leucine-rich repeat-containing G-protein-coupled receptor 5 (Lgr5) was an LCSC marker. They found that Lgr5 expression was correlated with the progression of HCC, and Lgr5$^+$ HCC cells exhibited properties of LCSCs [26]. Nevertheless, most of the LCSC markers identified so far are also markers for liver stem/progenitor cells. For example, CD133 is a biomarker of normal stem cells such as hematopoietic stem cells and neuronal stem cells. Therefore, most of the LCSC markers can also be detected in LPCs.

2.2 CSC Function-based Isolation

Aldehyde dehydrogenase (ALDH) is a widely expressed enzyme required for the oxidation of intracellular aldehydes. There has been experimental evidence showing that sorted cancer cells with potent ALDH activity possess CSC-like characteristics [27, 28]. Ma et al. reported that CD133$^+$ALDH$^+$ LCSCs exhibit a greater tumor initiating ability than CD133$^+$ALDH$^-$ cells [28].

Chiba et al. reported that SP cells sorted from HCC cells possessed potent tumorigenicity and CSC properties [29]. Their study also demonstrated that Bmi1, an essential stemness-associated transcription factor, was responsible for the self-renewal of SP cells [30]. SP cells possess the capability to efflux Hoechst 33342 dye. Based on this unique feature, LCSCs could be sorted by FACS. The active efflux of intracellular drugs is consistent with the chemoresistance properties of CSCs.

Sphere formation is a particular characteristic of stem cells, which is also widely used to enrich CSCs. For those tumors with no specific CSC marker identified, sphere formation is an ideal approach to obtain CSCs. We and others showed that non-adherent spheres from human HCC cell lines cultured in stem cell culture medium exhibit HCC initiating capability [31, 32]. Since cancer cells differentiated from CSCs also exist in the cultured spheres, the application of sphere culture to enrich for CSCs has been widely used in numerous CSC-related studies.

Asymmetric cell division is a distinct feature of stem cells as well as CSCs. The identification and isolation of lung CSCs could be achieved by a qualitative analysis of chromosomal co-segregation during mitosis [33]. For LCSCs, the expression of CSC markers has been found to co-segregate with the template DNA, while the expression of markers of differentiated cells has been observed in the other daughter HCC cells. These findings implied that LCSCs could be isolated based on their characteristic asymmetric cell division.

Recently, Heeschen and co-workers reported the intrinsic autofluorescence of CSCs in various epithelial tumors including HCC. By using this distinct feature of CSCs, they established a new functional approach to identify CSCs and successfully isolated them. Later, they found that the autofluorescence was due to the accumulation of the fluorescent vitamin riboflavin in ABCG2-coated vesicles, which are mainly distributed in the cytoplasm of cancer stem cells [34]. LCSCs could thus be isolated by FACS due to their intrinsic autofluorescence. Muramatsu et al. observed that LCSCs display a lower proteasome activity and possess reduced level of reactive oxygen species (ROS) in comparison with non-CSC cells through a visualization system [35]. A visualized subset of liver cancer cells exhibited asymmetric cell division and displayed potent tumorigenicity.

3 MOLECULAR MECHANISM UNDERLYING REGULATION OF LCSCs

3.1 Transcription Factors

It has been reported that the maintenance and expansion of CSCs are controlled by unique stemness-associated transcription factors, which include Sox2, Oct4, c-Myc, and Klf4. It has been demonstrated that these transcription factors are pivotal regulators of stem cell pluripotency, and dysregulation of these factors can lead to propagation of CSCs and tumorigenesis [36]. Meanwhile, an increasing amount of evidence has indicated that the expression of these transcription factors in HCC is associated with the amount of LCSCs and usually predicts a poor outcome [37–39].

The activation of c-Myc is a common feature in human malignancies, especially in HCC. LCSCs have been detected in Myc-driven rather than Akt/Ras-driven HCCs in mice, indicating the important role of Myc in the maintenance of LCSCs [40]. Thorgeirsson et al. revealed that the activation of c-Myc was indispensable in the process of reprogramming differentiated hepatocytes into LCSCs [41]. Recently, they revealed a close correlation between c-Myc activity and properties of CSCs, indicating that the expression of c-Myc has a distinct impact on characteristics of LCSCs [42].

Nanog is a stemness-associated transcriptional regulator that is activated in embryonic stem cells, and it has been implicated in the self-renewal of CD24$^+$ or CD133$^+$ LCSCs [22, 43]. In addition, hypomethylated Nanog promoter was observed in CD133$^+$ LCSCs, and Nanog overexpression elevated the proportion of CD133$^+$ cells [43]. Nanog is also involved in promoting chemoresistance and invasion in HCC [44]. Shan et al. found that the inhibition of Nanog in Nanog$^+$ LCSCs attenuated the expression of stemness-associated genes and augmented the expression of hepatocyte-related genes. Forced expression of Nanog in Nanog$^-$ hepatoma cells re-established the self-renewal ability of these cells [45]. Furthermore, there are also studies implicating the stemness-associated factors Sox2 [6], Oct4 [46], and Lin28 [47] in the regulation of LCSCs.

SALL4 is encoded by the human homologue of the Drosophila homeotic gene spalt, and it is a C2H2 zinc-finger transcription factor [48]. SALL4 has been implicated in the regulation of stemness in embryonic cells and hematopoietic stem cells, and elevated SALL4 expression has also been detected in stem cell-like HCCs. A recent report on the analysis of large cohorts indicated the importance of SALL4 as an indicator of LPC-derived HCCs with aggressive phenotypes. Microarray analysis revealed that SALL4$^+$ HCCs displayed a progenitor cell-associated expression pattern [37]. Zeng et al. showed that the activation of SALL4 enhanced LCSC-related gene expression and reinforced the self-renewal and invasive capability of hepatoma cells. Additionally, the expression of SALL4 correlated with the activity of histone deacetylase (HDAC), and an HDAC inhibitor significantly prevented the propagation of the SALL4$^+$ HCC cells [49]. Moreover, Marquardt et al. revealed that perturbation of NF-κB decreases the expansion of LCSCs and suggested that the dual suppression of NF-κB and HDAC pathway might be a feasible approach for HCC therapy [50].

3.2 Non-coding RNAs

Non-coding RNAs (ncRNAs) are modulatory RNAs that have been widely investigated in cancer research in recent years. A growing number of ncRNAs have been shown to display aberrant expression patterns in many human malignancies including HCC. It is important to screen for and identify ncRNAs associated with the regulation of LCSCs for early detection and therapy of HCC.

3.2.1 MicroRNAs

MicroRNAs (miRNAs) are small ncRNAs that regulate gene expression at the post-transcriptional level. A growing number of studies have demonstrated important role of miRNAs in the regulation of distinct biological processes, including stemness and differentiation. Accumulating data have implicated miRNAs in the modulation of CSCs including LCSCs. Wang et al. revealed that the expression of miR-181 was elevated in sorted EpCAM$^+$ LCSCs. miR-181 facilitated the self-renewal of EpCAM$^+$ LCSCs via targeting NLK to augment the activity of β-catenin and by targeting CDX2 and GATA6 to repress cell differentiation [51]. Through miRNA screening analysis, Ma et al. have revealed the preferential expression of miR-130b in CD133$^+$ LCSCs. Overexpression of miR-130b enhanced the self-renewal capability of LCSCs and augmented the onset and chemoresistance of liver cancer. Furthermore, miR-130b directly targets the tumor suppressor gene TP53INP1 to regulate self-renewal and tumorigenicity of CD133$^+$ LCSCs [52]. Additionally, the levels of miR-150 were significantly decreased in CD133$^+$ LCSCs in comparison with CD133$^-$ cells. Forced expression of miR-150 significantly decreased the size of the CD133$^+$ subpopulation and prevented self-renewal by suppressing c-Myc [53]. Chai et al. showed that miR-142-3p directly regulated CD133 to prevent its function in the regulation of LCSCs. Recently, our lab found that the expression of miR-429 was elevated in primary LCSCs sorted from clinical HCC samples, and miR-429 enhanced features of LCSCs by inhibiting Rb binding protein 4 (RBBP4)/E2F transcription factor 1 (E2F1) signaling [54].

3.2.2 lncRNAs

In the past decade, substantial effort has been made to investigate the function of long ncRNAs (lncRNAs) in human, including HCC [55]. Dysregulated expression of certain lncRNAs was implicated in the recurrence and metastasis of HCC [56, 57]. The underlying mechanism of the function lncRNAs remains the least clarified aspect of lncRNA studies. Recently, Wang et al. carried out a transcriptome microarray analysis and identified lnc-TCF7, a lncRNA that is highly expressed in LCSCs. Functional studies revealed that lnc-TCF7 was essential for the self-renewal and expansion of LCSCs. Regarding the mechanism, lnc-TCF7 was found to modulate the expression of TCF7 by recruiting the SWI/SNF complex to the promoter of TCF7, subsequently promoting the Wnt signaling [58]. Recently, Sun et al. found that the level of lnc-DANCR was elevated in LCSCs and was associated with patient outcome. They also demonstrated that lnc-DANCR facilitated the self-renewal of LCSCs by promoting the activation of β-catenin in a microRNA-dependent manner [59]. Our group also discovered a lncRNA named lnc-DILC that is downregulated in LCSCs. Depletion of lnc-DILC significantly promoted the expansion of LCSCs and facilitated the initiation and progression of HCC, whereas the ectopic expression of lnc-DILC dramatically inhibited the expansion of LCSCs. lnc-DILC has been shown to mediate the crosstalk between TNF-α/NF-κB signaling and autocrine IL-6/STAT3 cascade and to connect hepatic inflammation with expansion of LCSCs, suggesting that lnc-DILC could not only be a potential prognostic biomarker but also be a possible therapeutic target against LCSCs [32].

3.3 Epigenetic Factors

It is well accepted that epigenetic regulation to give rise to profound epigenomic changes is important, and epigenetic regulation includes DNA methylation, histone modifications, and chromatin remodeling without alteration of the genome from the zygote to somatic tissues. Aberrant epigenetic regulation has been considered an essential mechanism involved in tumorigenesis and expansion of CSCs [60].

It was reported that the expression of histone deacetylase 3 (HDAC3) was elevated in clinical specimens of HCC and was associated with poor outcomes of patients [61]. Marquardt et al. found that inhibiting DNA methylation in human hepatoma cells could result in an increase in the number of SP cells [62]. HDAC SIRT1 has been revealed to be essential for the self-renewal of LCSCs, and it has been shown to transcriptionally regulate SOX2 via DNA methylation [63]. By utilizing shRNA and pharmacological inhibitors, Chiba et al. elucidated that EZH2, a core component of PRC2, regulates the maintenance of LCSCs, and therefore inhibition of EZH2 is an ideal therapeutic strategy for the elimination of LCSCs [64]. Moreover, CHD1L, a chromatin remodeling factor, has been correlated with malignant phenotypes of HCC, and it has been shown to maintain chromatins at the promoter regions of genes involved in self-renewal and differentiation regulation of HCC in an open configuration [65]. Bmi-1 is a member of the polycomb group gene (PcG) family, which are highly evolutionary conserved. The PcG family is involved in the regulation of self-renewal of embryonic and adult stem cells. Upregulation of Bmi-1 has been observed in sorted $CD133^+$ [14], $CD90^+$ [18], $EpCAM^+$ [5], and $CD24^+$ [22] LCSCs, demonstrating the pivotal role of Bmi-1 in LCSCs. Moreover, Chiba et al. have reported the preferential expression of Bmi-1 in the SP of hepatoma cells compared with the non-SP cells [30]. Furthermore, overexpression of Bmi-1 could lead to liver propagation of stem cells and carcinogenesis in both p14/Arf-dependent and p14/Arf-independent manners [66].

3.4 Stemness-related Cascades

Wnt/β-catenin signaling is essential in development and differentiation. Using a lineage tracing mice model, Wang et al. identified a Wnt-responsive cell population adjacent to central vein in liver lobule. These diploid cells proliferate over the lifespan and can produce mature polyploid hepatocytes, which populate the entire liver lobule [67]. Lgr5 was reported as a Wnt target gene in numerous cell types. Huch et al. showed that only one single $Lgr5^+$ liver stem cell can expand and gradually form epithelial organoids in vitro and even could be differentiated into functional hepatocytes in vivo [68]. In another study, they also emphasized that Wnt signals were essential for long-term expansion of liver stem cell [69]. Aberrant hyperactivation of Wnt/β-catenin signaling has been observed in 20%–40% of human HCC [19]. Recent findings by using sequencing technologies have further confirmed aberrant somatic mutations in genes involved in Wnt/β-catenin signaling, including mutations in β-catenin and axin, which are common events in liver cancer [70, 71]. Mokkapati et al. found that Wnt/β-catenin signaling controls the initiation of HCC by using liver stem/progenitor cell-specific β-catenin-overexpressing transgenic mice. In addition, hepatomas display activated Ras/Raf/MAPK and PI3K/AKT/mTOR signaling and upregulated expression LCSC markers [72]. An increasing number of studies have demonstrated that Wnt/β-catenin signaling is hyperactivated in distinct LCSC populations including EpCAM$^+$ [5], CD133$^+$ [14], and OV6$^+$ [19] LCSCs. Among the recently identified LCSC markers, EpCAM was confirmed to be a direct downstream target of β-catenin. Consistently, Wang et al. found that Wnt/β-catenin signaling regulated the highly tumorigenic and invasive EpCAM and AFP dual positive LCSCs [5]. Moreover, our previous study also revealed that Wnt/β-catenin signaling was involved in the activation of transformed LPCs [19]. In our recent study, we found that Shp2, a non-receptor protein tyrosine phosphatase that contains two Src-homology 2 domains, enhanced the accumulation of β-catenin through the canonical Wnt/β-catenin pathway in LCSCs, leading to the enhanced self-renewal of LCSCs [73].

A growing number of studies have demonstrated that Notch is pivotal for the propagation and maintenance of CSCs in various cancers including HCC. Several studies showed that the liver-specific activation of Notch was able to reproduce the continuous stages of clinical hepatocarcinogenesis and correlated with HCC metastasis [74, 75]. Consistently, upregulated expression of Notch and Notch ligand was observed in clinical HCC samples and in CD133$^+$ LCSCs [14, 76]. Moreover, RUNX3 was found to suppress hepatocarcinogenesis by preventing Jagged-mediated expansion of CSCs [77].

Activation of IL-6/STAT3 signaling is essential for self-renewal and maintenance of stem cells [78]. In addition, IL-6/STAT3 signaling is implicated in liver inflammation and regeneration and shown to collaborate with TGF-β signaling to modulate the expansion of LCSCs [79, 80]. Mishra et al. found that hyperactivation of STAT3 in liver stem cells transformed them to HCC in an IL-6-dependent manner [14]. These studies were further validated by our lab; we revealed that hepatitis B virus X protein (HBx) activated IL-6/STAT3 signaling, inducing the malignant transformation of LPCs, which in turn promoted murine hepatocarcinogenesis [81]. In addition, a recent report showed that tumor-associated macrophages (TAMs) secreted IL-6, promoting the propagation of CD44$^+$ LCSCs and carcinogenesis via STAT3 signaling [82]. Ng et al. revealed that STAT3 activation increased the expression of Nanog, thereby enhancing the self-renewal of CD24$^+$ LCSCs [22]. In our unpublished data, we demonstrated that macrophage-secreted TNF-α but not IL-6 could enhance the self-renewal of LPCs through the TNFR1/Src/STAT3 pathway, which could induce the transformation of LPCs into LCSCs.

It is well accepted that the Hippo pathway exerts cell-autonomous functions in development and carcinogenesis [83]. In a recent study, Yimlamai et al.

reported that the activity of Hippo-pathway could influence liver cell fate. Acute inactivation of Hippo signaling in vivo is sufficient to de-differentiate the adult hepatocytes into progenitor cell-like cells. These hepatocyte-derived progenitor cells exhibited self-renewal and engraftment capacity at the single cell level [84]. Moreover, Lee et al. found that the mammalian Hippo–Salvador pathway could restrict the proliferation of liver oval cells and thus controlled liver size in development and prevents the occurrence of oval cell-derived tumors [85]. It has been reported that dysregulation of Hippo signaling or its downstream effector Yap/Taz is involved in various human diseases, including HCC [86]. In a previous study, Zhou et al. showed that Mst1 and Mst2 protein kinases, the mammalian Hippo orthologs, could phosphorylate Yap1 and thereby suppress its oncogenic activity in liver. Loss of Mst1 and Mst2 was sufficient to initiate hepatocyte overproliferation and HCC development [87]. A previous study has demonstrated that YAP can dedifferentiate hepatocytes into self-renewing progenitor cells, which could then differentiate and restore damaged livers. Recently, Zhao et al. found that activation of YAP efficiently prevents senescence and produces proliferative progenitor cell-like LCSCs [10].

3.5 Microenvironment

The tumor microenvironment/niche is complex, and it contains stromal cells, immune cells, endothelial cells, extracellular matrix (ECM) components, and various cytokines. CSCs have been speculated to reside in a unique niche. The niche maintains the CSCs in an undifferentiated state and enhances their self-renewing capabilities [88]. A growing amount of evidence has demonstrated that CSCs secrete various cytokines and growth factors into the niche to stimulate their self-renewal, more importantly to promote angiogenesis and recruit other stromal cells and immune cells that produce additional cytokines to reinforce the growth and chemoresistance of cancer. It has been reported that mesenchymal stem cells (MSCs) secrete CXCL12, IL-6, and IL-8 to support CSCs via the activation of NF-κB, and CSCs can recruit more MSCs by secreting IL-6 [89]. Another study reported that TAMs secreted TNF-α or TGF-β to support CSCs by inducing NF-κB- or TGF-β-dependent EMT [90]. For liver cancer, TAMs secrete TGF-β to facilitate the CSC-like features of Hepa1-6 cells and enhance their invasion [12]. Wan et al. showed that TAM-secreted IL6 facilitates the propagation of $CD44^+$ LCSCs and reinforces carcinogenesis via STAT3 signaling [82]. In addition, hypoxia could promote chemoresistance and radioresistance of CSCs via ROS response and ROS-triggered TGF-β and TNF-α signaling [91]. HGF/c-Met signaling is important in the differentiation of liver progenitors to hepatic/biliary cells [92] and is involved in the maintenance of LCSCs through EMT [93]. Recently, Lee et al. found that LCSCs could be enriched through the paracrine secretion of HGF by cancer-associated fibroblasts (CAFs), which activates FRA1 in an Erk1,2-dependent manner. They demonstrated that CAF-derived HGF plays a pivotal role in the regulation of

LCSCs, such as by increasing tumorigenicity, self-renewal, chemoresistance, and expression of LCSC markers [94]. Therefore, targeting the CSC niche could be a novel and important therapeutic strategy for HCC.

3.6 Metabolism

It is widely accepted that cancer cells display an altered metabolic phenotype, termed aerobic glycolysis, which may sustain the aggressive behavior of cancer cells. It is likely that CSCs might also employ unique metabolic strategies to survive. Recently, Zhang et al. found that low concentrations of glucose promoted the proliferation of LCSCs, which could preferentially take in glucose by the overexpression of glucose transporters (GLUT1 and GLUT3). These results suggested that LCSCs could survive via GLUT-mediated glucose uptake in a nutritionally deficient environment, which further promotes the development of HCC [95]. In addition, Chen et al. elucidated the metabolic reprogramming effect of Nanog in maintaining the properties of LCSCs. Their data showed that Nanog-mediated reprogramming of mitochondrial metabolism could lead to the amplification of oncogenic activity and chemoresistance of LCSCs [43].

4 PUTATIVE ORIGINS OF LIVER CANCER STEM CELLS

An increasing amount of evidence implies that only a subset of liver CSCs harbors genuine tumorigenicity; however, the accurate origin of liver CSCs remains obscure. It has been accepted that all cell types in the hepatic lineage including hepatoblasts, LPCs, hepatocytes, cholangiocytes might acquire stemness and convert into liver CSCs (Figure 14.2). In fact, the heterogenicity of hepatic CSCs is well accepted and may therefore contribute to the observed morphological and biological heterogeneity characteristic of HCC [41].

4.1 Liver Progenitor Cells

LPCs have the bipotential capacity of differentiating into both hepatocytes and cholangiocytes [79]. Hence, the transformation of LPCs might be a possible origin of LCSCs. Many types of liver cancers are induced by long-lasting chronic inflammation accompanied with the process of hepatocyte regeneration, such as HBV/HCV infection, alcohol consumption, or non-alcoholic fatty liver disease. This process might lead to the accumulation of genetic and/or epigenetic changes, which further promote the transformation of LPCs into LCSCs [80, 98]. We previously found that LPCs could be transformed into LCSCs during long-term TGF-β stimulation, which contribute to cirrhosis-elicited hepatocarcinogenesis [97]. We also found that macrophage-secreted TNF-α but not IL-6 could enhance the self-renewal of LPCs through the TNFR1/Src/STAT3 pathway, which could induce the transformation of LPCs into LCSCs [99].

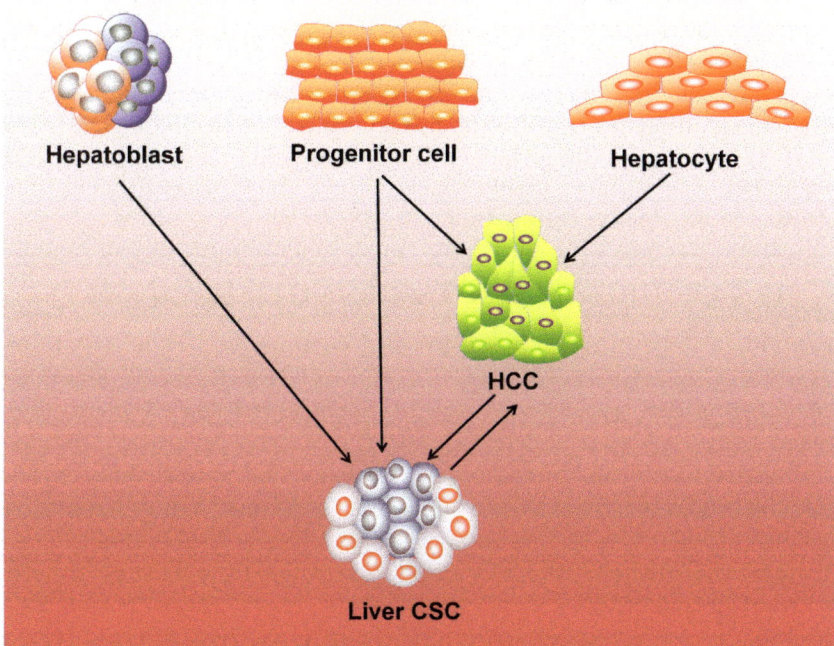

FIGURE 14.2 Putative cell types of origin for liver cancer stem cells.

4.2 Hepatocytes

Many studies have shown that hepatocytes have the ability to regenerate after liver injury and differentiate into both hepatic and biliary cells [98]. Due to genetic or epigenetic alteration, hepatocytes have a high risk of acquiring stemness properties and transform into LCSCs. By transducing the oncogene H-Ras in LPCs, hepatoblasts, and mature hepatocytes, Thorgeirsson et al. found that all of these hepatic cells could acquire CSC-like properties. These data indicated that LCSCs could originate from all types of hepatic cells after genetic/epigenetic alterations [41]. In addition, a few recent studies have shown that liver progenitor/stem cells could be derived from mature hepatocytes [100–102]. Therefore, hepatocytes could be converted into LPCs and then transformed into LCSC at certain condition as described above.

4.3 Liver Cancer Cells

Additionally, non-CSCs might be another source of CSCs via the dedifferentiation process [19]. In fact, an increasing amount of evidence has shown that non-CSCs can dedifferentiate into CSCs in liver cancer. Guan and coworkers demonstrated that the knockdown of ATOH8, a stemness regulator, could induce CD133⁻ liver cancer cells into CD133⁺ liver cancer cells, which facilitated CSC-like properties including self-renewal, differentiation, and chemoresistance [103]. The same group also found that genomic amplification of chromatin remodeling factor CHD1L can promote dedifferentiation of HCC cells and induce stemness properties by opening the chromatin [65]. Moreover, we observed a positive regulation of β-catenin by Shp2 in liver cancer cells, which should be at least partially responsible for the Shp2-mediated dedifferentiation of liver cancer cells into LCSCs [73].

5 THERAPEUTIC STRATEGY TARGETING LIVER CANCER STEM CELLS

Liver CSCs are known to initiate HCC in vivo, drive distant metastasis and tumor relapse after surgical treatment, and maintain resistance to conventional chemotherapy. All these harmful aspects of liver CSCs demonstrate that they are important therapeutic targets in HCC therapy. Given that different liver CSCs have been identified and sorted using their distinct surface markers and that their regulatory mechanisms have been partially clarified (Figure 14.1), the clinical significance of liver CSCs in the diagnosis and treatment of HCC still needs to be explored [104].

5.1 Potential in Diagnosis

As a conventional method, detection of the level of AFP in the blood of HCC patients has been used in HCC diagnosis worldwide. However, tumors found in this way are usually in an advanced stage. An increasing amount of evidence supports the existence of CTCs in many solid tumors, such as colorectal cancer, breast cancer, lung cancer, and liver cancer [105]. Due to the metastasis-initiating property of CSCs, some CTCs might possess CSC-like properties, which could be used to diagnose HCC at a very early stage. The FDA-approved CELLSEARCH system used EpCAM to capture circulating tumor cells and could successfully predict patient prognosis [106]. Yang et al. observed that CD45⁻CD90⁺ LCSCs were present in the blood of HCC patients [18]. Moreover, Liu et al. isolated ICAM-1⁺ CTCs in blood samples from HCC patients [23]. Therefore, the detection of circulating liver cancer cells expressing CSC markers might be a novel diagnostic approach for HCC.

Because of the properties of LCSCs, these cells might have a close correlation with patient prognosis. For example, the overexpression of stemness-associated surface markers, such as CD133, CD90, CD24, and CD44, indicates a worse clinical outcome in HCC patients. In our unpublished data, we found that the classic LPC marker OV6 was expressed in approximately 65% of CK19⁺ HCCs but rarely detected in CK19⁻ tumors. We thereby speculated that CK19⁺OV6⁺ HCCs might at least partially be derived from LPCs. Our data further demonstrated that CK19⁺OV6⁺ patients displayed a shorter time to recurrence and a shorter overall survival time than their counterparts. These findings suggest that CK19 and OV6 may serve as a novel combinatorial biomarker in the molecular classification of liver cancer, which is of clinical significance in the prediction of prognosis and individualized therapy. Clinical investigation revealed that CK19$^+$OV6$^+$ liver cancer patients display a worse prognosis and exhibit superior response to sorafenib therapy. With the application of gene-profiling methods, several stemness-associated genes could be detected to be in HCC, such as Sall4, Sox2, Oct4, and Nanog. These genes reflect the abundance of LCSCs and usually indicate a poor prognosis of HCC [37–39].

Apart from the surface markers of CSCs or stemness-associated genes, several other proteins might also have clinical correlation with HCC. We recently detected a remarkable increase in Shp2 in sorted EpCAM$^+$ or CD133$^+$ LCSCs and in CSC-enriched hepatoma spheroids of patients. HCC patients with low levels of Shp2 benefited from TACE (transcatheter arterial chemoembolization) or sorafenib treatment but the patients with high Shp2 expression did not, indicating the significance of Shp2 in personalized HCC treatment. These data suggest that Shp2 could serve as a biomarker for predicting patient response to chemotherapeutics, which is worthy of extensive clinical investigation [73].

5.2 Eliminating LCSCs

As described above, many stemness-related surface markers, such as CD90, CD133, EpCAM, CD44, $\alpha 2\delta 1$, and CD47, are highly expressed in LCSCs. Therefore, CSC-specific surface marker-targeted therapies might be an effective way to specifically eradicate LCSCs. Yang et al. administered a neutralizing anti-CD44 antibody to prevent CD90$^+$ LCSC-mediated tumor formation and metastasis [16]. Smith et al. reported that the use of an anti-CD133 antibody conjugated with a cytotoxic drug could inhibit the proliferation of CD133$^+$ LCSCs both in vitro and in vivo [107]. Wang et al. demonstrated that knockdown of EpCAM by RNA interference could dramatically inhibit both tumor-initiating and metastatic capacity of LCSCs [5]. Zhao et al. reported that a $\alpha 2\delta 1$-specific antibody (1B50-1) could induce apoptosis of LCSCs [108]. Lee et al. found that blockage of CD47 prevented the growth of HCC and promoted the sensitivity of HCC cells to chemotherapy drugs including sorafenib [109]. Ho et al. reported that the reduction of GEP levels by stable transfection or by blockage using an anti-GEP antibody inhibited cell proliferation and in vivo tumor growth of HCC [110].

5.3 Differentiation Strategy of LCSCs

Differentiation therapy for cancer has shown certain therapeutic potential in many types of cancers, especially leukemia. Differentiation therapy exerts its function by switching a malignant cancer cell into a benign phenotype as well as forcing CSCs to differentiate and lose their self-renewal properties. Similarly, the current treatment for APL, which consists of the combined use of all-trans retinoic acid (ATRA) and anthracycline-based chemotherapy, generally results in a high rate of complete cytogenetic remission (approximately 90%) and an overall survival rate of 80% in the initial phase [111]. It has been speculated that treatment with ATRA can induce apoptosis of leukemic cells due to its effect on the differentiation process [95]. There have been reports that ATRA could induce differentiation of EpCAM$^+$ LCSCs, which could be monitored by a notable reduction in the expression of CSC marker and the induction of expression of liver-specific genes. Furthermore, a more pronounced therapeutic effect could be observed when ATRA was combined with cisplatin compared with cisplatin treatment alone. This finding suggests that the combined treatment with differentiation therapy and conventional chemotherapy might be an effective strategy for HCC treatment [112]. Moreover, As_2O_3 has been shown to be able to induce differentiation of LCSCs by inhibiting the expression of stemness-related genes in vitro and to decrease recurrence and prolong survival of mice in vivo without apparent toxicity. Mechanistic studies have revealed that As_2O_3 promotes differentiation of LCSCs by targeting the expression of GLI1, further suggesting the clinical significance of As_2O_3 in LCSC-targeted therapy [113]. A recent study showed that overexpression of TFPI-2 markedly suppressed proliferation and induced apoptosis of hepatoma cells. Moreover, the expression of LCSC markers was significantly reduced, while the expression of hepatocyte markers was notably elevated in the TFPI-2-overexpressing hepatoma cells, suggesting that TFPI-2 might differentiate LCSCs into hepatocytes [114].

Bone morphogenetic proteins (BMPs), a subgroup of TGF-β superfamily members, play essential roles in embryonic liver development [115]. There have been reports showing that a high dose of exogenous BMP4 could promote the differentiation of CD133$^+$ LCSCs and inhibit their self-renewal, chemoresistance, and tumorigenicity. Oncostatin M (OSM) is a well-known cytokine for its function as an inducer of differentiation of hepatoblasts into hepatocytes via activating STAT3. New findings indicate that OSM can also induce hepatocyte differentiation of EpCAM$^+$ LCSCs in an OSMR-dependent manner. In addition, it has also been demonstrated that a combination of OSM and 5-FU exerts a synergistic effect on eliminating HCC cells by targeting LCSCs as well as non-CSC hepatoma cells [116].

Hepatocyte nuclear factor 4α (HNF-4α) is a liver-enriched transcription factor that plays a dual function in both regulating hepatocyte differentiation and maintaining hepatic function. By introducing HNF-4α into both hepatoma cells and DEN-treated rat liver by an adenoviral-based approach, the CD133$^+$/OV6$^+$ LCSC population and DEN-induced hepatocarcinogenesis could be reduced, while differentiation of LCSCs and non-CSC hepatoma cells could be induced [117, 120]. This suggests that differentiation therapy could be used with other hepatocyte-specific transcription factors, which was validated by further studies. Human fibroblasts can be reprogrammed into hepatocyte-like cells by introducing HNF1α, HNF4α, and FOXA3. A more recent study has found that

HCC cells can be converted into non-tumorigenic hepatocyte-like cells by the simultaneous expression of HNF1α, HNF4α, and FOXA3. In this process, the LCSCs can also be eradicated (unpublished data).

5.4 LCSC-targeted Immunotherapy

CSC-targeted immunotherapy has drawn substantial attention in recent years. Cytotoxic T lymphocyte (CTL)-based immunotherapy has been considered a promising CSC-targeted therapeutic strategy [119, 122]. It has been reported that glioblastoma multiforme (GBM) stem cells not only express high levels of tumor-associated antigens but also major histocompatibility complex molecules. Vaccination with dendritic cells (DCs) and GBM stem cell antigens activated CTLs against GBM stem cells and notably prolonged the survival of animals implanted with GBM stem cell-derived tumors [121]. In another study, by isolating ovarian cancer stem cells from patients and fusing them with DCs, T cells were activated to express elevated levels of IFN-γ and displayed enhanced eradication of the ovarian cancer stem cells [122]. Additionally, the DNAJB8-derived antigenic peptide could induce a DNAJB8-specific CTL response, leading to the specific killing of colorectal cancer stem cells since DNAJB8 is preferentially expressed in colorectal cancer stem cells [12]. Similarly, Annexin A3 (ANXA3) was found to be preferentially expressed in LCSCs and played a pivotal role in the promotion of stem cell-like features via dysregulation of the JNK pathway [124]. The use of ANXA3-transfected dendritic cells could effectively activate T cells, leading to specific killing of LCSCs, which suggested a potential role for ANXA3 in CSC-targeted immunotherapy [125]. Another group developed antibodies specifically targeting EpCAM, but none of those antibodies displayed anticancer activities in HCC. However, one of the anti-EpCAM antibodies, BiTE 1H8/CD3 triggered strong peripheral blood mononuclear cell-associated cytotoxicity particularly in EpCAM$^+$ hepatoma cells and entirely prevented the growth of xenografts derived from these cells in vivo, indicating that 1H8/CD3 may be an ideal agent for HCC therapy. Furthermore, these investigators also revealed that the overexpression of galectin-1 (Gal-1) in HCCs attenuated the 1H8/CD3-induced lymphocytotoxicity, suggesting that Gal-1 may confer resistance of HCC to 1H8/CD3 therapy [126]. A more recent study has demonstrated that blockade of CD47 could lead to a significant increase in the phagocytosis of HCC cell by macrophages as well as increased migration of macrophages into the HCC mass. In addition, in both heterotopic and orthotopic xenograft models of HCC, blockade of CD47 repressed tumor growth, which suggests that targeting CD47 with specific antibodies may exhibit potential immunotherapeutic efficacy in the treatment of HCC [127].

5.5 Targeting the LCSC Microenvironment

The tumor microenvironment has the ability to reinforce stem cell-associated programs and support plasticity of CSCs; therefore, targeting CSC niche factors has been proposed as a more effective modality in cancer therapy than directly targeting the CSCs. In certain malignancies, several attempts to target the CSC niche have already shown certain effects. There have been reports showing that using fibronectin and hyaluronic acid could facilitate the quiescent state of CSCs during treatment with a chemotherapeutic agent. There is also evidence showing that fibronectin receptor α4β1 integrin-specific antibody interrupts the interaction between CSCs and their niche [128]. Another mechanism to interfere with the niche of quiescent and drug-resistant CSCs is to target hypoxia; thus, HIF-1α and HIF-2α have been considered promising targets for the treatment of glioma [129, 132]. Moreover, blockage of the autocrine Akt/HIF-1α/PDGF-BB signaling could not only attenuate chemoresistance of liver cancer cells but also alleviate chemoresistance of LCSCs under hypoxic conditions [131]. It has also been demonstrated that anti-angiogenic therapy targeting VEGF could suppress the formation of tumor vasculature as well as deplete self-renewing CSCs, resulting in tumor regression [132]. In a recent study, activation of CAF-derived HGF and c-Met was efficiently blocked by an HGF-neutralizing antibody or the c-Met inhibitor PHA-665752, thereby suppressing the HGF/c-Met pathway, and their subsequent effects on properties of LCSCs could be abolished [94]. Another study investigated the effect of HSC line LX-2 CM (cultured medium) on the chemosensitivity of hepatoma cells to chemotherapeutic agents. The investigators found that HGF secreted by LX-2 activated the HGF receptor tyrosine kinase mesenchymal-epithelial transition factor (Met) signaling in hepatoma cells and promoted their EMT-associated and LCSC-like features, which could be effectively abolished by HGF neutralizing antibodies [133]. In addition, it was reported that TAM-secreted IL6 promotes the expansion of CD44+ LCSCs and their tumorigenesis. Blocking IL6 signaling with the FDA approved agent tocilizumab, indicated for rheumatoid arthritis therapy, prevents TAM-triggered expansion of CD44+ LCSCs [82]. Moreover, Govaere et al. detected the presence of laminin-332 in the ECM surrounding LCSCs, and laminin-332-targeting treatment could inhibit growth of HCC cells in vitro and in vivo but promote their resistance to sorafenib and doxorubicin, indicating the pivotal role of laminin-332 in supporting the LCSC niche and maintaining the chemoresistance and quiescence of LCSCs [134]. Taken together, the above evidence indicates that targeting the LCSC niche could be another therapeutic strategy for the treatment of HCC.

6 CURRENT CONTROVERSY AND FUTURE DIRECTION

Since liver CSCs have been regarded as key causative factors for recurrence and chemoresistance in HCC, therapeutic approaches are undoubtedly required to eradicate liver CSCs. New studies to identify and characterize liver CSCs have provided numerous strategies for the improvement of early diagnosis of HCC and prediction of patient prognoses. Biomarkers of liver CSCs play important roles in the classification of liver cancer and are critical for the personalized treatment of HCC patients. Due to the plasticity of liver CSCs, heterogeneity is frequently observed in liver cancer. Distinct origins of liver CSCs may contribute to their heterogeneity. Thus, a single marker might only identify a rather restricted subset of LCSCs, and thus an optimized combination of the identified markers is in demand. Emerging studies have utilized different combinations of liver CSC markers to identify the cells and have verified the increase in accuracy of these approaches. For example, we sorted CD133-positive and EpCAM-positive HCC cells and observed enhanced liver CSC-like properties [73]. However, whether all HCCs originate from CSCs or CSCs exist in all HCCs remains obscure. We and others would argue that CSCs could simply be extremely poorly differentiated HCC cells. The initiation of liver cancer by LPCs or transformation of hepatoblasts is consistent with the classical CSC theory. The liver CSCs are the originating cells for HCC, and they persist throughout the development of HCC. However, in the cases of HCC derived from the transformation of hepatocytes or cholangiocytes, whether these HCC cells can acquire stemness or not depends on intrinsic genetic and epigenetic alterations as well as extracellular circumstances. If these cells indeed acquire stemness and dedifferentiate into liver CSCs, this is more likely to be regarded as an outcome of cancer evolution rather than these cells being the originating cells of the development of cancer. Since we cannot obtain direct evidence for the origin of liver CSCs during the development of human HCC and since most of the animal models that are currently available possess certain limitations, this argument will need to be resolved in the future.

It is well accepted that the propagation of LCSCs requires comprehensive regulatory signaling. It has been reported that HBV can enhance stemness of liver CSCs by targeting EpCAM via the activation of β-catenin pathway and upregulation of miR-181 [135]. Additionally, studies have shown that DNA demethylation mediated by HBV is responsible for the upregulation of EpCAM [136]. Most recently, Chen et al. demonstrated that NANOG metabolically reprogrammed liver CSCs by inducing tumorigenic changes in oxidative phosphorylation and fatty acid metabolism, thereby providing insights into the mechanisms involved in the generation of liver CSCs involving the metabolic reprogramming of mitochondrial functions [43]. Moreover, metabolism-dependent epigenetic regulation contributes to the determination of the fate of cancer stem cells [137, 140]. A previous study has shown that the maintenance of CSCs could also be regulated by autophagy via the mediation of IL-6 secretion, implying that autophagy mediators might be potential targets for CSC-targeted cancer treatment [139]. Another recent breakthrough in the advancement of gene therapy has been the approval of alipogene tiparvovec (Glybera) in Europe as a treatment for familial lipoprotein lipase deficiency [140], which has given hope for gene therapy for cancer. Currently, the development of gene therapy to eradicate CSCs is gaining an increasing amount of interest. However, there are still challenges in identifying the liver CSC-specific targets.

Stem cell features are the main focus of current studies investigating regulatory mechanisms of CSCs. The most significant regulatory differences between liver stem cells and LCSCs are still unknown. The elimination of normal stem cells, when we are eradicating liver CSCs, would have a hazardous effect on HCC patients, particularly those who already have chronic liver disease. To develop an ideal strategy to eliminate CSCs without sacrificing normal stem cells, the unique regulatory mechanisms of CSCs need to be elucidated. Since the CSC niche is also important for liver CSCs, targeting specific liver CSC niche components might be effective.

Immunotherapy has already shown great promise in cancer therapy as demonstrated by inhibitors of programmed death 1 (PD-1). However, we should note that these approved therapies might not target liver CSCs since liver CSCs can express distinct surface antigens. Genetically modified T cells with chimeric antigen receptors (CARs) are considered the most effective approach to enrich for T cells that can specifically recognize tumor cells [141]. To recognize specific antigens on cancer stem cells, CAR T cells can also be genetically modified, making them distinct from normal T cells. Therefore, targeted immunotherapy offers a promising avenue to eliminate liver CSCs. Gao et al. constructed a subset of T cells that were redirected toward glypican-3 for the treatment of hepatocellular carcinoma and explored their therapeutic effects in vivo. Their data showed that GPC3-targeted CAR-T cells significantly inhibited the growth of liver cancer [142]. In recent years, Deng et al. developed new CAR-T cells targeting the liver CSC antigen EpCAM and verified its effect on certain solid tumors [143]. However, the complicated issue is that most of identified liver CSC markers as well as the essential signaling pathways in liver CSCs also exist in normal stem cells. Immunotherapy developed to eliminate liver CSCs by targeting these mutual markers or signaling molecules will undoubtedly affect the survival and function of normal stem cells. Therefore, the identification of liver CSC-specific targets remains an important challenge for the future.

REFERENCES

[1] Yoo YD, Kwon YT. Molecular mechanisms controlling asymmetric and symmetric self-renewal of cancer stem cells. J Anal Sci Technol 2015;6(1):28.

[2] Zhang F, et al. Combined hepatocellular cholangiocarcinoma originating from hepatic progenitor cells: immunohistochemical and double-fluorescence immunostaining evidence. Histopathology 2008;52(2):224–32.

[3] Cardinale V, et al. Mucin-producing cholangiocarcinoma might derive from biliary tree stem/progenitor cells located in peribiliary glands. Hepatology 2012;55(6):2041–2.

[4] Komuta M, et al. Clinicopathological study on cholangiolocellular carcinoma suggesting hepatic progenitor cell origin. Hepatology 2008;47(5):1544–56.

[5] Yamashita T, et al. EpCAM-positive hepatocellular carcinoma cells are tumor-initiating cells with stem/progenitor cell features. Gastroenterology 2009;136(3):1012–24.

[6] Wen W, et al. Cyclin G1 expands liver tumor-initiating cells by Sox2 induction via Akt/mTOR signaling. Mol Cancer Ther 2013;12(9):1796–804.

[7] Gottesman MM. Mechanisms of cancer drug resistance. Annu Rev Med 2002;53:615–27.
[8] Yamashita T, et al. Discrete nature of EpCAM+ and CD90+ cancer stem cells in human hepatocellular carcinoma. Hepatology 2013;57(4):1484–97.
[9] Calderaro J, et al. Programmed death ligand 1 expression in hepatocellular carcinoma: relationship with clinical and pathological features. Hepatology 2016;64(6):2038–46.
[10] Guo XC, et al. Single tumor-initiating cells evade immune clearance by recruiting type II macrophages. Genes Dev 2017;31(3):247–59.
[11] Salem AF, et al. Caveolin-1 promotes pancreatic cancer cell differentiation and restores membranous E-cadherin via suppression of the epithelial-mesenchymal transition. Cell Cycle 2011;10(21):3692–700.
[12] Fan QM, et al. Tumor-associated macrophages promote cancer stem cell-like properties via transforming growth factor-beta1-induced epithelial-mesenchymal transition in hepatocellular carcinoma. Cancer Lett 2014;352(2):160–8.
[13] Lee D, et al. Interaction of tetraspan(in) TM4SF5 with CD44 promotes self-renewal and circulating capacities of hepatocarcinoma cells. Hepatology 2015;61(6):1978–97.
[14] Ma S, et al. Identification and characterization of tumorigenic liver cancer stem/progenitor cells. Gastroenterology 2007;132(7):2542–56.
[15] Tang KH, et al. CD133(+) liver tumor-initiating cells promote tumor angiogenesis, growth, and self-renewal through neurotensin/interleukin-8/CXCL1 signaling. Hepatology 2012;55(3):807–20.
[16] Yang ZF, et al. Significance of CD90+ cancer stem cells in human liver cancer. Cancer Cell 2008;13(2):153–66.
[17] Zhu Z, et al. Cancer stem/progenitor cells are highly enriched in CD133+CD44+ population in hepatocellular carcinoma. Int J Cancer 2010;126(9):2067–78.
[18] Yang ZF, et al. Identification of local and circulating cancer stem cells in human liver cancer. Hepatology 2008;47(3):919–28.
[19] Yang W, et al. Wnt/beta-catenin signaling contributes to activation of normal and tumorigenic liver progenitor cells. Cancer Res 2008;68(11):4287–95.
[20] Haraguchi N, et al. CD13 is a therapeutic target in human liver cancer stem cells. J Clin Invest 2010;120(9):3326–39.
[21] Xu X, et al. DLK1 as a potential target against cancer stem/progenitor cells of hepatocellular carcinoma. Mol Cancer Ther 2012;11(3):629–38.
[22] Lee TK, et al. CD24(+) liver tumor-initiating cells drive self-renewal and tumor initiation through STAT3-mediated NANOG regulation. Cell Stem Cell 2011;9(1):50–63.
[23] Liu S, et al. Expression of intercellular adhesion molecule 1 by hepatocellular carcinoma stem cells and circulating tumor cells. Gastroenterology 2013;144(5):1031–41. e10.
[24] Ding J, Wang H. Multiple interactive factors in hepatocarcinogenesis. Cancer Lett 2014;346(1):17–23.
[25] Lee TK, et al. Blockade of CD47-mediated cathepsin S/protease-activated receptor 2 signaling provides a therapeutic target for hepatocellular carcinoma. Hepatology 2014;60(1):179–91.
[26] Lei ZJ, et al. Lysine-specific demethylase 1 promotes the stemness and chemoresistance of Lgr5(+) liver cancer initiating cells by suppressing negative regulators of beta-catenin signaling. Oncogene 2015;34(24):3188–98.
[27] Ginestier C, et al. ALDH1 is a marker of normal and malignant human mammary stem cells and a predictor of poor clinical outcome. Cell Stem Cell 2007;1(5):555–67.
[28] Ma S, et al. Aldehyde dehydrogenase discriminates the CD133 liver cancer stem cell populations. Mol Cancer Res 2008;6(7):1146–53.

[29] Chiba T, et al. Side population purified from hepatocellular carcinoma cells harbors cancer stem cell-like properties. Hepatology 2006;44(1):240–51.
[30] Chiba T, et al. The polycomb gene product BMI1 contributes to the maintenance of tumor-initiating side population cells in hepatocellular carcinoma. Cancer Res 2008;68(19):7742–9.
[31] Cao L, et al. Sphere-forming cell subpopulations with cancer stem cell properties in human hepatoma cell lines. BMC Gastroenterol 2011;11:71.
[32] Wang X, et al. Long non-coding RNA DILC regulates liver cancer stem cells via IL-6/STAT3 axis. J Hepatol 2016;64(6):1283–94.
[33] Pine SR, et al. Microenvironmental modulation of asymmetric cell division in human lung cancer cells. Proc Natl Acad Sci U S A 2010;107(5):2195–200.
[34] Miranda-Lorenzo I, et al. Intracellular autofluorescence: a biomarker for epithelial cancer stem cells. Nat Methods 2014;11(11):1161–9.
[35] Muramatsu S, et al. Visualization of stem cell features in human hepatocellular carcinoma reveals in vivo significance of tumor-host interaction and clinical course. Hepatology 2013;58(1):218–28.
[36] Takahashi K, Yamanaka S. Induction of pluripotent stem cells from mouse embryonic and adult fibroblast cultures by defined factors. Cell 2006;126(4):663–76.
[37] Yong KJ, Chai L, Tenen DG. Oncofetal gene SALL4 in aggressive hepatocellular carcinoma. N Engl J Med 2013;369(12):1171–2.
[38] Yin X, et al. Coexpression of stemness factors Oct4 and Nanog predict liver resection. Ann Surg Oncol 2012;19(9):2877–87.
[39] Huang P, et al. Role of Sox2 and Oct4 in predicting survival of hepatocellular carcinoma patients after hepatectomy. Clin Biochem 2011;44(8–9):582–9.
[40] Chow EK, et al. Oncogene-specific formation of chemoresistant murine hepatic cancer stem cells. Hepatology 2012;56(4):1331–41.
[41] Holczbauer A, et al. Modeling pathogenesis of primary liver cancer in lineage-specific mouse cell types. Gastroenterology 2013;145(1):221–31.
[42] Akita H, et al. MYC activates stem-like cell potential in hepatocarcinoma by a p53-dependent mechanism. Cancer Res 2014;74(20):5903–13.
[43] Chen CL, et al. NANOG metabolically reprograms tumor-initiating stem-like cells through tumorigenic changes in oxidative phosphorylation and fatty acid metabolism. Cell Metab 2016;23(1):206–19.
[44] Sun C, et al. NANOG promotes liver cancer cell invasion by inducing epithelial-mesenchymal transition through NODAL/SMAD3 signaling pathway. Int J Biochem Cell Biol 2013;45(6):1099–108.
[45] Shan J, et al. Nanog regulates self-renewal of cancer stem cells through the insulin-like growth factor pathway in human hepatocellular carcinoma. Hepatology 2012;56(3):1004–14.
[46] Wang XQ, et al. Octamer 4 (Oct4) mediates chemotherapeutic drug resistance in liver cancer cells through a potential Oct4-AKT-ATP-binding cassette G2 pathway. Hepatology 2010;52(2):528–39.
[47] Zhou J, Ng SB, Chng WJ. LIN28/LIN28B: an emerging oncogenic driver in cancer stem cells. Int J Biochem Cell Biol 2013;45(5):973–8.
[48] de Celis JF, Barrio R. Regulation and function of Spalt proteins during animal development. Int J Dev Biol 2009;53(8–10):1385–98.
[49] Zeng SS, et al. The transcription factor SALL4 regulates stemness of EpCAM-positive hepatocellular carcinoma. J Hepatol 2014;60(1):127–34.
[50] Marquardt JU, et al. Curcumin effectively inhibits oncogenic NF-κB signaling and restrains stemness features in liver cancer. J Hepatol 2015;63(3):661–9.

[51] Ji J, et al. Identification of microRNA-181 by genome-wide screening as a critical player in EpCAM-positive hepatic cancer stem cells. Hepatology 2009;50(2):472–80.
[52] Ma S, et al. miR-130b promotes CD133(+) liver tumor-initiating cell growth and self-renewal via tumor protein 53-induced nuclear protein 1. Cell Stem Cell 2010;7(6):694–707.
[53] Zhang J, et al. microRNA-150 inhibits human CD133-positive liver cancer stem cells through negative regulation of the transcription factor c-Myb. Int J Oncol 2012;40(3):747–56.
[54] Li L, et al. Epigenetic modification of MiR-429 promotes liver tumour-initiating cell properties by targeting Rb binding protein 4. Gut 2015;64(1):156–67.
[55] Prensner JR, Chinnaiyan AM. The emergence of lncRNAs in cancer biology. Cancer Discov 2011;1(5):391–407.
[56] Yuan SX, et al. Long noncoding RNA associated with microvascular invasion in hepatocellular carcinoma promotes angiogenesis and serves as a predictor for hepatocellular carcinoma patients' poor recurrence-free survival after hepatectomy. Hepatology 2012;56(6):2231–41.
[57] Huang JF, et al. Hepatitis B virus X protein (HBx)-related long noncoding RNA (lncRNA) down-regulated expression by HBx (Dreh) inhibits hepatocellular carcinoma metastasis by targeting the intermediate filament protein vimentin. Hepatology 2013;57(5):1882–92.
[58] Wang Y, et al. The long noncoding RNA lncTCF7 promotes self-renewal of human liver cancer stem cells through activation of Wnt signaling. Cell Stem Cell 2015;16(4):413–25.
[59] Yuan SX, et al. Long noncoding RNA DANCR increases stemness features of hepatocellular carcinoma via de-repression of CTNNB1. Hepatology 2016;63(2):499–511.
[60] Feinberg AP. Phenotypic plasticity and the epigenetics of human disease. Nature 2007;447(7143):433–40.
[61] Liu C, et al. Histone deacetylase 3 participates in self-renewal of liver cancer stem cells through histone modification. Cancer Lett 2013;339(1):60–9.
[62] Marquardt JU, et al. Human hepatic cancer stem cells are characterized by common stemness traits and diverse oncogenic pathways. Hepatology 2011;54(3):1031–42.
[63] Liu LM, et al. SIRT1-mediated transcriptional regulation of SOX2 is important for self-renewal of liver cancer stem cells. Hepatology 2016;64(3):814–27.
[64] Chiba T, et al. 3-Deazaneplanocin A is a promising therapeutic agent for the eradication of tumor-initiating hepatocellular carcinoma cells. Int J Cancer 2012;130(11):2557–67.
[65] Liu M, et al. CHD1L promotes lineage reversion of hepatocellular carcinoma through opening chromatin for key developmental transcription factors. Hepatology 2016;63(5):1544–59.
[66] Chiba T, et al. Bmi1 promotes hepatic stem cell expansion and tumorigenicity in both Ink4a/Arf-dependent and -independent manners in mice. Hepatology 2010;52(3):1111–23.
[67] Wang B, et al. Self-renewing diploid Axin2(+) cells fuel homeostatic renewal of the liver. Nature 2015;524(7564):180–5.
[68] Huch M, et al. In vitro expansion of single Lgr5+ liver stem cells induced by Wnt-driven regeneration. Nature 2013;494(7436):247–50.
[69] Huch M, et al. Long-term culture of genome-stable bipotent stem cells from adult human liver. Cell 2015;160(1–2):299–312.
[70] Fujimoto A, et al. Whole-genome sequencing of liver cancers identifies etiological influences on mutation patterns and recurrent mutations in chromatin regulators. Nat Genet 2012;44(7):760–4.
[71] Guichard C, et al. Integrated analysis of somatic mutations and focal copy-number changes identifies key genes and pathways in hepatocellular carcinoma. Nat Genet 2012;44(6):694–8.
[72] Mokkapati S, et al. β-Catenin activation in a novel liver progenitor cell type is sufficient to cause hepatocellular carcinoma and hepatoblastoma. Cancer Res 2014;74(16):4515–25.

[73] Xiang D, et al. Shp2 promotes liver cancer stem cell expansion by augmenting β-catenin signaling and predicts chemotherapeutic response of patients. Hepatology 2017;65(5):1566–80.
[74] Strazzabosco M, Fabris L. Notch signaling in hepatocellular carcinoma: guilty in association! Gastroenterology 2012;143(6):1430–4.
[75] Razumilava N, Gores GJ. Notch-driven carcinogenesis: the merging of hepatocellular cancer and cholangiocarcinoma into a common molecular liver cancer subtype. J Hepatol 2013;58(6):1244–5.
[76] Wang XQ, et al. Notch1-Snail1-E-cadherin pathway in metastatic hepatocellular carcinoma. Int J Cancer 2012;131(3):E163–72.
[77] Nishina S, et al. Restored expression of the tumor suppressor gene RUNX3 reduces cancer stem cells in hepatocellular carcinoma by suppressing Jagged1-Notch signaling. Oncol Rep 2011;26(3):523–31.
[78] Lin L, et al. The STAT3 inhibitor NSC 74859 is effective in hepatocellular cancers with disrupted TGF-beta signaling. Oncogene 2009;28(7):961–72.
[79] Mishra L, et al. Liver stem cells and hepatocellular carcinoma. Hepatology 2009;49(1):318–29.
[80] Tang Y, et al. Progenitor/stem cells give rise to liver cancer due to aberrant TGF-beta and IL-6 signaling. Proc Natl Acad Sci U S A 2008;105(7):2445–50.
[81] Wang C, et al. Hepatitis B virus X (HBx) induces tumorigenicity of hepatic progenitor cells in 3,5-diethoxycarbonyl-1,4-dihydrocollidine-treated HBx transgenic mice. Hepatology 2012;55(1):108–20.
[82] Wan S, et al. Tumor-associated macrophages produce interleukin 6 and signal via STAT3 to promote expansion of human hepatocellular carcinoma stem cells. Gastroenterology 2014;147(6):1393–404.
[83] Yu FX, Zhao B, Guan KL. Hippo pathway in organ size control, tissue homeostasis, and cancer. Cell 2015;163(4):811–28.
[84] Yimlamai D, et al. Hippo pathway activity influences liver cell fate. Cell 2014;157(6):1324–38.
[85] Lee KP, et al. The Hippo-Salvador pathway restrains hepatic oval cell proliferation, liver size, and liver tumorigenesis. Proc Natl Acad Sci U S A 2010;107(18):8248–53.
[86] Song H, et al. Mammalian Mst1 and Mst2 kinases play essential roles in organ size control and tumor suppression. Proc Natl Acad Sci U S A 2010;107(4):1431–6.
[87] Zhou D, et al. Mst1 and Mst2 maintain hepatocyte quiescence and suppress hepatocellular carcinoma development through inactivation of the Yap1 oncogene. Cancer Cell 2009;16(5):425–38.
[88] Plaks V, Kong N, Werb Z. The cancer stem cell niche: how essential is the niche in regulating stemness of tumor cells? Cell Stem Cell 2015;16(3):225–38.
[89] Cabarcas SM, Mathews LA, Farrar WL. The cancer stem cell niche—there goes the neighborhood? Int J Cancer 2011;129(10):2315–27.
[90] Smith AL, Robin TP, Ford HL. Molecular pathways: targeting the TGF-beta pathway for cancer therapy. Clin Cancer Res 2012;18(17):4514–21.
[91] Liu L, et al. Hypoxic reactive oxygen species regulate the integrated stress response and cell survival. J Biol Chem 2008;283(45):31153–62.
[92] Ishikawa T, et al. Hepatocyte growth factor/c-met signaling is required for stem-cell-mediated liver regeneration in mice. Hepatology 2012;55(4):1215–26.
[93] Ding W, et al. Epithelial-to-mesenchymal transition of murine liver tumor cells promotes invasion. Hepatology 2010;52(3):945–53.
[94] Lau EYT, et al. Cancer-associated fibroblasts regulate tumor-initiating cell plasticity in hepatocellular carcinoma through c-Met/FRA1/HEY1 signaling. Cell Rep 2016;15(6):1175–89.

[95] Zhang HL, et al. Blocking preferential glucose uptake sensitizes liver tumor-initiating cells to glucose restriction and sorafenib treatment. Cancer Lett 2017;388:1–11.
[96] He G, et al. Identification of liver cancer progenitors whose malignant progression depends on autocrine IL-6 signaling. Cell 2013;155(2):384–96.
[97] Wu K, et al. Hepatic transforming growth factor beta gives rise to tumor-initiating cells and promotes liver cancer development. Hepatology 2012;56(6):2255–67.
[98] Fausto N. Liver regeneration and repair: hepatocytes, progenitor cells, and stem cells. Hepatology 2004;39(6):1477–87.
[99] Li XF, et al. Inflammation-elicited Liver Progenitor Cell Conversion to Liver Cancer Stem Cell with Clinical Significance. Hepatology 2017;66(6):1934–51.
[100] Yanger K, et al. Adult hepatocytes are generated by self-duplication rather than stem cell differentiation. Cell Stem Cell 2014;15(3):340–9.
[101] Tarlow BD, Finegold MJ, Grompe M. Clonal tracing of Sox9+ liver progenitors in mouse oval cell injury. Hepatology 2014;60(1):278–89.
[102] Tarlow BD, et al. Bipotential adult liver progenitors are derived from chronically injured mature hepatocytes. Cell Stem Cell 2014;15(5):605–18.
[103] Song YY, et al. Loss of ATOH8 increases stem cell features of hepatocellular carcinoma cells. Gastroenterology 2015;149(4):1068–81.
[104] Nio K, Yamashita T, Kaneko S. The evolving concept of liver cancer stem cells. Mol Cancer 2017;16:4.
[105] Mavroudis D. Circulating cancer cells. Ann Oncol 2010;21(Suppl. 7). vii95–100.
[106] Maheswaran S, et al. Detection of mutations in EGFR in circulating lung-cancer cells. N Engl J Med 2008;359(4):366–77.
[107] Smith LM, et al. CD133/prominin-1 is a potential therapeutic target for antibody-drug conjugates in hepatocellular and gastric cancers. Br J Cancer 2008;99(1):100–9.
[108] Zhao W, et al. 1B50-1, a mAb raised against recurrent tumor cells, targets liver tumor-initiating cells by binding to the calcium channel alpha2delta1 subunit. Cancer Cell 2013;23(4):541–56.
[109] Lo J, et al. Nuclear factor kappa B-mediated CD47 up-regulation promotes sorafenib resistance and its blockade synergizes the effect of sorafenib in hepatocellular carcinoma in mice. Hepatology 2015;62(2):534–45.
[110] Ho JC, et al. Granulin-epithelin precursor as a therapeutic target for hepatocellular carcinoma. Hepatology 2008;47(5):1524–32.
[111] Chen Z, Chen SJ. Poisoning the devil. Cell 2017;168(4):556–60.
[112] Zhang Y, et al. All-trans retinoic acid potentiates the chemotherapeutic effect of cisplatin by inducing differentiation of tumor initiating cells in liver cancer. J Hepatol 2013;59(6):1255–63.
[113] Zhang KZ, et al. Arsenic trioxide induces differentiation of CD133(+) hepatocellular carcinoma cells and prolongs posthepatectomy survival by targeting GLI1 expression in a mouse model. J Hematol Oncol 2014;7.
[114] Li ZW, et al. Tissue factor pathway inhibitor-2 induced hepatocellular carcinoma cell differentiation. Saudi J Biol Sci 2017;24(1):95–102.
[115] Si-Tayeb K, Lemaigre FP, Duncan SA. Organogenesis and development of the liver. Dev Cell 2010;18(2):175–89.
[116] Yamashita T, et al. Oncostatin M renders epithelial cell adhesion molecule-positive liver cancer stem cells sensitive to 5-fluorouracil by inducing hepatocytic differentiation. Cancer Res 2010;70(11):4687–97.
[117] Yin C, et al. Differentiation therapy of hepatocellular carcinoma in mice with recombinant adenovirus carrying hepatocyte nuclear factor-4alpha gene. Hepatology 2008;48(5):1528–39.

[118] Ning BF, et al. Hepatocyte nuclear factor 4 alpha suppresses the development of hepatocellular carcinoma. Cancer Res 2010;70(19):7640–51.
[119] Saijo H, et al. Cytotoxic T lymphocytes: the future of cancer stem cell eradication? Immunotherapy 2013;5(6):549–51.
[120] Hirohashi Y, et al. Immune response against tumor antigens expressed on human cancer stem-like cells/tumor-initiating cells. Immunotherapy 2010;2(2):201–11.
[121] Xu Q, et al. Antigen-specific T-cell response from dendritic cell vaccination using cancer stem-like cell-associated antigens. Stem Cells 2009;27(8):1734–40.
[122] Weng D, et al. Induction of cytotoxic T lymphocytes against ovarian cancer-initiating cells. Int J Cancer 2011;129(8):1990–2001.
[123] Morita R, et al. Heat shock protein DNAJB8 is a novel target for immunotherapy of colon cancer-initiating cells. Cancer Sci 2014;105(4):389–95.
[124] Tong M, et al. ANXA3/JNK signaling promotes self-renewal and tumor growth, and its blockade provides a therapeutic target for hepatocellular carcinoma. Stem Cell Reports 2015;5(1):45–59.
[125] Pan QZ, et al. Annexin A3 as a potential target for immunotherapy of liver cancer stem-like cells. Stem Cells 2015;33(2):354–66.
[126] Zhang PF, et al. An EpCAM/CD3 bispecific antibody efficiently eliminates hepatocellular carcinoma cells with limited galectin-1 expression. Cancer Immunol Immunother 2014;63(2):121–32.
[127] Xiao Z, et al. Antibody mediated therapy targeting CD47 inhibits tumor progression of hepatocellular carcinoma. Cancer Lett 2015;360(2):302–9.
[128] Kaplan RN, et al. VEGFR1-positive haematopoietic bone marrow progenitors initiate the pre-metastatic niche. Nature 2005;438(7069):820–7.
[129] Gordan JD, et al. HIF-2alpha promotes hypoxic cell proliferation by enhancing c-myc transcriptional activity. Cancer Cell 2007;11(4):335–47.
[130] Li Z, et al. Hypoxia-inducible factors regulate tumorigenic capacity of glioma stem cells. Cancer Cell 2009;15(6):501–13.
[131] Lau CK, et al. An Akt/hypoxia-inducible factor-1alpha/platelet-derived growth factor-BB autocrine loop mediates hypoxia-induced chemoresistance in liver cancer cells and tumorigenic hepatic progenitor cells. Clin Cancer Res 2009;15(10):3462–71.
[132] Ye J, et al. The cancer stem cell niche: cross talk between cancer stem cells and their microenvironment. Tumour Biol 2014;35(5):3945–51.
[133] Yu GF, et al. Hepatic stellate cells secreted hepatocyte growth factor contributes to the chemoresistance of hepatocellular carcinoma. PLoS One 2013;8(9).
[134] Govaere O, et al. Laminin-332 sustains chemoresistance and quiescence as part of the human hepatic cancer stem cell niche. J Hepatol 2016;64(3):609–17.
[135] Arzumanyan A, et al. Does the hepatitis B antigen HBx promote the appearance of liver cancer stem cells? Cancer Res 2011;71(10):3701–8.
[136] Fan H, et al. Hepatitis B virus X protein induces EpCAM expression via active DNA demethylation directed by RelA in complex with EZH2 and TET2. Oncogene 2016;35(6):715–26.
[137] Menendez JA, Alarcon T. Metabostemness: a new cancer hallmark. Front Oncol 2014;4:262.
[138] Menendez JA, et al. Metabostemness: metaboloepigenetic reprogramming of cancer stem-cell functions. Oncoscience 2014;1(12):803–6.
[139] Maycotte P, et al. Autophagy supports breast cancer stem cell maintenance by regulating IL6 secretion. Mol Cancer Res 2015;13(4):651–8.

[140] Salmon F, Grosios K, Petry H. Safety profile of recombinant adeno-associated viral vectors: focus on alipogene tiparvovec (Glybera(R)). Expert Rev Clin Pharmacol 2014;7(1):53–65.
[141] Sadelain M, Brentjens R, Riviere I. The basic principles of chimeric antigen receptor design. Cancer Discov 2013;3(4):388–98.
[142] Gao HP, et al. Development of T cells redirected to glypican-3 for the treatment of hepatocellular carcinoma. Clin Cancer Res 2014;20(24):6418–28.
[143] Deng ZL, et al. Adoptive T-cell therapy of prostate cancer targeting the cancer stem cell antigen EpCAM. BMC Immunol 2015;16.

Chapter 15

Clinical Application of Stem Cells in Liver Diseases: From Bench to Bedside

Yan Xu*, Qi Zhang**
*Biotherapy Center, The Third Affiliated Hospital, Sun Yat-sen University, Guangzhou, PR China;
**Cell-gene Therapy Translational Medicine Research Center, The Third Affiliated Hospital, Sun Yat-sen University, Guangzhou, PR China

1 INTRODUCTION

Advances in stem cell biology bring remarkable revolution to human disease study and therapy, including various liver diseases. The establishment of pluripotent stem cells (PSCs), especially induced pluripotent stem cells (iPSCs), makes it possible to obtain infinite hepatocytes derived from human source for cell therapy, disease modeling, and drug testing. While the development of adult stem cells have made the prospect of stem cell therapy and tissue regeneration a clinical reality. Recently, stem cell therapy is widely applied in liver diseases, including acute liver failure, liver cirrhosis, later-stage chronic liver dysfunction, liver transplantation, and other liver diseases. Stem cell therapies hold great promise to repair, restore, or replace liver function. In this chapter, we will review recent progress in basic study and clinical application of different stem cells in diverse liver diseases as well as the existing questions to be answered.

2 PSCS IN LIVER DISEASE STUDY

Liver transplantation is the only therapeutically effective treatment for end-stage liver diseases. Unfortunately, the number of available donor organs is quite limited compared to the number of patients waiting for liver transplantation. Hepatocyte transplantation appears as a promising alternative strategy to treat a variety of liver diseases [1,2] as nearly 100 patients suffering from inherited metabolic liver disease, chronic liver disease, or acute liver failure have received donor hepatocytes to date [2]. Besides, primary hepatocytes are also intensively needed for bioartificial liver (BAL) system, drug development, and toxicology screening in pharmaceutical industry. However, all these applications are limited by the scarce number and poor quality of available donor cells [3]. In recent years, directed differentiation of hepatic lineages from stem cells and transdifferentiation from differentiated somatic cells open new possibility to obtain plenty hepatocytes in vitro.

PSCs, especially iPSCs generated by somatic cell reprogramming, hold great promise for individualized, cell-based regenerative therapies, disease modeling, and drug screening. By forced expression of four reprogramming transcription factors, Yamanaka et al. successfully converted differentiated fibroblasts to an embryonic stem cell (ESC)-like state [4]. The resulting iPSCs share the characteristics of infinite self-renewal and potential to differentiate into various cell types with ESCs while overcome the ethic issue of ESCs, making it possible to get inexhaustible autologous cells in vitro. Recently, researchers have developed numerous protocols to derive different liver cell populations (including hepatocytes and cholangiocytes) from PSCs.

Cells generated by most protocols in vitro showed encouraging results of cell function and repopulating efficiency after transplanted to hepatic injury animal models, supporting the feasibility of autologous liver cell therapies, though the safety and therapeutic effects need to be further tested on nonhuman primates and then clinical trials. These differentiation methods combined with gene editing technologies also paved the liver disease modeling both in a dish and in humanized animal models, including genetic diseases, metabolic diseases, and infectious diseases (Fig. 15.1), overcoming the limited availability of primary tissues, interspecies differences of animal models and immediately reduced proliferation and cell function of isolated primary cells cultured in vitro during human liver disease study in the past. Here we summarized the recent works on hepatocytes and cholangiocytes differentiation from PSCs, liver disease modeling in a dish and in humanized animal models using iPSCs and discuss the remaining problems and future directions in the field.

2.1 PSC-Derived HPCs/Hepatocytes and Cholangiocytes in Cell Therapy of Liver Disease

2.1.1 Hepatocyte Differentiation From PSCs

Hepatocytes occupy 80% of the liver volume and serve as the main metabolic cells of liver. Thus, hepatocyte differentiation is subject to most intensive research in recent years. Protocols for differentiating PSCs to hepatocytes were first developed using mouse ESCs. In 2002, Rambhatla et al. reported the differentiation of cells with features of hepatocytes from human ESCs for the first time [5]. By treating the attached embryonic bodies (EBs) with sodium butyrate and dimethyl sulfoxide, they can get a more homogeneous and multinucleated populations. This differentiation method is also effective for process without EB differentiation step [5]. However, sodium butyrate induced significant cell death and resulted in approximately only 10–15% yield of hepatocytes. By screening cytokines and extracellular matrix (ECM) proteins supporting differentiation of

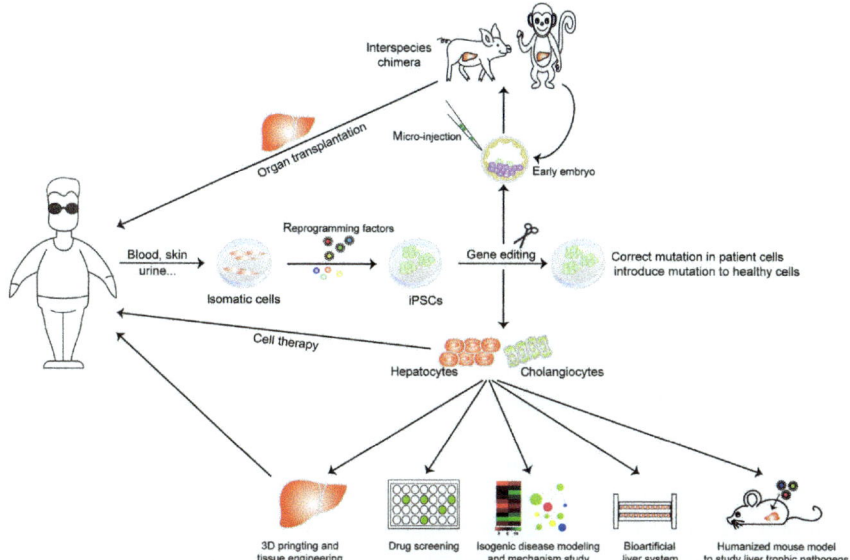

FIGURE 15.1 PSCs in liver disease study. Somatic cells from human blood, fibroblasts, urine et al. can be reprogrammed to a PSC state by forced expression of reprogramming factors. These induced pluripotent stem cells (iPSCs) resemble human embryonic stem cells (ESCs) and have the ability of infinite self-renewal and potential to differentiate into any cell types, including hepatocytes and cholangiocytes. Differentiated hepatocytes and cholangiocytes provide inexhaustible substitution of primary cells for cell therapy, drug screening, "seed cells" for 3D printing, tissue engineering, and bioartifical liver system. With the development of gene editing technologies, iPSC also provide a valuable tool for isogenic disease modeling and mechanism study in vitro and in humanized animal models. What's more, it is also possible to generate interspecies chimeras by injecting human PSCs to early embryos of other species (like pigs and monkeys) and get functional grafts all from iPSCs.

ESCs to hepatocytes, Shirahashi et al. found that type 1 collagen, human insulin and dexamethasone (Dex), while Schwarts et al. found type 1 collagen, fibroblast growth factor 4 (FGF4) and hepatocyte growth factor (HGF) can promote hepatic differentiation of hESC-derived EBs [6,7]. However, all these differentiation efficiency is low and most differentiation protocols involve EB formation, a highly inefficient and heterogeneous process. Also, most reports performed only limited phenotypic and functional tests on the differentiated cells.

In 2007, Cai et al. report a novel three-stage method to efficiently direct the differentiation of hESCs into hepatic cells in serum-free medium without EB formation [8]. The induction process includes stepwise induction of definitive endoderm (DE) by activin A, hepatic initiation by bone morphogenetic protein 4 and FGF4 and finally hepatocyte maturation by HGF, oncostatin M, and Dex. The differentiation process excludes the extraembryonic endoderm differentiation of ESCs and efficiency is much higher compared to previous EB-based protocols. Also, they did thorough characterization of the induced hepatocytes compared to previous reports, including susceptible to HIV-HCV pseudotype viruses and transplantation to carbon tetrachloride (CCl4)-treated liver-injured SCID mice [8]. In 2009, the same group reported the induction of hepatocytes from hiPSCs for the first time using an optimized protocol [9]. At the same time, several different groups reported new directed differentiation protocols of hepatocytes from hESCs and hiPSCs in adherent and 2 dimensional monolayer culture with progressive elimination of serum, the feeder cell layer and other undefined reagents, making the resulting cells more compatible with clinical applications (Table 15.1). By high-throughput screening for human hepatocyte expansion, Shan et al. found two chemical compounds, FH1 and FPH1, and their treatment during the maturation stage of iHep cell induction from hiPSCs can prompt the cells to a more mature phenotype [10]. Most of these protocols took a stepwise strategy (including endoderm induction, hepatic specification, hepatoblast expansion, and hepatocyte maturation) to induce the hepatic specification by titrating concentration and treatment duration of growth factors or cytokines important for liver development and small chemical compounds interfering signals involved in early development, such as BMP, FGF, HGF and WNT (Table 15.1). Moreover, with the development of biomaterials, more and more attention has been paid to the ECM [11] and the three-dimensional culture [12,13]. It has been shown that 3D cultures induce a more mature phenotype compared to 2D, with a significant decrease of fetal hepatocyte markers AFP and CYP3A7 as well as an increase in drug metabolism enzymes activity and cell polarity [13–15].

Cells generated by most protocols in vitro showed encouraging results of cell function including glycogen storage, LDL uptake, urine metabolism, and inducible cytochrome P450 activity test. And several protocols have generated cells with the ability to survive and repopulate the liver function after transplanted to hepatic injury animal models, supporting the feasibility of autologous liver cell therapies (Table 15.1). However, it should be noted that most of the cells differentiated so far are still far more immature compared to their primary counterparts, evidenced by remaining expression of fetal hepatocyte marker expression (like *AFP*) and limited ability of proliferation, reduced amount of ALB secretion, and especially the limited drug metabolism activity, which should be further improved in the future.

2.1.2 Cholangiocyte Differentiation From PSCs

Cholangiocytes, the epithelium lining the biliary tree, though representing only 3% of the total liver cell population, are nonparenchymal cells fundamental for normal liver function with the role for processing bile produced by hepatocytes [16]. Similar with hepatocytes, cholangiocytes are also originated from hepatoblasts [17], thus the differentiation process also includes induction of DE, hepatic specification, and hepatic progenitor induction. The first attempt of cholangiocyte differentiation from hPSCs was done by Hongkui Deng [18]. By culturing the hESC-derived hepatic progenitor cells (HPCs)

TABLE 15.1 Summary of Protocols for Hepatocyte Differentiation from hPSCs

			Functional testing in vitro							
Efficiency	Positive control	Cytochrome P450 metabolic assay	Albumin Secretion	Glycogen storage	Cholesterol uptake	Urea metabolism	Indocyanine green uptake	Murine transplantation	Refs.
10–15% of input cells	Primary hepatocytes	Yes	N/A	Yes	N/A	N/A	N/A	N/A	[5]
NA	Primary human hepatocytes, primary rodent hepatocytes or HepG2	N/A	Yes	N/A	N/A	Yes	N/A	N/A	[7]
2% ALB and CK18 positive	Primary rat hepatocytes or HepG2	Yes	Yes	N/A	N/A	Yes	Yes	N/A	[6]
NA	Human hepatocytes	Yes	Yes	Yes	N/A	Yes	N/A	N/A	[12]
16.2% of the total cell population	N/A	Yes	Yes	Yes	N/A	N/A	Yes	N/A	[144]
70% ALB positive	Hepatoma cell line Huh-7	Yes	Yes	Yes	Yes	N/A	Yes	Yes	[8]
67.4% ALB positive	Hepatoma cell line HepG2	N/A	Yes	Yes	N/A	N/A	Yes	Yes	[145]
70–90% ALB positive	HepG2	Yes	Yes	Yes	N/A	N/A	N/A	N/A	[146–148]
55% ALB positive	Primary porcine hepatocytes	Yes	Yes	N/A	N/A	Yes	N/A	Yes	[149]
60% AFP and Alb positive	Primary human hepatocytes	Yes	Yes	Yes	N/A	Yes	N/A	N/A	[9]
35% ASGR1, LDLR, c-met and α6 integrin positive	Thawed fetal hepatocytes	Yes	Yes	Yes	Yes	N/A	Yes	Yes	[150]

(Continued)

TABLE 15.1 Summary of Protocols for Hepatocyte Differentiation from hPSCs (cont.)

Efficiency	Positive control	Functional testing in vitro							Refs.
		Cytochrome P450 metabolic assay	Albumin Secretion	Glycogen storage	Cholesterol uptake	Urea metabolism	Indocyanine green uptake	Murine transplantation	
80% ALB positive	N/A	Yes	Yes	Yes	Yes	N/A	N/A	N/A	[28]
80% ALB positive	Cadaveric liver samples	N/A	N/A	Yes	Yes	Yes	Yes	Yes	[48,151]
N/A	Human primary hepatocyte	Yes	Yes	N/A	N/A	N/A	N/A	Yes	[35,152]
N/A	Primary human hepatocytes	Yes	N/A	Yes	N/A	Yes	N/A	Yes	[153]
53% ALB positive	Human adult liver tissue or HepG2	N/A	N/A	Yes	N/A	N/A	N/A	N/A	[154]
70% were double positive for ASGPR and DBOMF	Primary human hepatocytes	Yes	Yes	Yes	N/A	N/A	Yes	Yes	[155]
80% co-express AFP, ALB and A1AT	Primary hepatocytes	Yes	Yes	Yes	Yes	N/A	N/A	N/A	[3]
90% ALB positive	Fresh human hepatocytes	Yes	Yes	N/A	N/A	N/A	Yes	N/A	[156,157]
80% ALB positive	Primary human hepatocytes	N/A	N/A	Yes	Yes	N/A	Yes	Yes	[51]
N/A	Primary human hepatocytes	Yes	Yes	N/A	N/A	Yes	N/A	N/A	[158]
74.5% ± 5.9 ALB positive	Primary adult hepatocytes	Yes	Yes	Yes	N/A	N/A	Yes	N/A	[159]
90% ALB positive	Primary human hepatocytes	Yes	Yes	Yes	Yes	N/A	Yes	Yes	[57,160]

N/A, not available.

in previously reported William's E media on matrigel, they got KRT19- and KRT7-positive, AFP-negative cells and some cells acquired apicobasal polarity and possessed a secretory function. By stepwise treatment with growth hormone, epidermal growth factor, interleukin-6, and then sodium taurocholate, Dianat et al. induced cholangiocyte commitment of biopotent HepaRG-derived hepatoblasts and hepatic progenitors differentiated from hESCs or hiPSCs. The resulting cells express cholangiocyte markers (including SOX9, OPN, CK7, CK19, CK18, HNF1β, and HNF6) and cholangiocyte-specific transporters (SCTR, CFTR, ASBT, and TGR5), and VEGF receptor 2 (KDR) [19]. They formed primary cilia and also responded to hormonal stimulation by increasing intracellular Ca^{2+}; they developed epithelial/apicobasal polarity and formed functional cysts and biliary ducts in 3D matrix. However, the whole differentiation process from hPSCs takes as long as 23 days [19]. Another study used TGF-β to induce the cholangiocyte commitment from hESC-derived hepatic progenitors. The resulting cells showed similar structural and functional characteristics in vitro with those induced by Dianat et al. [19] and engrafted within mouse liver following transplantation [20]. And this process is accompanied with decrease of repressive histone mark H3K27me3 and its methyltransferase EZH2 [21]. However, most of these differentiation protocols are established as monolayers which is limited by the inability to accurately recapitulate the normal physiological conditions; thus the generated cholangiocytes have not been tested for key functions of bona fide cholangiocytes (Table 15.2), such as enzymatic activity and responses to hormonal stimuli.

In 2015, two laboratories reported different new methods for generating mature functional cholangiocyte organoids from human PSCs more efficiently in 3D culture systems on the same issue of *Nature Biotechnology* [22,23]. hPSCs was induced to hepatoblasts and then cholangiocytes in different ways. Ogawa et al. add growth factors including HGF, EGF, and TGF-β1 and activate NOTCH signaling pathway by co-culturing with OP9 stromal cells, which is similar to development in vivo [23]. While Sampaziotis et al. used activin, retinoic acid, and FGF10 to induce cholangiocyte progenitors and further mature the cholangiocyte-like cells in William's E medium supplemented with EGF in 3D culture [22]. Both methods lead to the formation of mature cholangiocytes expressing mature markers and organoids with tubular and cystic morphology and showed rhodamine efflux function. Ogawa et al. also showed in vivo formation of ductal structures of hPSCs-derived cholangiocytes after transplantation to immunodeficient mice [23]. Except for optimization of culture media and growth factors, improvement of ECM is another aspect to potentiate cellular differentiation. Using their previously described procedure, Takayama et al. found that Laminin 411 and 511 can promote the cholangiocyte differentiation of hiPSCs [24]. When cultured with Laminin 411 and 511, both the diameter and number of the cholangiocyte-like cells constituted cysts were upregulated.

TABLE 15.2 Summary of Protocols for Cholangiocyte Differentiation from hPSCs

Efficiency	Functional characterization in vitro							Murine transplantation	Refs.
	Cilia and calcium signaling	Polarization and tubulogenesis	Transport of rhodamine 123	ALP	γ-GGT	Responses to hormonal stimuli (secretin and somatostatin)	Chloride transfer through CFTR activity		
N/A	N/A	Yes	Yes	N/A	N/A	N/A	N/A	N/A	[18]
N/A	N/A	Yes	N/A	N/A	N/A	N/A	N/A	N/A	[62,161]
N/A	Yes	Yes	Yes	N/A	N/A	Yes	N/A	N/A	[19]
66.91% CK7+, 77.34% CFTR+ and 85.3% CK19+	Yes	Yes	N/A	N/A	N/A	N/A	N/A	Yes	[162]
74.5% both Sox9+ and CK7+	Yes	Yes	Yes	Yes	Yes	Yes	Yes	N/A	[22]
N/A	Yes	Yes	Yes	N/A	N/A	N/A	Yes	Yes	[23]

ALP, Alkaline phosphatase; GGT, glutamyl transferase; N/A, not available.

2.2 PSC in Liver Disease Modeling and Drug Screening

The in vitro differentiation of PSCs not only opens new possibility of individualized regenerative medicine, but also provides new tools to investigate disease mechanism. iPSCs have the advantage that they can be derived from distinct individuals with different genetic background and the parental somatic cells are accessible with minimal invasive methods, like from blood or urine [25,26]. Also, iPSCs expanded in vitro are amenable for genetic editing, making it possible to study specific genetic factors in the same genetic background. There is emerging evidence showing that iPSCs created either artificially by gene manipulation or by obtaining and reprogramming patient cells are an effective model to recapitulate the pathophysiology of many liver diseases (including genetic diseases, metabolic diseases and infectious diseases) and pharmacological screening as well [27].

2.2.1 Genetic Inherited Diseases

Rashid et al. is the first to report the generation of iPSCs from patients with genetic inherited liver disease (α1-antitrypsin deficiency (A1ATD), Glycogen storage disease type 1a (GSD1a) and familial hypercholesterolemia (FH) [28], showing it is feasible to recapitulate an inherited liver disease in a culture dish. Subsequently, a number of groups have described iPSCs from liver disease patients including Wilson's disease [29,30], familial transthyretin amyloidosis (ATTR) [31,32], Alpers-Huttenlocher syndrome [33], and so on (Table 15.3). After differentiated to hepatocytes, the patient iPSC-derived cells showed disease phenotype. For example, A1ATD patient iPSC-derived hepatocytes showed accumulation of α1-antitrypsin polymers within the ER [28] and the iPSC model can even recapitulate the personalized variations of disease with the same genetic defects [34]. While after gene correction by genetic targeting or drug treatment, the disease phenotype can be rescued [31,35,36]. Another advantage of iPSC disease model is that multisystem complexity of the disease can be recapitulate by simultaneous differentiation to several cell types using the same iPSC cell line [30,31,37]. By differentiating patient derived iPSC to neurons, cardiomyocytes, and hepatocytes, Leung et al. and Leung and Murphy demonstrated that mutant TTR produced by the hepatocytes can induce oxidative stress and cell death of neuronal and cardiac cells from patient-matched iPSCs [31,37].

Besides hepatocytes abnormal, cholangitic diseases, a set of disorders of the cholangiocytes lining the bile ducts, are another big group of liver diseases. Cholangiopathies not only are responsible for significant morbidity and mortality, but also can progress to end-stage liver disease or even malignant transformation and account for significant proportion of liver transplants, especially pediatric and young adult liver transplants [38]. Differentiation of cholangiocytes from iPSCs provides new models to study the cholangiopathies pathogenesis and finds new therapeutic ways. Both studies by Sampaziotis et al. and Ogawa

TABLE 15.3 Contribution of hiPSCs to Modeling Genetic Inherited Liver Diseases

Disease	Mutation	Origin of iPSCs used for comparison	Donor cells and reprogramming strategy	Differentiation protocol	Phenotype	Correction	Drug testing	Refs.
A1ATD	3, homozygous Glu342Lys of α1-antitrypsin	3 asymptomatic person	Dermal fibroblasts, retrovirus, OSKM	[3,28]	Accumulation of α1-antitrypsin polymers within the ER	N/A	N/A	[28]
GSD1a	1 patient absent hepatic glucose-6-Phosphatase enzyme				Accumulated substantially greater amounts of intracellular glycogen, excessive lipid accumulation and excessive production of lactic acid, induced expression of 3 canonical glucagon-responsive genes after glucagon stimulation	N/A	N/A	
FH	1 patient, autosomal-dominant mutation in LDL receptor				absence of the LDL receptor, impaired ability to incorporate LDL			
Crigler-Najjar syndrome	1 patient, homozygous for 13bp deletion, exon 2 of UGT1A1				N/A	N/A	N/A	
Hereditary tyrosinemia type 1	1 patient, 1 allele of FAH has Val166Gly mutation(553T > G), the other allele is unknown				N/A	N/A	N/A	
A1ATD	3 patients, homozygous Glu342Lys of α1-antitrypsin	Control human iPSCs	Fibroblasts, retrovirus, OSKM or Sendai virus	[3,28]	Accumulation of α1-antitrypsin polymers within the ER	Yes		[36]

A1ATD	Z mutation (Glu342Lys) of α1-antitrypsin	Nonpatient iPSCs	Fibroblasts, episomal vectors	[152]	Formation of intracellular globules that are formed by the polymers of mutant AAT proteins, CBZ could reduce the AAT accumulation	Yes	Yes	Yes	[35]
A1ATD	6 homozygous PiZZ ATD patients (Glu342Lys) of α1-antitrypsin with severe liver disease, 2 requiring liver transplantation during childhood and 4 with lung disease but no clinically overt liver disease	A wild type control patient	Fibroblasts, lentivirus of ONSLKM, or plasmids encoding OSKM(l)LE and p53 shRNA,; or hSTEMCCA-loxP lentivirus	[151]	Dilated rER, abnormal accumulation and processing of the ATZ molecule; marked delay in the rate of ATZ degradation in iHeps from severe liver disease patients compared to those from no liver disease patients; globular inclusions that are partially covered with ribosomes were observed only in iHeps from individuals with severe liver disease	N/A	N/A	N/A	[34]
Familial amyloidotic polyneuropathy	3 Japanese female patients with FAP ATTRVal30Met	Normal human fibroblasts BJ from neonatal foreskin (ATCC)	Fibroblasts, Sendai virus of OSKM	[163,164]	N/A	N/A	N/A	N/A	[32]
Familial transthyretin amyloidosis (ATTR)	A patient heterozygous for the leucine-55-proline (L55P) of TTR mutation	Normal control-iPSC	N/A	[146,148]	The ratio of ATTRL55P-to-TTRWT monomers in ATTR hs was calculated to be 1:2, extracellular matrix, protein folding and stress response and connective tissue genes were overrepresented, ATTRL55P hepatic cell supernatant had a deleterious effect on neuronal and cardiac cell survival	N/A	N/A	Yes	[31,37]
WD	1 ATP7B M769V homozygous mutant patient		Fibroblast line (FFF0101990), retrovirus, OSKM	[9,165]	N/A	N/A	N/A	N/A	[30]

(Continued)

TABLE 15.3 Contribution of hiPSCs to Modeling Genetic Inherited Liver Diseases (cont.)

Disease	Mutation	Origin of iPSCs used for comparison	Donor cells and reprogramming strategy	Differentiation protocol	Phenotype	Correction	Drug testing	Refs.
WD	1 Chinese patient bears the R778L of ATP7B Chinese hotspot mutation	iPSCs from control fibroblasts IMR90 (90 iPSC-2) and H9 ESCs	Skin fibroblasts, retroviruses OSKM	[9]	Abnormal cytoplasmic localization of mutated ATP7B and defective copper transport	N/A	Yes	[29]
CF	2 patients, homozygous for ΔF508 of CFTR	WT iPSC	Fibroblasts, retrovirus, OSKM	[23]	Cysts were not completely hollow and contained branched ductal structures even after forsklin induction, maturation of the mutant cells was delayed, VX-809 and Corr-4a did not improve cyst formation but incompletely rescue the trafficking defect of the mutant protein	N/A	Yes	[23]
CF	1 patient, homozygous for ΔF508 of CFTR	WT	Skin fibroblasts, retrovirus, OSKM	[22]	Minimal CFTR protein expression, no change intracellular chloride in response to media with varying chloride concentrations, CF drug VX809 increased CFTR function and increased organoid size	N/A	Yes	[22]

ATTR, Amyloidogenic transthyretin; CBZ, carbamazepine; CF, cystic fibrosis; FAH, fumarylacetoacetate hydrolase; FH, familial hypercholesterolemia; GSD1a, glycogen storage disease type 1a; TTR, transthyretin; WD, Wilson's disease.

et al. showed that cholangiocytes differentiated from hiPSCs derived from patients with cystic fibrosis (a disease caused by mutations in the CFTR gene, encoding a cell-surface chloride transporter) are successfully recapitulating the CFTR misfolding and testing effect of CF drug VX809 [22,23].

Except for mimicking the disease phenotype and facilitating mechanism study in a dish, hepatocytes derived from patients also provide an valuable platform for drug screening and toxicity testing [39–41]. A proof-of-concept study by Chio et al. found five clinical drugs were identified to reduce AAT accumulation in A1ATD patient iPSC-derived hepatocyte-like cells through a blind large-scale chemical compound screening [35].

2.2.2 Metabolic Diseases

Besides genetic inherited diseases, patient-specific iPSC-derived hepatocytes also have been used for disease-in-a-dish models for metabolic diseases, such as nonalcoholic fatty liver disease (NAFLD) [42,43] and alcohol-induced liver injury [44]. In contrast to primary hepatocytes, the amount of iPSC-derived HLCs is not limited, which enables larger and more detailed mimicking of different nutritional conditions. Graffmann et al. showed that iPSC derived hepatocyte-like cells are a reliable cellular tool for modeling NAFLD by inducing fat storage with oleic acid [43]. In combination with gene editing technologies such as CRISPR/cas9, this model would be valuable for studying the role of genetic factors, such as single-nucleotide polymorphisms on *PNPLA3* and *TM6SF2* identified by GWAS studies [45], during disease development.

2.2.3 Infectious Diseases

Liver tropic viruses (such as hepatitis C virus (HCV) and hepatitis B virus (HBV)) are big health problems worldwide since HCV is the leading cause of hepatocellular carcinoma (HCC) particularly in Western countries and HBV mainly in China [46]. However, hepatitis virus infection has never been fully understood because the culprit virus mainly preferably infects human. HLC from hiPSC or hESC provides an important novel tool for studying the replication cycle of HCV and HBV. Using HCV pseudotype virus (HCVpv) and HCV subgenomic replicons, Cai et al. and Yoshida et al. showed that HCV can enter and replicate in iHeps [8,47]. Schwartz et al. found that the iPSC-derived iHLCs support the complete HCV life cycle, including replication and release of infectious virions, responding to antiviral drugs and inducing an antiviral inflammatory response using a genotype 2a HCV reporter virus expressing secreted Gaussia luciferase [48]. The permissiveness of HCV infection is stage dependent, as PSCs and DE cells were not permissive for HCV infection whereas HPCs and hepatocytes were persistently infected by both HCV derived in cell culture (HCVcc) and patient-derived virus (HCVser) [49], and a recent work by Yan et al. showed hHBs proved to have the highest permissiveness and infectivity compared with all other stages [50]. After engrafted to immune-deficient transgenic mice, the HLCs can also be infected with HCV-positive sera and supported long-term infection of multiple HCV genotypes [51]. This work provided a wonderful model to study chronic HCV infection in vivo. However, since the recipient mice are immune-deficient, it is difficult to study the immune system's contribution to disease progression during hepatitis virus infection. Humanized mouse model engrafted with both human immune and human liver cells will be a valuable tool to recapitulate hepatotropic pathogens infection, human immune response, chronic hepatitis, and associated immunopathogenesis [52,53], although the pioneer work was using HSC and liver progenitor cells from human fetal liver instead of cells from PSCs [54,55].

In 2014, Shlomai et al. showed for the first time that HBV can also infect iPSC-derived HLCs [56]. More recently, Xia et al. confirmed and extended this finding by utilizing human ESC-derived and human iPSC-derived HLCs as a robust and manageable in vitro model for HBV using an optimized differentiation protocol [57]. However, it is difficult to maintain virus infected iHLCs in vitro for long time to monitor the pathogenesis. To overcome this shortage, Kaneko et al. constructed a sodium taurocholate cotransporting polypeptide (NTCP) overexpressing HPCs derived from PSC [58]. These findings demonstrated the potential of human iPSC-derived HLC for in vitro studies of HBV biology and opened the door in generating HBV/HCV-susceptible cells from individual groups of patients and individuals with certain genetic polymorphism. Besides hepatotropic virus infection, iHLCs also support plasmodium liver-stage infection in vitro [59]. And more recently, Xia et al. identified two novel antivirals drugs with the HBV infected iHLCs model [57], further demonstrating the invaluable iPSC model in liver disease studying.

2.3 Perspectives

An alternative approach to hPSC-derived hepatocyte and cholangiocyte differentiation is the generation and capture of HPCs from hPSCs [18,60–62]. These cells can proliferate for extended periods while maintaining the bipotential, offering a simplified route for generating large numbers of hepatocytes or cholangiocytes within relatively short time in vitro by starting at an intermediate step and may improve the engraft efficiency in animal models.

Although the clinical studies of PSCs just started, experiments on animal models showed extremely encouraging results [27]. Subcutaneously implanted BAL support system fueled with hepatocytes differentiated from mouse ESCs improved liver function and prolonged survival of mice treated of 90% hepatectomization, suggesting that hepatocytes derived from mouse ESCs are comparable primary hepatocytes [63]. Besides serving as a cell model in regenerative medicine, disease modeling and drug development, iPSC also brings new hope for organ transplantation. With the development of tissue engineering [64], 3D printing [65] and interspecies chimera technologies

[66], functional grafts all from iPSCs are also expected in the close future (Fig. 15.1).

3 CLINICAL APPLICATION OF MSCS IN LIVER DISEASES

Besides the necrosis of parenchyma cells, dysregulation of nonparenchymal cells and subsequent abnormal changes in the microenvironment is another important pathogenic factor of liver diseases. Recent studies have shown that stem cell-based therapies may reduce liver inflammation, and subsequently improve scarring and replenish hepatocytes, thus could be a promising alternative strategy for patients with liver failure and cirrhosis [67]. By March 11, 2017, there are 273 clinical trials have been registered on ClinicalTrials.gov when searching for the terms "stem cells AND liver diseases". Most of these studies use allogeneic or autologous mesenchymal stem cells (MSCs) to treat various liver disease, especially liver cirrhosis. In this part, we summarize the role of MSC in basic and clinical studies of liver diseases.

3.1 General Properties of MSCs

Pioneered by Friedenstein et al. in 1968, MSCs were described as adherent cells located in the bone marrow (BM) and exhibiting a fibroblast-like morphology with strong proliferative capacities during ex vivo expansion and potential to differentiate into a range of cell types including adipocytes, chondroblasts, myocytes and osteoblasts [68,69]. It was originally isolated from BM, and later extended to a variety of tissues including adipose tissue, placenta, dental pulp, Wharton jelly of the umbilical cord, peripheral blood, brain, lung, liver, dermis, and skeletal muscle [70,71]. In 2006, the International Society for Cellular Therapy proposed minimal criteria to define human MSCs [71]. These cells are characterized by positive for CD73, CD90 and CD105 while negative for the endothelial marker CD31 and hematopoietic marker CD45 [70,71]. The real origin of MSCs is still controversial. It was widely believed that MSCs are originated from pericytes in the past decades. However, this traditional view is challenged by the study of Guimaraes-Camboa et al. recently [72]. Using lineage tracing, they showed that pericytes in vivo do not behave like stem cells both in aging and diverse pathological settings [72].

At beginning, research of MSCs focused on their role as multipotent cells capable to differentiate into various mesenchymal lineages, while more recently, their paracrine and immunoregulatory features have been noticed and resulted in their extensive application in clinical trials. MSCs are of low immunogenicity since they are characterized by low expression of human leukocyte antigen class I molecules and the absence of major histocompatibility complex class II antigens, Fas ligand and the co-stimulatory molecules B7-1, mB7-2, CD40, and CD40L [73], and thus both allogeneic and autologous MSCs are widely used in animal models and clinical trials. The safety of MSC transplantation has also been tested in clinical trials of various diseases with encouraging results [73]. Moreover, they are amendable to routine cryogenic storage and several commercial companies are actively developing allogeneic MSC-based products [74].

3.2 Clinical Studies of MSC Therapy in Liver Diseases

MSCs have been applied in clinical trials of various liver diseases, including liver inflammation, fibrosis, cirrhosis, liver cancer, and liver failure. Pilot studies in this field showed promising results of MSC transplantation and no serious side effects or complications have been reported [73,75].

At beginning, autologous bone marrow cells were mainly used. Autologous bone marrow cell infusion therapy from the peripheral vein significantly improved serum albumin, Child-Pugh scores, and AFP and proliferating cell nuclear antigen expression in liver biopsy tissue in nine patients with liver cirrhosis [76]. Another study with 8 patients with end-stage liver disease showed that autologous mesenchymal stem cell injection through peripheral or the portal vein improved liver function as verified by the Model for End-Stage Liver Disease (MELD) score evaluated at baseline and 1, 2, 4, 8, and 24 weeks after injection [77]. Decreased serum bilirubin and INR levels and increased albumin levels were also observed after autologous bone marrow mononuclear cell transplantation in 10 patients with advanced chronic liver disease [78]. And intravenous autologous bone marrow infusion in patients with alcoholic liver cirrhosis improves the serum levels of albumin and total protein and the prothrombin time in five patients [79].

However, a pilot randomized controlled study with 30 patients showed that the improvement mainly happens in the first 90 days [80], suggesting the requirement of repeated infusion to obtain long-term effects. Consistently, using 53 subject and 105 control of liver failure patients caused by hepatitis B, Peng et al. showed that single transplantation with autologous MMSC through hepatic artery resulted in markedly improved levels of ALB, TBIL, and PT and MELD score from 2–3 weeks after transplantation but no dramatic differences in incidence of HCC or mortality was observed at 192 weeks of follow-up [81]. Also, peripheral vein infusion of autologous mesenchymal stem cells in Egyptian HCV-positive patients with end-stage liver disease showed improved s-albumin within the first 2 weeks and prothrombin concentration and alanine transaminases after 1 month [82], strengthening the idea that repeated infusion is necessary to achieve a long-term improvement of pathology.

The positive effect of MSC transplantation is confirmed by a randomized, placebo-controlled trial with 15 patients received MSC and 12 patients received placebo [83]. However, recently, a multicenter, randomized, open-label phase 2 clinical trial with 72 patients showed that either one-time or two-time hepatic arterial injections of 5×10^7 autologous bone marrow-derived MSCs transplantation safely improved histologic fibrosis and liver function in patients with alchoholic cirrhosis [84], challenging the idea that repeated infusion is necessary. In 2017, Lin et al. showed that peripheral infusion of allogeneic bone marrow-derived MSCs (BM-MSCs) is safe for patients with HBV-related Acute-on-Chronic Liver Failure (ACLF) and significantly improves the 24-week survival rate, due to liver function improvement and a decrease in the rate of severe infections. One hundred and ten patients with HBV-related ACLF were enrolled in this randomized controlled study from 2010 to 2013. The control group was treated with standard medical therapy (SMT) only. The experiment group was infused weekly for 4 weeks with $1.0–10 \times 10^5$ cells/kg allogeneic BM-MSCs. During 24 weeks of follow-up, none of the MSC recipients experienced toxic side effects. No carcinoma occurred in any trial patient. The cumulated survival rate of MSC group patients was 73.2% (41/56) versus 55.6% (30/54) of the SMT group ($p = 0.026$). Compared with the control group, allogeneic BM-MSC treatment markedly improved clinical laboratory measurements, including serum total bilirubin, and MELD scores. The incidences of severe infection and intestinal paralysis in the MSC group were much lower than those in the SMT group. The patients died from multiple organ failure and severe infection in the SMT group were much more than those in the MSC group [84a].

Besides BM-MSC, umbilical cord-derived MSC (UC-MSC) is also widely used in clinical trials since its easy availability and abundance of inexpensive raw material. Shi et al. conducted the infusion three times at 4-week intervals and found that UC-MSC transfusions significant increased liver function and survival rates during the 48-week or 72-week follow-up period of ACLF patients associated with HBV infection [85]. The same group also showed that UC-MSC is effective in improving liver function and ascites in decompensated liver cirrhosis patients [86]. In 2016, Li et al. showed that UC-MSC transplantation is effective in improving the hepatic function and survival of HBV-ACLF patients treated with plasma exchange and Entecavir with 11 patients [87]. Interestingly, it showed that either undifferentiated or differentiated MSCs demonstrated partial improvement of liver function tests with elevation of prothrombin concentration and serum albumin levels, decline of elevated bilirubin and MELD score in MSCs group liver cirrhosis following chronic HCV infection [88].

The effectiveness of differentiated MSCs is confirmed by Amer et al., Autologous bone marrow-derived hepatocytes transplantation through intrahepatic and intrasplenic injection is safe and effective to improve limb edema, serum albumin, Child score, MELD score, fatigue scale, and performance status in patients with end-stage liver cell failure due to chronic hepatitis C using 40 patients (20 vs. 20). No difference was observed between intrahepatic and intrasplenic groups [89].

Besides disease caused by hepatocytes dysfunction, bone marrow mesenchymal stem cell and UC-MSC transplantation were also reported to be safe and effective in patients with autoimmune liver diseases like ursodeoxycholic acid (UDCA)-resistant primary biliary cirrhosis (PBC) or in patients undergoing liver transplantation [90,91]. A single-arm survey consisted of seven UDCA-resistant

PBC patients received UC-MSCs three times at 4-week intervals and regular UCDA as well. During the follow-up period, the symptoms of fatigue were largely alleviated, and some of the patients underwent remission of pruritus [91]. These results not only indicate a safe and feasible way to administrate UC-MSCs, but also demonstrate the improvement of liver function and quality of life in these patients. Pan et al. found that systemic infusion of MSCs led to a significant prevention of liver enzyme release and an improvement in the histology of the acute injured liver and may alleviate hepatic ischemia/reperfusion injuries after liver transplantation via inactivation of the MEK/ERK signaling pathway [92]. Recently, the same research group reported that systemic administration of umbilical cord mesenchymal stem cells was clinically safe and short-term favorable in patients with ischemic-type biliary lesions (ITBL) after liver transplantation. In their phase I, prospective, single-center clinical study, 12 ITBL patients were enrolled in the UC-MSCs group and received six doses of UC-MSCs (1.0×10^6 cells/kg) through peripheral intravenous infusion [93]. Compared with the traditional therapeutic group, interventional therapies were performed in 64.3% (45/70) patients in control group and 33.3% (4/12) patients in MSCs groups, respectively. MSCs therapy significantly decrease the need of interventional therapies ($p = 0.046$). The 1- and 2-year graft survival rates were higher in MSCs group (100% and 83.3) than in control group (72.9% and 68.6%). In a more recent study reported by Detry et al., ten liver transplant recipients under standard immunosuppression received $1.5–3 \times 10^6$/kg third-party unrelated MSCs on postoperative day 3 ± 2, and were prospectively compared to a control group of 10 liver transplant recipients [94]. In this study, no side effect of MSC infusion at day 3 after liver transplant could be detected, but this infusion did not promote tolerance, which needs further MSCs based trials in liver transplantation recipients [94].

3.3 Potential Mechanisms of MSCs Therapy in Liver Disease

3.3.1 Replenish Functional Hepatocytes by Transdifferentiation

Though the effectiveness of MSC therapy is widely demonstrated, the underlining mechanism is not fully demonstrated so far. In 2000, two studies showed that hepatocyte and cholangiocyte could derive from adult human bone marrow stem cells in vivo by examining archival autopsy and biopsy liver specimens from patients after cross-gender bone-marrow transplant [95,96]. The engraftment of hepatocyte and cholangiocyte ranged from 4% to 43% and from 4% to 38%, respectively in the study by Theise et al. [95], but much lower (about 0.5–2%) in the study by Alison et al. [96]. This phenomenon is also observed in animal models [97–100]. However, it is arguable whether the BM-derived hepatocytes are generated through "transdifferentiation" or "spontaneous cell fusion"[101–105]. To clarify this issue, Jang et al. developed a co-culture system of BM cells and injured liver separated by a barrier and demonstrated microenvironmental cues rather than fusion are responsible for conversion [106]. This conclusion is confirmed by Sato et al. later in rat [107]. Hepatocytes can also be induced from bone marrow multipotent adult progenitor cells of various species in vitro using FGF-4 and HGF [108], or by co-culture with liver cells [109] and pellet culture [110]. Recently, more and more studies showed that hepatic cells can be induced from MSCs from various tissues, including human adipose-derived stem cells [111]. And more recently, it has been shown that sera from cardiac-failure-associated congestive/ischemic liver patients have the potential to induce hepatic transdifferentiation of hMSCs in vitro, indicating that pathologic local environment may support the transdifferentiation of hMSCs to hepatocytes [112].

Although it is firmly confirmed that hMSCs can be induced into hepatocytes in vitro, the transition efficiency in vivo is believed to be very low. However, this view may be challenged by a recent study by Barliga et al. Using a transgenic mouse expressing mutant human α1-antitrypsin, they showed that host hepatocytes can be replaced 40% and 13% by mouse BM progenitors and human MSCs, respectively [113]. The transdifferentiation efficiency may be affected by the route of administration, as studies in sheep showed that intrahepatic injection of human MSCs resulted in more efficient generation and widespread distribution of hepatocytes throughout the liver parenchyma compared to intraperitoneal injection [114].

3.3.2 Modulating the Microenvironment by Secretome and Immunoregulatory Effect

Although cell replacement is an essential component of MSC therapy for some diseases, published researches have shown that the therapeutic effect of MSCs is mainly a result of immunomodulation and this function is "licensed" by inflammation [115] (Fig. 15.2). In response to inflammatory mediators, MSCs produce ample amount of immunoregulatory factors, cell-mobilization factors and growth factors and thereby facilitate tissue repair by tissue-resident cells. Many studies have demonstrated that MSCs can suppress the activation and function of various cells of innate and adaptive immune system, including macrophages, neutrophils, natural killer cells, dendritic cells, T lymphocytes, and B lymphocytes [116]. A series of studies have shown that MSCs inhibit T cell proliferation and promote Treg cells generation while suppress the differentiation into the T_H1 and T_H17 subsets of helper T cells [117]. Deng et al. also found that MSCs could induce dendritic cells to acquire a tolerogenic phenotype, which in turn elicits the generation of Treg cells [118]. In addition, MSCs also suppress the activation of natural killers stimulated by IL-2 or IL-15 [119,120]. Switching of macrophages from a pro-inflammatory type 1 to an anti-inflammatory type 2 phenotype was also observed when using MSCs to treat sepsis [121,122]. These extraordinary immunomodulatory properties endow MSCs with great potential for treating diverse inflammatory disorders. The liver is a unique anatomical and immunological site in which antigen-rich blood from the gastrointestinal tract is pressed through a network of sinusoids and scanned by antigen-presenting cells and lymphocytes [123]. Therefore, Liver is regarded as a "unique immunological organ" and well equipped within it fenestrated vascular sinusoids, with

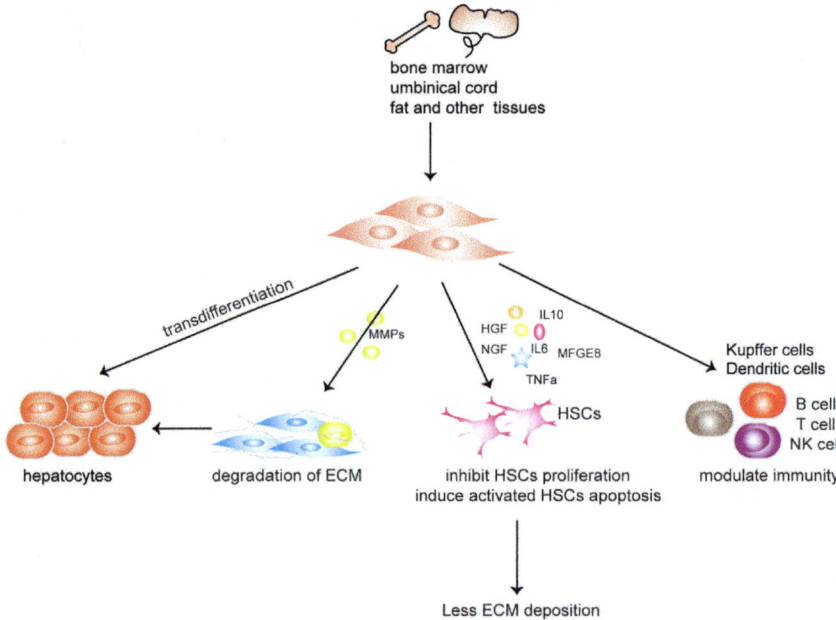

FIGURE 15.2 **Possible mechanisms of MSCs therapy in liver disease.** MSCs isolated from bone marrow, umbilical cord, adipose and other tissues has been widely used in treating various liver diseases. These cells has the potential to transdifferentiate to hepatocytes, reduce the deposition of extracellular matrix either by secreting matrix metalloproteinases (MMPs) or inhibit hepatic stellate cell (HSC) activation and induce HSCs apoptosis which is mediated by secreting cytokines, such as HGF, NGF. The immunoregulatory feature of MSCs also plays an essential role in liver disease therapy.

all cellular elements of innate immunity, macrophages, myeloid dendritic cells, Kuppfer cells and mucosal-associated invariant T cells, with all these complemented by barrier functions of liver sinusoidal epithelial cells. Acute or chronic liver injury resulting from various causes will build diverse microenvironments with different levels of inflammatory factors as well as recruitment of resident and circulating cells, which will definitely determine the immunoregulatory fate of MSCs and explain the inconsistency of efficacy of MSCs in diverse liver disease to some extent.

MSC can also improve pathology of liver disease by secretome [124] (Fig. 15.2). It has been increasingly observed that the transplanted MSCs could support the regeneration of hepatocytes, suppress the activation of HSCs and deposition of ECM and modulate immunologic activities by secreting various growth factors, cytokines through paracrine activity and change the local microenvironment. The mechanism studies and new therapy recipes are widely conducted in co-culture systems in vitro and different animal models in vivo, including mice [125], rat and pig [126]. Using mice after hematopoietic reconstitution with enhanced green fluorescent protein-expressing BM cells, Higashiyama et al. showed that autologous BM cells contribute to the spontaneous regression of liver fibrosis [127]. Similar phenomenon was observed after bone marrow-derived MSCs transplantation to CCl(4)-injured mouse liver [128]. The reduced deposition of ECM may be mediated by expression of matrix metalloproteinase (MMP)-13 and MMP-9 by MSCs [127] or inhibiting the phosphorylation of extracellular signal-regulated kinase (ERK) 1/2 and reduced expression of collagen type I and III by HSCs [129].

Besides directly inhibiting deposition of ECM, in vitro cultured MSCs also secret significant amount of vascular endothelial growth factor (VEGF), HGF, and transforming growth factor-beta (TGF-β) [129,130]. This effect is confirmed in clinical trials, as Jang et al. showed that both histological fibrosis and Child-Pugh score were improved while the levels of transforming growth factor-β1, type 1 collagen and α-smooth muscle actin (α-SMA) significantly decreased after BM-MSCs therapy in 12 patients [131]. Most cytokines secreted by MSCs not only are important growth factors for hepatocyte differentiation, but also can induce apoptosis of activated HSCs. Direct co-culture and indirect co-culture of MSCs with HSCs showed inhibited proliferation and induced apoptosis of HSCs [129,132–135]. In the direct co-culture system of BMSCs with HSCs and MSCs significantly suppressed the proliferation and α-SMA expression of HSCs, which was partially mediated by Notch pathway activation [132]. While in the indirect co-culture system, MSCs increased the production of HGF, IL-10, and NGF [133–135], leading to significant decrease in collagen synthesis and proliferation of activated HSCs, as well as elevated apoptosis of activated HSCs. UC-MSCs also suppress monocyte differentiation into DCs by secreting IL6 and HGF [136]. Interestingly, IL-6 secretion from activated HSCs will induce IL-10 secretion from MSCs, and the crosstalk between MSCs and HSCs may enhance the proapoptotic effect of MSCs on HSCs [134]. While HGF secreted by MSCs was responsible for the marked induction of apoptosis in HSCs [129,134]. Moreover, HGF is an important growth factor for hepatocytes development. Recently, An et al. found a new component of MSC secretome, milk fat globule-EGF factor 8 secreted by MSCs from umbilical cord, teeth and bone marrow and more strongly secreted by MSC-derived hepatocyte-like cells, can strongly inhibit the activation of human primary HSCs and has anti-fibrotic on liver fibrosis in mice by inhibiting TGF-β signaling and reduces ECM deposition [137].

The release of extracellular vesicles (EV) from cells has been implicated in intercellular communication, and may contribute to beneficial paracrine effects of stem cell-based therapies. Yan et al. showed that tail vein or oral gavage administration of human umbilical cord MSC-derived exosome has antioxidant and anti-apoptotic effects and could rescue the liver failure in CCl4 treating mouse. The protective effect is mainly mediated by through the delivery of GPX1 [138]. Similar results were obtained using both fresh and cryopreserved EV derived from BM-MSC and it seems that noncoding RNA Y-RNA-1 enriched in the extracellular vesicles also plays an important role [139].

3.4 Perspectives

Despite encouraging results from the most studies of MSCs transplantation so far, some studies got negative results. Nikeghbalian et al. showed no significant alterations of liver function parameters, liver enzymes, serum albumin, creatinine, serum bilirubin, and/or liver volume after intraportal administration of autologous bone marrow-derived CD133(+) or mononuclear cells of autologous cells in patients with end-stage liver disease. However, they used just in six patients [140]. Moreover, Yang et al. showed that human umbilical cord mesenchymal stem cells can be induced into cancer-associated mesenchymal stem cells (CA-MSCs) after incubation with condition medium from liver cancer cell line HepG-2 cells and show obtaining the proliferation and migration characteristic of CA-MSCs [141], indicating that the tumor microenvironment can remodel MSC cell fate.

Thus, the long-term clinical benefits and safety of MSC transplantation should be further confirmed in a large-sized randomized controlled trial, including the fibrogenic potential of MSCs and their ability to promote preexisting tumor cell growth. And many studies enrolled patients with chronic liver disease on the waiting list for liver transplantation [78,80] who have received various kinds of therapies before, which may interfere the effect of MSC therapy. Also, the type of transfused MSCs, the interval time, the dose of stem cells, the best route of administration (intravenous, intraperitoneal, intrahepatic, intrasplenic, or portal-venous injection) need to be systematically examined. Finally, the detailed mechanism of MSC function needs to be further elucidated, including the signals controlling homing and migration of MSCs into injury organs [142], factors determining engraft efficiency, survival period, and cell fate of transplanted cells, which will be benefited from lineage tracing and labeling technologies [143] and will facilitate the engineering of MSC to get better clinical efficacy in the future.

REFERENCES

[1] Fox IJ, et al. Treatment of the Crigler-Najjar syndrome type I with hepatocyte transplantation. N Engl J Med 1998;338:1422–6.

[2] Fisher RA, Strom SC. Human hepatocyte transplantation: worldwide results. Transplantation 2006;82:441–9.

[3] Hannan NR, Segeritz CP, Touboul T, Vallier L. Production of hepatocyte-like cells from human pluripotent stem cells. Nat Protoc 2013;8:430–7.

[4] Takahashi K, Yamanaka S. Induction of pluripotent stem cells from mouse embryonic and adult fibroblast cultures by defined factors. Cell 2006;126:663–76.

[5] Rambhatla L, Chiu CP, Kundu P, Peng Y, Carpenter MK. Generation of hepatocyte-like cells from human embryonic stem cells. Cell Transplant 2003;12:1–11.

[6] Schwartz RE, et al. Defined conditions for development of functional hepatic cells from human embryonic stem cells. Stem Cells Dev 2005;14:643–55.

[7] Shirahashi H, et al. Differentiation of human and mouse embryonic stem cells along a hepatocyte lineage. Cell Transplant 2004;13:197–211.

[8] Cai J, et al. Directed differentiation of human embryonic stem cells into functional hepatic cells. Hepatology 2007;45:1229–39.
[9] Song Z, et al. Efficient generation of hepatocyte-like cells from human induced pluripotent stem cells. Cell Res 2009;19:1233–42.
[10] Shan J, et al. Identification of small molecules for human hepatocyte expansion and iPS differentiation. Nat Chem Biol 2013;9:514–20.
[11] Ishii T, et al. Effects of extracellular matrixes and growth factors on the hepatic differentiation of human embryonic stem cells. Am J Physiol Gastrointest Liver Physiol 2008;295:G313–21.
[12] Soto-Gutierrez A, et al. Differentiation of human embryonic stem cells to hepatocytes using deleted variant of HGF and poly-amino-urethane-coated nonwoven polytetrafluoroethylene fabric. Cell Transplant 2006;15:335–41.
[13] Gieseck RL 3rd, et al. Maturation of induced pluripotent stem cell derived hepatocytes by 3D-culture. PLoS One 2014;9:e86372.
[14] Sivertsson L, Synnergren J, Jensen J, Bjorquist P, Ingelman-Sundberg M. Hepatic differentiation and maturation of human embryonic stem cells cultured in a perfused three-dimensional bioreactor. Stem Cells Dev 2013;22:581–94.
[15] Wang B, et al. Functional maturation of induced pluripotent stem cell hepatocytes in extracellular matrix-A comparative analysis of bioartificial liver microenvironments. Stem Cells Transl Med 2016;5:1257–67.
[16] Cervantes-Alvarez E, et al. Current strategies to generate mature human induced pluripotent stem cells derived cholangiocytes and future applications. Organogenesis 2017;13:1–15.
[17] Tanimizu N, Miyajima A, Mostov KE. Liver progenitor cells develop cholangiocyte-type epithelial polarity in three-dimensional culture. Mol Biol Cell 2007;18:1472–9.
[18] Zhao D, et al. Derivation and characterization of hepatic progenitor cells from human embryonic stem cells. PLoS One 2009;4:e6468.
[19] Dianat N, et al. Generation of functional cholangiocyte-like cells from human pluripotent stem cells and HepaRG cells. Hepatology 2014;60:700–14.
[20] De Assuncao TM, et al. Development and characterization of human-induced pluripotent stem cell-derived cholangiocytes. Lab Invest 2015;95:684–96.
[21] Jalan-Sakrikar N, et al. Hedgehog signaling overcomes an EZH2-dependent epigenetic barrier to promote cholangiocyte expansion. PLoS One 2016;11:e0168266.
[22] Sampaziotis F, et al. Cholangiocytes derived from human induced pluripotent stem cells for disease modeling and drug validation. Nat Biotechnol 2015;33:845–52.
[23] Ogawa M, et al. Directed differentiation of cholangiocytes from human pluripotent stem cells. Nat Biotechnol 2015;33:853–61.
[24] Takayama K, et al. Laminin 411 and 511 promote the cholangiocyte differentiation of human induced pluripotent stem cells. Biochem Biophys Res Commun 2016;474:91–6.
[25] Loh YH, et al. Generation of induced pluripotent stem cells from human blood. Blood 2009;113:5476–9.
[26] Zhou T, et al. Generation of induced pluripotent stem cells from urine. J Am Soc Nephrol 2011;22:1221–8.
[27] Shi Y, Inoue H, Wu JC, Yamanaka S. Induced pluripotent stem cell technology: a decade of progress. Nat Rev Drug Discov 2017;16:115–30.
[28] Rashid ST, et al. Modeling inherited metabolic disorders of the liver using human induced pluripotent stem cells. J Clin Investig 2010;120:3127–36.
[29] Zhang S, et al. Rescue of ATP7B function in hepatocyte-like cells from Wilson's disease induced pluripotent stem cells using gene therapy or the chaperone drug curcumin. Hum Mol Genet 2011;20:3176–87.

[30] Yi F, et al. Establishment of hepatic and neural differentiation platforms of Wilson's disease specific induced pluripotent stem cells. Protein Cell 2012;3:855–63.
[31] Leung A, et al. Induced pluripotent stem cell modeling of multisystemic, hereditary transthyretin amyloidosis. Stem Cell Rep 2013;1:451–63.
[32] Isono K, et al. Generation of familial amyloidotic polyneuropathy-specific induced pluripotent stem cells. Stem Cell Res 2014;12:574–83.
[33] Li S, et al. Valproic acid-induced hepatotoxicity in Alpers syndrome is associated with mitochondrial permeability transition pore opening-dependent apoptotic sensitivity in an induced pluripotent stem cell model. Hepatology 2015;61:1730–9.
[34] Tafaleng EN, et al. Induced pluripotent stem cells model personalized variations in liver disease resulting from alpha1-antitrypsin deficiency. Hepatology 2015;62:147–57.
[35] Choi SM, et al. Efficient drug screening and gene correction for treating liver disease using patient-specific stem cells. Hepatology 2013;57:2458–68.
[36] Yusa K, et al. Targeted gene correction of alpha1-antitrypsin deficiency in induced pluripotent stem cells. Nature 2011;478:391–4.
[37] Leung A, Murphy GJ. Multisystemic disease modeling of liver-derived protein folding disorders using induced pluripotent stem cells (iPSCs). Methods Mol Biol 2016;1353:261–70.
[38] Lazaridis KN, LaRusso NF. The Cholangiopathies. Mayo Clin Proc 2015;90:791–800.
[39] Lu J, et al. Morphological and functional characterization and assessment of iPSC-derived hepatocytes for in vitro toxicity testing. Toxicol Sci 2015;147:39–54.
[40] Szkolnicka D, et al. Accurate prediction of drug-induced liver injury using stem cell-derived populations. Stem Cells Transl Med 2014;3:141–8.
[41] Sirenko O, Hesley J, Rusyn I, Cromwell EF. High-content assays for hepatotoxicity using induced pluripotent stem cell-derived cells. Assay Drug Dev Technol 2014;12:43–54.
[42] Wruck W, Graffmann N, Kawala MA, Adjaye J. Concise review: current status and future directions on research related to nonalcoholic fatty liver disease. Stem Cells 2017;35:89–96.
[43] Graffmann N, et al. Modeling nonalcoholic fatty liver disease with human pluripotent stem cell-derived immature hepatocyte-like cells reveals activation of PLIN2 and confirms regulatory functions of peroxisome proliferator-activated receptor alpha. Stem Cells Dev 2016;25:1119–33.
[44] Tian L, Deshmukh A, Prasad N, Jang YY. Alcohol increases liver progenitor populations and induces disease phenotypes in human IPSC-derived mature stage hepatic cells. Int J Biol Sci 2016;12:1052–62.
[45] Dongiovanni P, Romeo S, Valenti L. Genetic factors in the pathogenesis of nonalcoholic fatty liver and steatohepatitis. Biomed Res Int 2015;2015:10. 460190.
[46] Alter MJ. Epidemiology of hepatitis C virus infection. World J Gastroenterol 2007;13:2436–41.
[47] Yoshida T, et al. Use of human hepatocyte-like cells derived from induced pluripotent stem cells as a model for hepatocytes in hepatitis C virus infection. Biochem Biophys Res Commun 2011;416:119–24.
[48] Schwartz RE, et al. Modeling hepatitis C virus infection using human induced pluripotent stem cells. Proc Natl Acad Sci U S A 2012;109:2544–8.
[49] Wu X, et al. Productive hepatitis C virus infection of stem cell-derived hepatocytes reveals a critical transition to viral permissiveness during differentiation. PLoS Pathog 2012;8:e1002617.
[50] Yan F, et al. Human ES cell-derived hepatoblasts are an optimal lineage stage for HCV infection. Hepatology 2017;66(3):717–35.
[51] Carpentier A, et al. Engrafted human stem cell-derived hepatocytes establish an infectious HCV murine model. J Clin Investig 2014;124:4953–64.

[52] Bility MT, et al. Hepatitis B virus infection and immunopathogenesis in a humanized mouse model: induction of human-specific liver fibrosis and M2-like macrophages. PLoS Pathog 2014;10:e1004032.
[53] Bility MT, et al. Chronic hepatitis C infection-induced liver fibrogenesis is associated with M2 macrophage activation. Sci Rep 2016;6:39520.
[54] Bility MT, et al. Generation of a humanized mouse model with both human immune system and liver cells to model hepatitis C virus infection and liver immunopathogenesis. Nat Protoc 2012;7:1608–17.
[55] Washburn ML, et al. A humanized mouse model to study hepatitis C virus infection, immune response, and liver disease. Gastroenterology 2011;140:1334–44.
[56] Shlomai A, et al. Modeling host interactions with hepatitis B virus using primary and induced pluripotent stem cell-derived hepatocellular systems. Proc Natl Acad Sci U S A 2014;111:12193–8.
[57] Xia Y, et al. Human stem cell-derived hepatocytes as a model for hepatitis B virus infection, spreading and virus-host interactions. J Hepatol 2017;66:494–503.
[58] Kaneko S, et al. Human induced pluripotent stem cell-derived hepatic cell lines as a new model for host interaction with hepatitis B virus. Sci Rep 2016;6:29358.
[59] Ng S, et al. Human iPSC-derived hepatocyte-like cells support Plasmodium liver-stage infection in vitro. Stem Cell Rep 2015;4:348–59.
[60] Li F, et al. Hepatoblast-like progenitor cells derived from embryonic stem cells can repopulate livers of mice. Gastroenterology 2010;139:2158–2169.e8.
[61] Yanagida A, Nakauchi H, Kamiya A. Generation and in vitro expansion of hepatic progenitor cells from human iPS cells. Methods Mol Biol 2016;1357:295–310.
[62] Takayama K, et al. Long-term self-renewal of human ES/iPS-derived hepatoblast-like cells on human laminin 111-coated dishes. Stem Cell Rep 2013;1:322–35.
[63] Soto-Gutierrez A, et al. Reversal of mouse hepatic failure using an implanted liver-assist device containing ES cell-derived hepatocytes. Nat Biotechnol 2006;24:1412–9.
[64] Uygun BE, et al. Organ reengineering through development of a transplantable recellularized liver graft using decellularized liver matrix. Nat Med 2010;16:814–20.
[65] Ma X, et al. Deterministically patterned biomimetic human iPSC-derived hepatic model via rapid 3D bioprinting. Proc Natl Acad Sci U S A 2016;113:2206–11.
[66] Wu J, Izpisua Belmonte JC. Interspecies chimeric complementation for the generation of functional human tissues and organs in large animal hosts. Transgenic Res 2016;25:375–84.
[67] Zhang Z, Wang FS. Stem cell therapies for liver failure and cirrhosis. J Hepatol 2013;59:183–5.
[68] Friedenstein AJ, Petrakova KV, Kurolesova AI, Frolova GP. Heterotopic of bone marrow. Analysis of precursor cells for osteogenic and hematopoietic tissues. Transplantation 1968;6:230–47.
[69] Pittenger MF, et al. Multilineage potential of adult human mesenchymal stem cells. Science 1999;284:143–7.
[70] Bianco P. "Mesenchymal" stem cells. Annu Rev Cell Dev Biol 2014;30:677–704.
[71] Frenette PS, Pinho S, Lucas D, Scheiermann C. Mesenchymal stem cell: keystone of the hematopoietic stem cell niche and a stepping-stone for regenerative medicine. Annu Rev Immunol 2013;31:285–316.
[72] Guimaraes-Camboa N, et al. Pericytes of multiple organs do not behave as mesenchymal stem cells in vivo. Cell Stem Cell 2017;20:345–359.e5.
[73] Eom YW, Shim KY, Baik SK. Mesenchymal stem cell therapy for liver fibrosis. Korean J Intern Med 2015;30:580–9.
[74] Johnson CL, Soeder Y, Dahlke MH. Mesenchymal stromal cells for immunoregulation after liver transplantation: the scene in 2016. Curr Opin Organ Transplant 2016;21:541–9.

[75] Eom YW, Kim G, Baik SK. Mesenchymal stem cell therapy for cirrhosis: present and future perspectives. World J Gastroenterol 2015;21:10253–61.
[76] Terai S, et al. Improved liver function in patients with liver cirrhosis after autologous bone marrow cell infusion therapy. Stem Cells 2006;24:2292–8.
[77] Kharaziha P, et al. Improvement of liver function in liver cirrhosis patients after autologous mesenchymal stem cell injection: a phase I-II clinical trial. Eur J Gastroenterol Hepatol 2009;21:1199–205.
[78] Lyra AC, et al. Feasibility and safety of autologous bone marrow mononuclear cell transplantation in patients with advanced chronic liver disease. World J Gastroenterol 2007;13:1067–73.
[79] Saito T, et al. Potential therapeutic application of intravenous autologous bone marrow infusion in patients with alcoholic liver cirrhosis. Stem Cells Dev 2011;20:1503–10.
[80] Lyra AC, et al. Infusion of autologous bone marrow mononuclear cells through hepatic artery results in a short-term improvement of liver function in patients with chronic liver disease: a pilot randomized controlled study. Eur J Gastroenterol Hepatol 2010;22:33–42.
[81] Peng L, et al. Autologous bone marrow mesenchymal stem cell transplantation in liver failure patients caused by hepatitis B: short-term and long-term outcomes. Hepatology 2011;54:820–8.
[82] Salama H, et al. Peripheral vein infusion of autologous mesenchymal stem cells in Egyptian HCV-positive patients with end-stage liver disease. Stem Cell Res Ther 2014;5:70.
[83] Mohamadnejad M, et al. Randomized placebo-controlled trial of mesenchymal stem cell transplantation in decompensated cirrhosis. Liver Int 2013;33:1490–6.
[84] Suk KT, et al. Transplantation with autologous bone marrow-derived mesenchymal stem cells for alcoholic cirrhosis: phase 2 trial. Hepatology 2016;64:2185–97.
[84a] https://www.ncbi.nlm.nih.gov/pubmed/28370357h
[85] Shi M, et al. Human mesenchymal stem cell transfusion is safe and improves liver function in acute-on-chronic liver failure patients. Stem Cells Transl Med 2012;1:725–31.
[86] Zhang Z, et al. Human umbilical cord mesenchymal stem cells improve liver function and ascites in decompensated liver cirrhosis patients. J Gastroenterol Hepatol 2012;27(Suppl. 2):112–20.
[87] Li YH, et al. Umbilical cord-derived mesenchymal stem cell transplantation in hepatitis B virus related acute-on-chronic liver failure treated with plasma exchange and entecavir: a 24-month prospective study. Stem Cell Rev 2016;12:645–53.
[88] El-Ansary M, et al. Phase II trial: undifferentiated versus differentiated autologous mesenchymal stem cells transplantation in Egyptian patients with HCV induced liver cirrhosis. Stem Cell Rev 2012;8:972–81.
[89] Amer ME, et al. Clinical and laboratory evaluation of patients with end-stage liver cell failure injected with bone marrow-derived hepatocyte-like cells. Eur J Gastroenterol Hepatol 2011;23:936–41.
[90] Wang L, et al. Allogeneic bone marrow mesenchymal stem cell transplantation in patients with UDCA-resistant primary biliary cirrhosis. Stem Cells Dev 2014;23:2482–9.
[91] Wang L, et al. Pilot study of umbilical cord-derived mesenchymal stem cell transfusion in patients with primary biliary cirrhosis. J Gastroenterol Hepatol 2013;28(Suppl. 1):85–92.
[92] Pan GZ, et al. Bone marrow mesenchymal stem cells ameliorate hepatic ischemia/reperfusion injuries via inactivation of the MEK/ERK signaling pathway in rats. J Surg Res 2012;178:935–48.
[93] Zhang YC, et al. Therapeutic potentials of umbilical cord-derived mesenchymal stromal cells for ischemic-type biliary lesions following liver transplantation. Cytotherapy 2017;19:194–9.
[94] Detry O, et al. Infusion of mesenchymal stromal cells after deceased liver transplantation: a phase I-II, open-label, clinical study. J Hepatol 2017;67(1):47–55.
[95] Theise ND, et al. Liver from bone marrow in humans. Hepatology 2000;32:11–6.
[96] Alison MR, et al. Hepatocytes from non-hepatic adult stem cells. Nature 2000;406:257.

[97] Theise ND, et al. Derivation of hepatocytes from bone marrow cells in mice after radiation-induced myeloablation. Hepatology 2000;31:235–40.
[98] Almeida-Porada G, Porada CD, Chamberlain J, Torabi A, Zanjani ED. Formation of human hepatocytes by human hematopoietic stem cells in sheep. Blood 2004;104:2582–90.
[99] Shu SN, et al. Hepatic differentiation capability of rat bone marrow-derived mesenchymal stem cells and hematopoietic stem cells. World J Gastroenterol 2004;10:2818–22.
[100] Terai S, et al. An in vivo model for monitoring trans-differentiation of bone marrow cells into functional hepatocytes. J Biochem 2003;134:551–8.
[101] Ying QL, Nichols J, Evans EP, Smith AG. Changing potency by spontaneous fusion. Nature 2002;416:545–8.
[102] Terada N, et al. Bone marrow cells adopt the phenotype of other cells by spontaneous cell fusion. Nature 2002;416:542–5.
[103] Vassilopoulos G, Wang PR, Russell DW. Transplanted bone marrow regenerates liver by cell fusion. Nature 2003;422:901–4.
[104] Wang X, et al. Cell fusion is the principal source of bone-marrow-derived hepatocytes. Nature 2003;422:897–901.
[105] Quintana-Bustamante O, et al. Hematopoietic mobilization in mice increases the presence of bone marrow-derived hepatocytes via in vivo cell fusion. Hepatology 2006;43:108–16.
[106] Jang YY, Collector MI, Baylin SB, Diehl AM, Sharkis SJ. Hematopoietic stem cells convert into liver cells within days without fusion. Nat Cell Biol 2004;6:532–9.
[107] Sato Y, et al. Human mesenchymal stem cells xenografted directly to rat liver are differentiated into human hepatocytes without fusion. Blood 2005;106:756–63.
[108] Schwartz RE, et al. Multipotent adult progenitor cells from bone marrow differentiate into functional hepatocyte-like cells. J Clin Investig 2002;109:1291–302.
[109] Lange C, et al. Liver-specific gene expression in mesenchymal stem cells is induced by liver cells. World J Gastroenterol 2005;11:4497–504.
[110] Ong SY, Dai H, Leong KW. Inducing hepatic differentiation of human mesenchymal stem cells in pellet culture. Biomaterials 2006;27:4087–97.
[111] Fu Y, et al. Rapid generation of functional hepatocyte-like cells from human adipose-derived stem cells. Stem Cell Res Ther 2016;7:105.
[112] Bishi DK, Mathapati S, Cherian KM, Guhathakurta S, Verma RS. In vitro hepatic trans-differentiation of human mesenchymal stem cells using sera from congestive/ischemic liver during cardiac failure. PLoS One 2014;9:e92397.
[113] Baligar P, et al. Bone marrow stem cell therapy partially ameliorates pathological consequences in livers of mice expressing mutant human alpha1-antitrypsin. Hepatology 2017;65(4):1319–35.
[114] Chamberlain J, et al. Efficient generation of human hepatocytes by the intrahepatic delivery of clonal human mesenchymal stem cells in fetal sheep. Hepatology 2007;46:1935–45.
[115] Bernardo ME, Fibbe WE. Mesenchymal stromal cells: sensors and switchers of inflammation. Cell Stem Cell 2013;13:392–402.
[116] Wang Y, Chen X, Cao W, Shi Y. Plasticity of mesenchymal stem cells in immunomodulation: pathological and therapeutic implications. Nat Immunol 2014;15:1009–16.
[117] Luz-Crawford P, et al. Mesenchymal stem cells generate a CD4 + CD25 + Foxp3+ regulatory T cell population during the differentiation process of Th1 and Th17 cells. Stem Cell Res Ther 2013;4:65.
[118] Deng Y, et al. Umbilical cord-derived mesenchymal stem cells instruct dendritic cells to acquire tolerogenic phenotypes through the IL-6-mediated upregulation of SOCS1. Stem Cells Dev 2014;23:2080–92.

[119] Spaggiari GM, Capobianco A, Becchetti S, Mingari MC, Moretta L. Mesenchymal stem cell-natural killer cell interactions: evidence that activated NK cells are capable of killing MSCs, whereas MSCs can inhibit IL-2-induced NK-cell proliferation. Blood 2006;107:1484–90.

[120] Spaggiari GM, et al. Mesenchymal stem cells inhibit natural killer-cell proliferation, cytotoxicity, and cytokine production: role of indoleamine 2,3-dioxygenase and prostaglandin E2. Blood 2008;111:1327–33.

[121] Nemeth K, et al. Bone marrow stromal cells attenuate sepsis via prostaglandin E(2)-dependent reprogramming of host macrophages to increase their interleukin-10 production. Nat Med 2009;15:42–9.

[122] Abumaree MH, et al. Human placental mesenchymal stem cells (pMSCs) play a role as immune suppressive cells by shifting macrophage differentiation from inflammatory M1 to anti-inflammatory M2 macrophages. Stem Cell Rev 2013;9:620–41.

[123] Racanelli V, Rehermann B. The liver as an immunological organ. Hepatology 2006;43:S54–62.

[124] Parekkadan B, et al. Mesenchymal stem cell-derived molecules reverse fulminant hepatic failure. PLoS One 2007;2:e941.

[125] Tan L, et al. Contribution of dermal-derived mesenchymal cells during liver repair in two different experimental models. Sci Rep 2016;6:25314.

[126] Sang JF, et al. Combined mesenchymal stem cell transplantation and interleukin-1 receptor antagonism after partial hepatectomy. World J Gastroenterol 2016;22:4120–35.

[127] Higashiyama R, et al. Bone marrow-derived cells express matrix metalloproteinases and contribute to regression of liver fibrosis in mice. Hepatology 2007;45:213–22.

[128] Rabani V, et al. Mesenchymal stem cell infusion therapy in a carbon tetrachloride-induced liver fibrosis model affects matrix metalloproteinase expression. Cell Biol Int 2010;34:601–5.

[129] Wang J, et al. Inhibition of hepatic stellate cells proliferation by mesenchymal stem cells and the possible mechanisms. Hepatol Res 2009;39:1219–28.

[130] Rehman J, et al. Secretion of angiogenic and antiapoptotic factors by human adipose stromal cells. Circulation 2004;109:1292–8.

[131] Jang YO, et al. Histological improvement following administration of autologous bone marrow-derived mesenchymal stem cells for alcoholic cirrhosis: a pilot study. Liver Int 2014;34:33–41.

[132] Chen S, et al. Activation of Notch1 signaling by marrow-derived mesenchymal stem cells through cell-cell contact inhibits proliferation of hepatic stellate cells. Life Sci 2011;89:975–81.

[133] Jang YO, Jun BG, Baik SK, Kim MY, Kwon SO. Inhibition of hepatic stellate cells by bone marrow-derived mesenchymal stem cells in hepatic fibrosis. Clin Mol Hepatol 2015;21:141–9.

[134] Parekkadan B, et al. Immunomodulation of activated hepatic stellate cells by mesenchymal stem cells. Biochem Biophys Res Commun 2007;363:247–52.

[135] Lin N, et al. Nerve growth factor-mediated paracrine regulation of hepatic stellate cells by multipotent mesenchymal stromal cells. Life Sci 2009;85:291–5.

[136] Deng Y, et al. Umbilical cord-derived mesenchymal stem cells instruct monocytes towards an IL10-producing phenotype by secreting IL6 and HGF. Sci Rep 2016;6:37566.

[137] An SY, et al. Milk fat globule-EGF factor 8, secreted by mesenchymal stem cells, protects against liver fibrosis in mice. Gastroenterology 2016;152(5):1174–86.

[138] Yan Y, et al. hucMSC exosome-derived GPX1 is required for the recovery of hepatic oxidant injury. Mol Ther 2017;25:465–79.

[139] Haga H, Yan IK, Takahashi K, Matsuda A, Patel T. Extracellular vesicles from bone marrow-derived mesenchymal stem cells improve survival from lethal hepatic failure in mice. Stem Cells Transl Med 2017;6(4):1262–2127.

[140] Nikeghbalian S, et al. Autologous transplantation of bone marrow-derived mononuclear and CD133(+) cells in patients with decompensated cirrhosis. Arch Iran Med 2011;14:12–7.
[141] Yang J, et al. Condition medium of HepG-2 cells induces the transdifferentiation of human umbilical cord mesenchymal stem cells into cancerous mesenchymal stem cells. Am J Transl Res 2016;8:3429–38.
[142] Xiao Ling K, et al. Stromal derived factor-1/CXCR4 axis involved in bone marrow mesenchymal stem cells recruitment to injured liver. Stem Cells Int 2016;2016:8906945.
[143] Hu SL, et al. In vitro labeling of human umbilical cord mesenchymal stem cells with superparamagnetic iron oxide nanoparticles. J Cell Biochem 2009;108:529–35.
[144] Hay DC, et al. Direct differentiation of human embryonic stem cells to hepatocyte-like cells exhibiting functional activities. Cloning Stem Cells 2007;9:51–62.
[145] Agarwal S, Holton KL, Lanza R. Efficient differentiation of functional hepatocytes from human embryonic stem cells. Stem Cells 2008;26:1117–27.
[146] Hay DC, et al. Highly efficient differentiation of hESCs to functional hepatic endoderm requires ActivinA and Wnt3a signaling. Proc Natl Acad Sci U S A 2008;105:12301–6.
[147] Hay DC, et al. Efficient differentiation of hepatocytes from human embryonic stem cells exhibiting markers recapitulating liver development in vivo. Stem Cells 2008;26:894–902.
[148] Sullivan GJ, et al. Generation of functional human hepatic endoderm from human induced pluripotent stem cells. Hepatology 2010;51:329–35.
[149] Basma H, et al. Differentiation and transplantation of human embryonic stem cell-derived hepatocytes. Gastroenterology 2009;136:990–9.
[150] Touboul T, et al. Generation of functional hepatocytes from human embryonic stem cells under chemically defined conditions that recapitulate liver development. Hepatology 2010;51:1754–65.
[151] Si-Tayeb K, et al. Highly efficient generation of human hepatocyte-like cells from induced pluripotent stem cells. Hepatology 2010;51:297–305.
[152] Liu H, Kim Y, Sharkis S, Marchionni L, Jang YY. In vivo liver regeneration potential of human induced pluripotent stem cells from diverse origins. Sci Transl Med 2011;3:82ra39.
[153] Chen YF, et al. Rapid generation of mature hepatocyte-like cells from human induced pluripotent stem cells by an efficient three-step protocol. Hepatology 2012;55:1193–203.
[154] Pal R, Mamidi MK, Das AK, Gupta PK, Bhonde R. A simple and economical route to generate functional hepatocyte-like cells from hESCs and their application in evaluating alcohol induced liver damage. J Cell Biochem 2012;113:19–30.
[155] Zhao D, et al. Promotion of the efficient metabolic maturation of human pluripotent stem cell-derived hepatocytes by correcting specification defects. Cell Res 2013;23:157–61.
[156] Ma X, et al. Highly efficient differentiation of functional hepatocytes from human induced pluripotent stem cells. Stem Cells Transl Med 2013;2:409–19.
[157] Duan Y, et al. Differentiation and characterization of metabolically functioning hepatocytes from human embryonic stem cells. Stem Cells 2010;28:674–86.
[158] Tasnim F, Phan D, Toh YC, Yu H. Cost-effective differentiation of hepatocyte-like cells from human pluripotent stem cells using small molecules. Biomaterials 2015;70:115–25.
[159] Siller R, Greenhough S, Naumovska E, Sullivan GJ. Small-molecule-driven hepatocyte differentiation of human pluripotent stem cells. Stem Cell Rep 2015;4:939–52.
[160] Carpentier A, et al. Hepatic differentiation of human pluripotent stem cells in miniaturized format suitable for high-throughput screen. Stem Cell Res 2016;16:640–50.
[161] Takayama K, et al. CCAAT/enhancer binding protein-mediated regulation of TGFbeta receptor 2 expression determines the hepatoblast fate decision. Development 2014;141: 91–100.

[162] De Assuncao TM, et al. Development and characterization of human-induced pluripotent stem cell-derived cholangiocytes. Lab Invest 2015;95:1218.
[163] Shiraki N, et al. Efficient differentiation of embryonic stem cells into hepatic cells in vitro using a feeder-free basement membrane substratum. PLoS One 2011;6:e24228.
[164] Shiraki N, et al. Differentiation of mouse and human embryonic stem cells into hepatic lineages. Genes Cells 2008;13:731–46.
[165] Li W, et al. Rapid induction and long-term self-renewal of primitive neural precursors from human embryonic stem cells by small molecule inhibitors. Proc Natl Acad Sci U S A 2011;108:8299–304.

Index

A

ABC subfamily G, member 2 (ABCG2), 224
Acellular scaffolds, 130
2-Acetylaminofluorene (AAF) model, 260
Adeno-associated viruses (AAVs) integration, 65
Adipose-derived stem cells (ADSCs), 131
Adipose tissue-derived mesenchymal stem cells (AT-MSCs), 85
Alagille syndrome, 14
Aldehyde dehydrogenase (ALDH) activity, 227
Alginate hydrogel beads, 149
α1-antitrypsin deficiency (A1ATD), 325
Annexin A3 (ANXA3), 305
ATPase copper transporting beta (ATP7b), 13

B

Basic helix-loop-helix transcription factor (hand2), 9
Bile acids, 2
Bile canaliculus-like structure, 156
Bile duct
  dynamic remodeling, 196
  as stem cells, 57
Bile duct ligation (BDL), 39
Bioartificial liver (BAL) system, 317
Blood glucose level maintenance, 1
Bone marrow cells, 57
Bone marrow-derived MSCs (BM-MSCs), 85
Bone morphogenetic protein (BMP), 126
Brain-dead liver transplantation, 77
Bromodomain and extraterminal domain (BET), 40
Budd–Chiari syndrome, 182
Bud formation, liver development
  endoderm and mesoderm, 3
  extracellular and intracellular signaling network dissection, 3
  Forkhead box protein A, 4
  hepatoblast differentiation, 3
  MKK4 and MKK7 expression, 6
  Prox1 expression, 6
  Tbx3 expression, 5

C

Canal of Hering, 57
Cancer, platelets
  development, 160
  metastasis, 160
  platelet activating factors, 160
  progression, 160
Cancer stem cells (CSCs). *See also* Liver cancer stem cells (LCSCs)
  aldehyde dehydrogenase activity, 227
  CD13, 231
  CD24, 230
  CD44, 229
  CD47, 231
  CD90, 228
  CD274, 233
  CD133 antigen, 227
  cell connection and polarity, 59
  cholangiocarcinoma, 279
  α2δ1, 233
  drug repositioning, 237
  EpCAM, 229
  epithelial-mesenchymal transition, 58
  hepatocellular carcinoma
    controversies, 280
    heterogeneity, 277
    markers, 277
  hierarchical CSC model, 223
  immunotherapy, 237
  K19, 231
  markers, 58, 59, 225
  mesenchymal-epithelial transition, 58
  OV6, 228
  prognostic factors, 233
  properties, 58
  self-renewal, 273
  side population cell sorting, 224
  SOX9, 232
  TLR4, 232
  tumor initiating cells, 58
CD13, 231
CD24, 230
CD44, 229
CD47, 231
CD90, 228

CD274, 233
CD133 antigen, 227
Celecoxib, 238
Cell transplantation
   candidate cell sources, 78, 79
   hepatocytes
      ESC-derived hepatocytes, 80
      induced hepatocyte-like cells, 83
      iPSCs-derived hepatocytes, 81
      liver progenitor/stem cells. *See* Liver progenitor/stem cells (LPCs)
      mature hepatic functionality, 77
Centrilobular ductular reaction, 182, 185
Chemically induced liver progenitors (CLiPs)
   diploid hepatocytes, 92
   liver regenerative therapy, 92
   stable long-term expansion, 91
   in vivo repopulation capacity, 91
Chemoresistance, 290
Cholangiocarcinoma (CCA)
   characteristics, 279
   controversies in stem cells, 280
   CSC markers, 279
   therapy, 280
Cholangiocytes, 273
   differentiation from pluripotent stem cells, 320
   heterogeneity, 38
   lineage plasticity
      adult cholangiocytes, 38
      fetal and neonatal cholangiocytes, 37
      molecular mechanisms, 40
   transdifferentiation, 276
Cholestasis, ductular reaction, 184
Choline-deficient ethionine-supplemented (CDE) diet model, 260
Chromodomain-helicase-deoxyribonucleic acid (DNA)-binding protein 4 (CHD4), 229
Chronic liver disease (CLD)
   bile duct system remodeling, 196
   chronic inflammatory responses, 180
   etiologies, 163
   fibrotic expansion, 180
   hepatocellular carcinoma, 163
   liver cirrhosis
      ductular reaction, 181, 182
      hepatocellular carcinoma, 182
      histological features, 181
      regenerative nodules sizes, 181
   liver fibrosis, 163
   liver transplantation, 163
   platelet transfusions on, 168
   thrombocytopenia, 164
   TPO receptor agonist effect, 168
   viral hepatitis, 180
   zonal necrosis of parenchyma, 180
Chronic liver injury
   abnormal regeneration, 255
   causative agents, 255
   fluorescence-activated cell sorting-based isolation, 256
   liver stem/progenitor cell
      human liver diseases, 260
      mouse model, 260
      rat model, 260
   PHx model, 255
Claudin 4 (CLDN4), 40
Constant latency, 290
CRISPR/Cas9 technology, 135
Cytotoxic T lymphocyte (CTL)-based immunotherapy, 305

D

Decellularized scaffolds, 130
Delta-like 1 homolog (DLK1) positive cells, 214
4,4'-Diaminodiphenylmethane (DAPM)-injured livers, 39
3,5-Diethoxycarbonyl-1,4-dihydro-collidine (DDC) model, 260
Dipeptidyl peptidase IV (DPPIV) positive rats, 41
Disulfiram, 237
DNA methylation
   DNA methyltransferases, 61
   hypomethylation, 61
   methyl binding domain, 61
   TET-mediated demethylation, 61
   tumor suppressor genes, 62
Drosophila Hippo pathway, 12
Ductal plates, 22
Ductular reaction
   atypical ductular reaction, 182
   in biliary obstruction, 182
   centrilobular ductular reaction, 182, 185
   cholestasis, 184
   definition, 182
   ductular metaplasia, 184, 186
   in fulminant hepatitis, 182
   hepatocyte differentiation, 188, 194
   liver cirrhosis, 181, 182
   periportal ductular reaction, 182, 184
   Sox9-CreER/ROSA26R cell lineage tracing system, 187
   stem/progenitor-like cells, 187
   Wnt-Frizzled signaling, 187
Ductular reaction (DR), 39

E

Eltrombopag, chronic liver disease, 168
Embryonic bodies (EBs), 80, 318
Endothelial cells, 126
Epigenetic factors, 297
Episomal vectors, 106
Epithelial cell adhesion molecule (EpCAM), 214, 229
Epithelial cells, 35
Epithelial-mesenchymal transition (EMT), 58
ESC-derived hepatocytes, 80
Extracellular matrix (ECM)
    cell-friendly 3D ECM condition, 130
    2D culture systems, 129
    loading, multicellular spheroids
        albumin secretion activity, 153, 155
        cell polarity formation, 155
        ECM-loaded MCS, 154
        emulsion method, 151
        functions, 153
        matrigel solution, 153, 154
        methylcellulose medium, 151, 152
Extrahepatic bile ducts (EHBD), 38

F

Fibroblast growth factor 19 (FGF19), 60
Fibroblasts induced hepatocyte transdifferentiation, 106
Fluorescence-activated cell sorting (FACS), 214
Forkhead box protein A (Foxa) family, 4
Forward-genetics analyses, 14
FoxM1, 79
Fumarylacetoacetate hydrolase (Fah) deficient mouse, 56
Future liver remnant volume (FLRV), 211

G

Galectin-1 (Gal-1), 305
Gallbladder cancer (GBC), 224
Genome-wide hypomethylation, 61
Glioblastoma multiforme (GBM) stem cells, 305
Grainyhead-like 2 (Grhl2), 40

H

Hematopoietic cells, hepatic morphogenesis, 126
Hemoglobin-bilirubin metabolism, in medaka, 11
Hepatic colony-forming unit in culture (H-CFU-C), 214
Hepatic lobule, 257, 258
Hepatic progenitor cells
    from human pluripotent stem cells, 28
    proliferation and maturation
        bile ducts, 22
        cell–cell interactions, 22
        oncostatin M, 21
        signaling pathways, 26
        transcription factors, 23
        transforming growth factor signaling, 22
Hepatic progenitor markers, 29
Hepatic stellate cell differentiation, 8
Hepatitis B virus (HBV) infection, 134
Hepatobiliary cancers
    cancer stem cells. *See* Cancer stem cells (CSCs)
    cholangiocarcinoma, 222
    chronic hepatitis, 222
    genetic lineage tracing, 223
    hepatocellular carcinoma, 222
    hepatocellular–cholangiocarcinoma, 222
    long-term label-retaining assay, 223
    origins and prognosis, 224
Hepatobiliary system
    anatomy and development, 212
    organogenesis, 212
    portal vein embolization, 211
    radical liver resection, 211
    stem cells
        adult liver, 215
        extrahepatic biliary tract, 217
        hepatobiliary surgery, 218
        liver development, 214
Hepatoblast, 212, 274
Hepatocarcinogenesis, 202
Hepatocellular carcinoma (HCC), 13
    adeno-associated virus integration, 65
    cancer stem cell
        controversies, 280
        heterogeneity, 277
        markers, 277
    characteristics, 276
    chromatin structure, 62
    DNA methylation, 61
    genetic alterations, 60
    histone modification, 62
    liver cirrhosis, 182
    noncoding RNAs, 64
    oncostatin M, 278
    therapy, 278

Hepatocellular carcinoma–cholangiocellular
    carcinoma (HCC-CCA), 289
Hepatocyte nuclear factor 4α (HNF-4α), 304
Hepatocytes
  cell transplantation therapeutics
    ESC-derived hepatocytes, 80
    induced hepatocyte-like cells, 83
    iPSCs-derived hepatocytes, 81
  differentiation, 6, 318
  ductular differentiation, in vitro
    collagen gel matrix, 188, 190
    hepatic stellate cells, 188
    reversibility, 189, 191
    ROSA26R mice, 190
    three-dimensional culture technique, 189, 192
    two-step collagenase perfusion method, 188
  ductular differentiation, in vivo
    centrilobular ductular reaction, 194
    lineage-tracing experiments, 191, 193
    Mx1-Cre × ROSA26R mice, 194
    periportal ductular reaction, 194
    tamoxifen-treated CK19-CreER mice, 191
  expansion in vitro
    coculture techniques, 78
    3D culture technique, 78
    immortalization, 79
    long-term in vitro culture, 78
    small molecules, 80
  experimental model, 103
  functional, 103
  heterogeneity, 43
  humanized liver animal models, 132
  lineage plasticity
    AAV8-Cre:ROSA-YFP mice, 41
    biphenotypic hepatocytes, 42
    dedifferentiation and lineage conversion, 41, 42
    dipeptidyl peptidase IV positive rats, 41
    FGF signal, 46
    Hippo-signaling pathway, 46
    Notch signaling pathway, 45
    SOX9⁺ hepatocytes, 43
    TGFβ signal, 47
    Wnt-β-catenin pathway, 47
  liver cancer stem cells origin, 301
  partial hepatectomy, 78
  stem cells
    benefits, 55
    fluorescence activated cell sorting, 57
    progenitor cells, 57
    transgenic mouse model, 56

3D configuration, 130
transdifferentiation, 276
transdifferentiation *vs.* dedifferentiation, 196
2D culture systems, 129
Hepatoma cell lines, 127
Herpes simplex virus type 1 thymidine kinase (HSVtk) transgene, 132
Heterogeneity
  cholangiocytes, 38
  hepatocytes, 43
  liver cancer stem cells, 290
hiohgi (hio) mutant, 12
Hippo-signaling pathway, 46, 60
hirame (hir) mutant, 12
Histone deacetylase (HDAC), 63
Humanized liver animal models, 132
Human liver stem cells (HLSCs), 128
Human telomerase reverse transcriptase (hTERT), 79
Human umbilical vein endothelial cells (HUVECs), 218, 219
Hybrid multicellular spheroids, 148

I

Immortalization, 79
Immune clearance, 290
Immunotherapy
  cancer stem cells, 237
  liver cancer stem cells, 305
Induced hepatic stem cells (iHepSCs), 83
Induced hepatocyte-like cells (iHeps), 83
Induced pluripotent stem cells (iPSCs)
  differentiation, 115
  somatic cell reprogramming, 318
Induced pluripotent stem cells-derived hepatocytes (iPSC-Heps), 81
  advantages, 128
  clinical trials, 128
  gene expression pattern, 128
  hepatic functions, 128
Insertional mutagenesis, 66
Integrative delivery approaches, 105
Integrin-associated protein (IAP), 231
Intermediate hepatobiliary cells, 184
Intracellular domain (ICD) cleavage, 6
Intrahepatic (IHBD) bile ducts, 38

J

JAGGED1 mutations, 14
Jagged-Notch signaling pathway, 6

Index   351

K
K19, 231
Kamifusen mutant medaka embryos, 11
Kendama mutant medaka, 11
Kupffer cells, 165, 273

L
Leucine-rich repeat-containing G protein-
    coupled receptor 5 (Lgr5), 187
Lineage plasticity
  cholangiocytes
    adult cholangiocytes, 38
    fetal and neonatal cholangiocytes, 37
    molecular mechanisms, 40
  hepatocytes
    AAV8-Cre:ROSA-YFP mice, 41
    biphenotypic hepatocytes, 42
    dedifferentiation and lineage conversion,
      41, 42
    dipeptidyl peptidase IV positive rats, 41
    FGF signal, 46
    Hippo-signaling pathway, 46
    Notch signaling pathway, 45
    SOX9$^+$ hepatocytes, 43
    TGFβ signal, 47
    Wnt-β-catenin pathway, 47
Lipid metabolism, in medaka, 11
Liver
  blood supply, 1
  cells, 21
  cell types, 1
  development and microenvironment, 125
  functions, 1
  hematopoiesis, 2
  metabolic functions, 21
  physical exposure, 55
  regenerative capability, 36
  regenerative capacity, 179
  size regulation in mice, 12
Liver cancer stem cells (LCSCs)
  chemoresistance, 290
  constant latency, 290
  controversy, 306
  CSC function-based isolation, 293
  diagnosis, 302
  differentiation strategy, 303
  elimination, 303
  evasion of immune clearance, 290
  heterogeneity, 290
  immunotherapy, 305
  liver cancer stem cells origin, 301
  multiple differentiation capacity, 289

origins of, 300
regulation
  epigenetic factors, 297
  metabolism, 300
  microenvironment, 299
  non-coding RNAs, 295
  stemness-related cascades, 297
  transcription factors, 294
self-renewal, 289
surface biomarkers, 291
tumor microenvironment, 305
Liver cirrhosis
  ductular reaction, 181, 182
  hepatocarcinogenesis, 202
  hepatocellular carcinoma, 182
  hepatocyte dedifferentiation, 202
  histological features, 181
  proliferation, 199
  regenerative nodules clonality, 198
  regenerative nodules sizes, 181
Liver development
  bud formation, 3
  hepatic stellate cell differentiation, 8
  hepatocyte and cholangiocyte
    differentiation, 6
  kupffer cells differentiation, 8
  lineage specification, 36
  LSEC differentiation, 8
  in medaka, 10
  mesothelial cell differentiation, 7
  pit cells differentiation, 8
  in zebrafish, 9
Liver fibrosis, 163
Liver progenitor/stem cells (LPCs), 36.
    See also Cancer stem cells (CSCs)
  adult, 275
  adult liver, 86
  characterization, 261
  chronic liver injury
    human liver diseases, 260
    mouse model, 260
    rat model, 260
  CLiPs
    diploid hepatocytes, 92
    liver regenerative therapy, 92
    stable long-term expansion, 91
    in vivo repopulation capacity, 91
  compensatory hypertrophy, 255
  deduced origin, 257
  fetal, 274
  fetal liver cells, 85, 86
  historical view, 257
  limitations, 87

Liver progenitor/stem cells (LPCs) (cont.)
  liver cancer stem cells origin, 300
  liver regeneration, 264, 275
  oval cells, 86
  partial hepatectomy, 255
  roles, 274
  signaling pathways, 263
  small hepatocyte culture, 87
  small molecule signaling inhibitors, 87
  in vitro reprogramming, 90
  YAC-induced proliferative cells
    bipotentiality, 90
    LPC-like characteristics, 89
Liver receptor homolog 1 (lrh1), 24
Liver regeneration, platelets, 164
Liver-related organoids
  cell–cell communication, 126
  CRISPR/Cas9 technology, 135
  direct reprogramming, 127
  3D vs. 2D culture
    coculture system, 131
    conventional monolayer culture, 129
    extracellular matrix, 129, 130
    multiple cellular community, 131
    quality, 129
    scaffold, 130
    self-assembly generation, 130
  embryonic development, 125
  ESCs, 128
  function and application, 116
  hematopoietic stem cells, 125
  hepatic bud formation, 126
  hepatoma cell lines, 127
  humanized liver animal models, 132
  iPSC-Heps, 128
  liver functions, 115
  mesenchyme-driven self-condensation, 125
  primary human hepatocytes, 127
  in vivo 3D environment, 125
Liver sinusoidal endothelial cells (LSECs), 8, 164
Liver transplantation
  brain-dead liver transplantation, 77
  living-donor liver transplantation, 77

M

Matrigel-loaded MCS, 154
Medaka mutations
  complementation rate, 10
  hemoglobin-bilirubin metabolism, 11
  hepatoblast development, 11
  hepatoblast proliferation, 11
  liver formation and function, 10
  liver laterality, 11
  mutants analysis, 12
  vs. zebrafish, 10
Megakaryocyte, platelet production
  differentiation, 160, 162
  maturation process, 161
  proplatelets, 161
Mesenchymal-epithelial transition (MET), 58
Mesenchymal stem cells (MSCs), 84
  cell transplantation therapeutics, 84
  hepatocytes transdifferentiation, 334
  induced hepatocyte transdifferentiation, 106
  in liver diseases, 332
  microenvironment modulation, 335
  properties, 331
Mesenchyme, liver expansion and differentiation, 126
Mesothelial cell differentiation, 7
Mesothelin (MSLN), 7
Metabolic disorders, 13
Metformin, 237
Methyl binding domain (MBD), 61
MicroRNA 122 (miR122), 40
MicroRNAs (miRNAs), 64, 295
Multicellular spheroids (MCSs)
  3D culture, 145
  ECM loading
    albumin secretion activity, 153, 155
    cell polarity formation, 155
    ECM-loaded MCS, 154
    emulsion method, 151
    functions, 153
    matrigel solution, 153, 154
    methylcellulose medium, 151, 152
  internal structures, 145
  methylcellulose medium
    cell aggregation process, 146
    colony formation assay, 146
    gyratory culture, 147
    multistep aggregation of cells, 148
    optical tweezers, 147
    viscosity, 147
  microchannel formation
    alginate hydrogel beads, 149
    drug metabolism, 148
    epithelial-mesenchymal transition, 150
    hybrid MCS, 148, 150
    longterm culture system, 152
    luminal structure, 148
    methylcellulose medium, 149
    microchannel-like structures, 149

PDMS membrane plate, 151
  three-dimensional culture, 148
  sponge-like scaffolds, 145
Multiple differentiation capacity, LCSCs, 289
Multiplicity of infection (MOI), 105
Multipotent stem cells, 56

N

Nanog, 295
2-(2-Nitro-4-trifluoro-methylbenzyol)-1, 3 cyclohexanedione (NTBC), 56
Nonalcoholic fatty liver disease (NAFLD), 329
Nonalcoholic steatohepatitis (NASH) model, 260
Non-coding RNAs (ncRNAs), 64
  lncRNAs, 296
  microRNAs, 295
Nonintegrating viral systems, 106
Nonintegrative delivery approaches, 106
Nonviral gene editing techniques, 79
Nonviral systems, 106
Notch2 (N2) signaling, 7
Notch signaling pathway, 45
Nuclear factor, erythroid 2 (NF-E2), 161

O

Oncostatin M (OSM), 6
  differentiation of hepatoblasts, 304
  hepatocellular carcinoma, 278
  hepatocytic differentiation, 188
Optical tweezers, 147
OV6, 228
Oval cells (OCs), 86

P

Periportal ductular reaction, 182, 184
Plasmid-based strategies, 106
Platelet-derived growth factor receptor (PDGFR), 23
Platelets
  blood transfusion, 161
  and cancer, 160
  chronic liver disease, 168
  degranulation, 159
  iPS cell-based regenerative medicine, 162
  in liver cirrhosis, 166, 168
  in liver regeneration, 164
  megakaryocyte
    differentiation, 160, 162
    maturation process, 161
    proplatelets, 161

platelet-rich plasma, 159
  secretary granules, 159
  thrombopoietin, 163
  from TPO-stimulated megakaryocytes, 159
Pluripotent stem cells (PSCs), 56
  animal models, 330
  bioartificial liver system, 317
  cholangiocyte differentiation, 320
  genetic inherited diseases, 325
  hepatocyte differentiation, 318, 321
  hepatocyte transplantation, 317
  infectious diseases, 329
  metabolic diseases, 329
  somatic cell reprogramming, 318
Podocalyxin-like protein 1 (PCLP1), 7
Polycomb group gene (PcG) family, 297
Polydimethylsiloxane (PDMS) membrane plate, 151
Portal triad, 257
Portal vein embolization, 211
p53 pathway, 104
Precursor miRNAs (pre-miRNA), 64
Primary human hepatocytes (PHHs), 127, 129
Programmed cell death ligand 1(PD-L1), 233
Proplatelets, 161
Prospero-related homeobox 1 (prox1), 24
Prox1 expression, 6

R

Recombinant activation gene-2 (RAG-2) mice, 134
Recombinant adeno associated virus (rAAV)
  clinical trials, 65
  gene therapy, 65
  genotoxicity, 66
  hemophilia B treatment, 66
  insertional mutagenesis, 66
  for lysosomal storage disease treatment, 65
Retroviral and lentiviral expression systems, 105
"Reversed lobulation" cirrhosis, 182
Romiplostim, 168

S

Sakura mutant medaka embryos, 11
SALL4, 278
Sal-like protein 4 (Sall4), 25, 229
Secretary granules, 159
Sendai virus, 106
Septum transversum mesenchyme (STM), 212
Sex-determining region Y-box 9 (SOX9), 216, 232

Side population cell sorting, 224
Simian virus 40T (SV40T), 79
"Small-for-size" syndrome, 179
Small-molecule-compounds-based transdifferentiation, 105
Small molecule signaling inhibitors, 87
Sodium taurocholate cotransporting polypeptide (NTCP), 330
Somatic cell reprogramming, 318
Somatic stem cells, 56
Space of Disse, 22
Sponge-like scaffolds, 145
Stellate cells
  liver expansion and differentiation, 126
  as stem cells, 57
Stem cells
  and hepatocellular carcinoma
    adeno-associated virus integration, 65
    chromatin structure, 62
    DNA methylation, 61
    genetic alterations, 60
    histone modification, 62
    noncoding RNAs, 64
  liver function and architecture, 273
  liver size control, 59
  multipotent, 56
  pluripotent, 56
  somatic, 56
  totipotent, 56
  in vitro and in vivo differentiation, 56
Stemness-related cascades, 297
Stromal-derived factor 1 (SDF-1), 161
Sulfasalazine, 238
Suou mutation in medaka, 11

T

Tbx3 expression, 5
Thioacetamide (TAA)-induced chronic injury, 39
Thrombocytopenia, 164, 168
Thrombopoietin (TPO)
  platelet production, 163
  stimulated megakaryocytes, 159
Tissue-specific stem/progenitor cells, 35
TNF-Related WEAK Inducer of Apotosis (TWEAK), 263
Toll-like receptor 4 (TLR4), 232
Totipotent stem cells, 56

Transcription factors
  cholangiocyte plasticity, 40
  c-Myc activation, 294
  hepatic progenitor cell proliferation and maturation, 23
  Nanog, 295
  SALL4, 295
Transdifferentiation
  induced hepatocytes
    BAL system, 109
    cell quality evaluation, 107
    donor cell types, 106
    generation, 104
    integrative delivery approaches, 105
    large-scale expansion, 109
    maturation and maintenance, 109
    nonintegrative delivery approaches, 106
    small-molecule-compounds-based, 105
Trop2 (Tacstd-2), 261
Tumor initiating cells, 58
Tumor initiating cells (TIC), 55
Tyrosine aminotransferase (TAT) expression, 6
Tyrosinemia, 56

U

Ukon mutation, 11
Urea cycle, 13
Urinary cells induced hepatocyte transdifferentiation, 107
Ursodeoxycholic acid (UDCA)-resistant primary biliary cirrhosis (PBC), 333

W

Wilms Tumor 1 (WT1) protein, 7
Wnt-β-catenin pathway, 47
"Writer" and "eraser" enzymes, 62

Y

Yes-associated protein (YAP), 12

Z

Zebrafish
  cholangiocyte-driven regeneration model, 40
  development, 9
Zebularine, 278

CPI Antony Rowe
Chippenham, UK
2018-05-25 11:34